Introduction to Fungi

To Brom, Adam and Sarah

Introduction to Fungi

JOHN WEBSTER
Professor of Biological Sciences
University of Exeter

CAMBRIDGE
AT THE UNIVERSITY PRESS · 1970

Published by the Syndics of the Cambridge University Press
Bentley House, 200 Euston Road, London N.W.1
American Branch: 32 East 57th Street, New York, N.Y. 10022

Library of Congress Catalogue Card Number: 77–9314

Standard Book Number: 0 521 07640 4

Printed by offset in Great Britain by
Alden & Mowbray Ltd
at the Alden Press, Oxford

Contents

Preface

There are several available good text-books of mycology, and some justification is needed for publishing another. I have long been convinced that the best way to teach mycology, and indeed all biology, is to make use, wherever possible, of living material. Fortunately with fungi, provided one chooses the right time of the year, a wealth of material is readily available. Also by use of cultures and by infecting material of plant pathogens in the glasshouse or by maintaining pathological plots in the garden, it is possible to produce material at almost any time. I have therefore tried to write an introduction to fungi which are easily available in the living state, and have tried to give some indication of where they can be obtained. In this way I hope to encourage students to go into the field and look for fungi themselves. The best way to begin is to go with an expert, or to attend a Fungus Foray such as those organised in the Spring and Autumn by mycological and biological societies. I owe much of my own mycological education to such friendly gatherings. A second aim has been to produce original illustrations of the kind that a student could make for himself from simple preparations of living material, and to illustrate things which he can verify for himself. For this reason I have chosen not to use electron micrographs, but to make drawings based on them.

The problem of what to include has been decided on the criterion of ready availability. Where an uncommon fungus has been included this is because it has been used to establish some important fact or principle. A criticism which I must accept is that no attempt has been made to deal with Fungi Imperfecti as a group. This is not because they are not common or important, but that to have included them would have made the book much longer. To mitigate this shortcoming I have described the conidial states of some Ascomycotina rather fully, to include reference to some of the form-genera which have been linked with them. A more difficult problem has been to know which system of classification to adopt. I have finally chosen the 'General Purpose Classification' proposed by Ainsworth, which is adequate for the purpose of providing a framework of reference. I recognise that some might wish to classify fungi differently, but see no great merit in burdening the student with the arguments in favour of this or that system.

Because the evidence for the evolutionary origins of fungi is so meagre I have made only scant reference to the speculations which have been made on this topic. There are so many observations which can be verified, and for this reason I have preferred to leave aside those which never will.

The literature on fungi is enormous, and expanding rapidly. Many undergraduates do not have much time to check original publications. However, since the book is intended as an introduction I have tried to give references to some of the more recent literature, and at the same time to quote the origins of some of the statements made.

Exeter. 27 April, 1970 J.W.

Acknowledgements

It is a pleasure to thank all who have helped me in the production of this book; especially Mrs N. Barnsley (née Lomas) and Mr G. Bellany for excellent technical assistance and Mr G. Woods for help with the photography. I have also to thank numerous fellow mycologists who have made gifts of material and cultures, or who have generously allowed me to make use of their published figures and keys. These have been individually acknowledged in the text. I am particularly grateful to Miss E. M. Wells of the University of Sheffield Library who obtained on loan many publications. My wife and children have been very patient and understanding, and I dedicate the book to them.

Introduction

WHY STUDY FUNGI?

The absence of photosynthetic pigments enforces upon fungi a saprophytic or a parasitic existence. As saprophytes they share with bacteria and animals the role of decay of complex plant and animal remains in the soil, breaking them down into simpler forms which can be absorbed by further generations of plants. Without this essential process of decay, the growth of plants, upon which life is dependent, would eventually cease for lack of raw materials. Soil fertility is thus in part bound up with fungal activity. The roots of most green plants are infected with fungi and absorption of minerals may be enhanced following infection. Such infected root systems are termed mycorrhiza, and they are an example of a symbiotic relationship between green plants and fungi. In infertile natural soils the success of the higher plant may depend on infection (Harley, 1969). Harmful effects of saprophytic fungi on human economy are seen when food, timber and textiles are rotted. Fungi are also of importance in industrial fermentations as in brewing, production of antibiotics, or citric acid fermentation. Food processing such as baking, cheese-making, or wine fermentation is also dependent on fungi. Increasing use is made of fungi in carrying out chemical transformations in the pharmaceutical industry. These activities of fungi have been ably reviewed by Christensen (1965) and Gray (1959).

As parasites, fungi cause disease in plants and animals. Although fungal pests of crop plants have been known since human records began, it was the impact of potato blight on the population of Ireland in the mid nineteenth century which gave the impetus to the scientific study of plant pathology (Large, 1958). As agents of disease in animals and man, fungi are less severe than bacteria and viruses, but as the control of other diseases improves, the importance of fungal disease is being recognised (Ainsworth, 1952).

Apart from these applied aspects of the study of fungi, they have a claim to interest in their own right, and as tools for the physiologist, microbiologist, biochemist and geneticist, who often find them ideally suited for investigations of all kinds. Our general understanding of genetics owes much to investigations with *Neurospora*, and our understanding of respiration to studies on yeast. Investigations into the bakanae disease of rice caused by *Gibberella fujikuroi* led to the discovery of the group of plant growth hormones called gibberellins. These aspects of fungal biology will not be stressed in this book. They have been well-described by Cochrane (1958), Fincham & Day (1965) and Esser & Kuehnen (1967).

CLASSIFICATION

Organisms do not classify themselves. They are classified by man for convenience of reference. Ideally a scheme of classification should reflect natural relationships, but in considering relationships mycologists may not attach the same weight to the criteria available. It should therefore not be surprising that different authorities do not use the same scheme of classification. I have chosen to adopt the scheme proposed by Ainsworth (1966), which is set out below. In other schemes of classification the Mastigomycotina and Zygomycotina are sometimes collectively referred to as Phycomycetes.

A General Purpose Classification of the Fungi

Note: Divisions end in '-mycota'; Sub-divisions in '-mycotina'; Classes in '-mycetes'; Sub-classes in '-mycetidae'; Orders in '-ales'; Families in '-aceae'.

Families are excluded. The orders cited, which should be regarded as only representative, are listed alphabetically by classes.

A few common names and alternative names have been inserted.

FUNGI (Mycota)

I. MYXOMYCOTA (Myxobionta)
- (1) Acrasiomycetes
  - Acrasiales
- (2) Hydromyxomycetes
  - Hydromyxales
  - Labyrinthulales
- (3) Myxomycetes (slime moulds)
- (4) Plasmodiophoromycetes
  - Plasmodiophorales

II. EUMYCOTA (Mycobionta; eumycetes)
- 1. Mastigomycotina
  - (1) Chytridiomycetes
    - Blastocladiales
    - Chytridiales (chytrids)
    - Monoblepharidales
  - (2) Hyphochytridiomycetes
    - Hyphochytriales
  - (3) Oomycetes
    - Lagenidiales
    - Leptomitales

Peronosporales (downy mildews, etc.)
Saprolegniales

2. Zygomycotina
   (1) Zygomycetes
      Entomophthorales
      Mucorales
   (2) Trichomycetes

3. Ascomycotina (ascomycetes)
   (1) Hemiascomycetes
      Endomycetales
      Taphrinales
   (2) Plectomycetes
      Erysiphales (powdery mildews, etc.)
      Eurotiales
   (3) Pyrenomycetes
      Hypocreales
      Sphaeriales
   (4) Discomycetes
      (inoperculate discomycetes)
      Helotiales
      Phacidiales
      (operculate discomycetes)
      Pezizales
      (hypogaeous discomycetes)
      Tuberales (truffles)
   (5) Laboulbeniomycetes
      Laboulbeniales
   (6) Loculoascomycetes (bitunicate ascomycetes)
      Capnodiales
      Dothideales
      Hysteriales
      Microthyriales (syn. Hemisphaeriales)
      Myriangiales
      Pleosporales

4. Basidiomycotina (basidiomycetes)
   (1) Hemibasidiomycetes
      Uredinales (rusts)
      Ustilaginales (smuts)
   (2) Hymenomycetes
      Agaricales (agarics, boleti)
      Aphyllophorales (polypores, etc.)
      Tulasnellales

(3) Gasteromycetes
    Hymenogastrales
    Lycoperdales
    Nidulariales
    Phallales
    Sclerodermatales

5. Deuteromycotina (fungi imperfecti)
   (1) Coelomycetes
       Melanconiales
       Sphaeropsidales
   (2) Hyphomycetes
       Hyphales (syn. Moniliales)
   (3) Agonomycetes (mycelia sterilia)
       Agonomycetales (syn. Myceliales)

No attempt will be made to treat each group in equal detail. In the Myxomycota only the Plasmodiophorales have been described, because of their importance as plant pathogens.

PART ONE

Myxomycota

Slime Moulds and Similar Organisms

Whether the Myxomycota are closely related to the Eumycota is doubtful. Possibly they are more closely related to Protozoa. The vegetative phase consists of a **plasmodium** (a multinucleate mass of protoplasm lacking a cell wall) or a **pseudoplasmodium** (an aggregate of separate amoeboid cells).

1: ACRASIOMYCETES

The Acrasiomycetes are sometimes termed the cellular slime moulds (Bonner, 1967). Genera such as *Dictyostelium* and *Polysphondylium* are ubiquitous in soil where they feed on bacteria. In culture, provided with a supply of bacteria for food, they exist as amoeboid cells (myxamoebae) which engulf the bacteria and increase in number. As the bacterial food becomes exhausted the myxamoebae begin to flow towards each other forming stellate pseudoplasmodial masses. This stage is termed aggregation. In some species such as *D. discoideum* the aggregated mass of myxamoebae, forming a cylindrical 'slug', migrates across the surface of the culture (migration). Migration is followed by a process of cellular rearrangement and differentiation in which a multicellular stalk made up of thousands of myxamoebae becomes surmounted by a globose sorus of spores. The spores, on germination, give rise to myxamoebae.

2: HYDROMYXOMYCETES

The best-known representative of this group is *Labyrinthula*, species of which are parasitic on marine algae and angiosperms. *Labyrinthula macrocystis* is a destructive parasite of eel-grass, *Zostera marina*. Air spaces of *Zostera* suffering from wasting-disease contain spindle-shaped cells with terminal branched pseudopodia. Aggregations of spindle-shaped cells, attached to each other by their pseudopodia, are formed. Within these colonies a network of slime-tracks is secreted, and the amoeboid cells are capable of moving within the labyrinth. Reproduction by means of biflagellate zoospores has been described for some species, and some have been grown in culture (see Alexopoulos, 1962; Bonner, 1967).

3: MYXOMYCETES

Slime moulds grow on decaying wood and bark, on fruit-bodies of fungi and decaying vegetation. Elaborately stalked or sessile, often brightly coloured fruit-bodies release spores which are dispersed by wind. The spores may germinate by giving rise to uniflagellate or unequally biflagellate zoospores. The flagella are of the whip type, a smooth cylindrical shaft lacking lateral appendages. After swimming for a period the zoospores withdraw their flagella and the cells assume an amoeboid form. Bacteria, yeast cells, fungus spores or other organic matter are ingested. In some species the spores germinate by forming an amoeboid stage directly: i.e. motile zoospores are not formed. The amoeboid cells are termed myxamoebae.

Fusion between myxamoebae or between the flagellated zoospores has been reported. The fusion cell or zygote undergoes nuclear division to form a multicellular plasmodium. The plasmodium often forms a network of veins, and within the veins the cytoplasm flows rhythmically, moving first in one direction, and then in reverse. As it flows over its substratum the plasmodium differentiates into fruit-bodies containing spores. The nuclear division preceding spore formation is meiotic, and uninucleate haploid spores are formed.

The Myxomycetes contain several hundred species. For a general account of the group see Alexopoulos (1962, 1963).

4: PLASMODIOPHOROMYCETES

PLASMODIOPHORALES

The Plasmodiophorales are obligate parasites. The best-known examples attack higher plants causing economically important diseases such as club-root of brassicas (*Plasmodiophora brassicae*), powdery scab of potato (*Spongospora subterranea*) and crook-root disease of watercress (*S. subterranea* f. sp. *nasturtii*). Others attack roots and shoots of non-cultivated plants, especially aquatic plants. Algae and fungi are also attacked. About eight genera are recognised, and they are separated from each other largely on

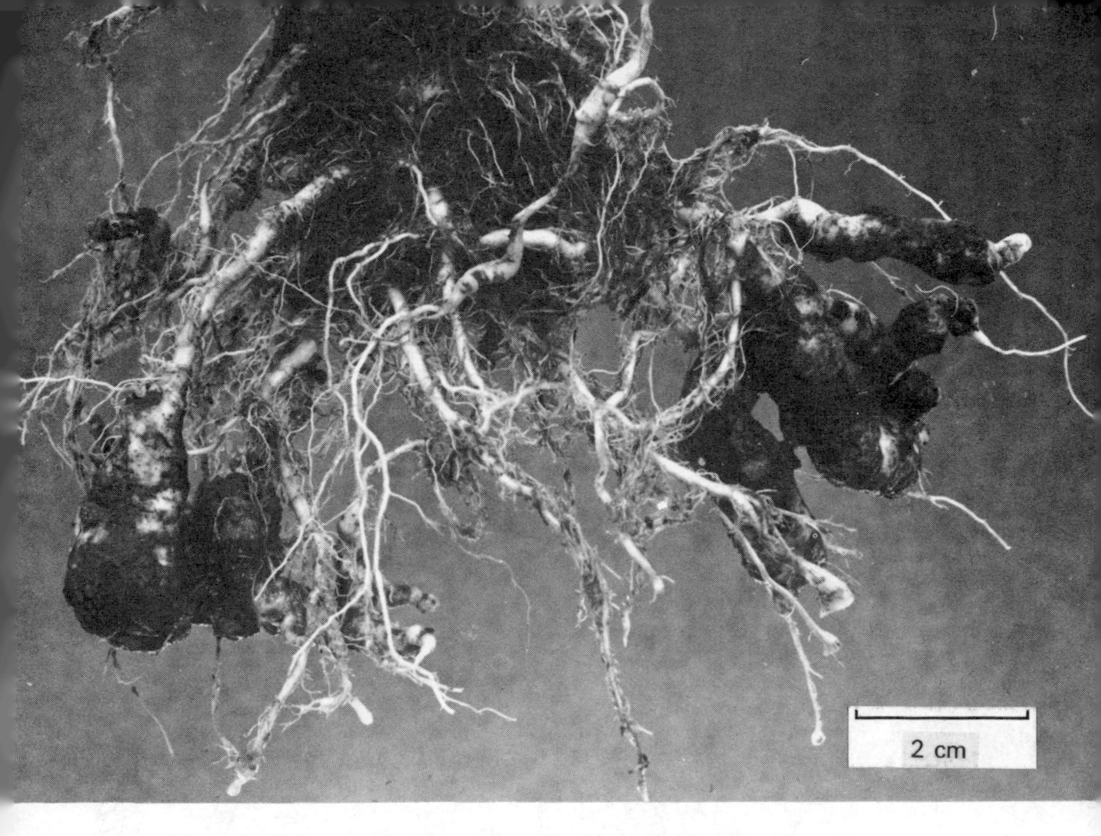

Figure 1. Club-root of cabbage caused by *Plasmodiophora brassicae*.

the way in which the resting spores are arranged in the host cell. Accounts of the group have been given by Karling (1942) and Sparrow (1960).

The zoospore in the Plasmodiophorales is biflagellate. The flagella are of unequal length and there is clear evidence from electron micrographs that the two flagella are both of the whip type. The wall of the resting spore in *P. brassicae* is believed to contain chitin and although cellulose has occasionally been reported in some species, it has been stated not to occur in *Woronina polycystis* and *Octomyxa brevilegniae*, both parasitic on Saprolegniaceae (Goldie-Smith, 1954; Pendergrass, 1950).

The details of the life cycle of many members of the Plasmodiophorales are still uncertain. In most genera there are two distinct plasmodial phases. The first, usually resulting from infection by a zoospore derived from a resting spore, gives rise to thin-walled zoosporangia. The second, which in its early development may be indistinguishable from the zoosporangial plasmodium, gives rise to resting spores. Although there have been suggestions that the development of the resting spore plasmodium may be preceded by a sexual fusion in some species, this cannot be regarded as proven nor may it be true of all species. In some cases the plasmodium giving rise to resting spores appears to arise merely at a later stage of infection of the host than the zoosporangial phase.

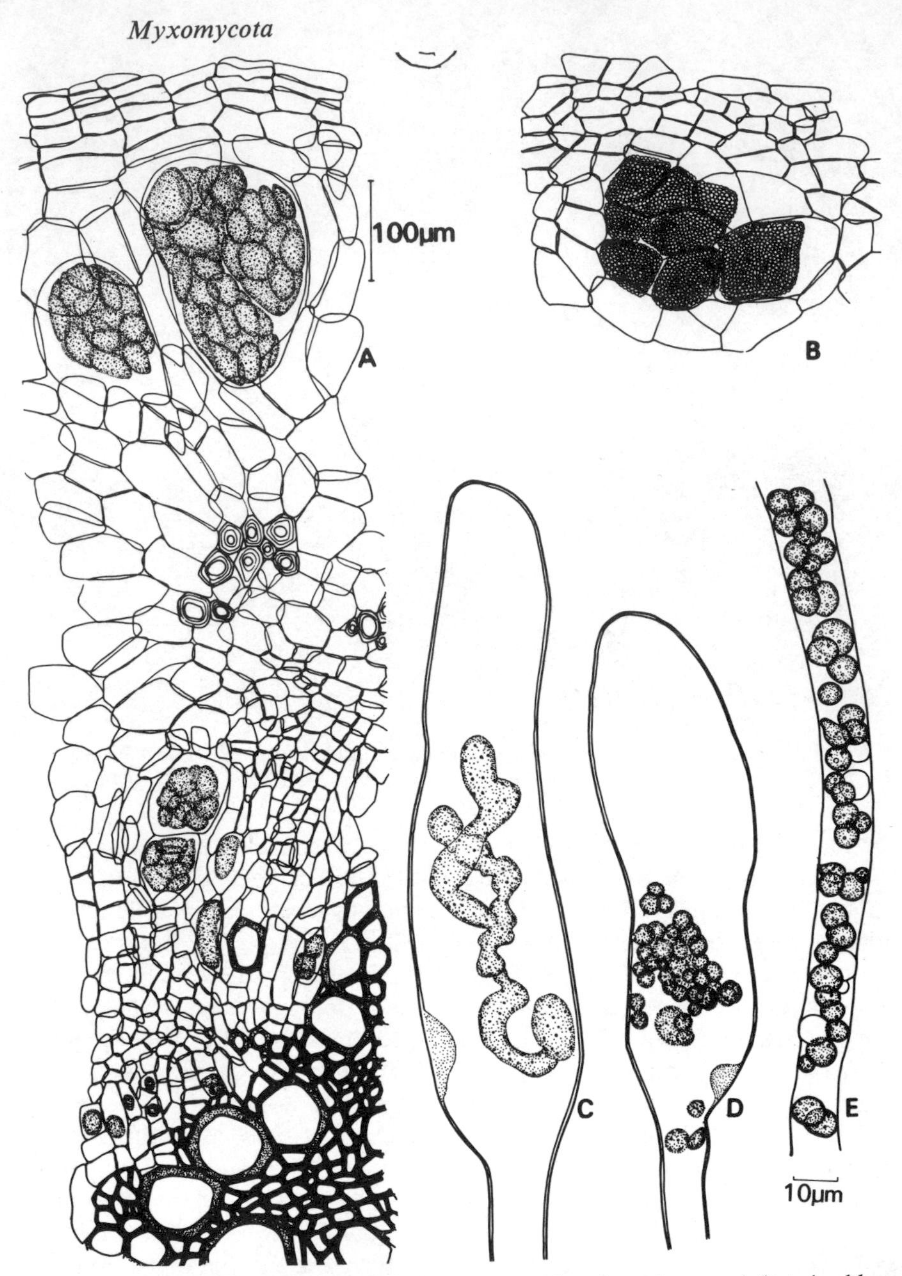

Figure 2. *Plasmodiophora brassicae*. A. Transverse section through young infected cabbage root showing plasmodia in the cortex. Note the hypertrophy of some of the host cells containing plasmodia, and the presence of young plasmodia in cells immediately outside the xylem. B. Transverse section of a cabbage root at a later stage of infection showing the formation of resting spores. C. Zoosporangial plasmodium in cabbage root hair 4 days after planting in a heavily contaminated soil. D. Young zoosporangia in root hair. Note the club-shaped swelling of the infected root hair. E. Mature and discharged zoosporangia. A and B to same scale. C, D and E to same scale.

Plasmodiophora

P. brassicae is the causal organism of club-root or finger-and-toe disease of brassicas (Fig. 1). The disease is common in gardens and allotments where cabbages are frequently grown, especially if the soil is acid and poorly drained. A wide range of cruciferous hosts is attacked and root hair infection of some non-cruciferous hosts can also occur. The disease is widely distributed throughout the world (Colhoun, 1958).

Infected crucifers usually have much-swollen roots. Both tap roots and lateral roots may be affected. Occasionally infection results in the formation of adventitious root buds which give rise to swollen stunted shoots. Above ground, however, infected plants may be difficult to distinguish from healthy ones. The first symptom is wilting of the leaves in warm weather, although often such wilted leaves recover at night. Later the rate of growth of infected plants is retarded so that they appear yellow and stunted. When plants are attacked in the seedling stage they can be killed, but if infection is delayed the effect is much less severe and well-developed heads of cabbage, cauliflower, etc. can form on plants with quite extensive root hypertrophy. Commonly, infected root hairs are hypertrophied, expanding at their tips to form club-shaped swellings which are sometimes lobed and branched (Fig. 2). MacFarlane & Last (1959) followed the growth of cabbage seedlings either uninfected or infected with *Plasmodiophora* at various times after germination. Within 35 days of infection of seedlings significant retardation of the weight-increase of the tops occurs as compared with healthy controls. In infected plants the root/shoot ratio is appreciably higher suggesting a diversion of materials to the clubbed roots. Swollen roots contain large numbers of sperical small resting spores and when these roots decay, the spores are released into the soil. Electron micrographs show that the resting spores have spiny walls (Williams & McNabola, 1967). The resting spore germinates to produce a single zoospore with two flagella of unequal length, both of the whiplash type (see Fig. 3; Kole & Gielink, 1962). Germination is stimulated by substances diffusing from cabbage roots (MacFarlane, 1952, 1959).

The primary zoospores (i.e. the first motile stage released from the resting spore) swim at first by means of their flagella, the long flagellum trailing and the short flagellum pointing forwards. Before infection of the host the zoospores become amoeboid. The amoeba becomes applied to a root hair, and penetration occurs, the entire body of the amoeba entering the host. Young epidermal cells are also infected. Within the root hair or epidermal cell a small uninucleate stage is first seen. Later the nucleus divides and a small plasmodium or thallus is formed with numerous nuclei (numbers varying from up to 30 to over 100 have been reported). The plasmodium divides up to from a variable number of roughly spherical thin-walled zoosporangia lying packed together in the host cell (Fig. 2). The production of zoosporangia may take place within four days. Each zoosporangium finally contains four to eight uninucleate zoospores. The mature zoosporangium becomes attached to the host-cell wall and a pore develops at this point through which the zoospores escape. Occasionally zoospores are released into the lumen of the host cell. The behaviour of the released zoospores is not completely known, but it is possible that they function as gametes and fuse in pairs. Quadriflagellate, binucleate swarmers have indeed been observed (Kole & Gielink, 1961) but swarmers with six flagella have also been seen and whether such swarmers result from fusion of separate biflagellate swarmers or from incomplete separation of zoospore initials has not been established.

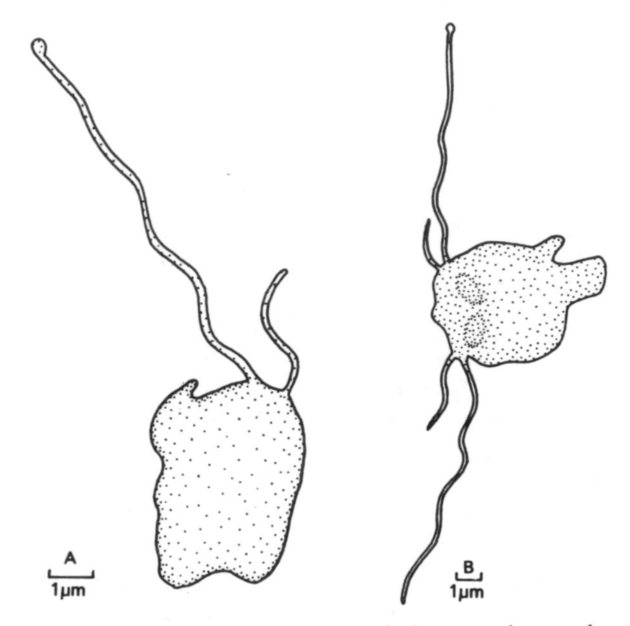

Figure 3. *Plasmodiophora brassicae*. Tracings of electron micrographs of zoospores. A. Zoospore (after Kole & Gielink, 1961). B. Quadriflagellate zoospore with two separate nuclei (after Kole & Gielink, 1961).

Nuclear fusion has not been observed at this point, nor are chromosome counts available to support the view that nuclear fusion occurs here. Cook & Schwartz (1930) believed that the zoospores (after reinfection of the root hairs) migrated to the epidermal and cortical cells where they fused to form binucleate zygotes but the details of the process were not figured. The behaviour of the zoosporangial zoospores in infection and their possible role as gametes are in need of reinvestigation. In *Woronina polycystis* and *Octomyxa brevilegniae* parasitic on Saprolegniaceae, and in *Sorosphaera veronicae* and *Ligniera verrucosa* both parasitic on roots of *Veronica*, although zoosporangia and resting spores (cysts) are known there is as yet no evidence of sexual fusion preceding their formation (Goldie-Smith, 1954; Pendergrass, 1950;

Miller, 1959). In *L. verrucosa*, Miller (1959) is of the opinion that plasmodia produced early in the infection develop into zoosporangia, while those produced later develop into resting spores. According to Cook & Schwartz the zygote of *P. brassicae* becomes spherical and develops into a plasmodium. It has been claimed by some workers that the plasmodium has the power of penetrating the cell walls of the host, but others claim that as the host cell divides the plasmodium is passively distributed to daughter cells. Possibly both mechanisms are involved. The plasmodium has no specialised feeding structures such as haustoria. It is immersed in the host cytoplasm surrounded by a thin plasmodial envelope (Williams & Yukawa, 1967). The plasmodium enlarges, repeated nuclear divisions take place, and the cells containing them become hypertrophied (Fig. 2A), although the host nucleus remains active. Hypertrophy of host cells is apparently brought about by blocking of the mechanism for cell division, and is accompanied by enhanced DNA synthesis. Starch accumulates in infected cells (Williams, 1966). At first only cortical cells of the young root are infected but later small plasmodia can be found in the medullary ray cells and in the vascular cambium. Subsequently, tissues derived from the cambium are infected as they are formed. In large swollen roots extensive wedge-shaped masses of hypertrophied medullary ray tissue may cause the xylem to be split. In this stage the root tissue shows a distinctly mottled appearance. When the growth of the plasmodia is complete they are transformed into masses of resting spores. According to Cook & Schwartz (1930) spore formation is preceded by meiosis, but this opinion is based only on comparison of nuclear size. The resting spores are at first naked but later become surrounded by a thin cell wall. They are closely packed together inside the host cell and are released into the soil as the root tissues decay.

This account of the possible life cycle of *P. brassicae* differs from that suggested by Heim (1955). She has described young plasmodia in the cortical and medullary parenchyma, forming pseudopodia which actively penetrate from an infected cell to neighbouring cells. In such plasmodia most of the nuclei become associated in pairs and then fuse to form diploid nuclei around which the cytoplasm condenses. Following nuclear fusion the plasmodia lose their ability to invade cells. Their contours become regular and they no longer put out pseudopodia. The diploid nuclei undergo a meiosis followed by a mitosis; then the haploid nuclei and adjacent cytoplasm become surrounded by cell walls to form spores. According to Heim, the spores can germinate within the infected root. The resting spores lose their spherical shape, elongate and acquire the appearance of young plasmodia, putting out pseudopodia into uninfected neighbouring cells. Fusion of amoeboid parasite cells occurs within the host before the vegetative plasmodium develops. Heim did not study the process of infection following resting spore germination in the soil and she ascribes no role to the zoosporangia.

Spongospora

The life cycle of *S. subterranea*, the cause of powdery scab of potato, is possibly similar to that of *P. brassicae* (see Kole, 1954; Kole & Gielink, 1963; Piard-Douchez, 1949). Diseased tubers show powdery pustules at their surface, containing masses of resting spores clumped into hollow balls. The resting spores release unequally biflagellate zoospores which can infect root hairs of potatoes or tomatoes. In the root hairs plasmodia form which develop into zoosporangia. Zoospores from such zoosporangia are capable of infection resulting in a further crop of zoosporangia. Zoospores released from the

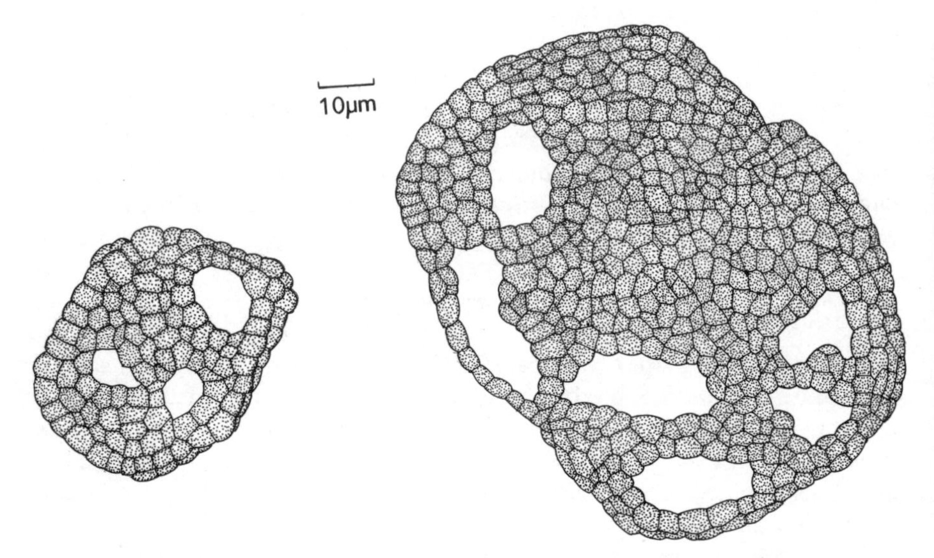

Figure 4. *Spongospora subterranea* f. sp. *nasturtii*. Spore balls from watercress roots with crook-root disease.

zoosporangia have also been observed to fuse in pairs or occasionally in groups of three to form quadri- or hexaflagellate swarmers (Kole, 1954) but whether this represents a true sexual fusion is uncertain. *S. subterranea* f. sp. *nasturtii* causes a disease of watercress in which the most obvious symptom is a coiling or bending of the roots. Zoosporangia and spore balls are found in infected root cells (Fig. 4; Tomlinson, 1958*b*). Heim (1960) has described a life cycle for this organism which resembles that which she has proposed for *P. brassicae* involving the fusion of myxamoebae within the infected host cells. Nuclear fusion is believed to initiate a short diploid phase which is ended by meiosis and a mitosis preceding spore formation.

CONTROL OF CLUB-ROOT, POWDERY SCAB AND CROOK-ROOT

The control of club-root disease is difficult. Because resting spores remain viable in soil for many years short-term crop rotation may not eradicate the disease. The fact that *Plasmodiophora* can infect cruciferous weeds such as Shepherd's purse (*Capsella bursa-pastoris*) suggests that the disease may be carried over on such hosts and that weed control may be important. Moreover, it is known that root hair infection can also occur on non-cruciferous hosts such as *Agrostis*, *Dactylis*, *Holcus*, *Lolium*, *Papaver* and *Rumex*, all of which are common garden or field weeds. Whether such infections play any part in maintaining the disease in the absence of a cruciferous host is not known. General measures aimed at mitigating the incidence of the disease include improved drainage and the application of lime. Lime probably inhibits the germination of the resting spores, but if the spore load in the soil is sufficiently high, lime will not prevent an outbreak of the disease. Since the effect of liming on the soil does not persist it is possible that it may delay the germination of the resting spores and thus prolong their persistence in the soil (MacFarlane, 1952). Early infection of seedlings can result in severe symptoms, so it is important to raise seedlings in non-infected or steam-sterilised soil. The young plants can then be transplanted to infested soil. Infection can be minimised or delayed by the application of mercurous chloride (calomel) or mercuric chloride (corrosive sublimate) to the soil at the time of transplantation. Since it is known that some resting spores survive animal digestion, manure from animals fed with diseased material should not be used for growing brassicas. Certain kinds of brassicas seem to have a natural resistance to the disease, e.g. some swede varieties, and plant breeders have also bred and selected resistant strains of other kinds of brassicas. The nature of the resistance is not known, but root hair infection of resistant and susceptible varieties occurs with equal facility so it is possible that it is during later stages of the cycle of infection that host resistance becomes effective. There are several physiological races of *P. brassicae* which vary in their ability to infect different kinds of brassicas and this complicates the problem of breeding resistant varieties. Varieties bred for club-root resistance in one country may be completely susceptible to strains of the pathogen derived from other countries. It is clearly desirable to test selected brassica varieties against material of *P. brassicae* obtained from many different sources. It is probable that resistance to club-root is a polygenic character.

Powdery scab of potatoes is normally of relatively slight economic importance and amelioration of the incidence of the disease can be brought about by good drainage. Crook-root of watercress can be controlled by application of zinc to the water supply. The zinc can be applied by dripping zinc sulphate into the water supply to the watercress beds to give a final concentration of about 0·5 parts per million, or by the use of finely powdered glass containing zinc oxide (zinc frit) to the watercress beds. The slow release of zinc from the frit maintains a sufficiently high concentration to inhibit infection (Tomlinson, 1958*a*).

The affinities of the Plasmodiophorales are not clear. They resemble some other Myxomycota in having a plasmodium and zoospores with flagella of unequal length. They are sometimes classified with zoosporic fungi (Mastigomycotina), although not clearly related to any other group. Sparrow (1958) has proposed that they be regarded as a separate group the Plasmodiophoromycetes. Possibly they are more closely related to Protozoa than to other fungi (Karling, 1944).

PART TWO

Eumycota

Introduction

The great majority of fungi belong to the Eumycota, distinguished from the Myxomycota by the absence of a plasmodium or pseudoplasmodium. The vegetative phase of Eumycota may consist of a single cell, as in yeasts or some chytrids. More usually it consists of branched filaments or *hyphae*, forming a *mycelium*. The wall of the hypha is generally composed of microfibrils of chitin, but in some fungi the wall is made up of cellulose or other glucans. Occasionally cellulose and chitin have been reported to occur together. The hyphae may be *coenocytic*, i.e. contain numerous nuclei not separated by cell walls, or may be divided by transverse walls (*septa*) into uninucleate or multinucleate segments. Asexual reproduction is by spores of various kinds. In the Mastigomycotina motile zoospores are found, but in the Zygomycotina reproduction is by non-motile *aplanospores* passively carried by wind or other agencies, or sometimes within sporangia which are violently projected. Asexual reproduction by means of passively dispersed *conidia* is common in Ascomycotina, but is less common in the Basidiomycotina.

Sexual reproduction (a process involving nuclear fusion and meiosis at some point in the life cycle) occurs by a variety of methods. In some Mastigomycotina there is fusion of motile gametes (*planogametes*) which may be morphologically indistinguishable (*isogamy*), or fusion of a smaller active motile male gamete with a larger, more sluggish, but motile female gamete (*anisogamy*). Or the male gamete only may be motile, and the larger female gamete stationary (*oogamy*). Fusion between sexually differentiated hyphal branches (*gametangia*) is found in many fungi. In the Oomycetes there is marked disparity between the size of the male gametangia (*antheridia*) and the female gametangia (*oogonia*) showing another type of oogamy. Here the oogonia contain one to several eggs (*oospheres*), which following fertilisation become thick-walled *oospores*. In the Zygomycotina there is fusion (conjugation) between morphologically indistinguishable gametangia, and the fusion cell becomes a thick-walled *zygospore*. In some Ascomycotina and Basidiomycotina sexual fusion is initiated by transfer of a nonmotile *microconidium* or a *spermatium* to a receptive hypha, but in many others no specially differentiated sexual structures are found, and the sexual process is initiated by fusion between non-differentiated hyphae. Nuclear fusion in Ascomycotina is followed by meiosis and usually mitosis so that eight haploid nuclei are formed. Around these nuclei spores develop inside a sac or *ascus*. The eight haploid *ascospores* are usually violently projected from the ascus. In the

Basidiomycotina there is a corresponding stage of nuclear fusion in a specialised cell, the *basidium*. The nuclei are pushed into four outgrowths of the basidium and around each a spore (*basidiospore*) develops. The individual basidiospores are projected violently from the basidium.

In the Deuteromycotina sexual reproduction has not been reported. Whilst it is possible that some of these fungi may prove to have sexual states, it is likely that many Deuteromycotina have lost the ability to reproduce sexually. This does not mean that they are incapable of genetical recombination. It has been shown that many of them can recombine genetical properties through unconventional methods such as parasexual recombination, in which nuclear fusion occurs, but no proper meiosis.

In many fungi a single spore can germinate to give rise to a mycelium or thallus capable of sexual reproduction. Such fungi are *homothallic*. In others sexual reproduction only occurs following interaction of two differing thalli. Such fungi are *heterothallic*. The two thalli often show no morphological differences which amount to a difference in sex and they are said to differ in *mating type*.

1

Mastigomycotina

The Mastigomycotina are zoosporic fungi. Amongst the Eumycota three distinct types of zoospore have been described (Fig. 5).

1. CHYTRIDIOMYCETES: Posteriorly uniflagellate zoospores with the flagellum of the whiplash type found in the Chytridiales, Blastocladiales and Monoblepharidales. For details of the fine structure of uniflagellate zoospores see Fuller (1966).

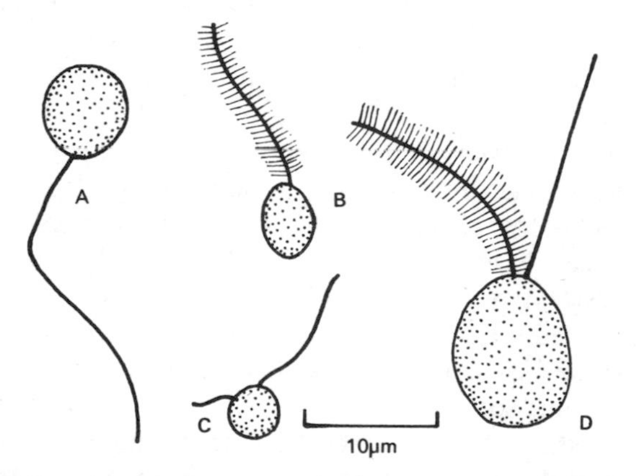

Figure 5. Types of zoospore. A. Posteriorly uniflagellate zoospore characteristic of Chytridiomycetes. B. Anteriorly flagellate zoospore characteristic of Hyphochytridiomycetes. C. Biflagellate zoospore characteristic of Plasmodiophoromycetes. D. Biflagellate zoospore characteristic of Oomycetes.

2. HYPHOCHYTRIDIOMYCETES: Anteriorly uniflagellate zoospores with the flagellum bearing lateral appendages (the so-called 'tinsel-type flagellum' or 'Flimmer-geissel'), characteristic of the Hyphochytriales (Fuller, 1966). This small group of aquatic fungi is not considered here in detail (but see Sparrow, 1960).

3. OOMYCETES: Biflagellate zoospores with subapically or laterally attached flagella; an anterior tinsel-type and a posterior whiplash-type flagellum characteristic of the Saprolegniales, Peronosporales, and two other orders (Colhoun, 1966).

The different structure of the zoospore suggests that the Mastigomycotina are polyphyletic, that is their evolutionary origin is to be found in unrelated groups of organisms. This means that the group is not a natural one but is merely an assemblage of organisms which have certain features in common, notably reproduction by zoospores.

CHYTRIDIOMYCETES

The Chytridiomycetes have one feature in common, the posteriorly uniflagellate zoospore of the whiplash type, and it has been suggested (Sparrow, 1958) that the ancestry of these organisms lies in the posteriorly uniflagellate Monads. The characters by which the group is separated into three orders are shown below (Sparrow, 1960).

(*a*) Thallus either lacking a vegetative system and converted as a whole into reproductive structures (**holocarpic**) or with a specialised rhizoidal vegetative system (**eucarpic**) and one (**monocentric**) or more (**polycentric**) reproductive structures; zoospore usually bearing a single conspicuous oil globule, germination monopolar CHYTRIDIALES

(*b*) Thallus nearly always differentiated into a well-developed vegetative system, often hypha-like, on which are borne numerous reproductive organs: zoospore without a conspicuous globule, germination bipolar.

(i) Thallus usually having a well-defined basal cell anchored in the substratum by a system of tapering rhizoids; resting structure an asexually formed, thick-walled, often punctate, resting spore; sexual reproduction by means of isogamous or anisogamous planogametes; alternation of generations present in some species BLASTOCLADIALES

(ii) Thallus without a well-defined basal cell; composed of delicate much-branched hyphae; resting structure an oospore; sexual reproduction oogamous, the male gamete always free-swimming, the female devoid of a flagellum MONOBLEPHARIDALES

CHYTRIDIALES

Members of this group are mostly aquatic, growing saprophytically on plant and animal remains in water or parasitically in the cells of algae and small aquatic animals. Some are terrestrial, apparently growing saprophytically on various plant and animal substrata in soil, whilst some attack the underground parts and also the aerial shoots of higher plants, occasionally causing diseases which are of economic significance, as in the case of *Synchytrium endobioticum* which causes black wart disease in potato. The chytrids which parasitise algae may cause severe depletion in the population level of the host alga. Some of the soil and mud-inhibiting chytrids can decompose cellulose, chitin and keratin, and 'baits' composed largely of these materials such as cellophane, shrimp exoskeleton, snake skin and hair, if floated in water to which soil has been added frequently become colonised by chytrid zoospores which give rise to mature thalli. Details of such methods for isolating these organisms have been given by Sparrow (1957, 1960), Willoughby (1956, 1958), and Emerson (1958). Some of the saprophytic forms have been grown in pure culture and information is accumulating about their nutrition and physiology. The nutritional requirements of most species which have been investigated are simple, and media containing mineral salts and carbohydrate in the form of sugar, starch or cellulose will support growth. As a group the chytrids have the capacity to reduce sulphate and to utilise both nitrate and ammonium salts for growth, and it has been argued that these are primitive nutritional characteristics (Cantino & Turian, 1959). A further 'primitive' feature is the ability of some chytrids to synthesise essential vitamins for growth, so that growth occurs on unsupplemented media, but it is also known that some members of the group are heterotrophic for vitamins, and their growth is stimulated by the addition of thiamine and other vitamins (Goldstein, 1960*a*,*b*; 1961).

The cell walls of some chytrid thalli have been examined microchemically and by X-ray diffraction, and chitin has been detected. The composition of the wall is of interest because chitin, a polymer of N-acetylglucosamine is also present in the walls of the Blastocladiales, Monoblepharidales, Zygomycotina, Ascomycotina, Basidiomycotina and Deuteromycotina whilst the cell walls of members of the Oomycetes are composed of cellulose, a linear polymer of D-glucose. Cellulose and chitin have been detected by X-ray diffraction methods occurring together in the walls of a species of *Rhizidiomyces*, a member of the Hyphochytridiomycetes (Aronson, 1965).

The form of the thallus in the Chytridiales is very variable. In the morphologically simpler types, such as *Olpidium* and *Synchytrium*, the mature thallus is a spherical or cylindrical sac surrounded by a wall. There are no rhizoids and the entire cytoplasmic contents of the thallus become transformed into reproductive structures, zoospores or gametes. Thalli of this type in which the whole structure is reproductive in function are described as **holocarpic**. In many other chytrids the thallus is differentiated into a vegetative part, concerned primarily with the collection of nutrients from the substratum, and a reproductive part which gives rise to zoospores or gametes. This type of thallus construction, which is the most common type, is termed **eucarpic**. The relationship between the rhizoidal system, the sporangium and the substratum is also variable. In some chytrids such as *Rhizophydium* the rhizoidal system only penetrates the host cell (often an algal cell or a pollen grain) and the sporangium is superficial or **epibiotic**. In others, such as *Diplophlyctis*, the whole thallus, rhizoids and sporangium are formed within the host cell, and are described as **endobiotic**. In *Physoderma* both epibiotic and endobiotic sporangia are found. Whilst in many types the zoospore on germination gives rise to a rhizoidal system bearing a single sporangium or resting spore (e.g. in *Entophlyctis*, *Rhizophlyctis* and *Diplophlyctis*), in others such as *Cladochytrium* and *Nowakowskiella* a more extensive rhizoidal system, sometimes termed the rhizomycelium, is established, on which numerous sporangia develop. Such thalli are **polycentric**, that is they form several reproductive centres instead of the single one where the thallus is termed **monocentric**. These types of thallus structure are illustrated in Fig. 6. When a

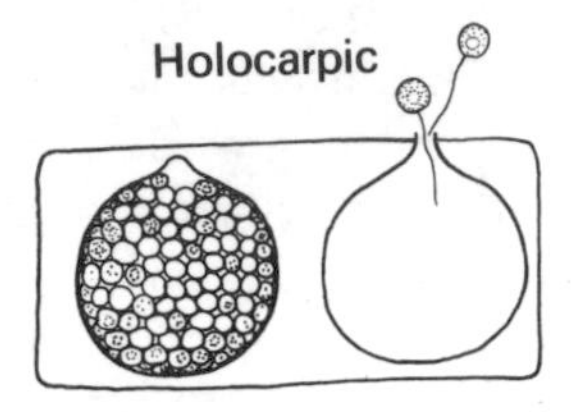

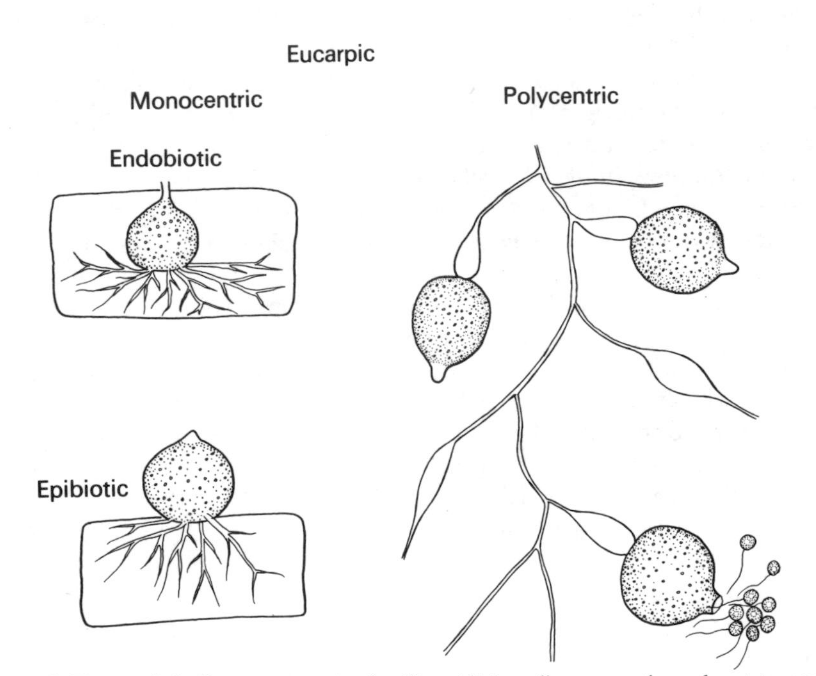

Figure 6. Types of thallus structure in the Chytridiales, diagrammatic and not to scale.

range of isolates of certain fungi are studied, it may be found that some are monocentric, whilst others are polycentric. Monocentric and polycentric forms of *Rhizophlyctis rosea* have been described. Thus the terms monocentric and polycentric are useful in a descriptive sense, but the distinction between the two conditions may not always be clear-cut.

The zoosporangium is usually a spherical or pear-shaped sac bearing one or more discharge tubes or exit papillae. The method by which zoospore release is achieved is used in classification. In the **inoperculate** chytrids such as *Olpidium*, *Diplophlyctis* and *Cladochytrium* the sporangium forms a discharge tube which penetrates to the exterior of the host cell and its tip becomes gelatinous and dissolves away. In **operculate** chytrids such as *Chytridium* and *Nowakowskiella* the tip of the discharge tube breaks open at a special line of weakness and becomes detached as a special cap or operculum or in some forms may remain attached at one side of the discharge tube and fold back like a hinged lid to allow the zoospores to escape.

The numbers of zoospores formed inside zoosporangia varies with the size of the zoospore and the zoosporangium. Although the size of the zoospore is roughly constant for a given species the size of the sporangium may be very variable. In *Rhizophlyctis rosea* tiny sporangia containing only one or two zoospores have been reported from culture media deficient in carbohydrate but on cellulose-rich media large sporangia containing many thousands of spores are commonly formed. The release of zoospores is brought about by internal pressure which causes the exit papillae to burst open. In studies of the fine structure of mature sporangia of *R. rosea* and *Nowakowskiella profusa* (Chambers & Willoughby, 1964; Chambers *et al.*, 1967), it has been shown that the single flagellum is coiled round the zoospore like a watch-spring. The zoospores are separated by a matrix of spongy material which may absorb water and swell rapidly at the final stages of sporangial maturation. When the internal pressure has been relieved by the escape of some zoospores those remaining inside the sporangium may escape by swimming or wriggling through the exit tube. In some species the spores are discharged in a mass which later separates into single zoospores, but in others the zoospores make their escape individually. The external form of the zoospore is similar in all chytrids, but in the details of internal structure there is considerable variation (see Fig. 7). There is a spherical body, which in some form is capable of plastic changes in shape, and a long trailing whiplash type of flagellum. When swimming the zoospores glide through the water often making abrupt changes in direction and also showing characteristic jerky or 'hopping' movements. In some species amoeboid crawling of zoospores has been reported. The internal structure of the zoospore as revealed by the light microscope and the electron microscope is somewhat variable (see Koch, 1956, 1958, 1961). The tip of the flagellum is often tapered into a narrower portion which may end in a bulbous enlargement. The detailed structure of the flagellum is revealed in electron micrographs of preparations in which the flagellum has become dismembered or in sections. It is then found to be composed of 11 fibrils, two thin central strands surrounded by nine slightly thicker ones, although the narrower tip of the flagellum may contain fewer strands. The individual fibrils may themselves be composed of two or three subfibrils, so that the whole flagellum may represent a complex bundle of about 30 subfibrils. The 11 major strands of flagellar structure

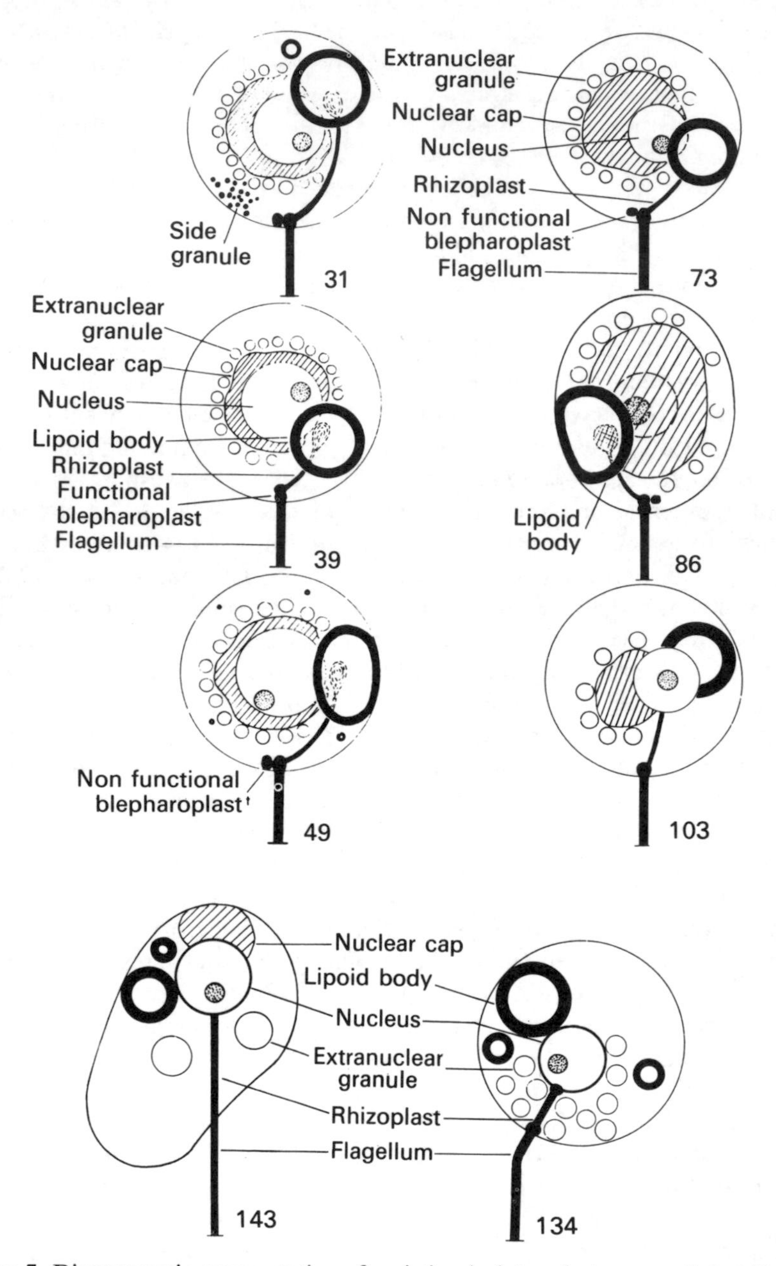

Figure 7. Diagrammatic representation of variation in internal structure of chytrid zoospores (after Koch, 1958). 31, *Nowakowskiella* sp.; 39, *Catenochytridium carolinianum*; 49, *Septochytrium variabile*; 73, *Rhizophydium sphaerotheca*; 86, *Chytridium* sp.; 103, *Phlyctochytrium irregulare*; 134, *Phlyctochytrium punctatum*; 143, *Rozella allomycis.*

are a feature common in a wide range of flagellated plant and animal cells of diverse relationships. Within the body of the zoospore in most chytrids the flagellum is attached to a basal granule or blepharoplast which is itself fibrillar in structure. An interesting feature of certain (but not all) chytrid zoospores is the presence of a second blepharoplast, attached to the one bearing the flagellum. Koch has termed the second blepharoplast 'non-functional'. Its presence has aroused speculation that the ancestors of these chytrids may have been biflagellate, and that the 'non-functional' blepharoplast is a relic of the second flagellum which has disappeared. In some zoospores the 'functional' blepharoplast is connected to a fibrillar rhizoplast which may end in a flattened grid-like disc probably attached to a lipoid body close to the nucleus. Possibly the disc is a photoreceptor (Chambers *et al.*, 1967). The nucleus itself is single and is surrounded in many cases (but not all) by a 'nuclear cap' of uneven thickness. The nuclear cap is usually thinner near the point of attachment of the rhizoplast to the nucleus. Beyond the nuclear cap in chytrid zoospores there is a closely apposed group of mitochondria. The most conspicuous feature within the zoospore is a large refringent globule which is a lipoid body, and in some zoospores other smaller lipoid bodies may be found. It has been suggested that the lipoid body forms a concentrated food reserve to provide energy for zoospore movement, or that it may function as a lens, concentrating light on the grid-like plate.

The period of zoospore movement varies. Some flagellate zoospores seem to be incapable of active swimming, and amoeboid crawling may take place instead, or swimming may last for only a few minutes. In others motility may be prolonged for several hours. On germination the zoospore comes to rest and encysts. The flagellum may contract, it may be completely withdrawn or it may be cast off, but the precise details are often difficult to follow. The subsequent behaviour differs in different species. In holocarpic parasites the zoospore encysts on a host cell and the cyst wall and host-cell wall are dissolved and the cytoplasmic contents of the zoospore enter the host cell. In many monocentric chytrids rhizoids develop from one point on the zoospore cyst and the cyst itself enlarges to form the zoosporangium, but there are variants of this type of development in which the cyst enlarges into a prosporangium from which the zoosporangium later develops. In the polycentric types the zoospore on germination may form a limited rhizomycelium on which a swollen cell arises, giving off further branches of rhizomycelium. It has been claimed that germination from a single point on the wall of the zoospore cyst (monopolar germination) as distinct from two points, enabling growth to take place in two directions (bipolar germination), is an important character distinguishing the Chytridiales (monopolar) and the Blastocladiales (bipolar), but this distinction is not absolute. Indeed the distinction between the two groups is becoming less clear. The morphological resemblance between certain forms is certainly most striking (compare *Rhizophlyctis rosea* and *Blastocladiella emersonii*).

CLASSIFICATION

Sparrow (1960) has provided a key separating the Chytridiales into families. The primary separation is based on the presence or absence of an operculum. Whiffen (1944) has objected to this method of classifying chytrids on the grounds that this character is of minor significance. She has emphasised instead the development of the thallus in relation to zoospore germination. For example, in *Rhizophydium* and *Rhizidium* the encysted zoospore enlarges into a zoosporangium, whilst in *Polyphagus* the encysted zoospore enlarges into a prosporangium from which the zoosporangium develops. In *Entophlyctis* the zoosporangium develops from an enlargement of the germ tube whilst in *Diplophlyctis* the germ tube enlarges to form a prosporangium from which the zoosporangium develops. On the basis of these and other differences Whiffen has proposed a division of the Chytridiales which she claims results in a more natural taxonomic grouping. However, the detailed classification need not concern us, and we shall study only selected examples.

Olpidiaceae

In this family the thallus is endobiotic and holocarpic and becomes entirely converted into a zoosporangium or resting sporangium. Sexual reproduction occurs by fusion of motile isogametes to form a biflagellate zygote which penetrates the host to form an endobiotic resting sporangium, which on germination gives rise to zoospores. The members of the family are mostly aquatic, but species of *Olpidium* are parasitic on the roots of higher plants.

Olpidium

Olpidium is a good example of the holocarpic type of thallus. About 30 species are known, most of them parasitic on aquatic algae, microscopic aquatic plants or the spores of various plants which fall into water or the soil. Other species are parasitic on moss protonemata and on leaves and roots of higher plants. *O. brassicae* is common on the roots of cabbages, especially when growing in wet soils, but is also found on a wide range of unrelated hosts. On lettuce a fungus morphologically similar to *O. brassicae* has been shown to be associated with symptoms of big-vein disease (yellow-vein banding sometimes associated with leaf puckering). However, not all lettuces showing big-vein symptoms contain *Olpidium*, and plants heavily infected with *Olpidium* may lack big-vein symptoms. Because of the characteristic syndrome and because the disease can be transmitted by grafting of lettuce shoots in the absence of *Olpidium* infection it has been concluded that a virus is involved. At first the virus was thought to be tobacco necrosis virus but some workers now regard it as a distinct big-vein virus. There is now much evidence to suggest that *Olpidium* can serve as a vector of several viruses. The virus is possibly carried either on or inside zoospores, and although there is no evidence that the virus can multiply inside the *Olpidium* thallus it can probably survive several months within the thick-walled resting sporangia and subsequently give rise to infection (Hewitt & Grogan, 1967).

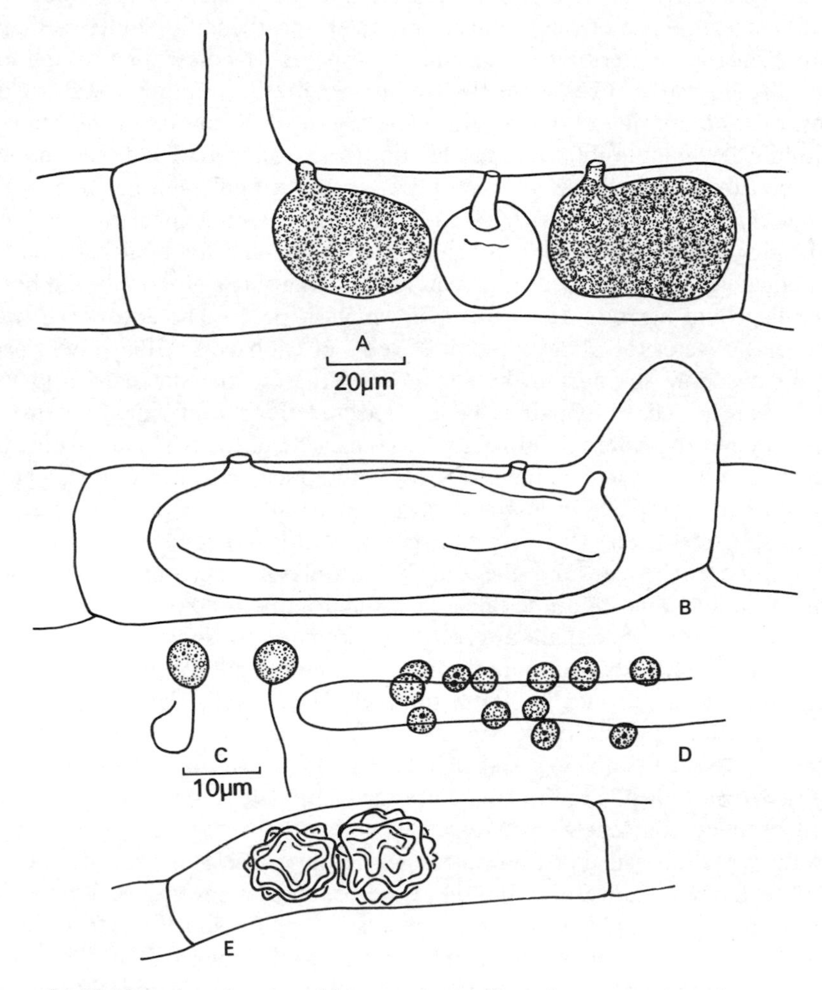

Figure 8. *Olpidium brassicae* in cabbage roots. A. Two ripe sporangia and one empty sporangium in an epidermal cell. Each sporangium has a single exit tube. B. Empty sporangium showing three exit tubes. C. Zoospores. D. Zoospore cysts on a root hair. Note that some cysts are uninucleate and some are binucleate. E. Resting sporangia. A, B, D, E to same scale.

The strain of *Olpidium* on lettuce is possibly not identical with that from cabbages since attempts to cross-inoculate the fungus from one host to the other have been unsuccessful. Sahtiyanci (1962) has described it as a distinct species. On cabbages infection seems to have little effect. When infected cabbage roots are washed in water the fungus can be seen under the low power of the microscope within the epidermal cells (and occasionally in the cortical cells) and root hairs (Fig. 8). The *Olpidium* thalli are spherical or cylindrical, and there may be one to several in a host cell. There are no rhizoids. Occasionally a large cylindrical thallus completely fills an epidermal cell. The cytoplasm of the thallus is granular and the entire contents divide into numerous posteriorly uniflagellate zoospores which escape through one or more discharge tubes from the thallus penetrating the outer wall of the host cell and opening to the exterior. The release of the zoospores takes place within a few minutes of washing the roots free from soil. The tip of the discharge tube breaks down and zoospores rush out and swim actively in the water. The zoospores are very small, tadpole-like, with a spherical head and a trailing flagellum. At high magnification under the light microscope the spherical body of the zoospore is seen to contain a single globose lipoid body. The flagellum narrows at its tip to a whiplash point. The zoospores swim actively in water for about 20 min. If roots of cabbage seedlings are placed in the zoospore suspension the zoospores settle on the root hairs and lose their flagella. The root hair is penetrated and the cytoplasmic contents of the encysted zoospore are transferred to the inside of the root hair, whilst the empty zoospore cyst remains attached to the outside. The process of penetration can take place in less than one hour. Within two days of infection small spherical thalli can be seen in the root hairs and epidermal cells of the root often carried around the cell by cytoplasmic streaming. The thalli enlarge and become multinucleate and within four to five days discharge tubes are developed and the thalli are ready to discharge zoospores.

In some infected roots, in addition to the smooth zoosporangia with their discharge tubes, stellate bodies with thick folded walls, lacking discharge tubes, are also found. These are resting sporangia and it seems likely that they are formed following a sexual fusion. In *O. viciae* and in *O. trifolii* Kusano has shown that the zoospores may copulate outside the host plant and produce biflagellate zygotes which infect their respective host plants, producing thick-walled resting sporangia. In these species the resting sporangia break open after several months to release further zoospores. The resting sporangia are binucleate and nuclear fusion occurs shortly before germination. Meiosis probably takes place during the division of the fusion nucleus before zoospore formation. In *O. brassicae* there is similarly evidence of zoospore fusion, indeed compound zoospores with up to six flagella have been seen. A proportion of the zoospore cysts found on root hairs are binucleate (see Fig. 7), and the resting sporangia are also binucleate (see Sampson, 1939). Sahtiyanci (1962) has shown that if cultures on cabbage roots were started from a single zoosporangium, resting sporangia were not formed. When cultures are started from zoospores derived from several zoosporangia resting spores were formed. By mixing zoospore suspensions from eight single zoosporangial lines in all possible combinations it was shown that the fungus exists in two distinct strains, and resting sporangia occurred only when opposite strains were mixed. The resting sporangia were capable of germination 7–10 days after they were mature and germinated by the formation of one of two exit papillae through which numerous zoospores were discharged. Figure 9 shows an outline of the probable life cycle of *Olpidium.*

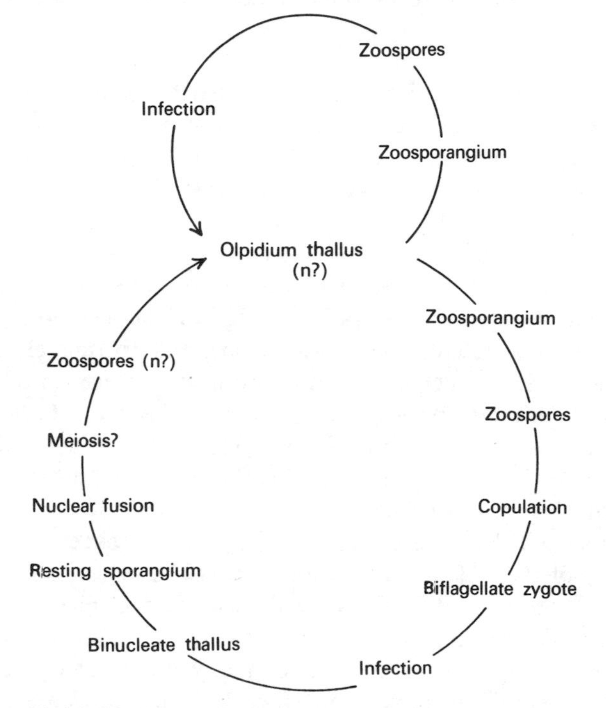

Figure 9. The probable life history of *Olpidium brassicae*. Certain information is lacking for this fungus, and this is indicated by question marks. This kind of life history has, however, been described in *O. viciae* and *O. trifolii*.

Synchytriaceae

In this family the thallus is endobiotic and holocarpic, and at reproduction it may become converted directly into a group (or sorus) of sporangia, or to a prosorus which later gives rise to a sorus of sporangia. Alternatively the thallus may become converted into a resting sporangium which can either function directly as a sporangium and give rise to zoospores, or can function as a prosorus, producing a vesicle whose contents cleave to form a sorus of sporangia. The zoospores are of the characteristic chytrid type. Sexual reproduction is by copulation of isogametes, resulting in the formation of thick-walled resting sporangia. Sparrow (1960) recognises three genera, *Synchytrium*, *Endodesmidium* and *Micromyces*. *Endodesmidium* and *Micromyces* are parasitic on green algae, but the largest genus is *Synchytrium*, with perhaps more than 100 species parasitic on flowering plants. Some species parasitise only a narrow range of hosts, e.g. *S. endobioticum* on Solanaceae, but others, e.g. *S. macrosporum*, may attack a wide range of hosts (Karling, 1964). Many species are not very destructive to the host plant, but result in the formation of galls on leaves, stems and fruits. The most serious parasite is *S. endobioticum*, the cause of black wart disease (or wart disease) of potato.

Synchytrium

Wart disease of potato (see Fig. 10) is now distributed throughout the main potato-growing regions of the world, especially in mountainous areas and those with a cool, moist climate. Diseased potato tubers when lifted bear dark-brown, warty cauliflower-like excrescences. Galls may also be formed on the aerial shoots, and they are then green with convoluted leaf-like masses of tissue. Heavily infected tubers may have a considerable proportion of their tissues converted to warts. The yield of saleable potatoes from a heavily infected crop may be less than the actual weight of the seed potatoes planted. The disease is thus potentially a serious one, but fortunately varieties of potatoes are available which are immune from the disease, so that control is practicable. The life history of the causal fungus has been studied by Curtis (1921), Köhler (1923, 1931*a,b*) and by Heim (1956*a,b*). Heim's account differs substantially from those of Curtis and Köhler and has not been generally accepted (see Lingappa, 1958*a,b*). The dark warts on the tubers are galls in which the host cells have been stimulated by the presence of the fungus to divide. Many of the cells contain resting sporangia which are more or less spherical cells with thick dark-brown walls with folded, plate-like extensions (see Fig. 11A). The resting sporangia are released by the decay of the warts and they may remain alive in the soil for many years. The outer wall (epispore) bursts open by an irregular aperture and the endospore balloons out to form a vesicle within which a single sporangium differentiates (Kole, 1965*a*). Thus the resting sporangium functions as a prosporangium on germination, and not directly as a sporangium as earlier workers had supposed.

The zoospores are capable of swimming for about two hours in the soil water. If they alight on the surface of a potato 'eye' or some other part of the potato shoot such as a stolon or a young tuber before its epidermis is suberised, they come to rest, and withdraw their flagella. Penetration of the epidermal cell of the host shoot occurs, and the contents of the zoospore cyst are transferred to the host cell, whilst the cyst membrane remains attached to the outside. When a dormant 'eye' is infected, dormancy may be broken and the tuber may begin to sprout at the infected 'eye'. If the potato variety is susceptible to the disease the small fungal thallus inside the host cell enlarges, and the host cell is stimulated to enlarge. Cells surrounding the infected cell also enlarge so that a rosette of hypertrophied cells surrounds a central infected cell (Fig. 11G). The walls of these cells adjacent to the infected cell are often thickened and assume a dark-brown colour. The infected cell remains

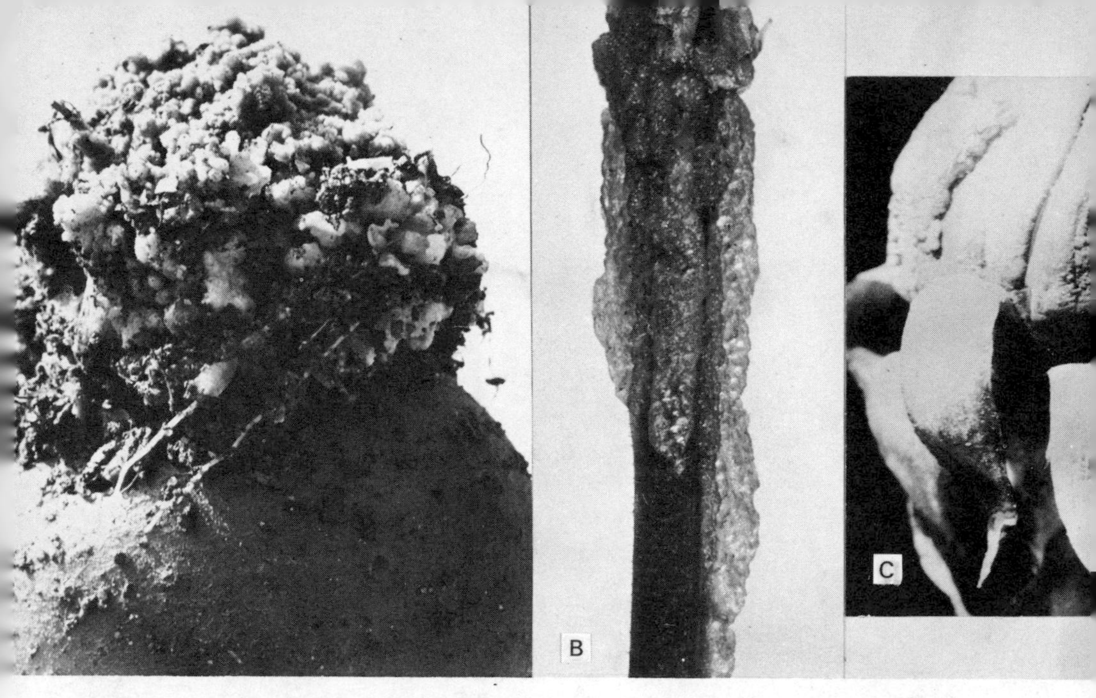

Figure 10. A. *Synchytrium endobioticum*. A potato tuber (variety Arran Chieftain) artificially infected with wart disease. The cauliflower-like excrescences are hypertrophied masses of host tissue resulting from abnormal growth and repeated reinfection of a shoot. The diseased tissue when growing near the surface of the soil is bright green and contains both prosori and resting sporangia. B. *S. mercurialis*. Stem of *Mercurialis perennis* showing hypertrophy of the epidermal cells surrounding resting sporangia of the fungus. C. *S. taraxaci*. Involucral bracts of *Taraxacum officinale* showing blisters of hypertrophied cells surrounding sporangial sori of the fungus.

alive for some time but eventually it dies. The parasite passes to the bottom of the host cell, enlarges, becomes spherical, and fills the lower part of the host cell. A double-layered chitinous wall which is golden-brown in colour is secreted around the thallus, and at this stage the thallus is termed a *prosorus* or summer spore. During its development until this stage the prosorus has remained uninucleate. Further development of the prosorus involves the protrusion of the inner wall through a pore in the outer wall, and the inner wall then expands as a vesicle which enlarges upwards and fills the upper half of the host cell (Fig. 11C). The cytoplasmic contents of the prosorus, including the nucleus are transferred to the vesicle. The process is quite rapid, and may be completed in four hours. During its passage into the vesicle the nucleus may divide, and division continues so that the vesicle contains about 32 nuclei, and at this stage the cytoplasmic contents of the vesicle become cleaved into a number of sporangia (Fig. 11D) forming a sorus. The number of sporangia varies from about four to nine. After the formation of the sporangial walls further nuclear divisions occur in each sporangium, and finally each nucleus with its surrounding mass of cytoplasm becomes differentiated to form a zoospore. As the sporangia ripen, they absorb water and swell, causing the host cell which contains them to burst open. Meanwhile, division

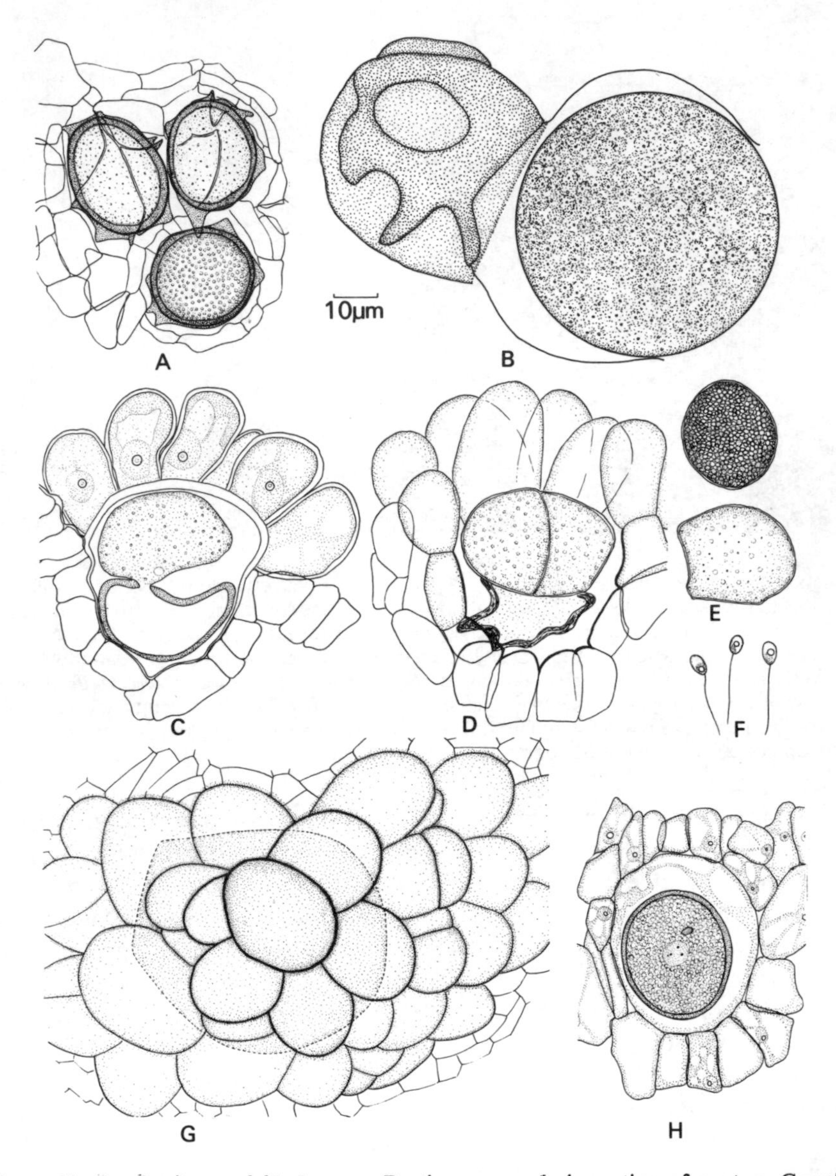

Figure 11. *Synchytrium endobioticum*. A. Resting sporangia in section of wart. B. Germinating resting sporangium showing the formation of a vesicle containing a single globose sporangium (after Kole, 1965*a*). C. Section of infected host cell containing a prosorus. The prosorus is extruding a vesicle. Note the hypertrophy of the infected cell and adjacent uninfected cells. D. Cleavage of vesicle contents to form zoosporangia. E. Two extruded zoosporangia. F. Zoospores. G. Rosette of hypertrophied potato cells as seen from the surface. The outline of the infected host cell is shown dotted. H. Young resting sporangium ersulting from infection by a zygote. Note that the infected cell lies beneath the epidermis due to division of the host cells.

of the host cells underlying the rosette has been taking place, and enlargement of these cells pushes the sporangia out on to the surface of the host tissue (Fig. 11E). The sporangia swell if water is available and burst open by means of a small slit through which the zoospores escape. There may be as many as 500–600 zoospores in large sporangia. The zoospores, which resemble those derived from resting sporangia, swim in the water film by a characteristic jerky, hopping movement and are capable of swimming for up to 20 hr. If suitable host tissues are available they encyst on the epidermis, and penetrate it within a few hours. Sometimes several zoospores succeed in penetrating a single cell so that it contains several fungal protoplasts. Within the host cell the thallus enlarges to form a prosorus, whilst the surrounding host cells enlarge to form the rosette. Eventually a further crop of sporangia is produced from which zoospores are released. This cycle of infection resulting in the formation of several generations of prosori can be continued throughout the spring and early summer.

According to Curtis, Köhler, and a number of other workers the resting sporangia of *S. endobioticum* are formed following copulation. These workers have noticed that the zoospores released from the soral sporangia may fuse in pairs (or occasionally in groups of three or four) to form zygotes which retain their flagella and swim actively for a time. The zoospores which function as gametes do not differ in size and shape, so copulation can be described as isogamous. There are however indications that the gametes may differ physiologically. Curtis has suggested that fusion may not occur between zoospores derived from a single sporangium, but only between zoospores from separate sporangia. Köhler (1956) has claimed that the zoospores are at first sexually neutral. Later they mature and become capable of copulation. Maturation may occur either outside the sporangia or within, so that in over-ripe sporangia the zoospores are capable of copulation on release. At first the zoospores are 'male', and swim actively. Later the swarmers become quiescent ('female') and probably secrete a substance which attracts 'male' gametes to them chemotactically. After swimming by means of the two flagella the zygote encysts on the surface of the host epidermis and penetration of the host cell may then follow by a process essentially similar to zoospore penetration. Multiple infections by several zygotes penetrating a single host cell can also occur. Nuclear fusion occurs in the young zygote, before penetration. The results of zygote infections differ from infection by azygotes (zoospores). When infection by an azygote occurs, the host cell reacts by undergoing *hypertrophy*, i.e. increase in cell volume, and adjacent cells also enlarge to form the characteristic rosette which surrounds the resulting prosorus. When a zygote infects, the host cell undergoes *hyperplasia*, i.e. repeated cell division. The parasite lies towards the bottom of the host cell, and division occurs in such a way that the fungal protoplast is transferred to the innermost daughter cell. As a result of repeated divisions of the host cells the fungal protoplasts may be buried several cell-layers deep beneath the epidermis (see Fig. 11H).

During these divisions of the host tissue the zygote thallus enlarges and becomes surrounded by a two-layered wall, a thick outer layer which eventually becomes dark-brown in colour and is thrown into folds or ridges which appear as spines in section, and a thin hyaline inner wall surrounding the granular cytoplasm. The host cell eventually dies and some of its contents may also be deposited on the outer wall of the resting sporangium. During its development the resting sporangium remains uninucleate. The resting sporangia are released into the soil and are capable of germination within about two months. Before germination the nucleus divides repeatedly to form the nuclei of the

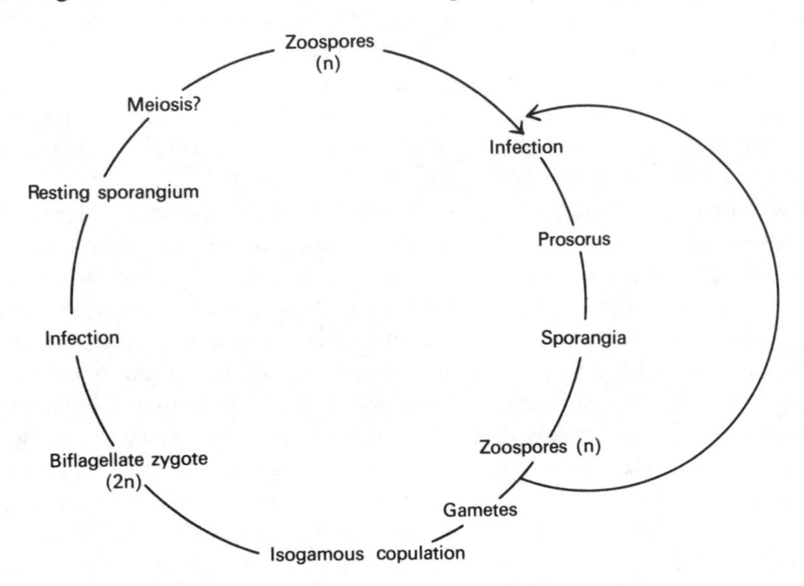

Figure 12. Probable life cycle of *Synchytrium endobioticum*.

zoospores whose further development has already been described. Some cytological details of the life history are still in doubt. Presumably the zygote and the young resting sporangium are diploid, and it has been assumed that meiosis occurs during germination of the resting sporangia prior to the formation of zoospores, so that these zoospores, the prosori and the soral zoospores are also believed to be haploid. These assumptions seem plausible in the light of knowledge of the life history and cytology of *Synchytrium fulgens*, a parasite of *Oenothera*, described by Kusano (1930*a,b*). In *S. fulgens* Kusano has described the occurrence of meiosis during the germination of resting sporangia, and Lingappa has also suggested that meiosis occurs at this point (Lingappa, 1958*b*; Karling, 1958). An outline of the probable life history of *S. endobioticum* is shown in Fig. 12.

CONTROL OF WART DISEASE

Control is based largely on the breeding of resistant varieties of potato. It was discovered that certain varieties of potato such as Snowdrop were immune from the disease and could be planted on land heavily infected with *Synchytrium* without developing warts. Following this discovery plant breeders have developed a number of immune varieties such as Arran Pilot, Arran Banner, Arran Peak and Majestic. A number of potato varieties susceptible to the disease are still widely grown, however, including King Edward, Eclipse and Duke of York. In most countries where wart disease occurs legislation has been introduced requiring that only approved immune varieties are planted on land where wart disease has been known to occur, and prohibiting the movement and sale of diseased material. In Britain the growing of immune varieties on infested land has prevented the spread of the disease, and it is confined to a small number of foci in the west Midlands, north-west England and mid- and south Scotland. The majority of the outbreaks are found in allotments, gardens and small-holdings. As new varieties of potatoes are developed by plant breeders they are tested for susceptibility to wart disease by the Ministry of Agriculture Plant Pathology Laboratory and most of those that are made available commercially are immune.

The reaction of immune varieties to infection varies. In some cases when 'immune' varieties are exposed to heavy infection in the laboratory they may become slightly infected, but infection is often confined to the superficial tissues which are soon sloughed off. In the field such slight infections would probably pass unnoticed. Occasionally infections of certain potato varieties may result in the formation of resting sporangia, but without the formation of noticeable galls. Penetration of the parasite seems to occur in all potato varieties, but when a cell of an immune variety is penetrated it may die within a few hours, and since the fungus is an obligate parasite, further development is checked. In other cases the parasite may persist in the host cell for up to two to three days, apparently showing normal development, but after this time the fungal thallus undergoes disorganisation and disappears from the host cell.

Other methods of control are less satisfactory. Attempts to kill the resting sporangia of the fungus in the soil have been made, but this is a costly and difficult process, and requires large applications of fungicides to the soil. In America applications of copper sulphate or ammonium thiocyanate at the rate of 1 ton/acre have been used, and local treatment with mercuric chloride has been used to eradicate foci of infection. Control measures based on the use of resistant varieties seem more satisfactory. Unfortunately, it has been discovered that new physiological races (or biotypes) of the parasite have arisen which are capable of attacking varieties previously considered immune. About six biotypes of *S. endobioticum* are now known. The implications of the discovery of these new races are obvious. It is vital to prevent their spread, and prompt action has been taken to do this. If the dispersal of the new races cannot be prevented, of if, as seems likely, new races arise independently elsewhere, much of the work of the potato breeder during the last 50 years may have to be started all over again. So far there are no reports of outbreaks of new races in Britain, but this is not to say that they do not exist, and intensive research in areas where wart disease has persisted may bring them to light.

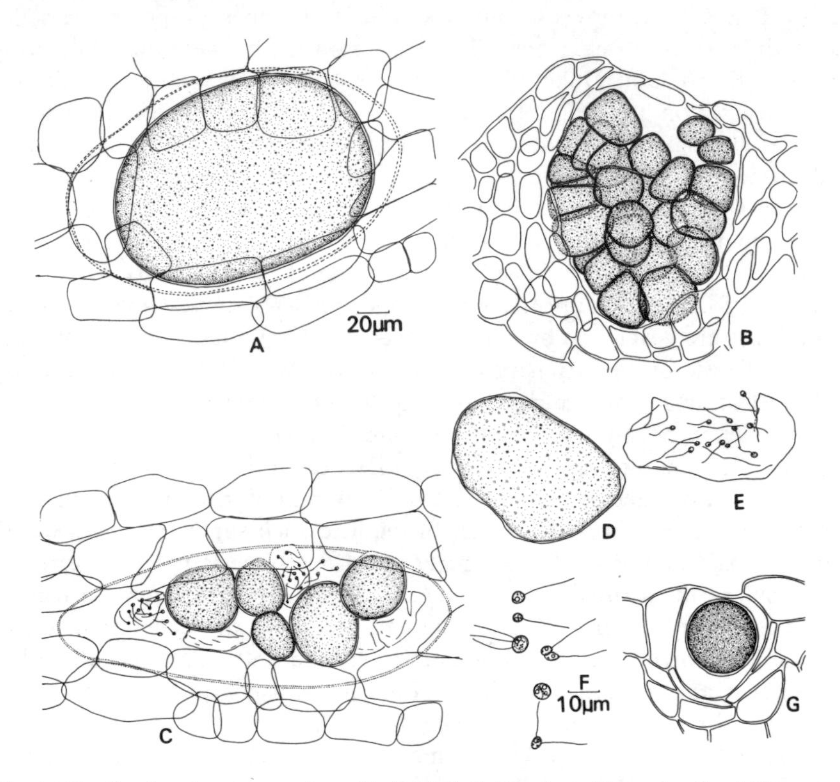

Figure 13. *Synchytrium taraxaci*. A. Undivided thallus in epidermal cell of *Taraxacum* scape. Outline of host cell shown dotted. B. Section of *Taraxacum* scape showing thallus divided into a sorus of sporangia. C. A sorus of sporangia seen from above. Two sporangia are releasing zoospores. D. A ripe sporangium. E. Sporangium releasing zoospores. F. Zoospores and zygotes. The triflagellate zoospore probably arose by incomplete separation of zoospore initials. G. Section of host leaf showing a resting sporangium. A, B, C, D, E, G to same scale.

OTHER SPECIES OF *Synchytrium*

Not all species of *Synchytrium* show the same kind of life history as that of *S. endobioticum*. *S. fulgens*, a parasite of *Oenothera*, resembles *S. endobioticum*. Both the summer spore and the resting sporangia function as prosori, extruding sporangia (Lingappa, 1958*a*,*b*). But in this species it has also been shown that the zoospores from resting sporangia can function as gametes and give rise to zygote infections from which further resting sporangia arise (Lingappa, 1958*b*; Kusano, 1930*a*). Köhler (1931*a*) has suggested that the same phenomenon may occasionally occur in *S. endobioticum*. In *S. taraxaci*, parasitic on *Taraxacum* (Fig. 13) and in a number of other species the mature thallus does not function as a prosorus but cleaves directly to form a sorus of sporangia and the resting sporangium also gives rise to zoospores directly. In some species, e.g. *S. aecidioides* resting sporangia are unknown whilst in others, e.g. *S. mercurialis*, a common parasite on leaves and stems of *Mercurialis perennis* (Fig. 14), only resting sporangia are known and summer sporangial sori do not occur. *Mercurialis* plants collected from March to June often show yellowish blisters on leaves and stems. The blisters are galls made up of one or two layers of hypertrophied cells, mostly lacking chlorophyll, which surround the *Synchytrium* thallus, which matures to form a

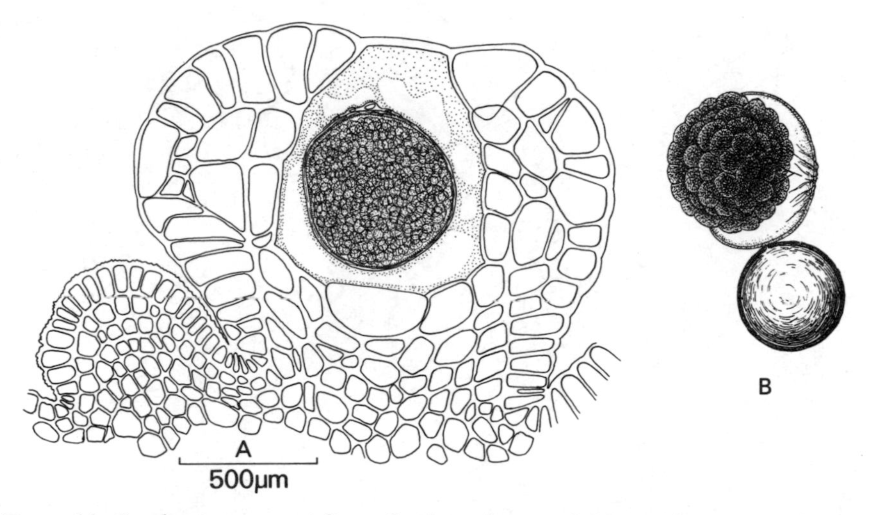

Figure 14. *Synchytrium mercurialis*. A. Section of stem of *Mercurialis perennis* showing hypertrophied cells surrounding a resting sporangium. B. Germination of the resting sporangia to release a sorus of zoosporangia. Thus in *S. mercurialis* the resting sporangium functions as a prosorus (after Fischer, 1892).

resting sporangium. In this species the resting sporangium functions as a prosorus during the following spring. The undivided contents are extruded into a spherical sac which becomes cleaved into a sorus containing as many as 120 sporangia, from which zoospores arise. The variations in the life histories of the various species of *Synchytrium* form a useful basis for classifying the genus (Karling, 1964).

Rhizidiaceae

In this family the thallus is typically monocentric, with a single, usually globose inoperculate sporangium or resting sporangium arising from an extensive richly-branched rhizoidal system. The resting sporangia may be formed asexually, or sexually by fusion of isogamous or anisogamous aplanogametes (i.e. non-motile gametes), and on germination the resting sporangia function either as sporangia or as prosporangia. The family includes forms parasitic on fresh-water algae and a large number saprophytic on the cast exoskeleton of insects and on plant debris in water and in soil.

Rhizophlyctis

Rhizophlyctis rosea grows on cellulose-rich substrata in a variety of habitats such as soil and lake mud, and it undoubtedly plays an active role in cellulose decay. It can readily be isolated and grown in culture. This has made possible

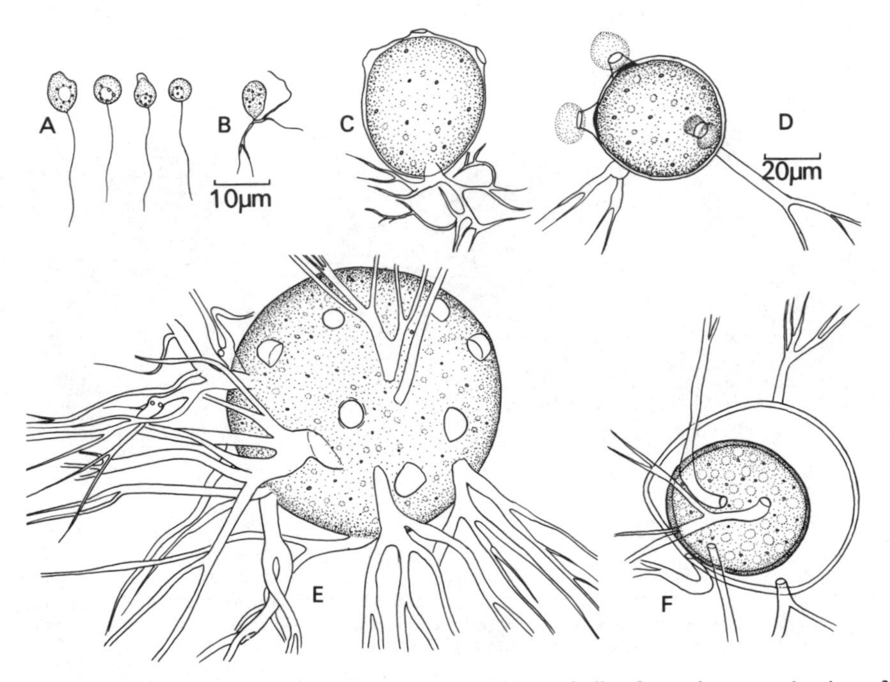

Figure 15. *Rhizophlyctis rosea.* A. Zoospores. B. Young thallus formed on germination of zoospore. The zoospore cyst has enlarged and will form the sporangium. C. An older plant showing three dehiscence papillae. D. A plant showing mucilage plugs at the tips of the dehiscence papillae and thickenings of the cell membrane at the bases of the papillae. Such thickenings are sometimes termed endo-opercula. E. A mature plant with a globose sporangium and seven visible papillae. F. A resting spore formed inside an empty zoosporangium. A, B to same scale; C, D, E, F, G to same scale.

studies on its nutrition and physiology (Stanier, 1942; Quantz, 1943; Haskins, 1939; Haskins & Weston, 1950; Cantino & Hyatt, 1953; Davies, 1961). Its nutritional requirements are simple and it shows vigorous growth on cellulose as the sole carbon source, although it can utilise a range of other carbohydrates such as glucose, cellobiose and starch. If boiled grass leaves or cellophane are floated in water to which soil or mud containing *R. rosea* has been added, the bright pink globose sporangia often up to 200 μ in diameter can be found after about seven days, and are easily visible under a dissecting microscope. The pink colour is due to the presence of carotenoid pigments such as γ-carotene, lycopene, and a xanthophyll. The sporangia are attached to coarse rhizoids which commonly arise at several points on the sporangial wall, and extend throughout the substratum, tapering to fine points. Although usually monocentric there are also records of some polycentric isolates. When ripe the sporangia have pink granular contents which differentiate into numerous uninucleate posteriorly uniflagellate zoospores (Fig. 15A). One to several discharge tubes are formed, and the tip of each tube contains a clear mucilaginous plug which prior to discharge is exuded in a mass from the tip of the tube (Fig. 15C). Whilst the plug of mucilage is being discharged the zoospores within the sporangium show active movement and then escape by swimming through the tube. In some specimens of *R. rosea* it has been found that a membrane may form over the cytoplasm at the base of the discharge tubes. If the sporangia do not discharge their spores immediately the membrane may thicken. When spore discharge occurs these thickened membranes can be seen often floating free within the sporangia, and the term endo-operculum has been applied to them. The genus *Karlingia* was erected for forms possessing such endo-opercula, including *R. rosea*, which is therefore sometimes referred to as *Karlingia rosea*, but the validity of this separation is questionable because the presence of absence of endo-opercula is a variable character. The zoospores are capable of swimming for several hours. The head of the zoospore is often globose, but can become pear-shaped or show amoeboid changes in shape. It contains a prominent lipoid body, several bright refrigent globules, and bears a single trailing flagellum. On coming to rest on a suitable substratum the flagellum is withdrawn and the body of the zoospore usually enlarges to form the rudiment of the sporangium, whilst rhizoids appear at various points on its surface.

Resting sporangia are also found. They are brown, globose or angular and have a thickened wall. Whether they are formed sexually in *R. rosea* is not known. Couch (1939) has, however, put forward evidence that the fungus is heterothallic (i.e. will only reproduce sexually following interaction of two differing thalli). Single isolates grown in culture failed to produce resting sporangia, but when certain cultures were paired resting sporangia were formed. Stanier (1942) has reported the occurrence of biflagellate zoospores, but whether these represented zygotes seemed doubtful. On germination the resting sporangium functions as a prosporangium.

Cladochytriaceae

In this family the thallus is eucarpic and usually polycentric, and the vegetative system may bear intercalary swellings, and septate turbinate cells (sometimes termed spindle organs). Many of its members are saprophytic on decaying plant remains, but some are parasitic on algae and on animal eggs.

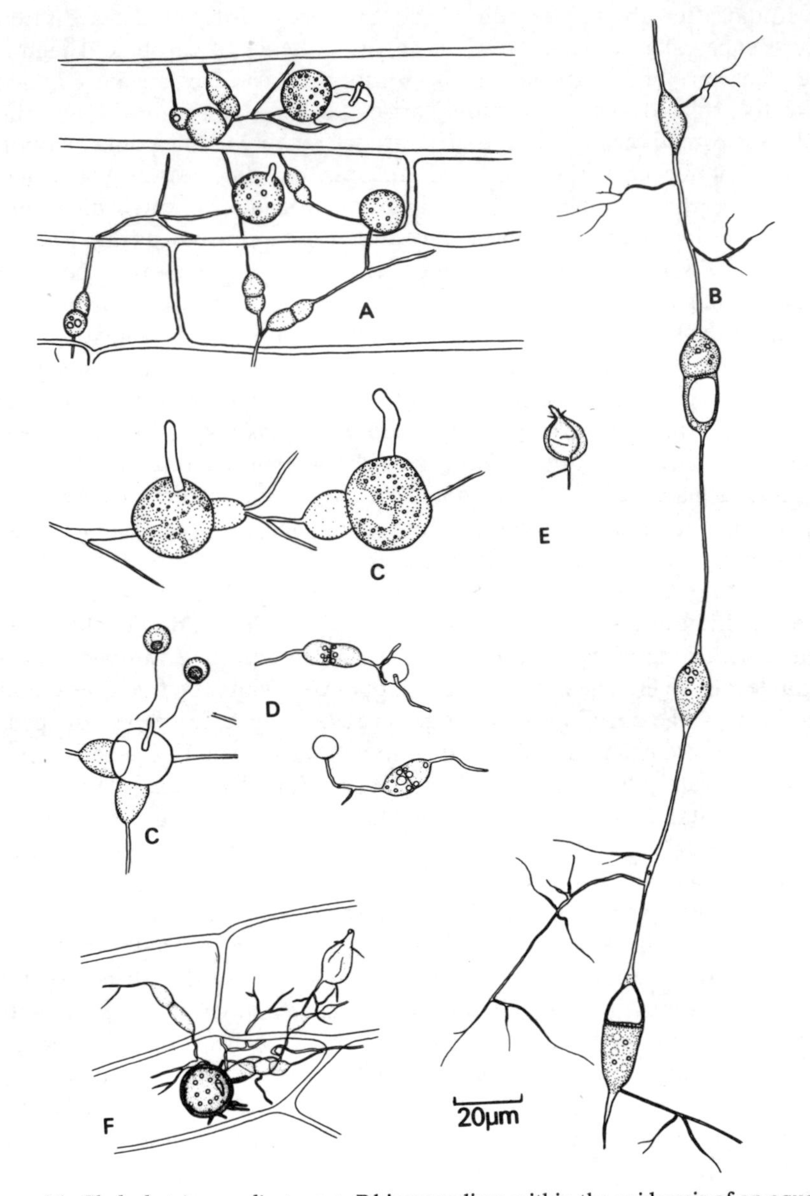

Figure 16. *Cladochytrium replicatum*. A. Rhizomycelium within the epidermis of an aquatic plant bearing the two-celled hyaline turbinate cells and globose orange zoosporangia. B. Rhizomycelium and turbinate cells from a culture. C. Zoosporangia from a two-week-old culture. One zoosporangium has released zoospores, each of which contains a bright orange coloured globule. D. Germinating zoospores on boiled wheat leaves. The empty zoospore cysts are spherical. The germ tubes have expanded to form turbinate cells. E. A zoosporangium which has proliferated internally to form a second sporangium. F. Rhizomycelium within a boiled wheat leaf bearing a thick-walled, spiny resting sporangium.

Cladochytrium

Cladochytrium replicatum is a common representative, which is to be found in decaying pieces of aquatic vegetation, and can be distinguished from other chytrids by the bright orange globules found in the sporangia. It can frequently be isolated during the winter months if moribund aquatic vegetation is placed in a dish of water and 'baited' with boiled grass leaves. The bright orange sporangia which are visible under a dissecting microscope appear within about five days arising from an extensively branched hyaline rhizomycelium bearing two-celled intercalary swellings. The zoospores on release from the sporangium each contain a single orange globule and bear a single posterior flagellum. After swimming for a short time the zoospore attaches itself to the surface of the substratum and puts out usually a single germ tube which can penetrate the tissues of the host plant. Within the cell of the host the germ tube expands to form an elliptical or cylindrical swollen cell which is often later divided into two by a transverse septum. This swollen cell is termed a spindle organ or turbinate cell. The zoospore is uninucleate and during germination the single nucleus is transferred to the swollen turbinate cell which becomes a vegetative centre from which further growth takes place. Nuclear division is apparently confined to the turbinate cells. From the turbinate cell rhizoids are put out which in turn produce further turbinate cells (see Fig. 16B,D). Nuclei are transported through the rhizoidal system but are not resident there. The thallus so established branches profusely and the characteristic two- or three-celled turbinate cells and the fine rhizoidal system may be widely distributed in the host tissues. At certain points on the thallus spherical zoosporangia form either terminally or in an intercalary position. Sometimes one of the cells of a pair of turbinate cells swells and becomes transformed into a sporangium, and in culture both cells may be modified in this way. The spherical zoosporangium undergoes progressive nuclear division and becomes multinucleate. Meanwhile the contents of the sporangium acquire a bright orange colour due to the accumulation of the globules containing the carotene lycopene which are later found in the zoospores. Cleavage of the cytoplasm to form uninucleate zoospore initials follows. The zoospores escape through a narrow exit tube which penetrates to the exterior of the host plant and becomes mucilaginous at the tip. There is no operculum. Sometimes the zoosporangia may proliferate internally, a new zoosporangium being formed inside the wall of an empty one. Resting sporangia with thicker walls and a more hyaline cytoplasm are also formed either terminally or in an intercalary position on the rhizomycelium. In some cases the wall of the resting sporangium is reported to be smooth and in others spiny, but it has been suggested (Sparrow, 1960) that the two kinds of resting sporangia may belong to different species. However, studies by Willoughby (1962*a*) of a number of single-spore isolates have shown that the presence or absence of spines is a variable character. The contents of the resting sporangia divide to form zoospores which also have a conspicuous orange granule, and escape by means of an exit tube as in the thin-walled zoosporangia. Whether the resting sporangia are formed as a result of a sexual process is not known. Pure cultures of *C. replicatum* have been studied by Willoughby (1962*a*) and Goldstein (1960*b*). The fungus is heterotrophic for thiamine. Biotin, while not required absolutely, stimulates growth. Nitrate and sulphate were utilised. A number of different carbohydrates were utilised, and a limited amount of growth took place on cellulose.

Chytridiaceae

The Chytridiaceae and the Megachytriaceae are distinguished from the preceding families in the method by which their sporangia discharge zoospores, by the removal of a circular lid or operculum. In the Chytridiaceae the thallus is monocentric, eucarpic and epi- or endobiotic, whilst in the Megachytriaceae the thallus is polycentric. Most Chytridiaceae are parasitic or saprophytic on fresh-water and marine algae, fungi, pollen and protozoa. There is a remarkable parallel in thallus organisation to the inoperculate chytrids. The details of sexual reproduction are known in only a few species, and usually involves fusion of adjacent 'male' and 'female' thalli, as in *Zygorhizidium* and *Chytridium*.

Chytridium

This is a large genus with about 40 species mostly parasitic on algae or on other fungi or protozoa. The sporangium is epibiotic, being formed by enlargement of the zoospore cyst, whilst the branched rhizoidal system penetrates the host. The thick-walled resting sporangium is endobiotic. *C. olla* is a parasite of oogonia and oospores of *Oedogonium*, but it can readily be grown in pure culture (Emerson, 1958). In culture the sporangia are globose, with granular pale-pink contents, and arise from an extensive rhizoidal system, with a series of main axes branching into very fine extremities. When mature sporangia (about three days old) are flooded with water they dehisce by lifting off a large operculum and numerous posteriorly uniflagellate zoospores escape and swim for a few hours (Fig. 17A). On coming to rest the flagellum is withdrawn and the zoospore develops a rhizoidal system from one end. The zoospore cyst meanwhile enlarges to form the body of the zoosporangium (Fig. 17B,C). In this species resting spores have been described on the endobiotic rhizoidal system within the algal host, but the details of their formation are not known. They are smooth and thick-walled, and after a long resting period they germinate to form a cylindrical germ tube which grows to the exterior of the *Oedogonium* oogonium and expands to form a pear-shaped epibiotic sporangium, resembling the normal epibiotic zoosporangia (Fig. 17D). Sexual fusion preceding the formation of resting spores has been described by Koch (1951) in *C. sexuale*. In this species which

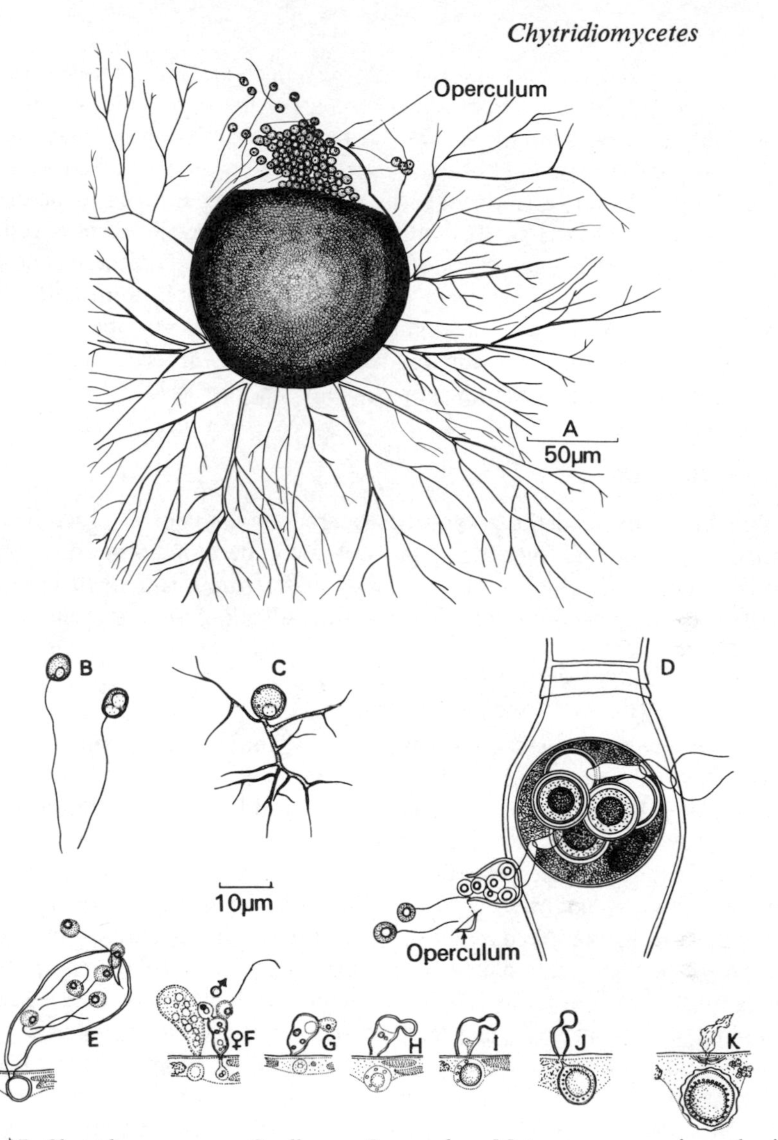

Figure 17. *Chytridium spp.* A–D, *C. olla* E–K, *C. sexuale*. A. Mature zoosporangium releasing zoospores through a wide operculum. B. Zoospores. C. Young plant showing the enlargement of the zoospore cyst to form the zoosporangium. D. Oogonium of *Oedogonium* with an immature oospore killed by the parasite. The oospore contains five resting spores of *Chytridium*, two of which have germinated to form a sporangium. Note the operculum cast off from the mouth of the lower sporangium. E. Zoosporangium of *C. sexuale* attached to a *Vaucheria* filament. Zoospores are being released. F–K. Stages in the formation of a resting sporangium. F. An epibiotic female filament (♀) bearing an attached encysted male thallus and a male motile cell. G, H, I, J. Stages in plasmogamy between male thallus and female thallus. The cytoplasm of the female thallus is gradually transferred from the epibiotic part to an endobiotic spherical cell, and in J the transfer is complete. K. Mature resting spore. The epibiotic part of the female thallus and the attached male thallus have collapsed. A, D to same scale; B, C, E–K to same scale. D, after de Bary, E–K, after Koch.

is parasitic on *Vaucheria* the motile cells may encyst on the host-cell wall and develop into a sporangial thallus, forming a further crop of zoospores from an epibiotic sporangium (Fig. 17E). Alternatively, the thalli may develop as 'female' thalli to which another zoospore may become attached, to form a 'male' thallus. Koch believes that 'the female thallus appears to be nothing more than a sporangial thallus with its exogenous development arrested at an early stage by the engagement of the male cells; however, there is no proof of this'. Later the encysted male cell empties its contents into the female thallus and the combined protoplasts move into an endobiotic swelling, increase in size and become surrounded by a thick warty wall (Fig. 17F–K). The germination of the resting spore has not been described.

Megachytriaceae

In this family most of the polycentric operculate chytrids are included. The thalli are epi- and endobiotic, and saprophytic in decaying remains of aquatic plants or occur in soil. The zoosporangia arise from terminal or intercalary swellings of the mycelium. Resting spores are also known, apparently formed asexually.

Nowakowskiella

Species of *Nowakowskiella* are widespread saprophytes in soil and in decaying aquatic plant debris, and can be obtained by 'baiting' aquatic plant remains in water with boiled grass leaves, cellophane and the like. *N. elegans* is often encountered in such cultures, and pure cultures can be obtained and grown on agar or on liquid culture media (Emerson, 1958). In boiled grass leaves the fungus forms an extensive rhizomycelium with intercalary swellings. Zoosporangia are formed terminally or in an intercalary position (see Fig. 18), and are globose or pear-shaped with a sub-sporangial swelling or apophysis, and granular or refractile hyaline contents. At maturity some sporangia develop a prominent beak, but in others this is not present. Sporangia dehisce by detaching an operculum (Fig. 18B,C) followed by escape of the zoospores which at first remain clumped together at the mouth of the sporangium. Thicker-walled, yellowish resting sporangia have been described (Fig. 18E) especially when liquid cultures are grown at slightly reduced temperatures (Emerson, 1958). There is no evidence that these structures are formed following sexual fusion, and details of their germination are not known. Goldstein (1961) has reported that the fungus requires thiamine, and can utilise nitrate, sulphate and a number of carbohydrates including cellulose, but cannot utilise starch.

In *N. profusa* three kinds of dehiscence have been described: exo-operculate in which the operculum breaks away to the outside of the sporangium; endo-operculate, in which the operculum remains within the sporangium; and inoperculate where the exit papilla opens without any clearly defined operculum (Chambers, Markus & Willoughby, 1967). This variation in dehiscence in a single strain of the fungus adds emphasis to criticisms of the value of dehiscence as a primary criterion in classification.

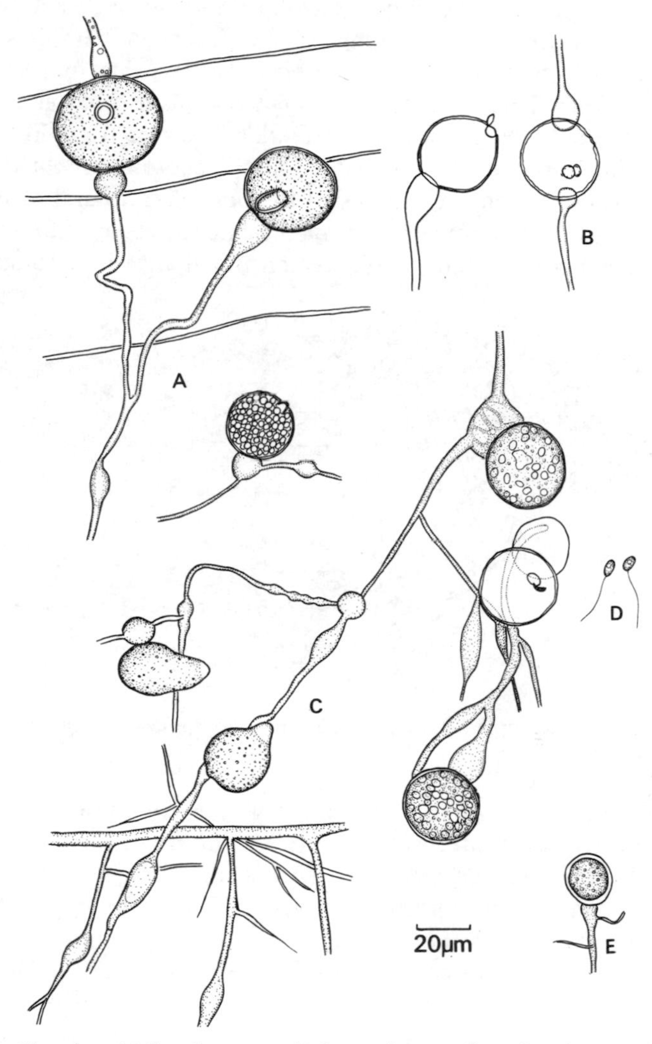

Figure 18. *Nowakowskiella elegans*. A. Polycentric mycelium bearing zoosporangia. B. Empty zoosporangia showing opercula. C. Mycelium showing turbinate cells and zoosporangia. D. Zoospores from culture. E. Resting spore from culture.

BLASTOCLADIALES

The Blastocladiales are mostly saprophytes in soil, water, mud or inhabiting plant and animal debris, but one genus *Coelomomyces* is made up of obligate parasites of insects, usually mosquito larvae. This genus is also unusual in having a naked plasmodium-like thallus lacking rhizoids. In the remaining genera the thallus is eucarpic. The morphologically simpler forms such as *Blastocladiella* are monocentric, with a spherical or sac-like zoosporangium or resting sporangium arising directly or on a short one-celled stalk from a tuft of radiating rhizoids. These simpler types show considerable similarity to the monocentric chytrids, and in the vegetative state they may be difficult to distinguish from them. The more complex organisms such as *Allomyces* are polycentric, and the thallus is differentiated into a trunk-like portion bearing rhizoids below, branching above, often dichotomously, and bearing sporangia of various kinds at the tips of the branches. Studies of the composition of the wall of *Allomyces* by microchemical, X-ray diffraction and electron microscope techniques have demonstrated that it is composed of

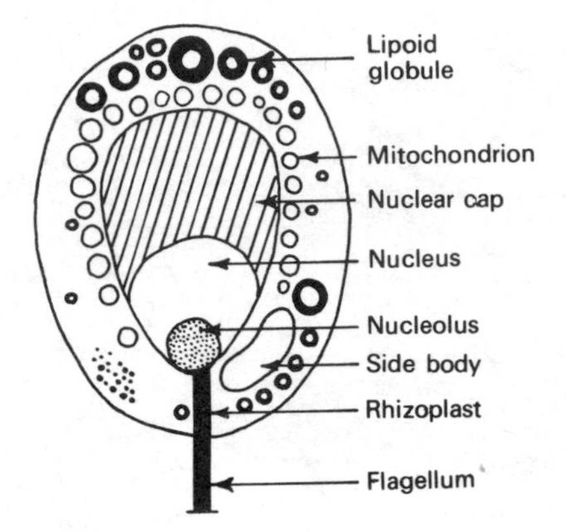

Figure 19. Diagrammatic representation of structure of zoospore of *Allomyces macrogynus* (after Koch, 1961).

microfibrils of chitin, and that appreciable quantities of glucan, ash and protein are intimately associated with the walls. Chitin has also been demonstrated in the walls of *Blastocladiella* (Aronson, 1965). The zoospores of Blastocladiales have a single posterior flagellum of the whip type. The flagellum in *Allomyces* shows the familiar eleven-strand structure found in numerous flagella (Kole, 1965*b*; Renaud & Swift, 1964; Fuller, 1966). The internal structure of the body of the zoospore in *Allomyces* is summarised in Fig. 19. Characteristic features are the large nuclear cap, and the presence of numerous small lipoid globules. In contrast the zoospores of the Chytridiales usually possess one or a few lipoid bodies, and whilst some possess nuclear caps, others have none. The nuclear cap of *Allomyces* consists largely of ribonucleic acid and protein and this has also been confirmed in *Blastocladiella*. The large side-body has been shown in *B. emersonii* to be a single mitochondrion, penetrated by a channel through which the flagellum-base passes, making contact with the nuclear membrane (Cantino, *et al.*, 1963, 1968; Reichle & Fuller, 1967; Lessie & Lovett, 1968). After swimming, the zoospore comes to rest, withdrawing its flagellum in *B. emersonii*. The flagellum is possibly wound in around the nuclear cap. In *Allomyces* the zoospore cyst produces a germ tube at one point, which branches to form the rhizoidal system. At the opposite pole the zoospore cyst forms a wider germ tube which gives rise to branches bearing sporangia. The biopolar method of germination is another point of difference between the Blastocladiales and the Chytridiales, in which germination is typically unipolar. A number of distinct life history patterns are found in the group. In *Allomyces arbuscula*, for example, an isomorphic alternation of haploid gametophyte and diploid sporophyte has been demonstrated. In *A. neo-moniliformis* (= *A. cystogenus*) there is no free-living sexual generation, but this stage is represented by a cyst (see below). In *A. anomalus* the asexual stage only has been found, and neither free-living sexual plants nor cysts have been reported. These types of life cycle have also been found in other genera, such as *Blastocladiella*. A characteristic feature of the asexual plants of the Blastocladiales is the presence of resting sporangia with dark brown pitted walls. The brown pigment in the resting sporangia of *Allomyces* is of the melanin group but γ-carotene is also present. Melanin has also been identified in the resting sporangia of *Blastocladiella*. The pits are inwardly directed conical pores in the wall. The inner ends of the pores abut against a smooth colourless inner layer of wall material surrounding the cytoplasm. Such pigmented pitted resting sporangia are not found in the Chytridiales.

The ease with which certain members of the group can be grown in culture has facilitated extensive studies of their nutrition and physiology, and the results of some of these investigations are discussed below.

Sparrow (1960) has recognised three families, but of these we shall study only representatives of the Blastocladiaceae.

Blastocladiaceae

Allomyces

Species of *Allomyces* are found most frequently on mud or soil from the tropics and subtropics, and if dried samples of soil are placed in water and 'baited' with boiled hemp seeds the baits may become colonised by zoospores. From such crude cultures it is possible to pipette zoospores on to suitable agar media and to follow the complete life history of these fungi in the laboratory. Emerson (1941) has analysed soil samples from all over the world for species of *Allomyces* and has distinguished three types of life history, represented by three subgenera.

1. *Eu-Allomyces*

The *Eu-Allomyces* type of life history is exemplified by *A. arbuscula* and *A. macrogynus*. In *A. arbuscula* the outer wall of the brown pitted resting sporan-

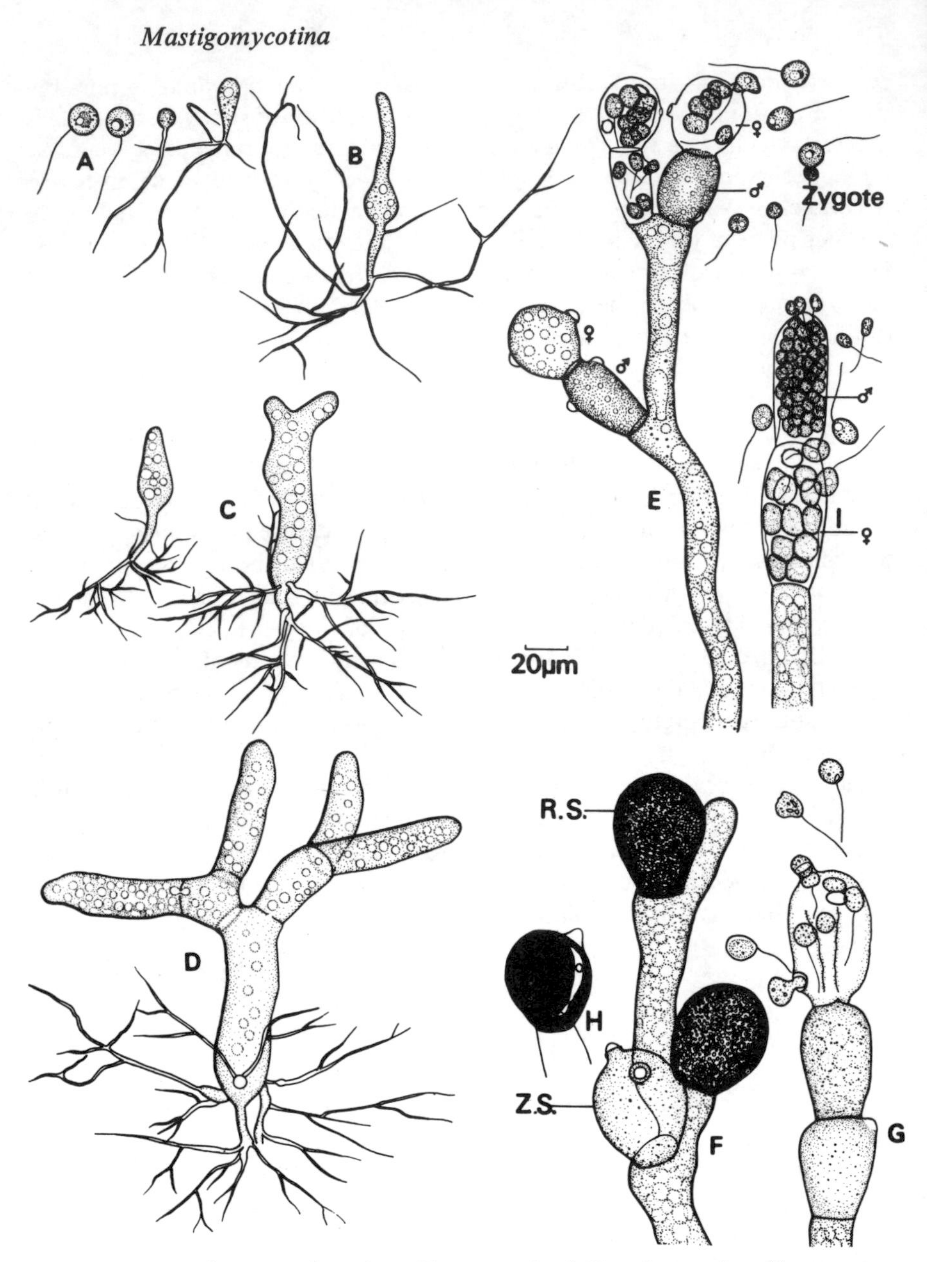

Figure 20. A–H. *Allomyces arbuscula*. A. Zoospores (haploid meiospores). B. Young gametophytes, 1 day old. C. Young sporophytes, 18 hr old. D. Sporophyte, 30 hr old. Note the perforations visible in some of the septa. E. Gametangia at tips of branches of gametophyte. Note the disparity in the size of the gametes. The smaller ♂ gametes are orange in colour whilst the larger ♀ gametes are colourless. Compare the arrangement of gametangia with those of *A. macrogynus* in I. F. R.S. Resting sporangia (or meiosporangia) and Z.S. zoosporangia (or mitosporangia) on sporophyte. G. Release of mitospores from mitosporangia of sporophyte. H. Rupture of meiosporangium. I. *Allomyces macrogynus*. Gametophyte showing gametangia. The ♂ gametangia are here terminal or epigynous.

gia cracks open by a slit and the inner wall balloons outwards, then opens by one or more pores, to release about 48 posteriorly uniflagellate swarmers. The resting sporangia are formed on asexual, diploid plants and cytological studies have shown that the resting sporangia contain about 12 diploid nuclei, which undergo meiotic division during the early stages of germination (Wilson, 1952). The cytoplasm cleaves around the 48 haploid nuclei to form the zoospores. Since meiosis occurs in the resting sporangia, they have been termed *meiosporangia*, and the haploid zoospores *meiospores*. The meiospores swim by movement of the trailing flagellum, and on coming to rest germinate as described above to form a rhizoidal system and a trunk-like region which bears dichotomous branches. Repeated nuclear division occurs to form a coenocytic structure, and ring-like ingrowths from the walls of the trunk-region and branches extend inwards to form incomplete septa with a pore in the centre through which cytoplasmic connection can be seen (Fig. 20D). The haploid plants which develop from the meiospores are gametophytic or sexual plants. They are monoecious, and the tips of their branches swell to form paired sacs—the male and female gametangia. The male gametangia can be distinguished from the female by the presence of a bright orange pigment, γ-carotene, whilst the female gametangia are colourless. In *A. arbuscula* the male gametangium is subterminal, or hypogynous, beneath the terminal female gametangium, but in *A. macrogynus* the positions are reversed and the male gametangium is terminal or epigynous, above the subterminal female (Fig. 20E,I). The gametangia bear a number of colourless papillae on their walls which eventually dissolve to form pores. The contents of the gametangia differentiate into uninucleate gametes which differ in size and pigmentation. The female gametangium forms colourless swarmers, whilst the male gametangium releases smaller, more active orange-coloured swarmers. After escape through the papillae in the walls of the gametangia the gametes swim for a time and then pair off. A female gamete which fails to pair can germinate to form a new sexual plant. There is evidence that a hormone, sirenin, secreted by the female gametangia during gametogenesis and by the released female gametes stimulates a chemotactic response in male gametes (Machlis, 1958*a*,*b*; Carlile & Machlis, 1965*a*,*b*). The paired gametes swim for a time, the zygotes showing two flagella. The cytoplasm of the two gametes becomes continuous, the zygote comes to rest, encysts, casts off the flagella, and nuclear fusion then follows. The zygote develops immediately into a diploid asexual plant resembling the sexual plants in general habit, but bearing two types of zoosporangia instead of gametangia. The first type is a thin-walled, papillate zoosporangium formed singly or in rows at the tips of the branches. Within these thin-walled sporangia the nuclei undergo mitosis and the cytoplasm cleaves around the nuclei to form diploid colourless zoospores which are released from the sporangia as plugs composed of a pectic substance blocking the papillae in the wall dissolve to form circular pores through which the swarmers can escape (Skucas, 1966).

Since nuclear division in the thin-walled sporangia is mitotic they are termed *mitosporangia*, and the swarmers they release are *mitospores*. The mitospores after a swimming phase encyst and are capable of immediate germination to form a further diploid asexual plant resembling the plant which bore them. The second type of zoosporangium is the dark-brown, thick-walled, pitted resting sporangium, or meiosporangium, formed at the tips of the branches. Meiotic divisions within these sporangia result in the formation of the haploid meiospores, which develop to form sexual plants. The life cycle of a member of the subgenous *Eu-Allomyces* is an isomorphic alternation of generations, the haploid sexual plant or gametophyte forming gametes, and the resulting zygotes develop into diploid sporophytes on which the mito- and meio-sporangia are formed (Fig. 22). Comparisons of the nutrition and physiology of the two generations show no essential distinction between them up to the point of production of gametangia or sporangia.

Emerson & Wilson (1954) and Emerson (1954) have made cytological and genetical studies of a number of collections of *Allomyces*. Interspecific hybrids between *A. arbuscula* and *A. macrogynus* have been produced in the laboratory, and it has been shown that the fungus earlier described as *A. javanicus* is a naturally occurring hybrid between these two species. Cytological examination of the two parent species and of artificial and natural hybrids showed a great variation in chromosome number. In *A. arbuscula* the basic haploid chromosome number is 8, but strains with 16, 24 and 32 chromosomes have been found. In *A. macrogynus* the lowest haploid number encountered was 14, but strains with 28 and 56 chromosomes are also known. The demonstration that these two species each represent a polyploid series was the first to be made in the fungi.

The behaviour of the hybrid strains is of considerable interest. The parent species differ in the arrangement of the primary pairs of gametangia, *A. arbuscula* being hypogynous, whilst *A. macrogynus* is epigynous. Zygotes formed following fusion of gametes derived from the different parents, germinated and gave rise to sporophyte plants. The meiospores from the hybrid sporophytes had a low viability (0·1–3·2% as compared with a viability of about 63% for *A. arbuscula* meiospores), but some germinated to form gametophytes. The arrangement of the gametangia on these F_1 gametophytes showed a complete range from 100% epigyny, to 100% hypogyny. Also in certain gametophytes the ratio of male to female gametangia (normally about one in the two parents) was very high with less than 1 female per 1,000 male gametangia. It was concluded from these experiments that since intermediate gametangial arrangements are found in hybrid haploids this arrangement is not under the control of a single pair of non-duplicated allelic genes, but that a fairly large number of genes must be involved. Hybridisation in some way upsets the mechanism which controls the arrangement of gametangia in the parental species.

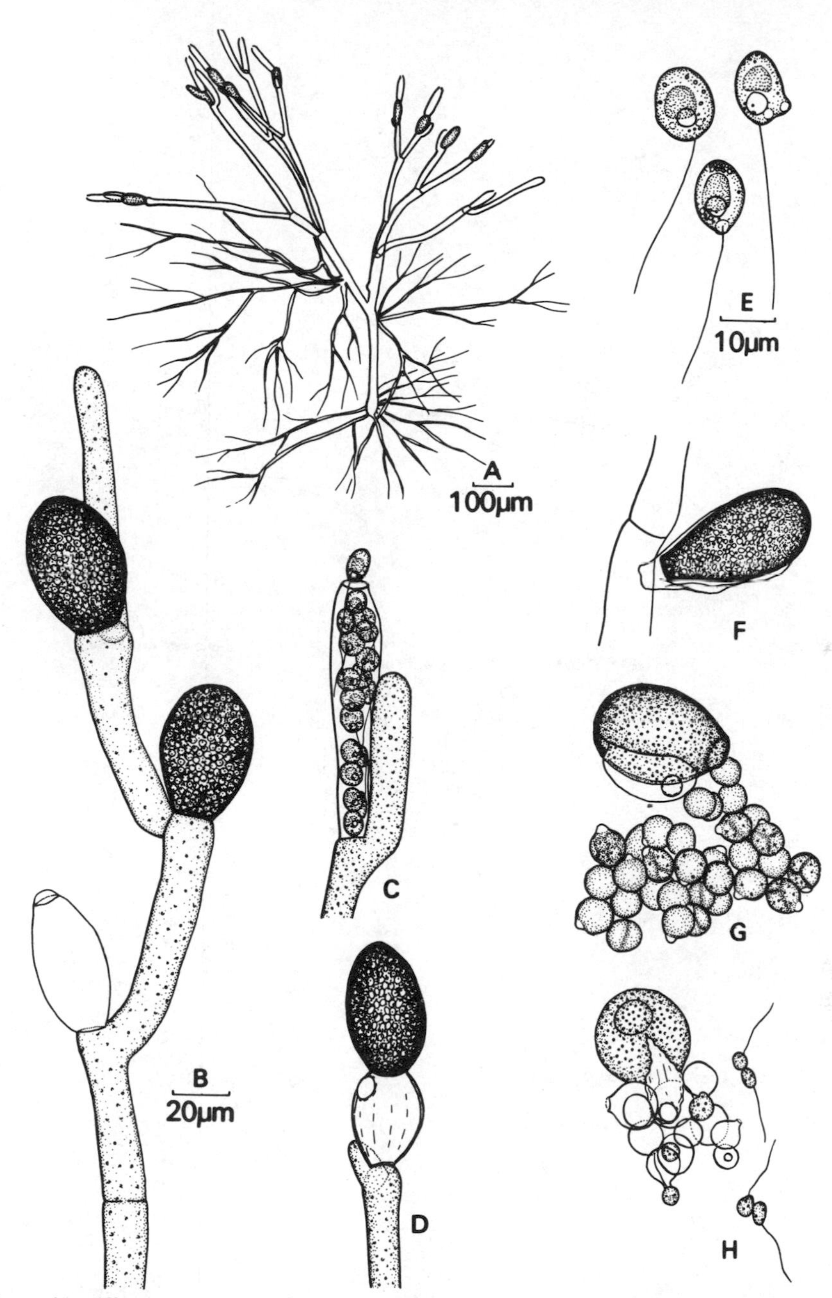

Figure 21. *Allomyces neo-moniliformis.* A. Whole plant. B. Branch tips showing resistant sporangia and zoosporangium (mitosporangium). C. Mitosporangium releasing mitospores. D. Branch tip showing a terminal resistant sporangium and a subterminal empty mitosporangium. E. Mitospores. F. Resistant sporangium escaping from its surrounding envelope. G. Germinating resistant sporangium. A thin-walled vesicle has ballooned out through the cracked wall. From the pore in the vesicle swarmers have escaped which have encysted immediately. H. Release of swarmers from the cysts. The swarmers pair to form biflagellate zygotes. B, C, D, E, G, H to same scale.

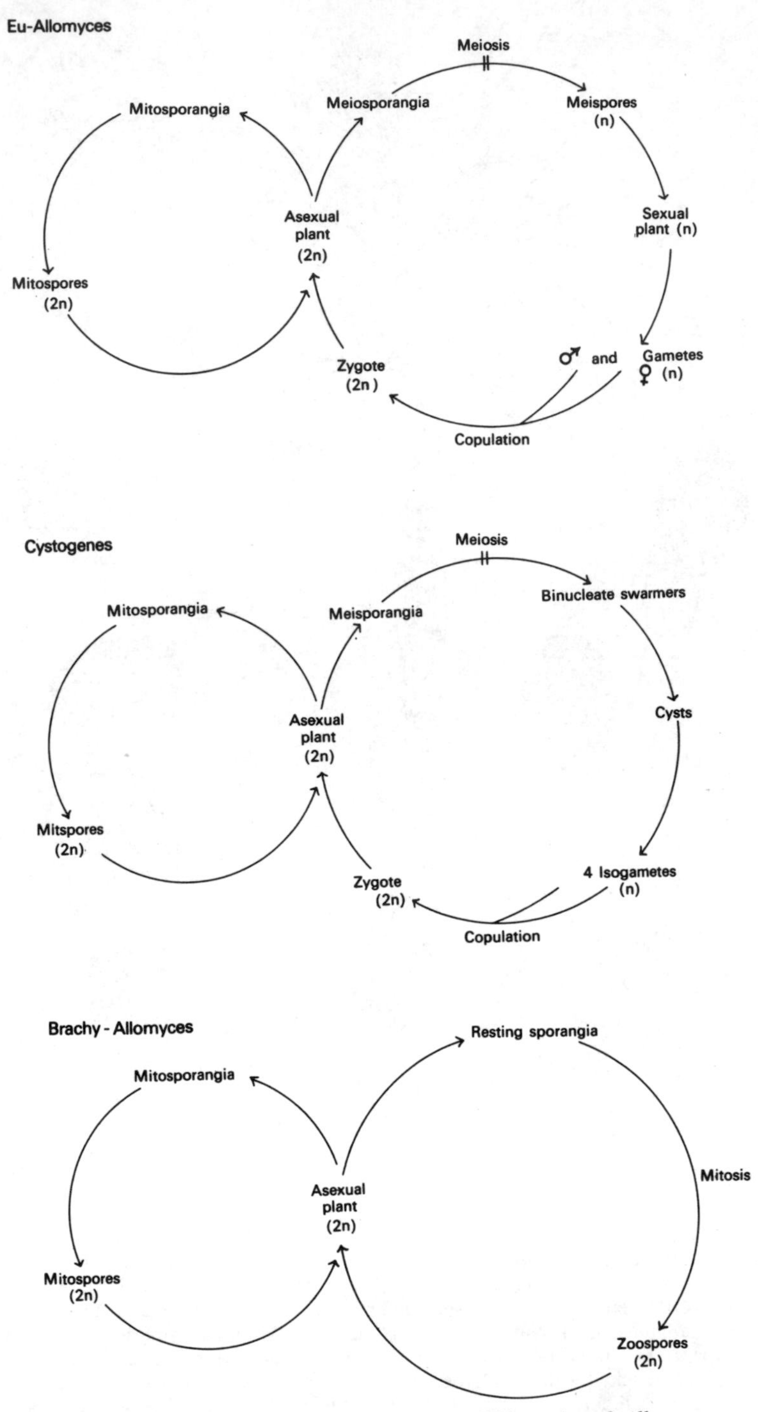

Figure 22. Diagrammatic outlines of life cycles of *Allomyces*.

2. *Cystogenes*

A life cycle differing from *Eu-Allomyces* is found in *Allomyces moniliformis* and *A. neo-moniliformis*. There is no independent gametophyte generation, but this stage is probably represented by a cyst. The asexual plants resemble those of *Eu-Allomyces*, bearing both mitosporangia and brown thick-walled punctate meiosporangia. The mitosporangia are thin-walled, and bear a single terminal papilla, through which discharge of the posteriorly uniflagellate mitospores takes place. These swarmers germinate to form asexual plants again. The development of the meiosporangia is, however, very distinctive. The thin sporangial wall surrounding the sporangium breaks open irregularly and the brown pitted sporangium slips out (Fig. 21F). It is capable of germination within about two to four days after its release. Before germination the sporangium tends to become more round in shape, then its outer brown wall cracks open. Through the crack the thin inner wall expands and one to four discharge pores develop. The contents of the resting sporangium have meanwhile divided to form up to 30 amoeboid bodies which escape through the pores. There are reports that the amoeboid bodies bear a pair of posterior flagella (Emerson, 1941; Wilson, 1952) or even as many as six (Teter, 1944; Wilson, 1952), but flagella are not always present (McCranie, 1942; Teter, 1944). The movement of these bodies is very sluggish, and they rarely move far from their parent resting sporangium. They encyst quickly, frequently all clumped together (Fig. 21G,H). The period of encystment is short, often only a few hours, and then the contents of the cysts divide to form four colourless posteriorly uniflagellate swarmers, which escape through a single papilla on the cyst wall. These swarmers behave as isogametes and copulate to form biflagellate zygotes. There is evidence that the swarmers from one cyst are capable of copulation. The zygote swims for a time by movements of both flagella, then comes to rest and develops to form an asexual plant (Fig. 21A). The cytological details of the life cycle in *Cystogenes* have been worked out by Wilson (1952). As in *Eu-Allomyces* the meiotic divisions occur in the resting sporangia. Before final cleavage of the cytoplasm to form swarmers the nuclei pair, being held together by a common nuclear cap. Cleavage thus results in the formation of binucleate cells, some of which bear two flagella, and some do not. It is these cells which form the cysts. During encystment a single mitotic division occurs in each cyst, and four separate haploid nuclei are formed, each with a distinct nuclear cap, and cleavage of the cytoplasm around these nuclei results in the formation of the four gametes. In the *Cystogenes* life cycle there is thus a diploid asexual generation, and the haploid phase is represented by the cyst and gametes. The life cycle is illustrated diagrammatically in Figure 22.

3. *Brachy-Allomyces*

In certain isolates of *Allomyces* which have been placed in a 'form-species'

A. anomalus there are neither sexual plants nor cysts. Asexual plants bear mitosporangia and brown, resting sporangia. The spores from the resting sporangia develop directly to give asexual plants again. The cytological explanation proposed by Wilson (1952) for this unusual behaviour is that due to complete or partial failure of chromosome pairing in the resting sporangia, meiosis does not occur and nuclear divisions are mitotic. Consequently the zoospores produced from resting sporangia are diploid, like their parent plants, and on germination, give rise to diploid asexual plants again. Similar failures in chromosome pairing were also encountered in the hybrids between *A. arbuscula* and *A. macrogynus* leading to very low meiospore viability from certain crosses. In view of this it seems possible that some of the forms of *A. anomalus* might have arisen through natural hybridisation.

Blastocladiella

Species of *Blastocladiella* have been isolated from soil and mud. Most of the isolates have been from soils from southerly latitudes, but two species have been found in Britain, one parasitic on the blue-green alga *Anabaena* (Canter & Willoughby, 1964). The form of the thallus is comparatively simple, resembling that of some monocentric chytrids. There is an extensive branched rhizoidal system which either arises directly from a sac-like sporangium, or from a cylindrical trunk-like region bearing a single sporangium at the tip. In some species there is an isomorphic alternation of generations, probably matching in essential features the *Eu-Allomyces* pattern, but cytological details are needed to confirm this. For example, in *Blastocladiella variabilis* two kinds of asexual plant are found. One bears thin-walled zoosporangia which release posteriorly uniflagellate swarmers. These swarmers may develop to form plants resembling their parents or may give rise to the second type of asexual plant bearing a thick-walled dark-brown sculptured resting sporangium within the terminal sac. The resting sporangial walls crack open on germination to allow papillae to protrude. The papillae dissolve to release posteriorly uniflagellate swarmers which, after swimming, germinate to form sexual plants of two kinds. In habit the sexual plants resemble the zoosporangial asexual plants, bearing a terminal club-shaped sac. About half of the sexual plants are colourless, and about half are orange-coloured. By analogy with *Allomyces* where the orange-coloured gametes are smaller or 'male', the orange-coloured plants of *Blastocladiella variabilis* are regarded as male. However in *Blastocladiella* there is no distinction in size between the gametes, in contrast with the anisogamy of *Eu-Allomyces*. The orange and colourless gametes pair to produce zygotes, which germinate directly to produce asexual plants. Because the orange and colourless plants were produced in approximately equal numbers from swarmers derived from resting sporangia it was believed that sexuality (i.e. colour) was probably genotypically determined at meiosis in the resting sporangia. A similar life cycle has been described for *B. stübenii*. In other species (e.g. *B. cystogena*) the life cycle is of the *Cystogenes* type. In this species there are no zoosporangial thalli but only resting sporangial thalli. The resting sporangia crack open to release posteriorly uniflagellate swarmers which encyst in an irregular mass. The cysts germinate to give rise to four uniflagellate isogamous gametes. These gametes pair to form a biflagellate zygote which develops into a plant bearing a resting sporangium. In some other species there is no clear evidence of sexual fusion. In *B. emersonii* (Fig. 23) the resting sporangial thallus contain a single globose, dark-brown, resting sporangium with a dimpled wall. After a resting period the wall cracks open, and one to four papillae protrude from which swarmers are released. The swarmers germinate to form two

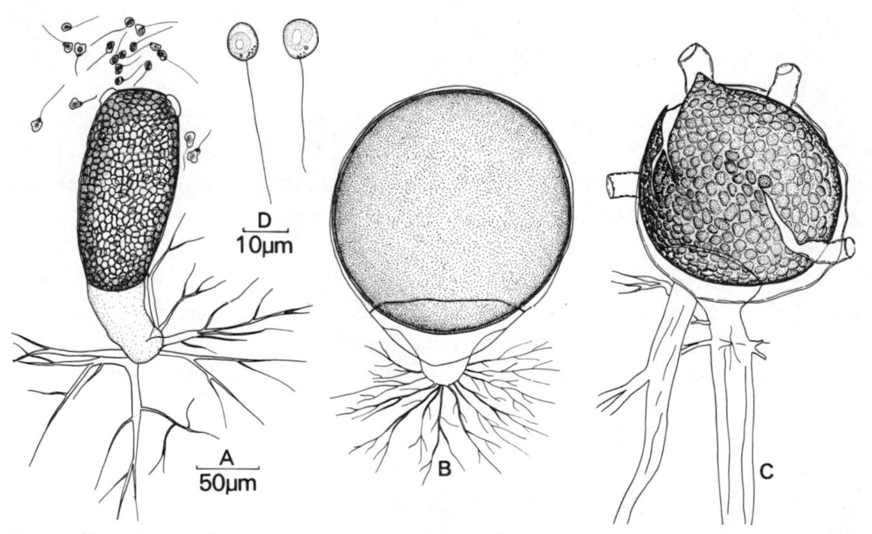

Figure 23. *Blastocladiella emersonii.* A. Mature thin-walled plant releasing zoospores, 3-day-old plant. B. Immature resting sporangium, 3 days old. C. Germinating resting sporangium showing the cracked wall and four exit tubes. D. Zoospores from a thin-walled plant.

types of plant bearing thin-walled zoosporangia. About 98% of the swarmers give rise to plants bearing colourless sporangia, and about 2% to plants with orange-coloured sporangia due to the presence of γ-carotene. The proportion of these two types of plant is, however, variable. Naturally occurring mutants are known in which the swarmers from resting sporangia develop almost exclusively into orange plants. Also, by growing plants in the presence of the antibiotic cycloheximide the ratio can be adjusted to 25% orange to 75% colourless plants.

Cantino & Horenstein (1956) have speculated that the incidence of colourless and orange plants was controlled by the random distribution of a hypothetical cytoplasmic factor *gamma*. On cleavage of the cytoplasm inside the resting sporangium prior to the differentiation of swarmers some zoospore initials would have a high concentration of *gamma*, and would give rise to colourless plants, whilst others would have a low concentration of *gamma*, giving rise on germination to orange-coloured plants. When the zoospores derived from colourless, wild-type orange-coloured, and mutant orange plants were stained with Nadi reagent (a 1% aqueous solution of p-amino-dimethylaniline hydrochloride mixed with an equal volume of 1% alcoholic *alpha*-naphthol) it was found that they contained a different distribution pattern of small blue-black particles. The swarmers from colourless plants had an average of 12·5 particles per spore, whilst the swarmers from the wild-type orange plants and mutant orange plants had averages of 7·5 and 8·0 particles per spore respectively. The distribution of these Nadi-positive particles thus corresponds with the predicted distribution of the hypothetical cytoplasmic factor *gamma*, although it should not too readily be assumed on the evidence so far available that *gamma* and the Nadi-positive particles are identical.

Despite careful and repeated examination there is no evidence of conventional sexual fusion between swarmers derived from orange and colourless plants. When swarmers from colourless and *mutant* orange plants are mixed together, however, transitory contact may be established by means of a cytoplasmic bridge, but the two cells eventually separate and nuclear fusion does not occur. Such contacts may occur several times, a colourless swarmer being visited in turn by more than one orange swarmer. There is some evidence that during such transitory contact exchange of cytoplasmic genetical determinants occurs. For example mutant zoospores show negligible viability on a purely synthetic medium, whilst the colourless zoospores possess good viability on the same medium. If zoospore suspensions of the two types are mixed, the number of viable plants derived from the suspension increases with the length of the length of time the suspensions remain mixed. This and other evidence has led Cantino & Horenstein (1954) to the view that in the absence of ordinary gametic copulation in *Blastocladiella emersonii* its motile swarmers appear capable of unilateral transfer of certain cytoplasmic materials.

B. emersonii has a number of other unusual features. If zoospore suspensions are sprayed on to Peptone Yeast Glucose Agar the majority of plants which develop will be of the thin-walled, colourless type. On the same medium containing 10^{-2} M, bicarbonate, resting sporangial plants with the thick, pitted brown walls will result. Thus by means of a simple manipulation of the environment it is possible to switch the metabolic activities of the fungus into one of two morphogenetic pathways. Cantino and his colleagues have analysed the differences in metabolism of plants grown in the presence and absence of bicarbonate and have shown that there are a number of important differences in the level of activity of certain enzymes (for reviews of literature see Cantino & Lovett, 1964). In the absence of bicarbonate there is evidence for the operation of a Krebs cycle. However, in the presence of bicarbonate part of the Krebs cycle is reversed, leading to alternative pathways of metabolism. In addition a polyphenol oxidase, absent in the thin-walled plant, replaces the normal cytochrome oxidase. There is also increased synthesis of melanin and of chitin in the presence of bicarbonate. Analysis of such changes will provide an insight at the enzymic level of differences between the two morphogenetic pathways.

The effect of bicarbonate can also be brought about by suitable levels of CO_2. However another unusual feature is that *B. emersonii* fixes CO_2 more rapidly in the light than in the dark. In the presence of CO_2, light-grown plants show a number of differences when compared with dark-grown controls. Illuminated plants take about three hours longer to mature, and are larger than dark-grown plants. They also have an increased rate of nuclear reproduction and a higher nucleic acid content. The most effective wavelengths for this increased CO_2 fixation (or lumisynthesis) lie between 400 and 500 nm, i.e. at the blue end of the spectrum. This suggests that the photoreceptor should be a yellowish substance. Attempts to identify the photoreceptor have, as yet, been unsuccessful, but it is known that it is not a carotenoid and not chlorophyll. Also the actual locus of the light-activated reaction has not yet been identified.

In *B. britannica* similarly no sexual fusion has been reported, but resting sporangial plants occur. The formation of resting sporangia in this species is stimulated by incubating cultures in white light. There is, however, no evidence of stimulation of formation of resting sporangia by bicarbonate.

MONOBLEPHARIDALES

This group is represented by three genera, *Monoblepharis*, *Gonapodya* and *Monoblepharella*. The first two are to be found on submerged twigs and fruits in fresh-water ponds and streams, especially in the spring, whilst *Monoblepharella* has been reported from hemp-seed baited water cultures from soil samples collected in the warmer parts of North and South America and from the West Indies. In all genera the thallus is eucarpic and filamentous. Studies of the wall of *Monoblepharella* show that it contains microfibrils of chitin. A characteristic feature is the frothy or alveolate appearance of the cytoplasm caused by the presence of numerous vacuoles often arranged in a regular fashion. Asexual reproduction is by posteriorly uniflagellate zoospores which are borne in terminal, cylindrical or flask-shaped sporangia. The zoospores have not yet received the same detailed examination as those of the Chytridiales and Blastocladiales. At the anterior end of the zoospore a crown of refractive granules (possibly lipoid globules) has been described. There is a large conical nuclear cap and usually a single refractive body at the point of insertion of the flagellum (Sparrow, 1960; Springer, 1945; Fuller, 1966). Germination may be bipolar or monopolar. Very little is known about the physiology of the group, probably because pure cultures of them have only been obtained comparatively recently (Springer, 1945; Perrott, 1955, 1958;

Aronson & Preston, 1960). Sexual reproduction is unusual for fungi in being oogamous with motile spermatozoids. After fertilisation has occurred the egg may move to the mouth of the oogonium by amoeboid crawling in some species of *Monoblepharis* or propelled by movement of the spermatozoid flagellum in *Monoblepharella* (Sparrow, 1939; Springer, 1945) and *Gonapodya* (Johns & Benjamin, 1954). Sparrow (1960) has separated the three genera into two families on the basis of this behavioral difference, the Monoblepharidaceae for *Monoblepharis*, and the Gonapodyaceae for *Gonapodya* and *Monoblepharella.*

Monoblepharis

Species of *Monoblepharis* can be collected on water-logged twigs on which the bark is still present, from quiet silt-free pools where the water is neutrally alkaline, i.e. pH 6·4 to 7·5. Twigs of birch, ash, elm and especially oak are suitable substrata, and although twigs collected at varying times throughout the year may yield growths of the fungus, there are two periods of vegetative growth, one in spring and another in autumn, with resting periods during the summer and winter months (Perrott, 1955). Low temperatures appear to favour development and good growths can be obtained on twigs incubated in dishes of distilled water at temperatures around 3°C. The mycelium is delicate and vacuolate. The hyphae are multinucleate. During the formation of a sporangium a multinucleate tip is cut off by a septum. The cytoplasm cleaves around the nuclei to form zoospore initials which are at first angular, later pear-shaped. The ripe sporangium is cylindrical or club-shaped and may not be much wider than the hypha bearing it. A pore is formed at the tip of the sporangium through which the zoospores escape by amoeboid crawling. The free zoospores have a single posterior flagellum and swim away. On coming to rest the zoospore germinates to form a germ tube. The single nucleus of the zoospore divides and further nuclear divisions occur as the germ tube elongates.

Sexual reproduction can be induced by incubating twigs at room temperature. In most species the sex organs may arise on the same plant which bears sporangia, although in two species *M. regignens* and *M. ovigera* no sexual stages have been reported. Whether these species are correctly classified in *Monoblepharis* is an open question, because it is possible that if they do reproduce sexually they might behave in the manner of *Monoblepharella.* In *Monoblepharis polymorpha* and related species the antheridia are epigynous. The antheridium is formed from the tip of the hypha which becomes cut off by a cross wall. Beneath the antheridium the hypha becomes swollen, somewhat asymmetrically so that the antheridium is displaced into a lateral position. The swollen subterminal part becomes spherical and is then cut off by a basal septum to form the oogonium. In *M. sphaerica* and some other species the arrangement of the sex organs is the reverse of that in *M. polymorpha* the oogonium being terminal and the antheridium subterminal. In both groups of species the antheridium has often discharged sperm before the adjacent oogonium is ripe. The release of sperm resembles zoospore release, and each antheridium releases about four to eight posteriorly uniflagellate swarmers which resemble, but are somewhat smaller than, the zoospores. The oogonium contains a single spherical uninucleate oosphere, and when this is mature an apical receptive papilla on the oogonial wall breaks down. A

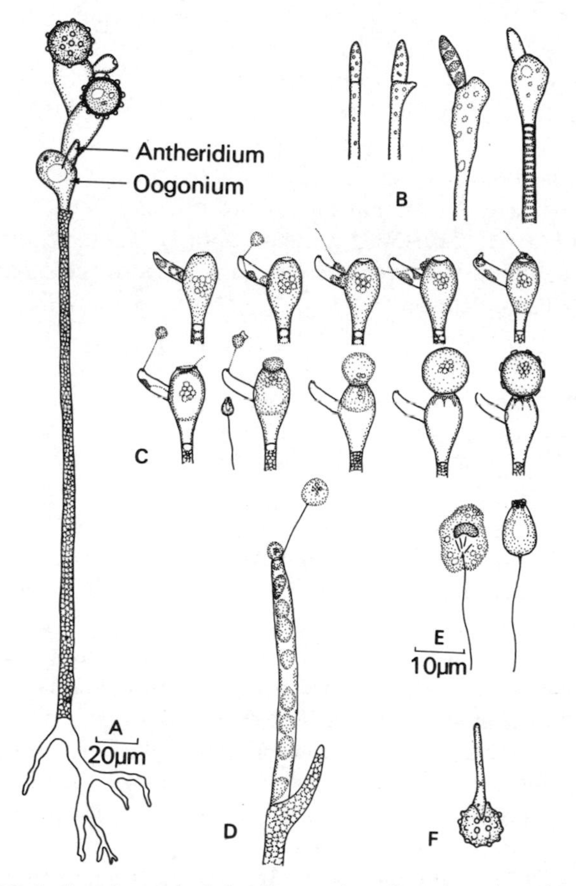

Figure 24. *Monoblepharis polymorpha* (after Sparrow). A. Complete plant showing rhizoidal system, antheridia, oogonia and mature oospores. B. Stages in the development of gametangia. C. Stages in fertilisation. D. Discharging sporangium. E. Zoospores. F. Germinating oospore. A, B, C, D, F to same scale.

spermatozoid approaching the receptive papilla of the oogonium becomes caught up in mucus and fusion with the oosphere then follows, the flagellum of the spermatozoid being absorbed within a few minutes. Following plasmogamy the oospore secretes a golden-brown membrane around itself and nuclear fusion later occurs. In some species, e.g. *M. sphaerica*, the oospore remains within the oogonium (endogenous) but in others, e.g. *M. polymorpha*, the oospore begins to move towards the mouth of the oogonium within a few minutes of fertilisation, and remains attached to it (exogenous)—Fig. 24. In the exogenous species nuclear fusion is delayed but finally fusions occur and the oospore becomes uninucleate. In some species the oospore wall remains smooth, but in others the wall may be ornamented by hemispherical warts or bullations. The oospore germinates after a resting period which coincides with frozen winter conditions or summer drought by producing a single hypha which branches to form a mycelium. The cytological details of the life cycle are not fully known but it seems likely that reduction division occurs during the germination of the over-wintered oospores (Laibach, 1927).

The Monoblepharidales are possibly related to the Blastocladiales, and the two groups have in common similar protoplasmic structure, uniflagellate swarmers with a well developed nuclear cap and bipolar zoospore germination (Sparrow, 1958). However, there is no parallel to the range of life cycles found in the Blastocladiales.

OOMYCETES

The Oomycetes are Mastigomycotina with a characteristic type of zoospore bearing two flagella, one of the whiplash type, and the other of the tinsel type. Most Oomycetes are aquatic, although some Saprolegniales and Peronosporales grow in soil. Although the possession of zoospores implies a dependence on water for dispersal, in some Peronosporales causing diseases of plants, the sporangia are detached from the sporangiophore and dispersed by wind. Subsequent germination may be by means of zoospores, or, as in the downy mildews, may be by means of germ tubes—i.e. zoospores may be lacking. The cell walls of Oomycetes are unusual for fungi, in that chitin is absent. Small amounts of cellulose are present, but the principal components are glucans with β-(1$\rightarrow$3) and β-(1$\rightarrow$6) glucosidic linkages (Aronson *et al.*, 1967). The form of the thallus varies. In the Lagenidiales (for the most part parasitic on algae, other aquatic fungi and microscopic aquatic animals) the thallus is holocarpic. In the remaining orders the thallus is predominantly eucarpic, often composed of coarse coenocytic hyphae. In the Leptomitales the hyphae are constricted at intervals. For a fuller account of these groups see Sparrow (1960). Sexual reproduction of the Oomycetes is oogamous, and results in the formation of a thick-walled resting spore or oospore. In the Lagenidiales two holocarpic thalli of different sizes may fuse, but in the Saprolegniales, Peronosporales and Leptomitales fusion is between a more or less globose oogonium containing one to several eggs, and an antheridium.

Confirmation of details of some life cycles is needed, but there is increasing evidence that the Saprolegniales and Peronosporales may be diploid organisms.

SAPROLEGNIALES

This is the best-known group of aquatic fungi, often termed the water-moulds. Members of the group are abundant in fresh-water, mainly as saprophytes on plant and animal debris, whilst some are marine. They are also common in soil. A few species of *Saprolegnia* and *Achlya* are economically important as parasites of fish and their eggs (Scott & O'Bier, 1962). *Aphanomyces euteiches* causes a root rot of peas and some other plants, whilst another species is a serious parasite of the crayfish *Astacus* (Unestam, 1965). Algae, fungi, rotifers and copepods may also be parasitised by members of the group and occasional epidemics of disease among zooplankton have been reported.

In the Ectrogellaceae, parasites of fresh-water and marine diatoms and Phaeophyceae, the thallus is holocarpic and endobiotic, resembling *Olpidium*. In the Thraustochytriaceae, saprophytic on marine algae, the thallus is eucarpic and resembles *Chytridium*; it consists of a globose sporangium and a rhizoidal vegetative part. In the largest family (the Saprolegniaceae) the thallus is eucarpic and coenocytic, often forming a vigorous, coarse, stiff mycelium. The cell wall of the Saprolegniaceae contains some cellulose.

The zoospores are biflagellate and when the flagella are attached laterally the anterior flagellum is of the tinsel type and the posterior is of the whip type (Manton *et al.*, 1951). It has been suggested that the posterior flagellum is concerned with propulsion of the zoospore whilst the anterior tinsel-type flagellum functions as a rudder (McKeen, 1962). Two types of zoospore are found in some genera. The first motile stage is usually pear-shaped with two apically attached flagella. After swimming the zoospore encysts, withdrawing its flagella. The cyst may germinate to produce a second type of zoospore, bean-shaped, with laterally attached flagella. The second stage zoospore after swimming also encysts and on germination may produce a further zoospore of the second type. Repeated encystment and motility may occur and this phenomenon is sometimes called repeated emergence. Alternatively the cyst may germinate to form a germ tube from which the filamentous thallus develops. In some genera the behaviour of the zoospores appears to be modified (see below).

The sexual reproduction of the Saprolegniales is oogamous, with a large, usually spherical oogonium containing one to many eggs. Antheridial branches apply themselves to the wall of the oogonium, penetrating the wall by fertilisation tubes through which a single male nucleus is introduced into each egg.

Most of the Saprolegniaceae will grow readily in pure culture, even on purely synthetic media, and extensive studies of their physiology have been made (Cantino, 1950, 1955; Papavizas & Davey, 1960; Barksdale, 1962). Most species examined have no requirements for vitamins. Organic forms of sulphur such as cysteine, cystine, glutathione and methionine are preferred, and most species are unable to reduce sulphate. Organic nitrogen sources such as amino-acids, peptone and casein are preferred to inorganic sources. Nitrate is generally not available although ammonium is utilised. Glucose, maltose, starch and glycogen are utilised by some species, but in others glucose is the only carbon source available. In liquid culture *Saprolegnia* can be maintained in the vegetative state indefinitely if supplied with organic nutrients in the form of broth. When the nutrients are replaced by water the hyphal tips quickly develop into zoosporangia. The formation of sexual organs can similarly be affected by manipulating the external conditions in some species, and the concentration of salts in the medium may play a decisive role. Deficiency of sulphur, phosphorus, calcium, potassium and magnesium may limit the formation of oogonia (Barksdale, 1962). In pure cultures of *Aphanomyces euteiches* oogonia are only produced in abundance when sulphur is supplied in the reduced form (Davey & Papavizas, 1962).

In the account which follows only the Saprolegniaceae will be considered. Taxonomic treatments of the Saprolegniaceae have been provided by Coker (1923), Coker & Matthews (1937), Chaudhuri *et al.* (1947), of *Achlya* by Johnson (1956) and of *Aphanomyces* by Scott (1961). Ecological aspects have been considered by Perott (1960), Dick & Newby (1961), Dick (1962, 1966), Hughes (1962) and Willoughby (1962*b*).

Saprolegniaceae

Members of this group can readily be obtained from water, mud and soil by floating boiled hemp seed in dishes containing pond water or soil samples or waterlogged twigs covered with water (for details see Johnson, 1956). Within about four days the fungi can be recognised by their stiff, radiating, coarse hyphae bearing terminal sporangia, and cultures can be prepared by transferring hyphal tips or zoospores to maize-extract agar or other suitable media. About 20 genera are recognised but the most commonly encountered are *Saprolegnia*, *Achlya*, *Dictyuchus* and *Thraustotheca*. In all genera the hyphae are coenocytic, containing within the wall a peripheral layer of cytoplasm surrounding a continuous central vacuole. In the peripheral cytoplasm streaming can be readily seen. Numerous nuclei are present un-separated by cell walls. The small size of the nuclei makes observation difficult, but nuclear division does not appear to be accompanied by the formation of a spindle nor have chromosomes been distinguished. Instead the nucleus appears to divide by constriction. Spindles and chromosomes have, however, been observed in the development of sex organs. Filamentous mitochondria and fatty globules can also be observed in vegetative hyphae. Fine structure of vegetative hyphae has been studied in *Aphanomyces* (Shatla *et al.*, 1966) and *Saprolegnia* (Gay & Greenwood, 1966). In *Aphanomyces* a number of organelles and inclusions of unknown function, e.g. crystals and randomly orientated microtubules have been described. A feature of many Saprolegniaceae, especially when grown in culture is the formation of thick-walled enlarged terminal or intercalary portions of hyphae which become packed with dense cytoplasm and are cut off from the rest of the mycelium by septa. These structures which may occur singly or in chains (see Fig. 29) are termed gemmae or chlamydospores, and their formation can be induced by manipulation of the culture conditions. It has been suggested that gemmae are homologous to oogonia, and indeed they occasionally function as female gametangia, but in some forms the gemmae may subsequently function as sporangia and release zoospores. Alternatively they may germinate by means of a germ tube. It is known that they cannot survive desiccation or prolonged freezing, but it is likely that they remain viable for long periods. Another feature of old cultures is the fragmentation of cylindrical pieces of mycelium cut off at each end by a septum.

ASEXUAL REPRODUCTION IN THE SAPROLEGNIACEAE

Saprolegnia

Species of *Saprolegnia* are common in soil and in fresh-water, saprophytic on plant and animal remains but a number of species such as *S. ferax* and *S. parasitica* have been implicated in disease of fish and their eggs. These species have been experimentally inoculated into wounded fish, resulting in death within 24 hr (Tiffney, 1939*a,b*; Vishniac & Nigrelli, 1957; Scott & O'Bier, 1962). In nature *S. parasitica* may be associated with severe epidemics of disease among fish (Stuart & Fuller, 1968). The sporangia of *Saprolegnia* develop at the tips of the hyphae. The hyphal tip which is pointed in the vegetative condition, swells and becomes club-shaped, rounded at the tip and accumulates denser cytoplasm around the still clearly visible central vacuole. A septum develops at the base of the sporangium and it is at first convex with respect to the sporangial tip, i.e. it bulges into the sporangium. The hyphal tip contains numerous nuclei and cleavage furrows separate the cytoplasm into uninucleate pieces each of which differentiates into a zoospore. As the zoospores are cleaved out the central vacuole is no longer visible. The tip of the cylindrical sporangium contains clearer cytoplasm and a flattened protuberance develops at the end. As the sporangium ripens and the zoospores become fully differentiated they show limited movement and change of shape (Fig. 25B–D). Evidence of a build-up of pressure within the sporangium can be seen from the shape of the basal septum, which shortly before zoospore discharge becomes concave, i.e. is pushed into the lumen of the hypha beneath the sporangium. How the pressure develops is not known, but it is possibly brought about by increase in osmotic pressure of the sporangial contents resulting in increased uptake of water and increased turgor. Earlier workers suggested that mucilaginous material within the sporangium might swell up due to water absorption, but such material has not been demonstrated. The sporangium undergoes a slight change in shape at this time and

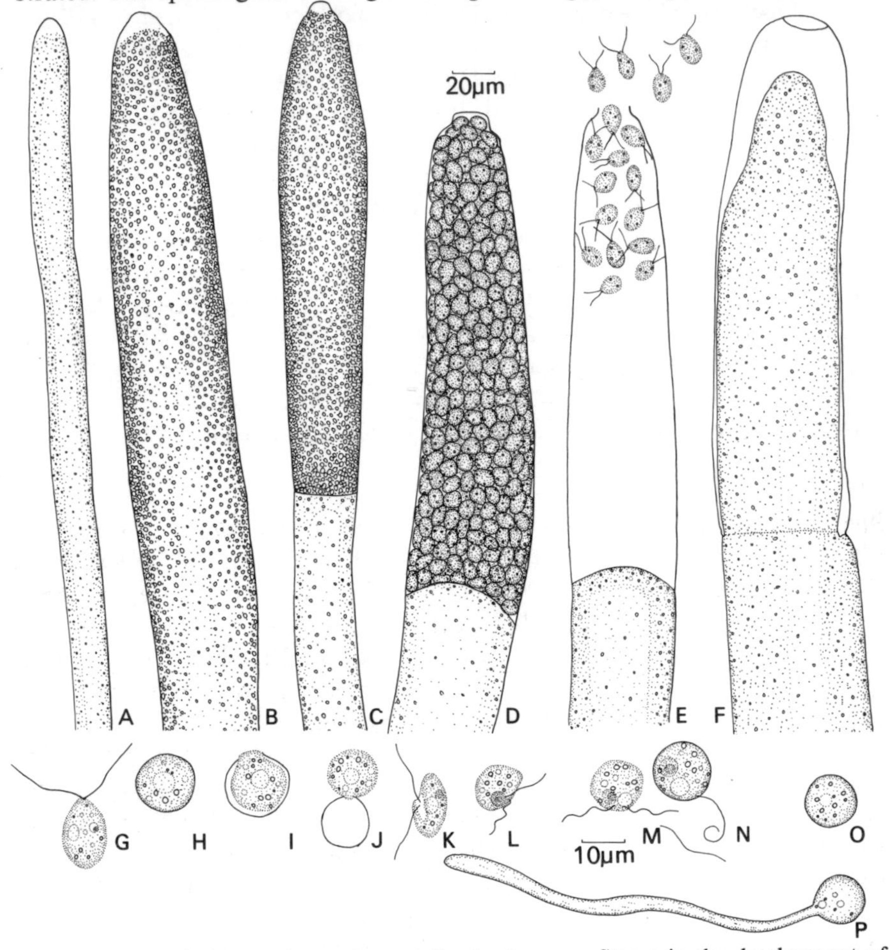

Figure 25. *Saprolegnia.* A. Apex of vegetative hypha. B–D. Stages in the development of zoosporangia. E. Release of zoospores. F. Proliferation of zoosporangium. A second zoosporangium is developing within the empty one. G. Primary zoospore (first motile stage). H. Cyst formed at end of first motile stage (primary cyst). I–J. Germination of primary cyst to release a second motile stage (secondary zoospore). K–M. Secondary zoospores. N. Secondary zoospore at moment of encystment. Note the cast-off flagella. O. Secondary cyst. P. Secondary cyst germinating by means of a germ tube. A–F to same scale; G–P to same scale.

the tip of the sporangium breaks down at the clear protuberant tip. The spores are released quickly, many zoospores escaping in a few seconds, moving as a column through the opening as though pushed out under pressure (Gay &

Greenwood, 1966) (Fig. 25E). The zoospores leave the sporangium backwards, with the blunt posterior emerging first. The size of the zoospore is sometimes slightly less than the diameter of the sporangial opening so that the zoospores are squeezed through it. If release of the zoospores is caused by a build-up of pressure the difference in size of the zoospore and the opening may be important in preventing sudden loss of pressure. However, an occasional zoospore is sometimes left behind and swims about in the empty sporangium finally making its way out backwards through the opening. A characteristic feature of *Saprolegnia* is that following the discharge of the zoosporangium growth is renewed from the septum at its base so that a new apex develops inside the old sporangial wall. This in turn may develop into a zoosporangium, discharging its spores through the old pore (Fig. 25F). The process may be repeated so that several empty zoosporangial walls may be found inside each other. In other genera proliferation usually does not take place in this way. Sometimes in *Saprolegnia* growth is renewed to one side of the old zoosporangium.

The zoospores on release turn round and swim with the pointed end directed forwards. They are pear-shaped and bear two apically attached flagella (Fig. 25G). Each spore also contains a nucleus and a contractile vacuole. Electron micrographs of *Saprolegnia* zoospores have been provided by Manton *et al.*, (1951). The zoospores swim for a time which may be less than a minute in *S. dioica*. However the zoospores from a single sporangium show variation in their period of motility, the majority encysting within about a minute but some remaining motile for over an hour. The zoospore then withdraws its flagella and encysts, i.e. the cytoplasm becomes surrounded by a distinct firm membrane (Fig. 25H). Following a period of rest (two to three hours in *S. dioica*) the cyst germinates to release a further zoospore (Fig. 25I–J). This zoospore differs in shape from the zoospore released from the sporangium, being bean-shaped (although amoeboid changes in shape occur) and bearing two laterally attached flagella which are inserted in a shallow groove running down one side of the zoospore (Fig. 25K–M). This laterally biflagellate bean-shaped zoospore is called the secondary zoospore or second motile stage, in contrast to the apically biflagellate primary zoospore or first motile stage. The secondary zoospore may swim vigorously for several hours before encysting. Salvin (1941) compared the rates of movement of primary and secondary zoospores in *Saprolegnia* and found that the secondary zoospores swam about three times more rapidly than the primary. He wrote 'as soon as there is a transition of the zoospore from a primary to a secondary type, not only is there a change in the general morphology, but in reality a transformation in the fundamental biochemical constitution'. This movement of secondary zoospores is chemotactic and zoospores can be stimulated to aggregate about parts of animal bodies such as the leg of a fly. Fischer & Werner (1958*a*) have substituted this chemotactic stimulus by glass capillaries containing 10^{-1} M sodium chloride or potassium chloride and traces of amino acids. They have also shown (1958*b*) that encystment can be stimulated by the presence of traces of nicotinamide (10^{-7} M), but that if the cysts are washed free of this substance then a further motile stage may be formed. This phenomenon of repeated encystment and motility is termed repeated emergence. The secondary cysts instead of forming further zoospores may germinate by means of a germ tube to give rise to vegetative filaments (Fig. 25P). Because there are two distinct motile stages in the asexual reproduction of *Saprolegnia* it is said to be diplanetic (the zoospore is sometimes termed a planospore, another term for a motile spore). However as we have seen there may be several periods of motility, and the term polyplanetic could equally well be applied. An alternative terminology has been suggested. Since the two motile stages differ morphologically the term dimorphic has been proposed. Electron micrographs of the cysts formed at the end of the primary and secondary motile stages in *Saprolegnia* have shown that in this genus the *cysts* may also be dimorphic. According to Meier & Webster (1954) the primary cysts are smooth but the secondary cysts bear small double-headed hooks somewhat resembling boat-hooks. In *S. ferax* the hooks are short and arise singly but in *S. parasitica* they are longer and arise in tufts (see Fig. 28). There is evidence that these hooks are effective in attaching the cysts to surfaces. Other workers have, however, claimed that the primary cysts of *Saprolegnia* bear hooks (e.g. Manton *et al.*, 1951; Nagai & Takahashi, 1962).

Achlya

Species of *Achlya* are also common in soil and in waterlogged plant debris such as twigs (Johnson, 1956). Certain species have also been reported as naturally occurring pathogens of fish, and experimental inoculations have resulted in fish mortality (Vishniac & Nigrelli, 1957; Scott & O'Bier, 1962). The zoosporangial development in *Achlya* is similar in all respects to that of *Saprolegnia*, and according to some earlier workers the zoospores develop flagella, but Johnson (1956) states that the question of primary zoospore flagellation is still unsolved. On discharge the zoospores do not swim away but cluster in a hollow ball at the mouth of the zoosporangium and encyst there (Fig. 26A). The primary cysts are termed cystospores by some authors. Protoplasmic connections linking the zoospores together have been described. The first motile stage in *Achlya* is thus very brief. In *Aphanomyces*, which is distinguished from *Achlya* by its delicate mycelium and narrow zoosporangia forming a single row of zoospores, encystment similarly occurs at the mouth of the sporangium. In *A. patersonii* Scott (1956) has shown that the motility of the primary zoospore could be controlled by variation in temperature. Below 20°C encystment of the primary zoospores at the mouth of the sporangium occurred in the manner typical of the genus, but above this temperature the primary zoospores swam away and encysted at a distance from the zoosporangium. In this respect its behaviour resembled that of a *Leptolegnia*.

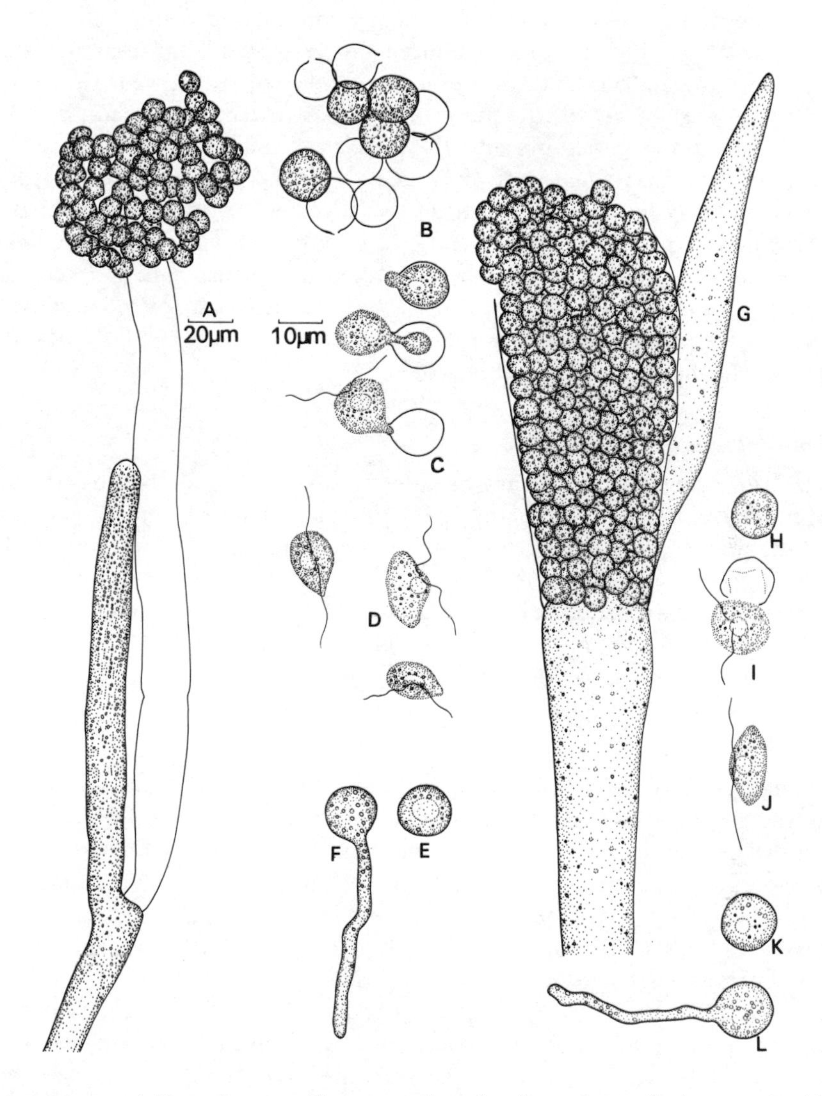

Figure 26. A–F. *Achlya colorata*. A. Zoosporangium showing a clump of primary cysts at the mouth. Note the lateral proliferation of the hypha from beneath the old sporangium. B. Full and empty primary cysts. C. Stages in the release of secondary zoospores from a primary cyst. D. Secondary zoospores. E. Secondary cyst. F. Secondary cyst germinating by means of a germ tube. G–L. *Thraustotheca clavata*. G. Zoosporangium showing encystment within the sporangium. The primary cysts are being released through breakdown of the sporangial wall. H. Primary cyst. I. Primary cyst germinating to release a secondary-type zoospore (here the first motile stage). J. Secondary-type zoospore. K. Secondary cyst. L. Secondary cyst germinating by means of a germ tube. A and G to same scale; B–F and H–L to same scale.

A similar phenomenon has been described in a *Saprolegnia*. Salvin (1941) has claimed that below 10°C the primary zoospores did not behave in a normal manner, but in some instances they encysted almost immediately after emerging and formed a loose mass of encysted spores near the mouth of the sporangium. In *Achlya* the primary cysts remain at the mouth of the sporangium for a few hours and then each cyst releases a secondary-type zoospore through a small pore (Fig. 26B,C). After swimming, the zoospore encysts and may germinate by a germ tube (Fig. 26E–F) or repeated emergence of secondary-type zoospores may occur. The cysts of *A. klebsiana* may remain viable for at least two months when stored aseptically at 5°C (Reischer, 1951*a*). When the zoosporangium of *Achlya* has released zoospores growth is usually renewed laterally by the outpushing of a new hyphal apex just beneath the first sporangium (Fig. 26A).

Thraustotheca

In *T. clavata* the sporangia are broadly club-shaped, and there is no free-swimming primary zoosporic stage. Encystment occurs within the sporangia and the primary cysts are released by irregular rupture of the sporangial wall (Fig. 26G). After release the cysts germinate to release bean-shaped zoospores with laterally attached flagella (Fig. 26I–J), and after a swimming stage further encystment occurs followed by germination by a germ tube (Fig. 26K–L), or by emergence of a further zoöspore.

Dictyuchus

In this genus there is again no free-swimming primary zoospore stage. Zoospore initials are cleaved out but encystment occurs within the cylindrical sporangium. The cysts are tightly packed together and release their secondary zoospores independently through separate pores in the sporangial wall (Fig. 27A). When zoospore release is complete a network made up of the polygonal walls of the primary cysts is left behind. After swimming, the laterally biflagellate zoospore encysts (Fig. 27B,C). Electron micrographs have shown that the wall of the secondary cyst of *D. sterile* bears a series of long spines looking somewhat like the fruit of a horse-chestnut (Meier & Webster, 1954). Following the formation of the first zoosporangium a second may be formed immediately beneath it by the formation of a septum cutting off a subterminal segment of the original hypha, or growth may be renewed laterally to the first sporangium. Commonly the entire sporangium in *Dictyuchus* may become detached from its parent hypha, but spore release proceeds normally.

Because there is only one motile stage in *Thraustotheca* and *Dictyuchus* (i.e. a zoospore of the secondary type) they are said to be monoplanetic or monomorphic. However, in *Pythiopsis* where there is again only one motile stage, the zoospore is of the primary type. After swimming the zoospore encysts and then germinates directly, i.e. there are no secondary zoospores (Fig. 27G–I).

APLANETIC FORMS

In certain cultures of Saprolegniaceae the zoosporangia produce zoospores which encyst within the sporangia but the cysts do not produce a motile stage. Instead, germ tubes are put out which penetrate the sporangial wall. It would

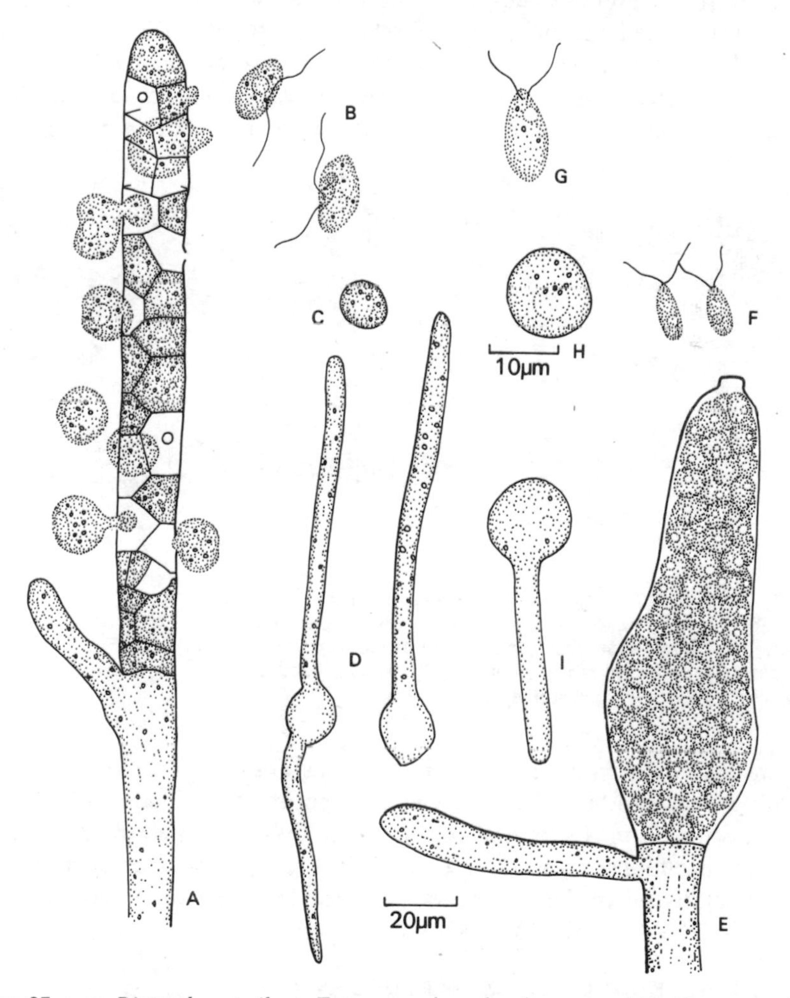

Figure 27. A–D. *Dictyuchus sterile*. A. Zoosporangium showing cysts within the sporangium and the release of secondary-type zoospores through separate pores in the sporangium wall. Note the network of primary cyst membranes. B. Secondary-type zoospores. C. Secondary cyst. D. Germination of secondary cysts by means of germ tubes. E–I. *Pythiopsis cymosa*. E. Zoosporangium. F, G. Primary zoospores. H. Primary cyst. I. Primary cyst germinating by means of a germ tube. Secondary-type zoospores have not been described. A, B, C, E, F to same scale; G, H, I, to same scale.

appear then that there is no motile stage and such forms are said to be aplanetic. The aplanetic condition is occasionally found in staling cultures of *Saprolegnia*, *Achlya*, *Thraustotheca* and *Dictyuchus*, but genera such as *Aplanes* have been erected for forms in which the aplanetic habit appears to be established. However doubt is being expressed about the validity of some of these genera since it is known that the aplanetic condition can be induced by heavy bacterial contamination. When studied in pure culture some of the aplanetic forms produce zoospores and can be assigned to other genera.

The usual methods of zoospore release in *Achlya*, *Thraustotheca* and *Dictyuchus* are referred to as Achlyoid, Thraustothecoid or Dictyuchoid respectively, and sometimes the sporangia of *Dictyuchus* are termed dictyosporangia or net-sporangia. However these methods are not exclusively

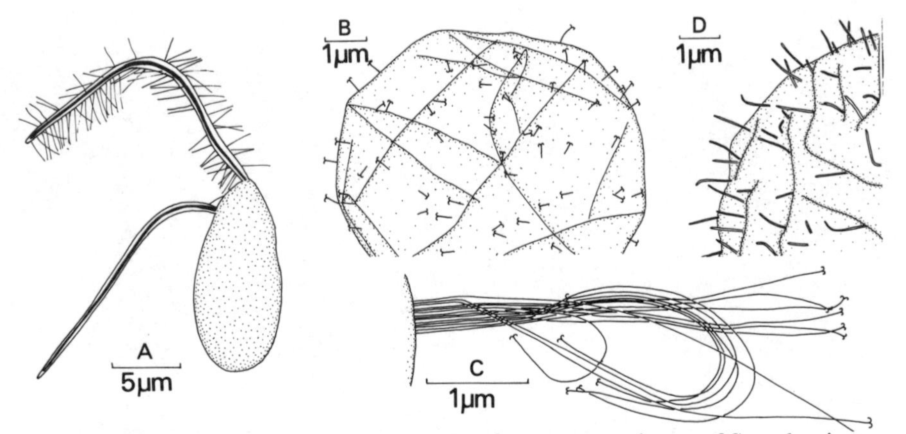

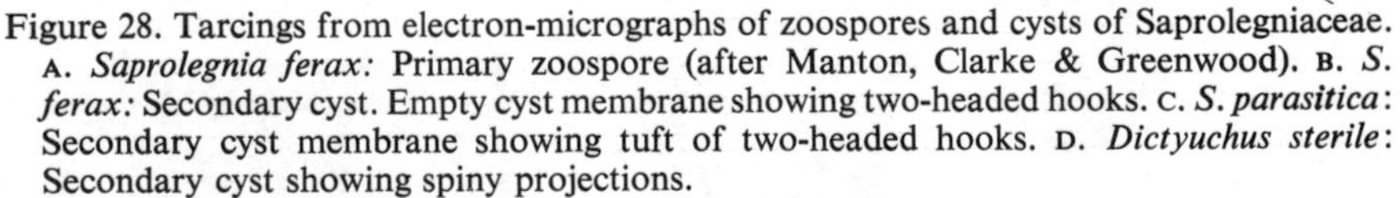
Figure 28. Tarcings from electron-micrographs of zoospores and cysts of Saprolegniaceae. A. *Saprolegnia ferax:* Primary zoospore (after Manton, Clarke & Greenwood). B. *S. ferax:* Secondary cyst. Empty cyst membrane showing two-headed hooks. C. *S. parasitica*: Secondary cyst membrane showing tuft of two-headed hooks. D. *Dictyuchus sterile*: Secondary cyst showing spiny projections.

confined to a particular genus, and numerous observations have been made showing Thraustothecoid or Dictyuchoid sporangia in *Saprolegnia*, Dictyuchoid sporangia in *Achlya* and so on. Salvin (1942*a*) has shown that staling products accumulating in cultures of *Thraustotheca* can affect the type of sporangial discharge. These observations have led some authors to doubt the validity of the distinction between these genera.

SEXUAL REPRODUCTION IN SAPROLEGNIACEAE

In all the Saprolegniaceae sexual reproduction follows a similar course. Oogonia containing one to several eggs are fertilised by antheridial branches. Fertilisation is accomplished by the penetration of fertilisation tubes into the eggs. The majority of species are homothallic (monoecious), that is a culture derived from a single zoospore will give rise to a mycelium forming both oogonia and antheridia. Some species are, however, heterothallic (dioecious) and sexual reproduction only occurs when two different strains are juxtaposed, one forming oogonia, the other antheridia. In some species the eggs develop apogamously (parthenogenetically) and in such forms ripe oogonia are found without any antheridia associated with them. There is a very extensive literature on the cytology of sexual reproduction (Olive, 1953). Light may inhibit the formation of oogonia (Szaniszlo, 1965).

The arrangement of oogonia and antheridia in monoecious forms is shown in Figs. 29 and 30, and in a dioecious form in Fig. 31. Where the antheridial

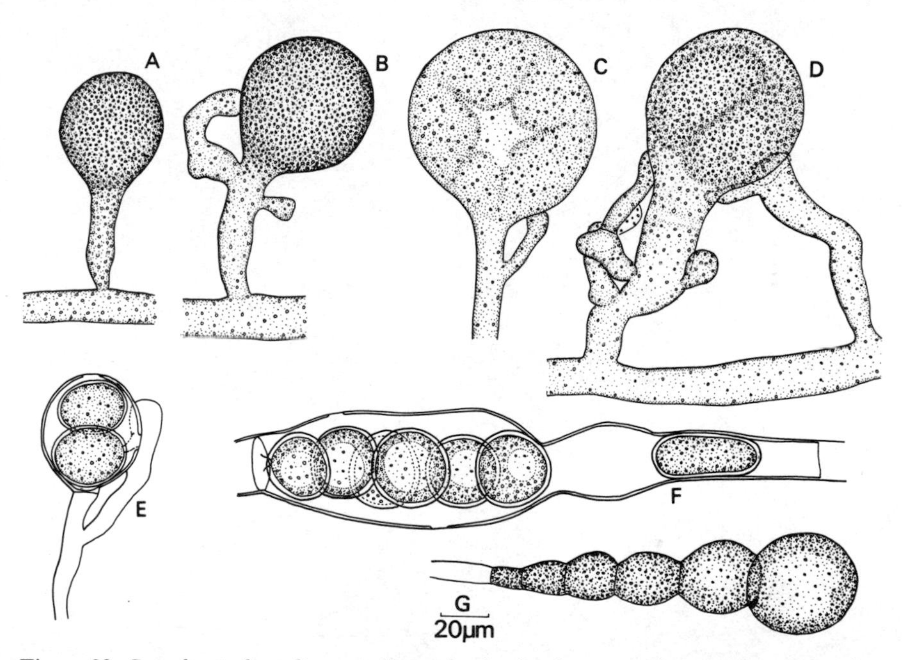

Figure 29. *Saprolegnia litoralis*. A–D. Stages in the development of oogonia. C. Oogonium showing furrowed cytoplasm. D. Outline of two oospheres visible. E. Oogonium with two mature oospores. F. Intercalary oogonium lacking antheridia. The oospores have developed apogamously. G. Chain of gemmae.

branches arise from the stalk of the oogonium they are said to be *androgynous*. If they originate on the same hypha as the oogonium they are said to be *monoclinous*, and when they originate on different hyphae they are *diclinous*. The oogonial initial is multinucleate and as it enlarges nuclear divisions may continue. Eventually some of the nuclei degenerate leaving only those nuclei which are included in the eggs. From the central vacuole within the oogonium cleavage furrows radiate outwards to divide the cytoplasm into uninucleate portions which round off to form eggs. The number of eggs is variable. In *Aphanomyces* there is only a single egg, but in other genera there may be one to several or occasionally as many as 30. The entire mass of cytoplasm within the oogonium is used up in the formation of eggs and there is no residual cytoplasm (periplasm) as in the related Peronosporales. The wall of the oogonium may be uniformly thin, but in some forms it shows uneven thickening, with thin areas or pits through which fertilisation tubes may penetrate (see Fig. 29E). In *Achlya colorata*, one of the commonest British species, the oogonial walls bear blunt rounded projections so that they appear somewhat

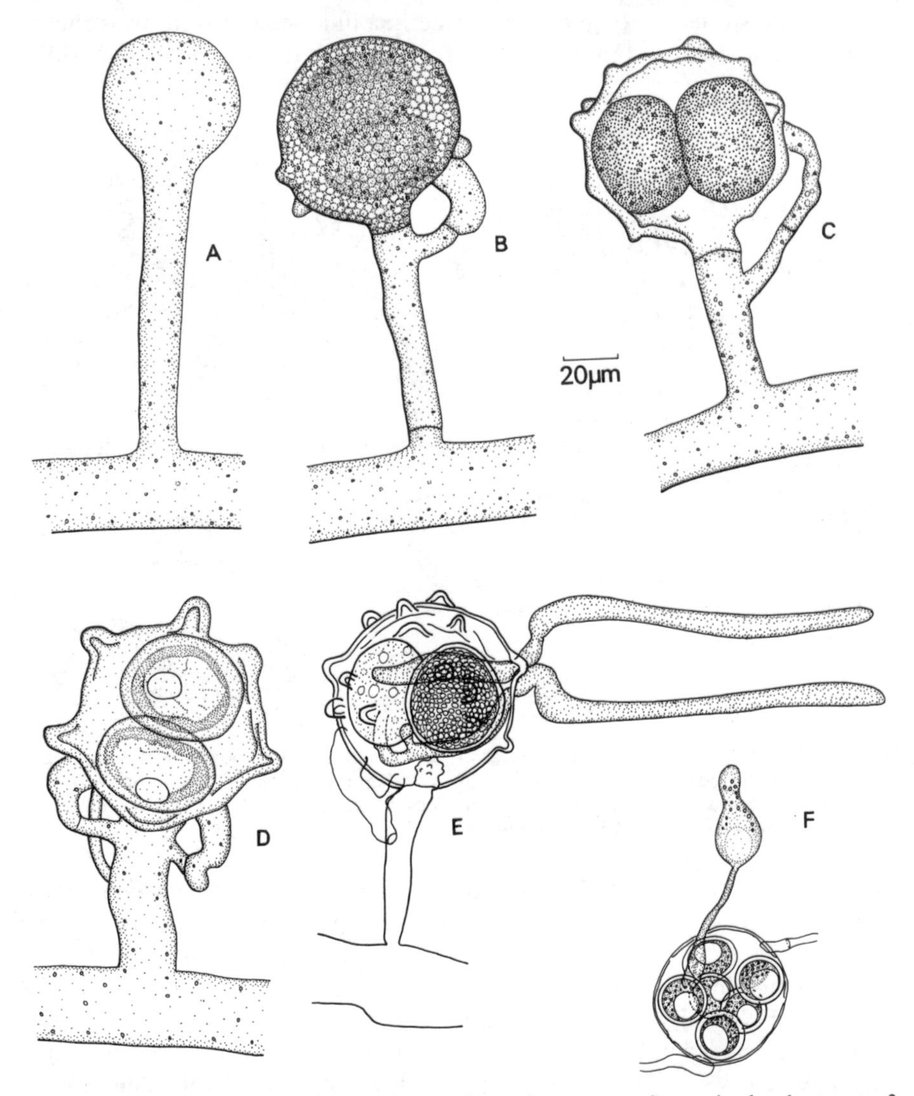

Figure 30. A–E, *Achlya colorata*; F, *Thraustotheca clavata*. A–D. Stages in development of oogonia. E. Six-month-old oospores germinating after 40 hr in charcoal water. F. *T. clavata*: six-month-old oospore germinating after 17 hr in charcoal water. The germ tube is terminated by a sporangium.

spiny (Fig. 30D). The antheridia are also multinucleate and although nuclear division may occur during development most of the nuclei degenerate. The antheridial branches grow chemotropically towards the oogonia and apply themselves to the oogonial wall. The tip of the antheridial branch is cut off by a septum and the resulting antheridium puts out a fertilisation tube which penetrates the oogonial wall. Within the oogonium the fertilisation tube contains a single male nucleus and after the tube has penetrated the oosphere wall the tip breaks down and the male nucleus enters the oosphere and eventually fuses with the single egg nucleus. The fertilised egg (or oospore) now undergoes a series of changes. Its wall thickens and oil globules become obvious. They tend to aggregate in a characteristic way for a given species.

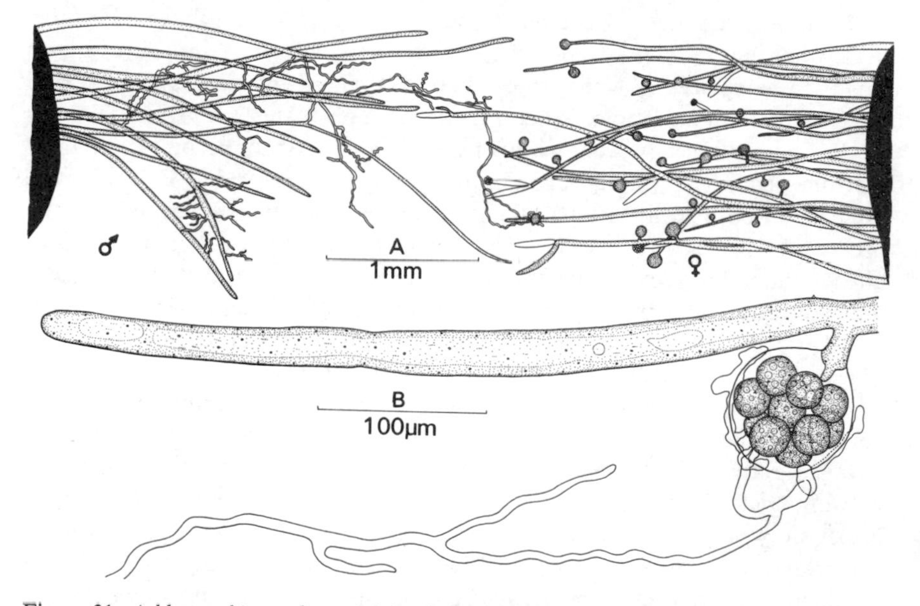

Figure 31. *Achlya ambisexualis*. A. Male plant and female plant grown on hemp seeds and placed together in water for 4 days. Note the formation of antheridial branches on the male and oogonial initials on the female. B. Fertilisation, showing the diclinous origin of the antheridial branch.

In some species the eggs contain a layer of peripheral oil globules surrounding a central mass of clear cytoplasm. Such eggs are described as *centric*. In others a single large oil globule appears to one side of the oospore; such eggs are *eccentric*. Other arrangements are also known. The eggs are rarely capable of immediate further development but seem to need a period of maturation, sometimes lasting weeks or months. Germination can be stimulated by transferring mature oogonia to freshly distilled water (preferably after shaking with charcoal and filtering) and occurs by means of a germ tube which grows out from the oospore, through the oogonial wall. Here it may continue growth as mycelium or may give rise to a sporangium (Fig. 30E,F). Ziegler (1953) and a number of other authors have claimed that meiosis occurs on germination of the oospore implying that the vegetative mycelium is haploid, but Sansome & Harris (1962) have claimed that in camphor-treated material of *Achlya* stained with aceto-orcein, meiosis takes place in the antheridia and oogonia, implying that the mycelium is diploid. The same conclusion had been reached by a number of earlier investigators. Some support for the view that the life cycle is diploid has been obtained by studying the mating reactions of single zoospore isolates derived from germinating oospores (see below).

Studies on heterothallic species of *Achlya* (J. R. Raper, 1952, 1954, 1957), have shown that the sequence of events in sexual reproduction is co-ordinated by hormones secreted by the male and female plants. If isolates of *Achlya bisexualis* and *A. ambisexualis* made from soil, water or mud are grown singly on hemp seed in water, reproduction is entirely asexual, and no sexual organs are distinguishable. When certain of the isolates are grown together in the same dish, however, within two to three days it becomes apparent that one strain is forming oogonia, and the other antheridia. Evidently the presence of both strains in the same dish causes a mutual stimulation for the production of sex organs. It is not necessary for the two strains to be in direct contact but they can be held apart in the water or separated by a membrane of cellophane or by agar, and development of antheridial and oogonial initials occurs as the mycelia of the two strains approach each other. This suggests that a diffusible substance (or substances) may be responsible for the phenomenon. The first observable reaction as the compatible hyphae approach each other is the production of fine lateral branches behind the advancing tips of the male hyphae. These are antheridial branches. By growing male plants in water in which female plants had been growing Raper showed that the vegetative female plants initiated the development of antheridial branches on the male. When female plants were grown in water in which vegetative male plants had previously been growing there was no noticeable reaction on the part of the female, i.e. no oogonial initials were produced. The role of the vegetative female plants as initiatiors of the sequence of events in sexual reproduction was ingeniously confirmed by experiments in micro-aquaria consisting of several consecutive chambers through which water flowed by means of small siphons. In adjacent chambers male and female plants were placed alternately so that water from a male plant would flow over a female plant and so on. If a female plant was placed in the first chamber the male in the second reacted by developing antheridial hyphae. If, however, a male plant was placed in the first chamber and a female in the second, the female failed to respond, and the male plant in the third chamber was the first to react. Raper postulated that the development of the antheridial branches was in response to the secretion of a hormone, Hormone A, secreted by the vegetative females. By further experiments of this kind Raper showed that the later steps in the sexual process were also regulated by means of diffusible substances. He postulated that the antheridial branches secreted a second substance, Hormone B, which resulted in the formation of oogonial initials on the female plants, the next observable step following the formation of antheridial hyphae. The oogonial initials in their turn secreted a further substance, Hormone C, which stimulated the antheridial initials to grow towards the oogonial initials and also resulted in the antheridia being delimited. Having made contact with the oogonial initials the antheridial branches secreted Hormone D which resulted in the formation of a septum cutting of the oogonium from its stalk and the formation of oospheres. If the interchange of diffusible material was prevented by removing male plants from close proximity to the female the sexual process could be interrupted. If the direct contact of antheridia and oogonia is prevented by a barrier of cellophane fertilisation does not occur. The outlines of the proposed scheme are shown below.

Effects of hormones on sexual reactions in *Achlya ambisexualis*, after Raper (1939)

Hormone	Produced by	Affecting	Specific action
A	Vegetative hyphae	Vegetative hyphae	Induces formation of antheridial branches
B	Antheridial branches	Vegetative hyphae	Initiates formation of oogonial initials
C	Oogonial initials	Antheridial branches	(1) Attracts antheridial branches (2) Induces thigmotropic response and delimitation of antheridia
D	Antheridia	Oogonial initials	Induces delimitation of oogonium by formation of basal wall

The original scheme necessitated four hormones but later work showed that Hormone A was in reality a complex of at least four hormones, some of which were secreted by male plants. Certain of the hormones of the 'A complex' augmented the hormonal activity of the vegetative female, and another inhibited it. Separation of the different hormones of the 'A complex' can be achieved by chemical means. Purification, and concentration and assay of Hormone A have been achieved but it has not been identified fully biochemically. The purified hormone of *A. bisexualis* is a colourless crystal having the empirical formula $C_{29}H_{42}O_5$, to which the name antheridiol has been given (McMorris & Barksdale, 1967).

Fischer & Werner (1955) have shown that vegetative hyphae and also antheridial branches can be stimulated chemotropically by mixtures of certain amino-acids, but this is not to say that these substances are identical to Hormone C.

Barksdale (1963*b*) has since shown that if hormone A is adsorbed on to small particles of the inert plastic polystyrene, the particles not only induce initiation of antheridial initials but also directional growth, and she has therefore questioned the necessity to postulate a hormone distinct from A which is exclusively concerned with directional growth.

The hormonal control of sexual reproduction is not confined to heterothallic forms: there is evidence that in homothallic species there is similar co-ordination. The fact that it is possible to initiate sexual reactions between homothallic and heterothallic species of *Achlya* shows that some of the hormones are probably common to more than one species although there is also evidence of some degree of specificity of the hormones of different species (Raper, 1950; Barksdale, 1960, 1965). Sexual reactions may also occur in intergeneric matings (Couch, 1926; Salvin, 1942*b*; Raper, 1950), which again raises the question of the validity of the separation of these genera.

One further interesting phenomenon which has been discovered in relation to the heterothallic *Achlyas* is relative sexuality. If a number of isolates of *A. bisexualis* and *A. ambisexualis* from widely separated sources are paired in all possible combinations it is found that certain strains show a capacity to react either as male or female depending on the particular partner to which they are apposed. Other strains remain invariably male or invariably female and these are referred to as true males or strong males, etc. The strains can be arranged in a series with strong males and strong females at the extremes, and intermediate strains whose reaction may be either male or female depending on the strength of their mate. Interspecific responses between strains of *A. bisexualis* and *A. ambisexualis* are also possible and the behaviour of certain of these strains may also show reversibility. Further, some of the strains which appear heterothallic at room temperature are homothallic at lower temperatures. On this and other evidence Barksdale (1960) has postulated that the heterothallic forms may have been derived from homothallic ones. She has argued that the most notable difference between strong males and strong females is in their production and response to Hormone A. Very little of this substance is found in male cultures, and male cultures are also much more sensitive in their response to the hormone than female cultures. Another important difference is in the uptake of Hormone A. Certain strains appear capable of absorbing Hormone A much more readily than others, and it is these strains which have the ability to absorb the hormone that produce antheridial branches during conjugation with other thalli (Barksdale, 1963*a*). If one assumes that heterothallic forms have been derived from homothallic this might have occurred by mutations leading to increased sensitivity to A and hence to maleness. Mutations leading to extracellular accumulation of A should lead to increasing femaleness.

Germination of the oospores of *A. ambisexualis* results in the formation of a multinucleate germ tube which develops into a germ sporangium if transferred to water, or to an extensive coenocytic mycelium in the presence of nutrients. This mycelium can be induced to form zoosporangia when transferred to water. From zoosporangia of either source single zoospore cultures can be obtained which can be mated with the parental ♂ or ♀ strains. All zoospores or germ tubes derived from a single oospore gave the same result in regard to their sexual interaction, and appeared to have the same tendency. These findings suggest that nuclear division on oospore germination is probably not meiotic, and thus are consistent with the idea that the life cycle is diploid (Mullins & Raper, 1965). Confirmation of these results, implying meiosis during gamete differentiation, have also been obtained with *A. bisexualis* (Barksdale, 1966).

RELATIONSHIPS

The Saprolegniales are probably closely related to the Peronosporales, both groups having walls containing cellulose, similar zoospore structure and oogamous reproduction. The Leptomitales is another group of aquatic fungi showing clear similarities with the Saprolegniales.

PERONOSPORALES

Many fungi in this group are parasites of higher plants and may cause diseases of economic importance such as late blight of potato, downy mildew of the grape-vine and blue-mould of tobacco. Some, particularly the Peronosporaceae and Albuginaceae, appear to be obligate parasites, but some members of the Pythiaceae are facultative parasites, and probably survive saprophytically, in soil, mud or on decaying vegetation.

The mycelium is coenocytic but delicate. The composition of the walls of species of *Pythium* and *Phytophthora* has been studied (Aronson *et al.*, 1967). They have a complex chemical structure consisting of polysaccharide, protein and lipid. Glucans constitute about 90% of the wall. Cellulose, a β-(1→4) linked glucan may make up a relatively small proportion of this glucan (up to about 36%), but in some forms, e.g. *Pythium butleri*, cellulose is replaced by a β-(1→2) linked glucan. The lack of chitin in the walls appears to be correlated with the absence of 'multi-vesicular bodies' found associated with the plasmalemma in other fungi which contain chitin, in regions of hyphae where active extension growth has ceased (Marchant *et al.*, 1967). In many of the obligate parasites haustoria are formed within the host cell, and they vary in form from minute spherical or cylindrical ingrowths to more extensively branched or lobed structures. Asexual reproduction is by means of sporangia which vary greatly in shape. In certain species of *Pythium* the sporangia are merely inflated lobes of mycelium, but more usually they are spherical or pear-shaped. In the Peronosporaceae they arise on differentiated and often characteristically branched sporangiophores, whilst in the Albuginaceae they occur in chains. The sporangia of foliar pathogens are detached from the sporangiophores and are dispersed by wind. In the water- and soil-inhabiting forms, the sporangia usually give rise to zoospores. Whilst zoospores are found in certain pathogens such as *Phytophthora*, *Albugo* and in *Plasmopara*, sporangial germination by means of a germ tube may also occur, especially in *Peronospora* where zoospore formation from sporangia does not occur. The zoospore is laterally biflagellate with one flagellum of the whip type and one of the tinsel type (Colhoun, 1966). It thus corresponds to the secondary type of zoospore found in the Saprolegniales with which this group may be related.

Sexual reproduction is oogamous. Each oogonium contains a single egg (except in *Pythium multisporum* where there are several). The antheridial and oogonial intials are commonly multinucleate at their inception and further nuclear divisions may occur during development. In many forms there is only one functional male and female nucleus, but in others multiple fusions occur. A feature of the oogonia of the Peronosporales is that the residual cytoplasm left after the differentiation of the central oosphere persists as the periplasm which may play a part in the deposition of wall material around the fertilised oospore. It now seems likely that meiosis occurs during the development of the gametangia implying that the soma is diploid. Oospore germination may be by means of a germ tube or by zoospores. Most species are homothallic, but heterothallism and relative sexuality have been reported.

Nutritional and physiological studies have been made on members of the Pythiaceae which can be grown in artificial culture (for references see Cantino, 1950, 1955; Cantino & Turian, 1959; Fothergill & Hide, 1962; Fothergill & Child, 1964). Most species of *Pythium* and *Phytophthora* are heterotrophic for thiamine. It has also been shown that sterols are essential for oospore production in *Pythium* and *Phytophthora* (Lilly, 1966). Sulphur requirements can be met in most forms by sulphate, but organic forms of sulphur are also utilised. Inorganic nitrogen compounds such as nitrates and ammonium salts may support growth, but species differ in their ability to utilise these sources. A range of sugars and sugar alcohols and other carbohydrates serve as sources of carbon.

Three families can be distinguished (Martin, 1961).

1. Sporangiophores differing little, if at all, from assimilative hyphae; mycelium saprophytic or parasitic, but if parasitic usually intracellular and without haustoria. **Pythiaceae**
2. Sporangiophores specialised; mycelium intercellular, with haustoria penetrating the host cells; obligate parasites of higher plants.
   (a) Sporangia borne singly or in clusters at the tips of branched sporangiophores which emerge through stomata; haustoria various **Peronosporaceae** (downy mildews)

Figure 32. A. Pot of cress (*Lepidium sativum*) 1 week old, showing symptoms of damping-off caused by *Pythium spp*. B. Symptoms of late blight caused by *Phytophthora infestans* on a terminal leaflet of potato. The white zone is due to the development of sporangiophores.

(b) Sporangia in chains on club-shaped sporangiophores borne in dense sori beneath the host epidermis; haustoria globose
Albuginaceae (white blisters)

Pythiaceae

Representatives of the Pythiaceae can be found in water and in soil growing saprophytically on plant and animal debris, or as parasites of animals (e.g. *Zoophagus*, parasitic on rotifers) and plants. Although about six genera can be distinguished (Middleton, 1952; Sparrow, 1960) we shall consider only *Pythium* and *Phytophthora*.

Pythium

Pythium grows in water and soil as a saprophyte, but under suitable conditions, e.g. where seedlings are grown crowded together in poorly drained soil, it can become parasitic causing diseases such as pre-emergence killing, damping-off, and foot-rot. Damping-off of cress (*Lepidium sativum*) can be demonstrated by sowing seeds densely on heavy garden soil which is kept liberally watered. Within five to seven days some of the seedlings may show brown lesions at the base of the hypocotyl, and the hypocotyl and cotyledons may become water-soaked and flaccid. In this condition the seedling collapses. A collapsed seedling may come into contact with other seedlings and so spread the disease (Fig. 32A). The host cells separate from each other easily due to the breakdown of the middle lamella, probably brought about by pectic and possibly cellulolytic enzymes secreted by the *Pythium*. The enzymes diffuse from the hyphal tips, so that softening of the host tissue actually occurs in advance of the mycelium. Pure culture studies *in vitro* suggest that species of *Pythium* may also secrete heat-stable substances which are toxic to plants. Within the host the mycelium is coarse and coenocytic, with granular cytoplasmic contents (Fig. 33). At first there are no septa, but later cross-walls may cut off empty portions of hyphae. Thick-walled chlamydospores may also be formed. There are no haustoria. Several species are known to cause damping-off, e.g. *P. debaryanum* and, perhaps more frequently, *P. ultimum*. *Pythium aphanidermatum* is associated with stem-rot and damping-off of

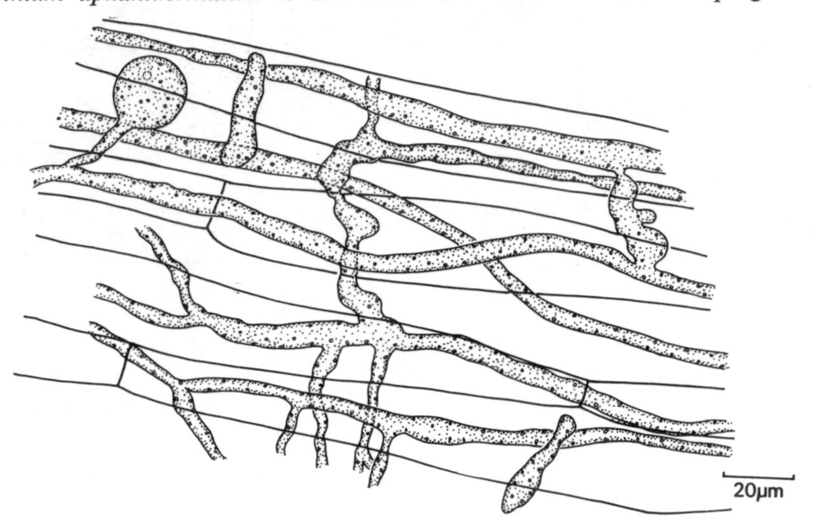

Figure 33. *Pythium* mycelium in the rotting tissue of a cress seedling hypocotyl. Note the absence of haustoria and the spherical sporangium initial.

cucumber, and the fungus may also cause rotting of mature cucumbers. *P. mamillatum* causes damping-off of mustard and beet seedlings, and is also associated with root-rot in *Viola*. A taxonomic account of *Pythium* has been given by Middleton (1943), and keys to species and original descriptions by Waterhouse (1967, 1968).

Asexual reproduction: the mycelium within the host tissue, or in culture, usually produces sporangia, but their form varies. In some species, e.g. *P. gracile*, the sporangia are filamentous and are scarcely distinguishable from vegetative hyphae. In *P. aphanidermatum* the sporangia are formed from inflated lobed hyphae (Fig. 34B). In many species however, e.g. *P. debaryanum*, the sporangia are globose (Fig. 34A). A terminal or intercalary portion of a hypha enlarges and assumes spherical shape, and becomes cut off from the mycelium by a cross-wall. The sporangia contain numerous nuclei. Zoospores do not differentiate within the sporangium but in a thin-walled vesicle formed at the tip of a fine tube which develops from the sporangium. The vesicle enlarges as cytoplasm from the sporangium is transferred to it, and during the next few minutes the cytoplasm cleaves into about 8–20 uninucleate zoospores which jostle about inside the sporangium, and cause the thin wall of the vesicle to bulge irregularly (Webster & Dennis, 1967). Finally, about 20 minutes after the inflation of the vesicle, its wall breaks down and the zoospores swim away. They are broadly bean-shaped with two laterally

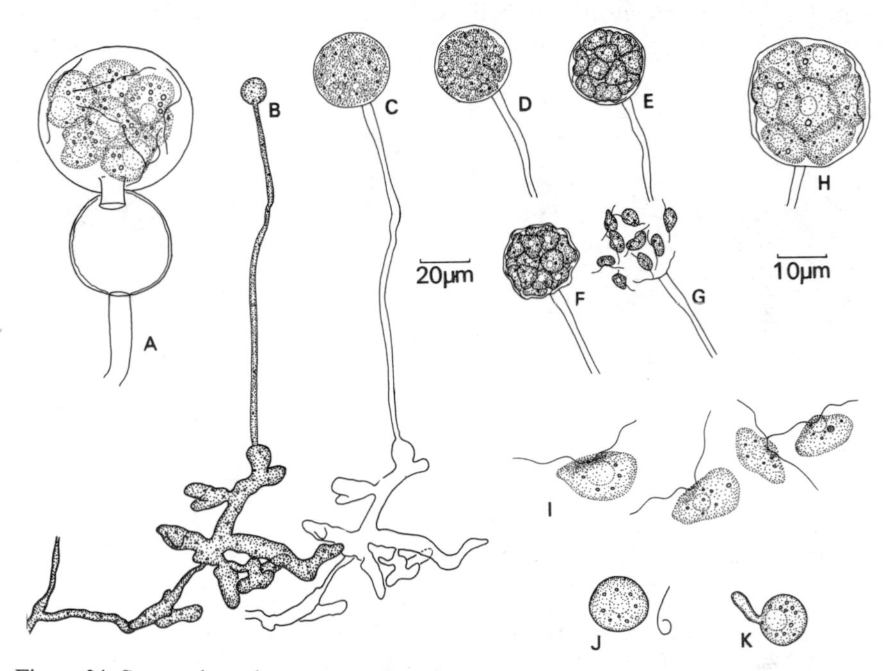

Figure 34. Sporangia and zoospores of *Pythium*. A. *P. debaryanum*. Spherical sporangium with short tube and a vesicle containing zoospores. B–K. *P. aphanidermatum*. B. Lobed sporangium showing a long tube and the vesicle beginning to expand. C–G. Further stages in the enlargement of the vesicle, and differentiation of zoospores. Note the transfer of cytoplasm from the sporangium to the vesicle in C. The stages illustrated in B–G took place in 25 min. H. Enlarged vesicle showing the zoospores. Flagella are also visible. I. Zoospores. J. Encystment of zoospore showing a cast-off flagellum. K. Germination of a zoospore cyst. B–G to same scale; A, H, I, J, K to same scale.

attached flagella. After swimming for a time the flagella are cast off, the spore encysts and germination takes place by means of a germ tube (Fig. 34K). Repeated emergence has also been reported. In some forms, e.g. *P. ultimum* var. *ultimum*, zoospore formation from sporangia does not occur; the sporangia germinate directly to form a germ tube. However Drechsler (1960) has described a variety, *P. ultimum* var. *sporangiferum*, in which zoospores develop from the zoosporangia. Sporangia showing direct germination are sometimes referred to as conidia. Zoospores are formed in *P. ultimum* var. *ultimum* when the oospore germinates, although direct germination of the oospores has also been reported (Fig. 41D,E). Sporangial proliferation occurs in certain species, e.g. *P. middletonii* and *P. undulatum*.

Sexual reproduction: the formation of oogonia and antheridia occurs readily in cultures derived from single zoospores, and it is therefore probable that most species of *Pythium* are homothallic. However certain species fail to form oospores in culture and it would be of interest to pair different isolates to see if sexual stimulation would result. Heterothallism has been demonstrated in *P. sylvaticum* (Campbell & Hendrix, 1967). Oogonia arise as terminal or intercalary spherical swellings which become cut off from the

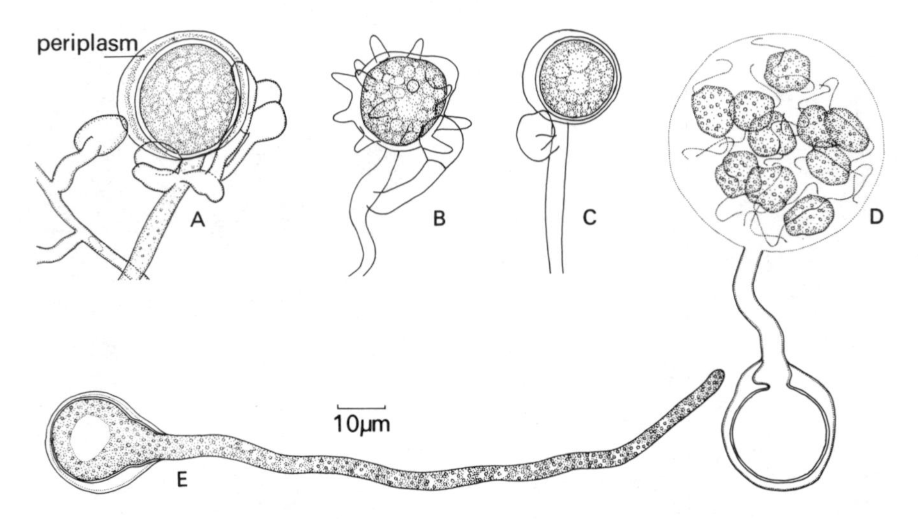

Figure 35. Oogonia and oospores of *Pythium*. A. *P. debaryanum*. Note that there are several antheridia. B. *P. mamillatum*. Oogonium showing spiny outgrowths of oogonial wall. C. *P. ultimum*. D, E. *P. ultimum* germinating oospores (after Drechsler).

adjacent mycelium by cross-wall formation. In some species, e.g. *P. mamillatum*, the oogonial wall is folded into long projections (Fig. 35B). The antheridia arise as club-shaped swollen hyphal tips, often as branches of the oogonial stalk, or sometimes from separate hyphae. In some species, e.g. *P. ultimum*, there is typically only a single antheridium to each oogonium, whilst in others, e.g. *P. debaryanum*, there may be several (Fig. 35). The young oogonium is multinucleate and the cytoplasm within it differentiates into a multinucleate central mass, the ooplasm, from which the egg develops, and a peripheral mass, the periplasm, also containing several nuclei. The periplasm does not contribute to the formation of the egg. There are conflicting accounts of the type of nuclear division which occurs. Many early workers, assuming that meiosis occurred on germination of the oospore and that the vegetative filaments were therefore haploid, have described only mitoses during the development of the egg. However Sansome (1961, 1963) has claimed that meiosis occurs in the oogonia and antheridia of *P. debaryanum* and also in *Phytophthora cactorum*. Meiosis during gametogenesis has also been reported in *Phytophthora drechsleri*, a finding supported by genetical evidence (Galindo & Zentmeyer, 1967). In *P. debaryanum* Sansome has described young oogonia with more numerous nuclei than older ones, due to degeneration of certain nuclei. One to eight nuclei survive and enlarge, undergoing meiosis. At a later stage of division up to 32 nuclei were counted. In this and other species only one egg nucleus survives in the centre of the ooplasm, while the remaining nuclei degenerate. The nuclei of the periplasm also degenerate. The young antheridia are multinucleate, but all nuclei except one degenerate. The surviving nucleus undergoes meiosis so that older antheridia have four nuclei. The antheridia are applied to the oogonial wall and penetrate it by means of a fertilisation tube. Following penetration only three nuclei were counted in the antheridium, suggesting that one had entered the oogonium. Later still, empty antheridia were found, and it is presumed that the three remaining nuclei enter the oogonium and join the nuclei degenerating in the periplasm. Fusion between a single antheridial and egg nucleus has been described. The fertilised egg secretes a double wall, and a globule of reserve material appears in the cytoplasm. Material derived from the periplasm may also be deposited on the outside of the egg. Such oospores may need a period of rest (after-ripening) of several weeks before they are capable of germinating. Germination may be by means of a germ tube, or by the formation of a vesicle in which zoospores are differentiated (Fig. 35D,E), or in some forms the developing oospore produces a short germ tube ending in a sporangium.

Pythium can live saprophytically, and can also survive in air-dry soil for several years. It is more common in cultivated than in natural soils, and appears to be intolerant of acid soils. As saprophytes, species of *Pythium* are important primary colonisers of virgin substrata, and probably gain initial advantage in colonisation by virtue of their rapid growth rate. They do not, however, compete well with other fungi which have already colonised a substrate and appear to be rather intolerant of antibiotics.

The control of diseases caused by *Pythium* is obviously rendered difficult by its ability to survive saprophytically in soil. Its wide host range means that it is not possible to control diseases by means of crop rotation. The effects of disease can be minimised by improving drainage and avoiding overcrowding of seedlings. In the greenhouse or nursery some measure of control can be achieved by partial sterilisation using steam or formalin to destroy the fungus, and recolonisation of the treated soil by *Pythium* is slow.

Phytophthora

The name *Phytophthora* is derived from the Greek (*phyton*, plant; *phthora*, destruction); and some species are certainly destructive parasites. The best-known is *P. infestans*, the cause of late blight of potatoes. This fungus is confined to Solanaceous hosts but others have a much wider range. For example *P. cactorum* has been recorded from over 40 families of flowering plants causing a variety of diseases such as damping-off, root-rot and fruit-rot. Other important pathogens are *P. erythroseptica* associated with a pink-rot of potato tubers, *P. fragariae* the cause of red-core of strawberries, and

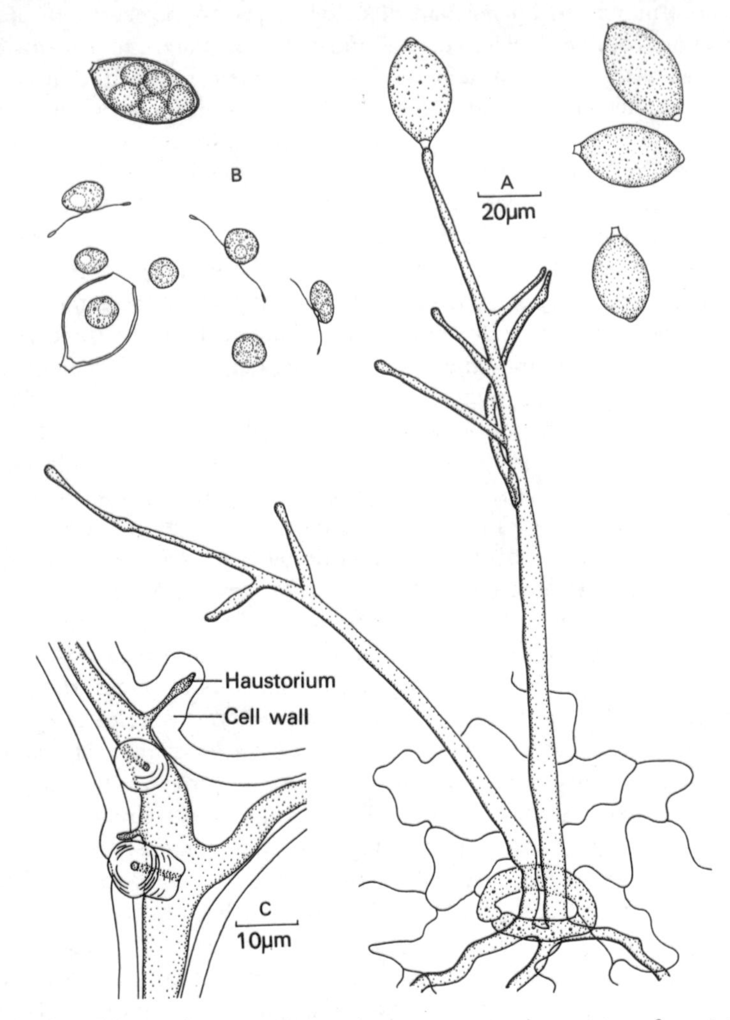

Figure 36. *Phytophthora infestans*. A. Sporangiophores penetrating a stoma of a potato leaf. B. Sporangial contents dividing and releasing zoospores. C. Intercellular mycelium from a potato tuber showing the finger-like haustoria penetrating the cell walls. Note the thickening of the cell wall around the haustorium.

P. palmivora causing pod-rot and canker of cacao. A taxonomic account of the genus has been given by Tucker (1931); an invaluable general account by Hickman (1958). Waterhouse (1956) has compiled a collection of original descriptions of species, and provided a key (1963). Whilst some species may live saprophytically as water-moulds the majority of pathogenic forms probably have no prolonged free-living saprophytic existence, but survive in the soil in the form of oospores, or within diseased host tissue. However, almost all the pathogenic forms can be isolated from their hosts at the appropriate season and can be grown in pure culture. Most species form a non-septate mycelium producing branches at right angles, often constricted at their point of origin. Septa may be present in old cultures. Within the host the mycelium is intercellular, but haustoria are formed which penetrate the host

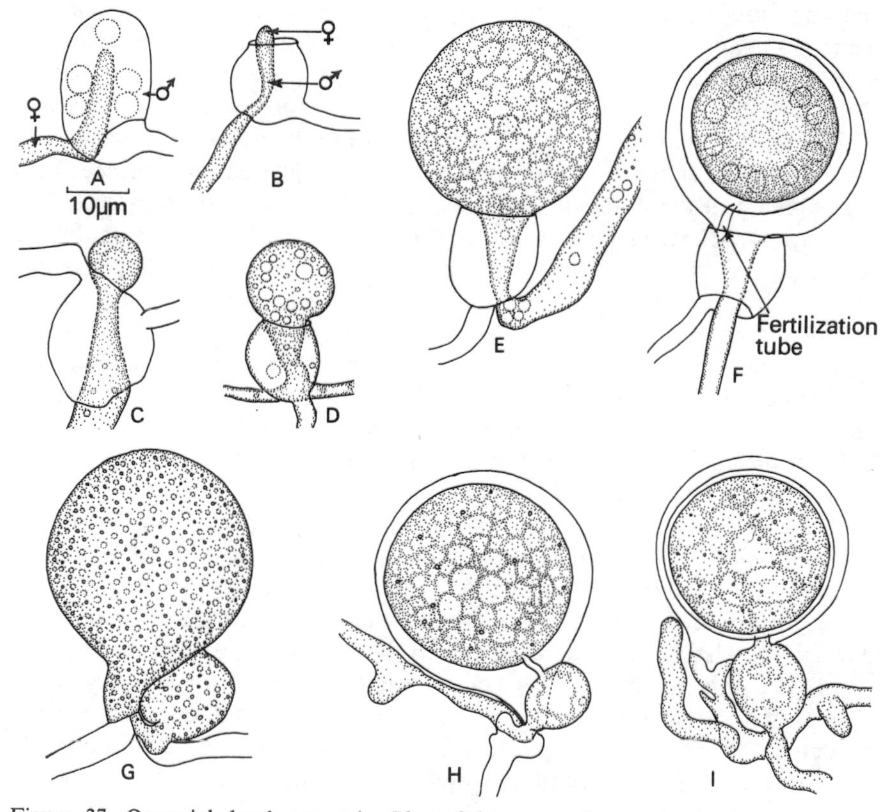

Figure 37. Oogonial development in *Phytophthora*. A–F. Stages in development of *P. erythroseptica*. G–I. Stages in development of *P. cactorum*.

cells. In *P. infestans* within potato tubers the haustoria are finger-like protuberances which may be in part surrounded by thickenings of host wall material (Fig. 36C). Electron micrographs of haustoria in potato leaves show that the haustoria are not surrounded by host cell wall. They are, however, surrounded by an encapsulation whose origin and function are not certain (Ehrlich & Ehrlich, 1966). Sporangia are usually pear-shaped and arise on well-differentiated simple or branched sporangiophores. On the host plant the sporangiophores may emerge through the stomata as in *P. infestans*

(Fig. 36A). The first sporangium is terminal, but the hypha bearing it may push it to one side and proceed to the formation of further sporangia. The mature sporangium has a terminal papilla. The fine structure of sporangia of *P. erythroseptica* has been described by Chapman & Vujičić (1965). In terrestrial forms the sporangia are detached, possibly aided by hygroscopic twisting of the sporangiophore on drying, and are dispersed by wind before germinating, but in aquatic forms zoospore release commonly occurs whilst the sporangia are still attached. In these aquatic species internal proliferation of sporangia may occur. Germination may be by means of zoospores, especially if water is available, or by germ tube under drier conditions. Direct germination (i.e. by germ tube) is stimulated by high temperature in *P. erythroseptica* (Vujičić & Colhoun, 1966). When zoospores are formed they are typically differentiated *within* the sporangium, and not in a vesicle as in *Pythium*, but this distinction is not absolute. An evanescent vesicle may be formed from the papilla and the differentiated zoospores pass into it before release, but in many cases the zoospores are released directly from the sporangium by the breakdown of the papilla. The uninucleate laterally biflagellate zoospores swim for a time, then usually encyst and germinate by means of a germ tube, but repeated emergence has occasionally been reported. In *P. cactorum* sporangia have been preserved for several months under moderately dry conditions. On return to suitable conditions such a sporangium may germinate by the formation of a vegetative hyphae, or a further sporangium. Thick-walled spherical chlamydospores have also been described.

Sexual reproduction in *Phytophthora* is similar in outline to that of *Pythium*, but two distinct types of antheridial arrangement are found. In *P. cactorum* and a number of other species antheridia are attached laterally to the oogonium and are described as *paragynous* (Fig. 37G–I). In some other species including *P. erythroseptica* and *P. infestans* the oogonia during their development actually penetrate and grow through the antheridium. The oogonial hypha after penetrating the antheridium emerges above it and inflates to form a spherical oogonium, with the antheridium persisting as a collar about its base (Fig. 37A–F). This arrangement of the antheridium is termed *amphigynous*. Both the oogonia and the antheridia are multi-

---

Figure 38. *Phytophthora cactorum.* Development of oogonium, antheridium and oospore. A. Initials of oogonium and antheridium. B. Oogonium and antheridium grown to full size: the oogonium has about 24 nuclei and the antheridium about nine. C. Development of a septum at the base of each, and degeneration of some nuclei in each until the oogonium has eight or nine nuclei and the antheridium four or five. D. A simultaneous division of the surviving nuclei in oogonium and antheridium. The protoplasm has large vacuoles. E. Separation of ooplasm from periplasm. Nuclei in division in periplasm prior to degeneration. Oogonium presses into antheridium. F. Entry of one antheridial nucleus by a fertilisation tube. The protoplasm and remaining nuclei of the antheridium degenerate. G. Development of oospore wall. H. Oospore enters on its dormant period with exospore formed from dead periplasm, endospore (of cellulose, protein, etc.) deposited upon it, and paired nuclei in association but not yet fused. A–H are composite drawings of eight stages in sequence (after Blackwell).

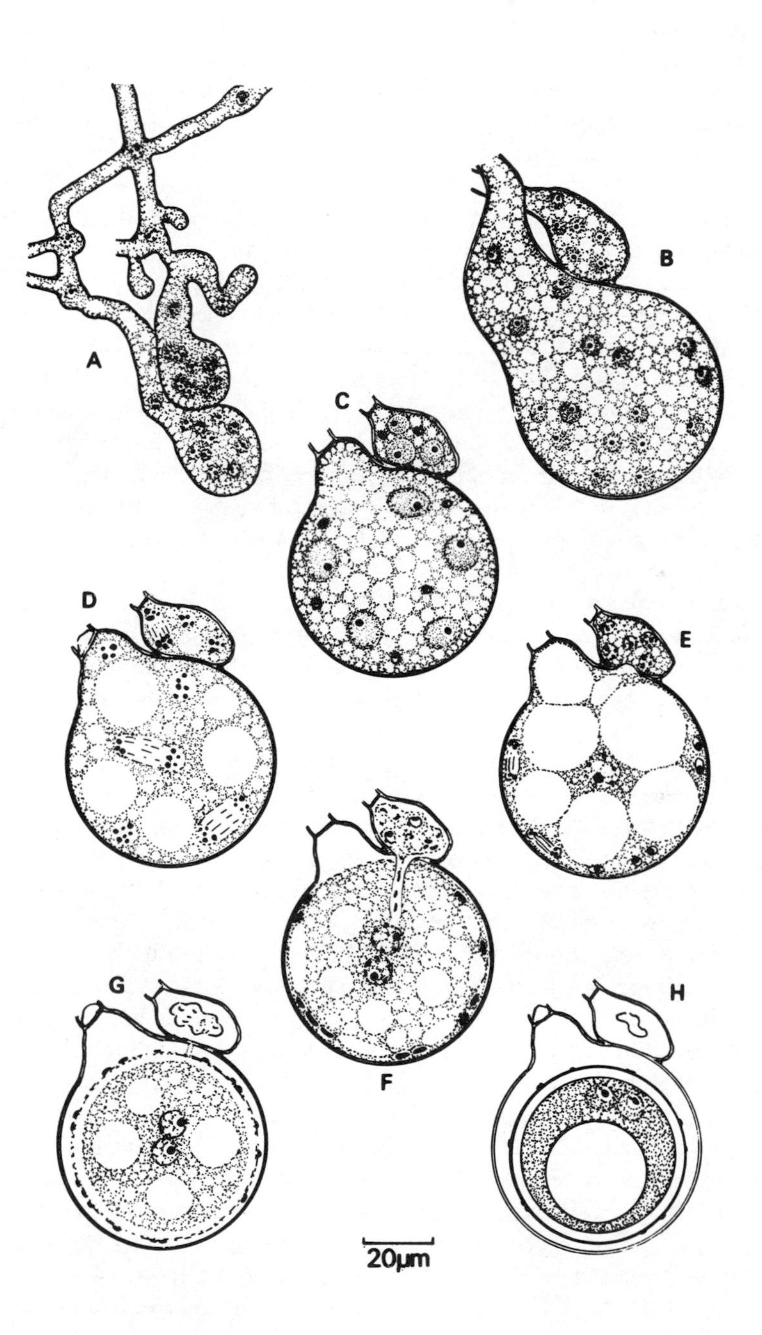
A
B
C
D
E
F
G
H
20µm

nucleate but as the oosphere matures only a single nucleus remains at the centre while the remaining nuclei are included in the periplasm. Fertilisation tubes have been observed and a single nucleus is introduced from the antheridium (Fig. 38). Fusion between the egg nucleus and antheridial nucleus is delayed until the oospore wall is mature. This wall has a thin exospore derived from the periplasm composed of pectic substances, and a thicker endospore of cellulose, protein and possibly other reserve substances. Within, the cytoplasm contains a large oil globule. Such oospores do not germinate immediately, but undergo a period of maturation lasting several weeks or months. Before germination, nuclear fusion will have occurred and the fusion nucleus will have divided several times. The egg swells and the exospore is stretched. The oil globule breaks up into a number of smaller globules and there are signs of digestion of the endospore from within. A germ tube penetrates the exospore and the oogonial wall and sporangia may be formed at the tip. There is now evidence that in *P. cactorum* and *P. drechsleri* meiosis occurs in the process of gamete formation (Sansome, 1963; Galindo & Zentmeyer, 1967).

Oospores may survive in soil for long periods. In any population of oospores there is considerable physiological heterogeneity, and no one set of environmental conditions will stimulate all the oospores to germinate (Blackwell, 1943*a*,*b*).

Species of *Phytophthora* such as *P. cactorum* and *P. erythroseptica* are homothallic and form oogonia readily in culture, but a number of others, e.g. *P. infestans*, often behave as though they were heterothallic, and in this species relative sexuality has been reported. Certain isolates of *P. infestans*, however, have been reported which are homothallic, forming oogonia in unpaired cultures. The behaviour of a number of isolates has shown that some strains are potentially bisexual. When mated with certain other isolates a given strain will form oogonia, but in different pairings the same strain will form antheridia. The behaviour of a given strain is thus dependent on the condition of the strain with which it is mated (Galindo & Gallegly, 1960). Although compatible mating types are present in Mexico where oospores occur in leaves and shoots of *Solanum*, in Europe only one mating type is prevalent, although there are earlier reports of oospores in culture and on potato tubers.

It has been noted that mixed cultures containing two different species of *Phytophthora* may form oogonia although when the individual species are grown singly oospores are formed only sparsely or not all all (Stamps, 1953). Cultures inoculated with *P. cinnamomi* and *P. cryptogea* behaved in this way, forming hybrid oospores and also oogonia and antheridia of *P. cinnamoni*. Although in her experiments Stamps could not demonstrate directly chemical stimulation of oospore production, Galloway (1936) induced oospore formation in single cultures of *P. meadii* and *P. colocasiae*, species not normally producing them, by addition to the medium of untreated filtrate from a paired culture.

LATE BLIGHT OF POTATO

Late blight of potato caused by *P. infestans* is a notorious disease. In the period between 1845 and 1847 it resulted in famine among the working-class population of Ireland who had come to depend on potatoes as their major source of food. 'The cost of the potato famine in terms of life and health of the people is very difficult to estimate. We have certain fixed points. The population of Ireland in 1841 was 8,175,124. In 1851 it was 6,552,385, a diminution of 1,622,739' (Salaman, 1949). Indirectly, however, the famine provided a stimulus to the study of plant pathology. The history of the Irish famine has been ably documented by Large (1958) and Woodham-Smith (1962). The present status of the disease has been assessed by Cox & Large (1960). It is found wherever the crop is grown, but it is not the major cause of lost production because certain virus diseases are more injurious. There is now an immense literature to the disease (Moore, 1959).

The fungus overwinters in tubers infected during the previous season, and a very low proportion of such tubers give rise to infected shoots. In experimental plots the proportion of infected plants developing from naturally and artificially infected tubers was found to be less than 1% (Hirst, 1955; Hirst & Stedman, 1960*a,b*). Nevertheless such infected shoots form foci within the crop from which the disease spreads. Sporangia formed on the diseased shoots are blown to healthy leaves, and there germinate either by the formation of germ tubes or zoospores. Zoospore production is favoured by temperatures between 9 and 15°C. After swimming for a time the zoospores encyst and then form germ tubes which usually penetrate the epidermal walls of the potato leaf, or occasionally enter the stomata. An appressorium is formed at the tip of the germ tube, attaching the zoospore cyst firmly to the leaf, and penetration of the cell wall is probably achieved by both mechanical and enzymatic action. Penetration can occur within two hours. Within the leaf tissues mycelium develops which grows out radially from the point of penetration. The resulting lesion had a dark green water-soaked appearance associated with tissue disintegration, possibly aided by toxin-secretion. Such lesions (see Fig. 32B) are visible within about three to five days of infection under suitable conditions of temperature and humidity. Around the margin of the advancing lesion, especially on the lower surface of the leaf, a zone of sporulation is found with sporangiophores emerging through the stomata. Sporulation is most prolific during periods of high humidity and commonly occurs at night following the deposition of dew. In potato crops as the leaf canopy closes over between the rows to cover the soil a humid microclimate is established which may result in extensive sporulation. As the foliage dries during the morning the sporangiophores undergo hygroscopic twisting which results in the flicking off of sporangia. Thus the concentration of sporangia in the air usually shows a characteristic diurnal fluctuation, with a peak at about 10 a.m.

The destructive action of *P. infestans* is directly associated with the killing of the foliage, which results in a reduction in the photosynthetic area of the potato plant and hence a reduction in the weight of tubers. When about 75% of the leaf tissue has been destroyed, further increase in weight of the crop ceases (Cox & Large, 1960). Thus an early epidemic of blight can result in a lowering of yield. This reduction tends to be offset by the fact that such epidemics are more common in rainy cool seasons which are conducive to high crop yields.

A more sinister cause of crop losses is occasioned by the fact that tubers can be infected by sporangia falling on to them, either during growth or lifting. Such infected tubers may rot in storage, and the diseased tissue is often invaded by secondary bacterial and fungal saprophytes.

Control of potato blight is achieved by several means:

1. Spraying

By spraying with suitable fungicides, which prevent sporangial germination, epidemic spread of the disease can be delayed. This can result in a prolongation of photosynthetic activity of the potato foliage and hence an increase in yield. Various types of fungicide have been developed, and among the most common is one containing copper sulphate and calcium oxide (Bordeaux mixture). However the damage caused by the copper on the potato leaves and by tractor wheels of the spraying apparatus to the potato haulms may cause a greater reduction in the potential crop yield, than the blight which might have resulted had the crop not been sprayed. To avoid unnecessary spraying and to ensure that timely spray applications are made it has proved possible to provide, for certain countries, forecasts of the incidence of potato blight epidemics. Analysis of the incidence of blight epidemics in south Devon by Beaumont (1947) established that a 'temperature–humidity rule' controlled the relationship between blight epidemics and weather. After a certain date (which varied with the locality) blight followed within 15–22 days, a period during which the minimum temperature was not less than 50°F (10°C) and the relative humidity was over 75% for two consecutive days. The warm humid weather during this period provides conditions suitable for sporulation and the initiation of new infections. Modified in the light of experience, this relationship, coupled with field observations on the first appearance of blight lesions in the field, has enabled the Plant Pathology Laboratory of the Ministry of Agriculture to issue accurate forecasts of the likelihood of potato blight epidemics (Cox & Large, 1960) and thus to increase the potential value of spraying.

2. Haulm destruction

The danger of infection of tubers by sporangia falling on to them from foliage at lifting time can be minimised by ensuring that all the foliage is destroyed beforehand. This is achieved by spraying the foliage two to three weeks before lifting time with such sprays as sulphuric acid, copper sulphate, tar acid compounds or sodium chlorate. The ridging of potato tubers also helps to protect the tubers from infection. Although sporangia may survive in soil for several weeks they do not penetrate deeply into it.

3. Use of disease-resistant varieties

A world-wide search for species of *Solanum* showed a number had natural resistance to *P. infestans*. One species which has proved to be an important source of resistance is *S. demissum*. Although this species is valueless in itself for commercial cultivation it is possible to cross it with *S. tuberosum* and some of the hybrids are resistant to the disease. *Solanum demissum* contains at least four major genes for resistance (R_1, R_2, R_3 and R_4), together with a number of minor genes which determine the degree of susceptibility in susceptible varieties (Black, 1952). The four genes may be absent from a particular host strain (O), present singly (e.g. R_1) in pairs (e.g. R_1, R_3), in threes (e.g. R_1, R_2, R_3), or all together (e.g. R_1, R_2, R_3, R_4) so that 16 host genotypes are possible representing different combinations of *R* genes. The identification of the *R* gene complex was dependent on the discovery that the fungus itself exists in a number of strains or physiologic races. On the assumption that the fungus itself carries genes which correspond to, and enable it to overcome the effect of, a host *R* gene 16 races of the fungus should theoretically be demonstrable, If the corresponding genes of the fungus are termed 1, 2, 3 and 4 then the different races can be labelled (0), (1), (2), etc., (1,2) (1,3) etc., (1,2,3), (1,3,4) etc., 1,2,3,4. By 1953, 13 of the 16 had been identified. The prevalent race commonly found, however, is Race 4. Later six major genes for blight resistance in potato were identified (Black, 1960). On this basis 2^6, i.e. 64, races of *P. infestans* should be demonstrable.

In addition to the major genes for resistance in potato a large number of genes probably also exists which, although individually of small effect, may, if present together, contribute to resistance. Resistance of this type is known as *field resistance*, and some potato breeding programmes aim at producing varieties possessing it.

The origin of the physiologic races is difficult to determine. If sexual reproduction is a rare phenomenon, then the oppprtunity which this presents for genetical recombination must be small. However, it is possible that some of the races arise by mutation followed by selection on a susceptible host (Fincham & Day, 1965). Another possibility is that the mycelium of *P. infestans* may be heterokaryotic, carrying nuclei of more than one race (Graham, 1955). There is also some evidence to suggest that the pathogenicity of a strain of the parasite can be increased by continued passage through resistant hosts (see Buxton, 1960).

Within one or two days of infection the tissues of a resistant host undergo necrosis so rapidly that sporulation of the fungus does not occur. Because of this rapid reaction of the host tissue further growth of the fungus becomes impossible. Such a reaction is sometimes termed hypersensitivity. According to Müller (1959) the function of the *R* genes is to accelerate this host reaction.

Peronosporaceae

The Peronosporaceae are in nature obligate parasites of higher plants and are responsible for a group of diseases known collectively as downy mildews. Most species are confined to particular host families or host genera. The mycelium in the host tissues is coenocytic and intercellular, with haustoria of various types penetrating the host cells (Fraymouth, 1957). The sporangia arise on well-differentiated branched sporangiophores which grow out from stomata. They are disseminated by wind. In some genera germination by means of biflagellate zoospores has been reported, but in most the sporangia germinate directly, by means of a germ tube. The term conidium is sometimes used for sporangia showing direct germination.

Representative genera of the family are *Peronospora*, *Plasmopara* and *Bremia*.

Peronospora

A number of diseases of economic importance are caused by species of *Peronospora*. *P. destructor* causes a serious disease of onions and shallots whilst *P. farinosa* causes downy mildew of sugar beet, beetroot and spinach, but can also be found on weeds such as *Atriplex* and *Chenopodium*. *Peronospora tabacina* causes blue-mould of tobacco. This name refers to the bluish-purple colour of the sporangia which is a feature of many species of *Peronospora*. *Peronospora parasitica* attacks members of the Cruciferae. Although many specific names have been applied to forms of this fungus on different host genera, it is now customary to regard them all as belonging to a single species (Yerkes & Shaw, 1959). Turnips, swede, cauliflower, brussels sprouts and wallflowers (*Cheiranthus*) are commonly attached, and the fungus is frequently found on shepherd's purse (*Capsella bursa-pastoris*). Diseased plants are detected by their swollen and distorted stems bearing a white 'fur' of sporangiophores (Fig. 39A). On leaves the fungus is associated with yellowish patches on the upper surface and the formation of white sporangiophores beneath. Sections of the diseased tissue show a wide coenocytic intercellular mycelium and branched lobed coarse haustoria in certain host cells (Fig. 40C). According to Fraymouth (1956) the haustoria of the Peronosporales do not actually penetrate the host protoplast. In plasmolysis experiments following staining of the protoplast with Neutral Red, she was able to show that the protoplast withdrew from the haustorium leaving it entirely free. Following penetration of the host cell reactions are set up between the host protoplasm and the invading fungus. The haustorium becomes ensheathed by a layer of callose which is often visible as a thickened collar around the base of the haustorium. Fraymouth's findings have been confirmed and extended by

Figure 39. A. Inflorescence of *Capsella bursa-pastoris* infected with *Peronospora parasitica*. The white 'fur' is due to development of sporangiophores. B. Stem of *Capsella* infected with *Albugo candida*. Note the distortion and hypertrophy of the infected stem. Compare the diameter of the infected stem with that of the uninfected stem behind it. The white blisters have not yet ruptured.

Peyton & Bowen (1963), who studied thin sections under the electron microscope of leaf tissue of soya bean infected with *P. manshurica*. Their interpretation of the haustorium is shown in Fig. 41. At the point of penetration of the host cell the haustorium is surrounded by a sheath of host wall material. It is enclosed for the whole of its length by a further sheath, the zone of apposition, possibly also derived from the host. The plasma membrane of the haustorium invaginates at certain points to form characteristic inclusions termed lomasomes, which have been seen in other fungi. Secretory bodies

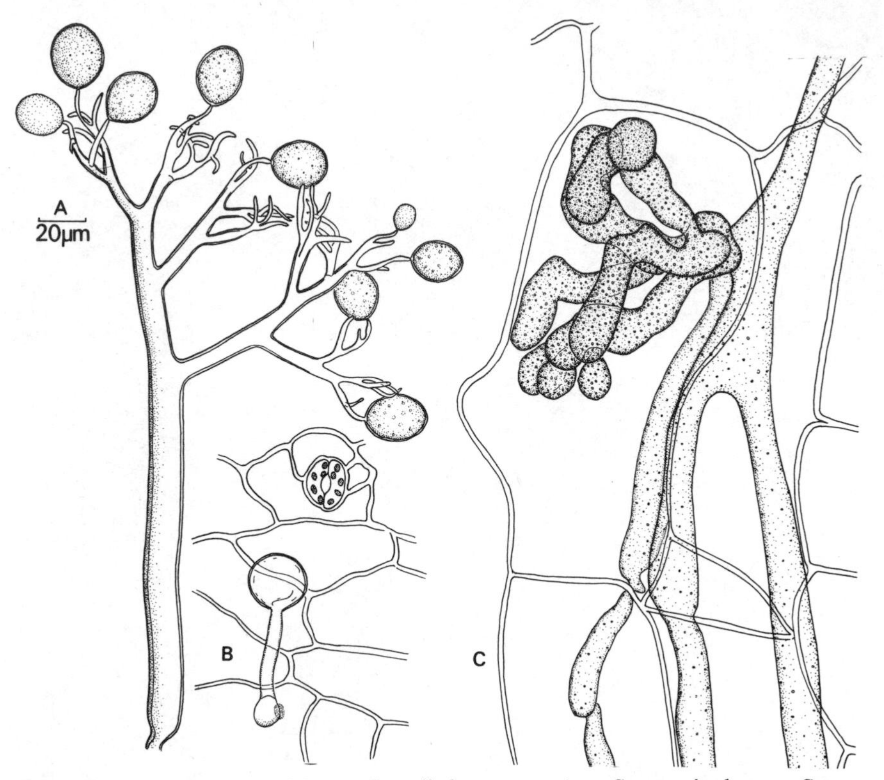

Figure 40. *Peronospora parasitica* on *Capsella bursa-pastoris*. A. Sporangiophore. B. Sporangium germinating by means of a germ tube. C. L.S. host stem showing intercellular mycelium and coarse lobed haustoria.

are apparently formed in the host cell around the haustoria, and the concentration of host ribosomes is also apparently increased. The sporangiophores emerge singly or in groups from stomata. There is a stout main axis which branches dichotomously to bear the oval sporangia at the tips of incurved branches (Fig. 40A). Detachment of the sporangia is possibly caused by hygroscopic twisting of the sporangiophores related to changes in humidity. In *P. tabacina*, however, it has been suggested that changes in turgor of the sporangiophores occur which parallel changes in the water content of the tobacco leaf. It has also been claimed for this fungus that the sporangia are discharged actively by energy applied at the point of attachment of the sporangia. Violent sporangial discharge occurs also in *Sclerospora philippinensis*, a downy mildew of maize in the Philippines. Here, a flat septum is formed between the sporangium and the sporangiophore, and separation is brought about by the rounding-off of the turgid cells on either side of the septum, so that the sporangia bounce off for a distance of 1–2 mm. In *P. parasitica* the sporangia, on alighting on a suitable host, germinate by the formation of a germ tube, and not by zoospores. The germ tube penetrates the wall of the epidermis (Fig. 40B).

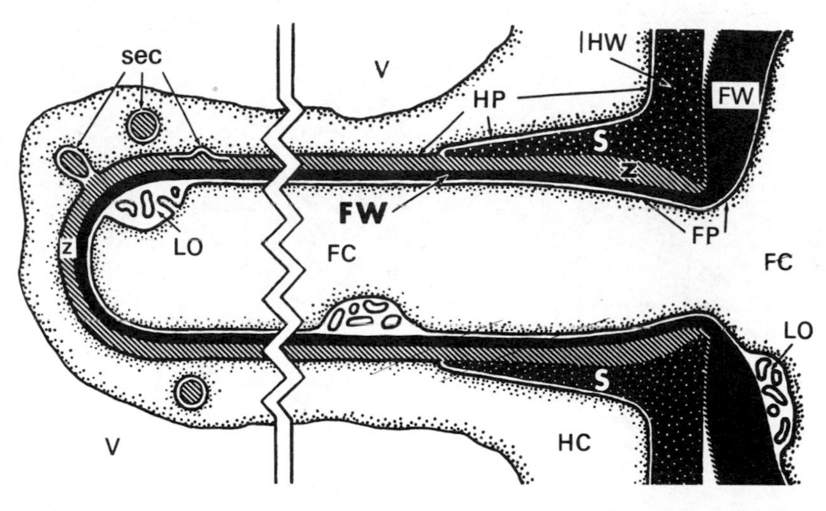

Figure 41. *Peronospora manshurica.* Diagram of host–parasite interface in haustorial region. Fungal cytoplasm (FC) is bounded by the fungal plasma membrane (FP), lomasomes (LO), and the fungal cell wall (FW) in both the intercellular hypha (right) and the haustorium (centre). The relative positions of the host cell vacuole (V), host cytoplasm (HC) and host plasma membrane (HP) are indicated. The host cell wall (HW) terminates in a sheath (S). The zone of apposition (Z) separates the haustorium from the host plasma membrane. Invaginations of the host plasma membrane and vesicular host cytoplasm are considered evidence for host secretory activity (sec.) (after Peyton & Bowen, 1963).

Oospores of *P. parasitica* are embedded in senescent tissues and are found throughout the season. There is evidence that some strains of the fungus are heterothallic, whilst others are homothallic (McMeekin, 1960). Both the antheridium and oogonium are at first multinucleate. Nuclear division precedes fertilisation, and chromosome reduction probably occurs in the oogonium and in the antheridium. The single functional gamete nuclei do not immediately fuse, and fusion is delayed until the oospore wall is partly formed.

The wall of the oospore in *P. parasitica* is very tough and it is difficult to induce germination. In *P. destructor* and some other species germination occurs by means of a germ tube but in *P. tabacina* zoospores have been described. It is probable that oospores overwinter in the soil and give rise to infection in subsequent seasons. The oospores of *P. destructor* have been germinated after 25 years, but it has not proved possible to infect onions from such germinating oospores. Possibly in this case the disease is carried over by means of systemic infection of onion bulbs (McKay, 1957). Although it

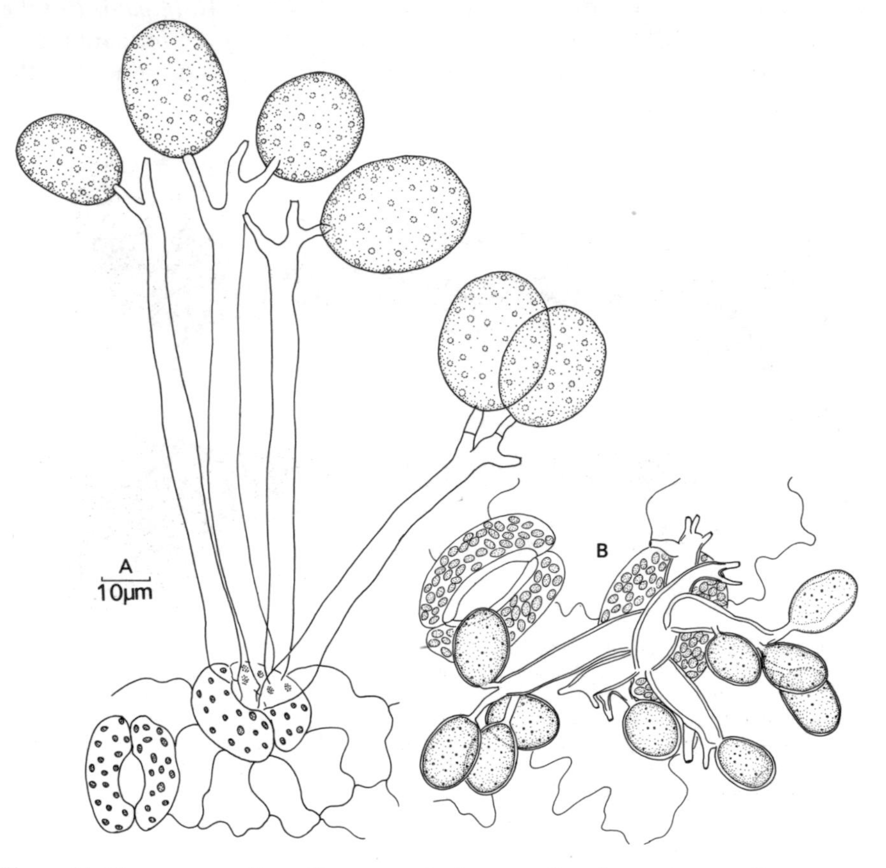

Figure 42. *Plasmopara*. A. *P. pusilla*; sporangiophores on *Geranium pratense*. B. *P. pygmaea*; sporangiophores on *Anemone nemorosa*.

has been found possible to grow *P. parasitica* in pure culture (Guttenberg & Schmoller, 1958) it is very unlikely that the fungus grows saprophytically under natural conditions.

Plasmopara

Although downy mildews caused by species of *Plasmopara* are rarely serious in Britain, *P. viticola* is potentially a very destructive pathogen of the grape vine. The disease which was endemic in America and not particularly destructive there was probably imported into France during the nineteenth century with disastrous results. Historically the disease is of great interest because experiments by Millardet to control the disease in France led to the formulation of Bordeaux mixtures which proved effective not only against this fungus but against *Phytophthora infestans*, and a number of other important foliar pathogens.

Plasmopara nivea is occasionally reported in Britain on Umbelliferous crops such as carrot and parsnip, and is also found on *Aegopodium podagraria*. *Plasmopara pygmaea* is found on yellowish patches on the leaves of *Anemone nemorosa*, whilst *P. pusilla* is similarly associated with *Geranium pratense*. The haustoria of *Plasmopara* are knob-like; the sporangiophores are branched modopodially and the sporangia are hyaline (Fig. 42). Two types of sporangial germination have been reported. In *P. pygmaea* there are no zoospores but the entire sporangial contents escape and later produce a germ tube. In other species the sporangia germinate by means of zoospores which penetrate the host stomata. Oospore germination in *P. viticola* is also by means of zoospores.

Bremia

Bremia lactucae causes downy mildew of lettuce (*Lactuca sativa*) and strains of it can also be found on a number of Composite hosts such as *Sonchus* and *Senecio*. Cross-inoculation experiments using sporangia from these hosts have failed to result in infection of lettuce and it seems that the fungus exists in a number of physiological races. Although wild species of *Lactuca* can carry strains capable of infecting lettuce, these hosts are not sufficiently common to provide a serious source of infection. The disease can be troublesome both in lettuce grown in the open and under frames, and in market gardens there may be sufficient overlap in the growing time of lettuce crops for the disease to be carried over from one sowing to the next. The damage to the crop caused by *Bremia* may not in itself be severe, but infected plants are prone to infection by the more serious grey mould, *Botrytis cinerea*. Systemic infection can occur. The intercellular mycelium is often coarse, and the haustoria are sac-shaped, often several in each host cell (Fig. 50D). The sporangiophores emerge singly or in small groups through the stomata and branch dichotomously. The tips of the branches expand to form a cup-shaped disc bearing short cylindrical sterigmata at the margin, and occasionally at the centre, and from these the hyaline sporangia arise (Fig. 43A,B). Germination of the sporangia is usually by means of a germ tube which penetrates an epidermal cell directly (Fig. 50C) or through a stoma, but zoospore formation has also been reported. Oogonia are apparently rare on lettuce.

Albuginaceae

This family has only a single genus, *Albugo*, with about 30 species of obligate parasites of flowering plants, causing diseases known as white blisters or white rusts. The commonest British species is *A. candida* (= *Cystopus candidus*) causing white blister of crucifers such as cabbage, turnip, swede, horse-radish, etc., but particularly common on shepherd's purse (*Capsella*

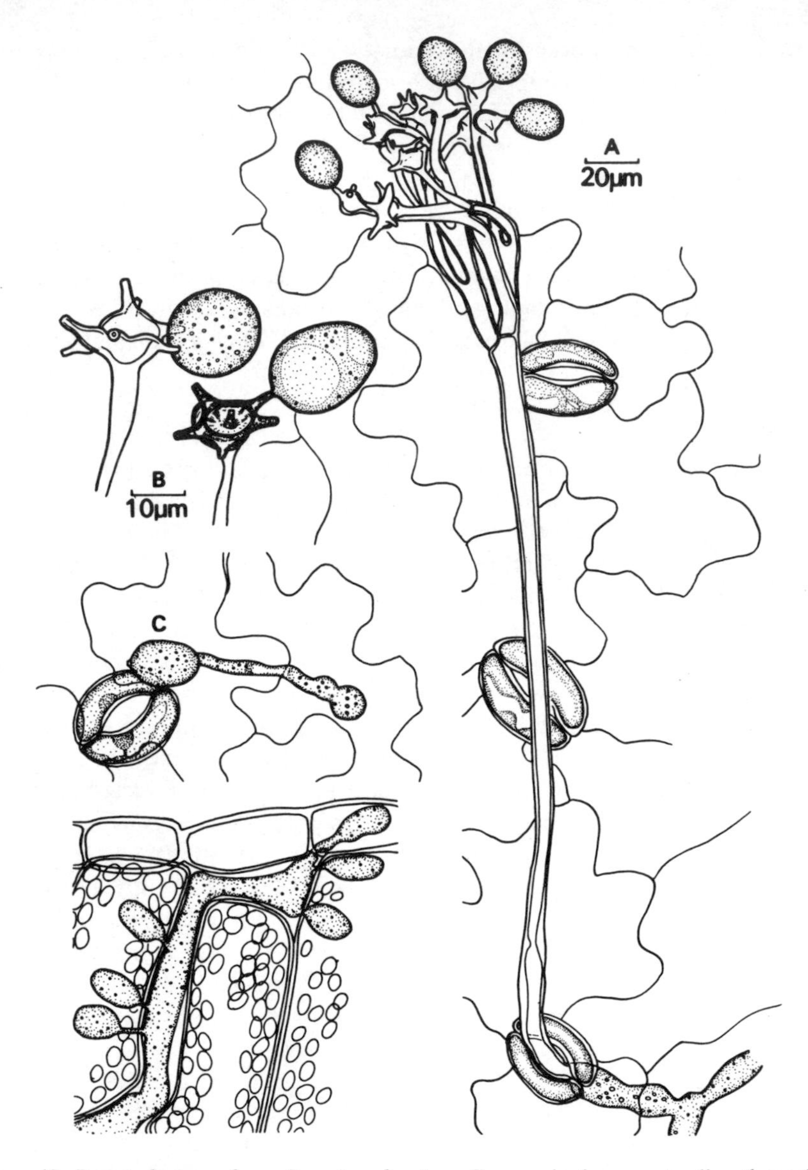

Figure 43. *Bremia lactucae* from *Senecio vulgaris*. A. Sporangiophore protruding through a stoma. B. Sporangiophore apex. C. Sporangium germinating by means of a germ tube. D. Cells of epidermis and palisade mesophyll, showing intercellular mycelium and haustoria. A, C, D to same scale.

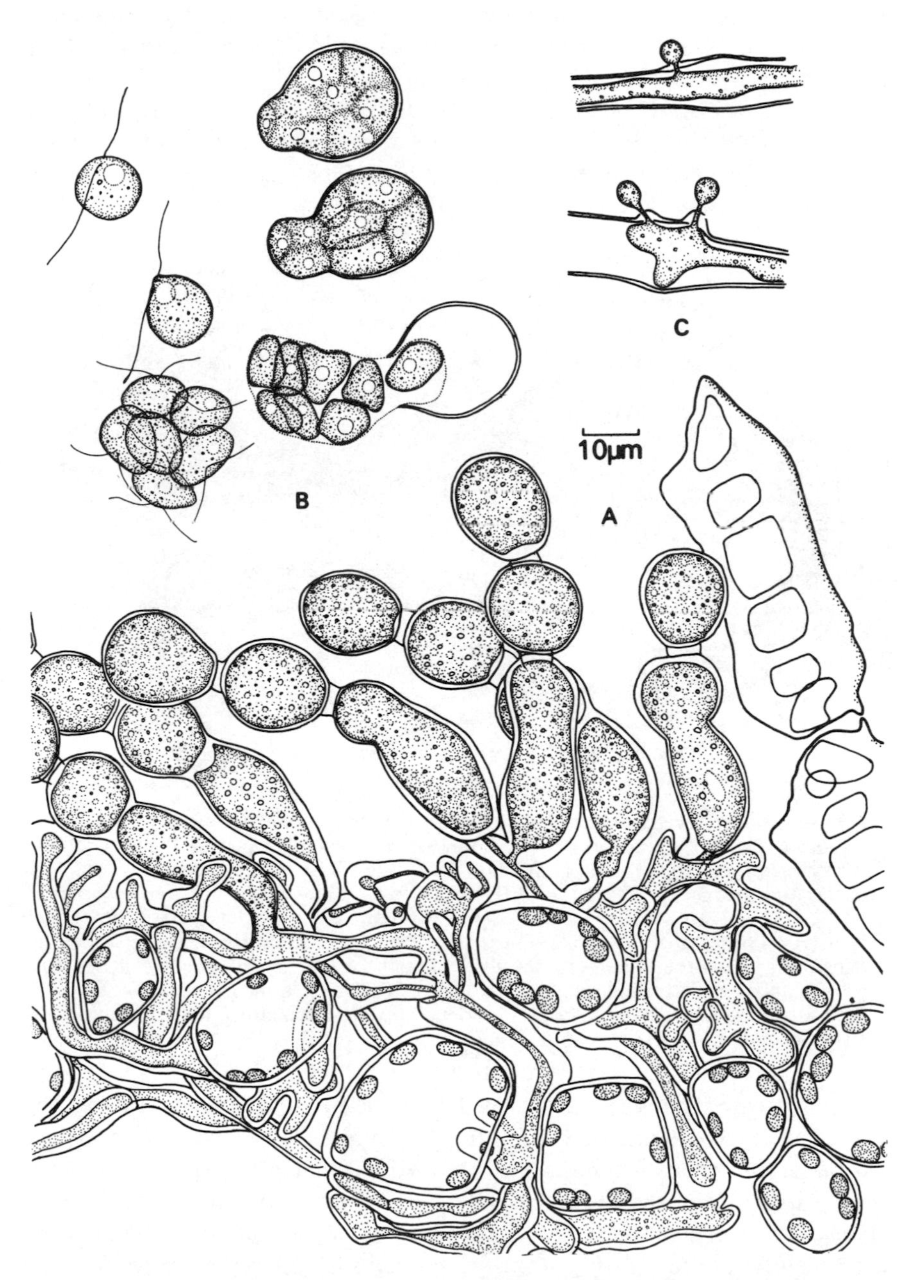

Figure 44. *Albugo candida* on *Capsella bursa-pastoris*. A. Mycelium, sporangiophores and chains of sporangia formed beneath the ruptured epidermis (right). B. Germination of sporangia showing the release of eight biflagellate zoospores. The stages illustrated took place within 2 min. C. Haustoria.

bursa-pastoris) (Fig. 39B). There is some degree of physiological specialisation in the races of this fungus on different host genera. Another less common species is *A. tragopogi* causing white blisters of salsify (*Tragopogon porrifolius*), goatbeard (*T. pratensis*) and *Senecio squalidus*.

In *A. candida* on shepherd's purse, diseased plants may be detected by the distorted stems (Fig. 39B) and the shining white raised blisters on the stem, leaves and fruits before the host epidermis is ruptured. Later, when the epidermis is burst open, a white powdery pustule is visible. The distortion is possibly

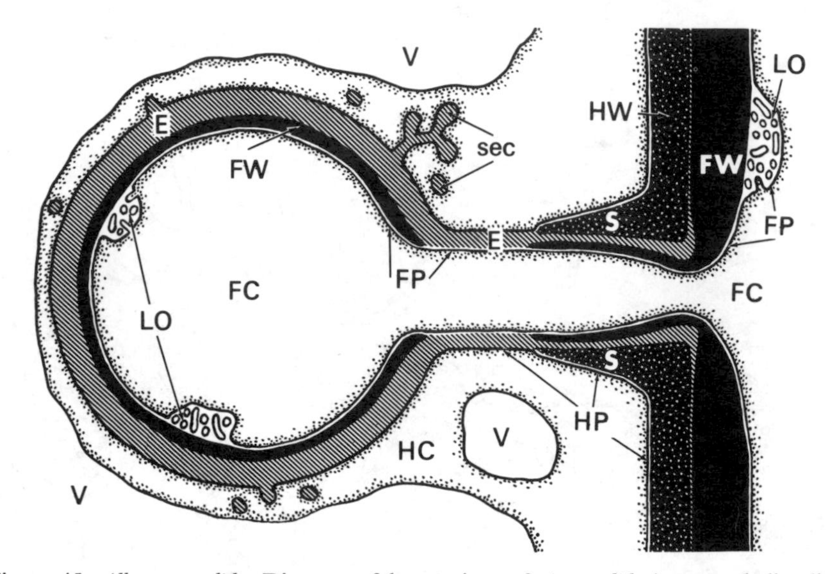

Figure 45. *Albugo candida*. Diagram of haustorium of *A. candida* in mesophyll cell of *Raphanus*. Fungal cytoplasm (FC) is bounded by the fungal plasma membrane (FP) with its lomasomes (LO) and the fungal cell wall (FW) in both the intercellular hypha (right) and the haustorial head (left). The relative positions of the host cell vacuole (v), host cytoplasm (HC) and the host plasma membrane (HP) are indicated. The host cell wall (HW) terminates in a collar-like sheath (S). The encapsulation (E) separates the haustorium from the host. Note the discontinuity of the fungal cell wall in the stalk proximal to the haustorial head. Invaginations of the host plasma membrane and vesicular host cytoplasm are suggested as evidence for host secretory activity (sec.) (after Berlin & Bowen, 1964).

associated with changes in auxin level of diseased as compared with healthy tissues. The host plant may be infected simultaneously with *Peronospora parasitica*, but the two fungi are easily distinguishable microscopically both on the structure of the sporangiophores and by their different haustoria. In *Albugo* the mycelium in the host tissues is intercellular and only small spherical haustoria are found (Fig. 44C), which contrast sharply with the coarsely lobed haustoria of *P. parasitica*. The fine structure of the haustorium of *Albugo candida* has been studied by Berlin & Bowen (1964) (see Fig. 45). The haustoria are spherical or somewhat flattened and about 4 μm in diameter, connected to the intercellular mycelium by a narrow stalk about 0·5 μm wide. Within the plasma membrane of the haustorium lomasomes, a system of unit-membranes and tubules, apparently derived from the plasma membrane, are more numerous than in the intercellular hyphae. The cytoplasm of the haustorial head is densely packed with mitochondria, ribosomes, endoplasmic reticulum and occasional lipoidal inclusions, but nuclei have not been observed. Since nuclei of *Albugo* are about 2·5 μm in diameter they may be unable to penetrate the constriction which joins the haustorium to the intercellular hypha. The base of the haustorium is surrounded by a collar-like sheath which is an extension of the host cell wall, but this wall does not normally completely surround the haustorium. Between the haustorium and the host plasma membrane is an encapsulation. Host cytoplasm reacts to infection by

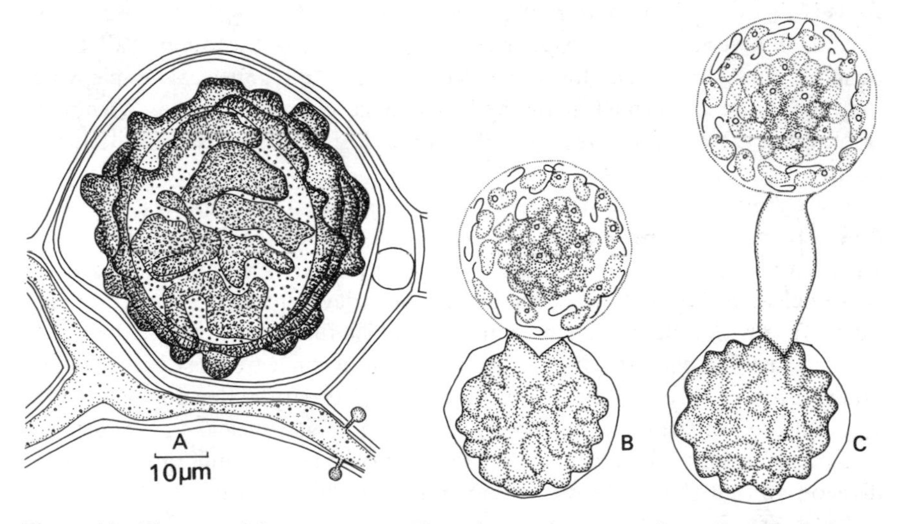

Figure 46. *Albugo candida* oospores. A. Oogonium and oospores from *Capsella* leaf. B, C. Two methods of oospore germination (after Vanterpool, 1959).

an increase in the number of ribosomes and Golgi complexes. In the vicinity of the haustorium the host cytoplasm contains numerous vesicular and tubular elements not found in uninfected cells. These structures have been interpreted as evidence of a secretory process induced in the host cell by the presence of the parasite. The intercellular mycelium masses beneath the host epidermis to form a palisade of cylindrical or skittle-shaped sporangiophores which give rise to chains of spherical sporangia in basipetal succession—i.e. new sporangia are formed at the base of the chain. The pressure of the developing chains of sporangia raises the host epidermis and finally ruptures it. The sporangia are then visible externally as a white powdery mass, dispersed by wind. If sporangia alight on a suitable host leaf they are capable of germinating within a few hours in films of water to form biflagellate zoospores, about eight per sporangium (Fig. 44B). After swimming for a time a zoospore encysts and then forms a germ tube which penetrates the host epidermis. A further crop of sporangia may be formed within ten days. Infection may be localised or systemic. Oogonia grow in the intercellular spaces of the stem and leaves. Both the antheridium and the oogonium are multinucleate at their inception, and during development two further nuclear divisions occur so that the oogonium may contain over 200 nuclei. However there is only one functional male and one functional female nucleus. In the oogonium all the nuclei except one migrate to the periphery and are included in the periplasm. Following nuclear fusion a thin membrane first develops around the oospore. Division of the zygote nucleus takes place, and is repeated, so that at maturity the oospore may contain as many as 32 nuclei. The mature oospore is surrounded by a brown warty exospore, thrown into folds (Fig. 46A). Germination of the oospores only takes place after a resting period of several months. Under suitable conditions the outer wall of the oospore bursts and the endospore is extruded as a thin spherical vesicle which may be sessile or formed at the end of a wide cylindrical tube. Within the thin vesicle 40–60 zoospores are differentiated and are released on its breakdown (Fig. 46B,C).

The cytology of oospore development in some other species of *Albugo* differs from that of *A. candida*. In *A. bliti* a parasite of *Portulaca* in North America and Europe, the oogonia and antheridia are also multinucleate and two nuclear divisions take place during their development. Numerous male nuclei fuse with numerous female nuclei and the fusion nuclei pass the winter without further change. In *A. tragopogi* a multinucleate oospore develops, and again there are two nuclear divisions involved in the development of the egg and the antheridium, but finally there is a single nuclear fusion between one male and one female nucleus. The fusion nucleus undergoes repeated division so that the overwintering oospore is multinucleate.

2

Zygomycotina

The Zygomycotina is an assemblage of fungi which reproduce asexually by non-motile aplanospores. The spores are contained in sporangia which may be violently projected, but more usually spores are passively dispersed by wind, rain or animals. Sexual reproduction is by gametangial copulation which is typically isogamous, and results in the formation of a *zygospore*. The mycelial organisation is coenocytic, and the cell wall contains chitin.

Two classes are included, the Zygomycetes and the Trichomycetes. The Zygomycetes comprise two orders, the Mucorales and Entomophthorales. Mucorales are ubiquitous in soil and dung, mostly as saprophytes, although a few are parasitic on plants and animals. Entomophthorales include a number of insect parasites, but some saprophytic forms exist. The Trichomycetes is a group of uncertain affinity, and are mostly parasitic in the guts of arthropods, e.g. millipedes and insect larvae. This group will not be considered further. For references see Lichtwardt (1960), Manier (1963, 1964), Trotter & Whisler (1965), Whisler (1966), Farr & Lichtwardt (1967).

MUCORALES

Unlike the fungi previously considered the Mucorales do not possess motile zoospores, but reproduce asexually by non-motile spores carried passively by wind, or dispersed by rain splash, insects or other animals. In many forms numerous spores are contained in globose sporangia surrounding a central core or columella. Some also possess few-spored sporangia, termed sporangiola, dispersed as a unit. In the so-called conidial types unicellular propagules are dispersed. Sporangia are also known in which there is no columella, or where the spores are arranged in a row inside a cylindrical sac termed a merosporangium. Sexual reproduction is by conjugation, the fusion of usually equal gametangia derived from branch tips, to form a warty zygospore.

The Mucorales are widely distributed in soil, and are mostly saprophytic. Some may cause spoilage of food. A few are weak parasites of fruits; some are parasitic on other fungi and some cause diseases of animals including man. A number of species have been used in fermentations for the production of alcohol. Extensive studies of their nutrition and physiology have been made (for references see Foster, 1949; Lilley & Barnett, 1951; Cochrane, 1958; Cantino & Turian, 1959). A wide variety of sugars can be used, and whilst starch can be decomposed by some species, cellulose is not utilised by most. Under anaerobic conditions ethyl alcohol and numerous organic acids are produced. Many Mucoraceae need an external supply of vitamins for growth in synthetic culture. Thiamine is a common requirement and since *Phycomyces* needs an external supply of this substance the amount of growth of the fungus can be used as a sensitive assay for the concentration of thiamine. Carotene synthesis by cultures of certain Mucorales may prove to be an economic proposition. Nine families have been recognised (Martin, 1961):

(a) Sporangia all columellate and alike — b

(a) Columellate sporangia lacking or, if present, accompanied by sporangioles or conidia — c

(b) Sporangial membrane thin, fugacious; sporangiospores liberated by breaking or dissolution of sporangial wall; suspensors rarely tong-like — **Mucoraceae**

(b) Sporangial wall densely cutinised above; sporangium violently discharged or passively separated as a unit from sporangiophore; suspensors always tong-like — **Pilobolaceae**

(c) Columellate sporangia usually present, accompanied by few-spored subglobose sporangioles in which spores are never formed in linear series, or by conidia — d

(c) Columellate sporangia never present; sporangioles 1-spored, conidium-like, or bearing spores in linear series, or modified non-columellate sporangia functioning as disseminules — e

(d) Sporangioles borne in clusters on lower part of sporangiophore, the latter usually tipped by a columellate sporangium or a spine — **Thamnidiaceae**

(d) Sporangioles or conidia never borne on same sporangiophores as columellate sporangia; zygospores with tong-like suspensors — **Choanephoraceae**

(e) Merosporangia borne on tips of sporangiophores, at first cylindrical, dividing into a single row of sporangiospores, simulating a chain of conidia — **Piptocephalidaceae**

(e) Merosporangia lacking, or not clearly defined — f

(f) Conidia or very short merosporangia borne unilaterally on special branches (sporocladia) of sporophore, in comb-like or fan-like aggregates — **Kickxellaceae**

(f) Non-columellate sporangia or conidia present, but latter not borne on sporocladia — g

(g) Conidia present borne on spicules on inflated tips of sporophores; sporangia lacking — **Cunninghamellaceae**

(g) Conidia lacking; sporangia non-columellate — h

(h) Sporangia borne freely on surface, entire sporangium sometimes modified to function as a disseminule; zygospores in hyphal matrix but not forming a sporocarp **Mortierellaceae**

(h) Sporocarp present, sclerotium-like, enclosing sporangia, zygospores or azygospores **Endogonaceae**

Representatives of all these families with the exception of the Kickxellaceae will be discussed.

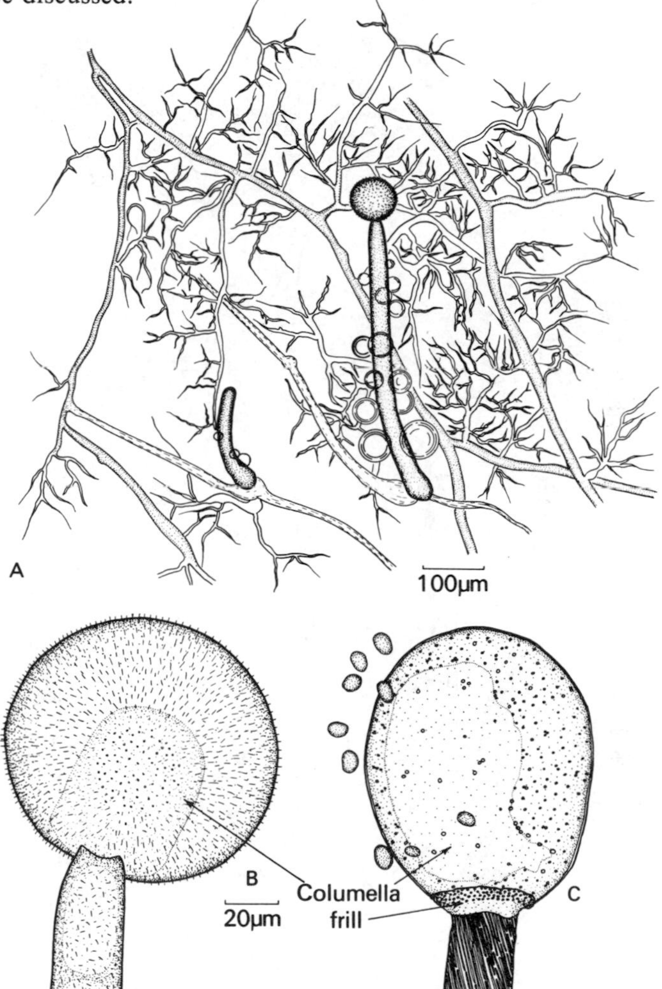

Figure 47. *Mucor mucedo*. A. Mycelium and young sporangiophores, with globules of liquid attached. B. Immature sporangium with the columella visible through the sporangial wall. C. Dehisced sporangium showing the columella, the frill representing the remains of the sporangial wall, and sporangiospores.

Mucoraceae

Members of this family are abundant in soil, on dung, and on moist fresh organic matter in contact with soil. For the most part they are saprophytic, and play an important role in the early colonisation of substrata in soil. Sometimes, however, they can behave as parasites of plant tissues, e.g. *Rhizopus stolonifer* can cause a rot of sweet potatoes. They are also known to cause fungal diseases of animals and man (mucormycosis), and such conditions seem to be particularly frequent in patients suffering from diabetes,

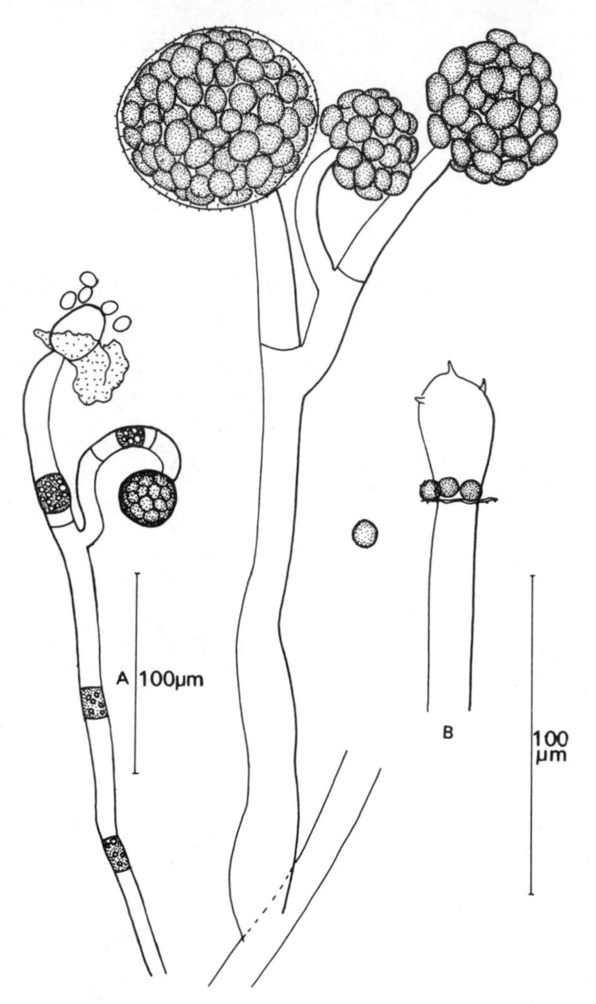

Figure 48. A. *Mucor racemosus*: branched sporangiophore containing chlamydospores. The remnants of the sporangial wall in the upper dehisced sporangium are clearly visible. B. *M. plumbeus*: branched sporangiophore. The sporangial wall has disappeared in two sporangia. Note the spiny process on the columella.

leukaemia and cancer. Lesions may be localised in the brain, lungs or other organs, or may be disseminated, e.g. at various points in the vascular system. Species of *Rhizopus* and *Mucor* are reported from human lesions, and these genera together with species of *Absidia* are associated with mucormycosis in domestic animals. Occasionally *Rhizopus* and *Mucor* cause spoilage of bread and other food. Certain species, e.g. *Mucor rouxii*, are also used industrially to break down starch to sugar before fermentation by yeasts to alcohols, since yeasts do not contain amylolytic enzymes for the initial breakdown.

The mycelium first established by a germinating spore on a solid substrate is coarse, coenocytic and richly branched; the branches usually taper to fine points (Fig. 47). Later, septa may appear. Thick-walled mycelial segments or chlamydospores may be cut off by such septa, and in certain species, e.g. *Mucor racemosus*, the presence of chlamydospores may be a useful diagnostic feature (Fig. 48A). In anaerobic liquid culture, especially in the presence of CO_2, *Mucor* may grow in a yeast-like instead of a filamentous form (Fig. 49) but reverts to filamentous growth in the presence of O_2 (Bartnicki-Garcia &

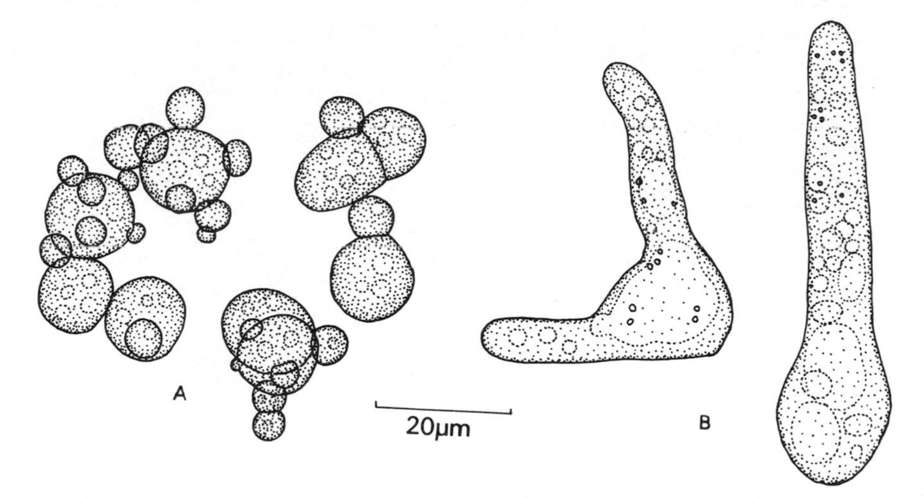

Figure 49. *Mucor rouxii*. A. Yeast-like growth in liquid medium under anaerobic conditions 24 hr after inoculation with spores. B. Filamentous growth from spores in liquid medium under aerobic conditions 4 hr after inoculation.

Nickerson, 1962*a*,*b*). The cell walls of Mucoraceae are complex chemically. Whilst chitin microfibrils are present, chitosan may be the most abundant component. Other polysaccharides such as glucosamine and galactose, proteins, lipids, purines, pyrimidines, magnesium and calcium have also been detected (Bartnicki-Garcia & Nickerson, 1962*c*; Kreger, 1954). Comparison of the structure and composition of yeast-like and filamentous cells of *Mucor rouxii* show that the yeast-like cells have much thicker walls. They also have a mannose content about five times as great as that of filamentous cell walls.

The synthesis of chitin microfibrils may be correlated with the presence of multi-vesicular bodies, which seem to arise from the endoplasmic reticulum. Such structures have been described from sporangiophores of *Phycomyces* and from a number of other fungi with chitin walls (Peat & Banbury, 1967; Marchant *et al.*, 1967).

The cytoplasm often shows rapid streaming, and phase contrast microscopy shows granules being carried in the stream, mitochondria which are short at the hyphal tips but longer and filamentous in older parts, sap vacuoles and nuclei. The nuclei are often irregular in shape. Conventional mitosis and spindle formation has not been seen, and the nuclei appear to divide by constriction, although chromosomes can be demonstrated (Cutter, 1942*a*,*b*; Robinow, 1957*a*,*b*; 1962). Electron micrographs of the mycelium of *Rhizopus* reveal no especially distinctive features (Hawker & Abbott, 1963*a*).

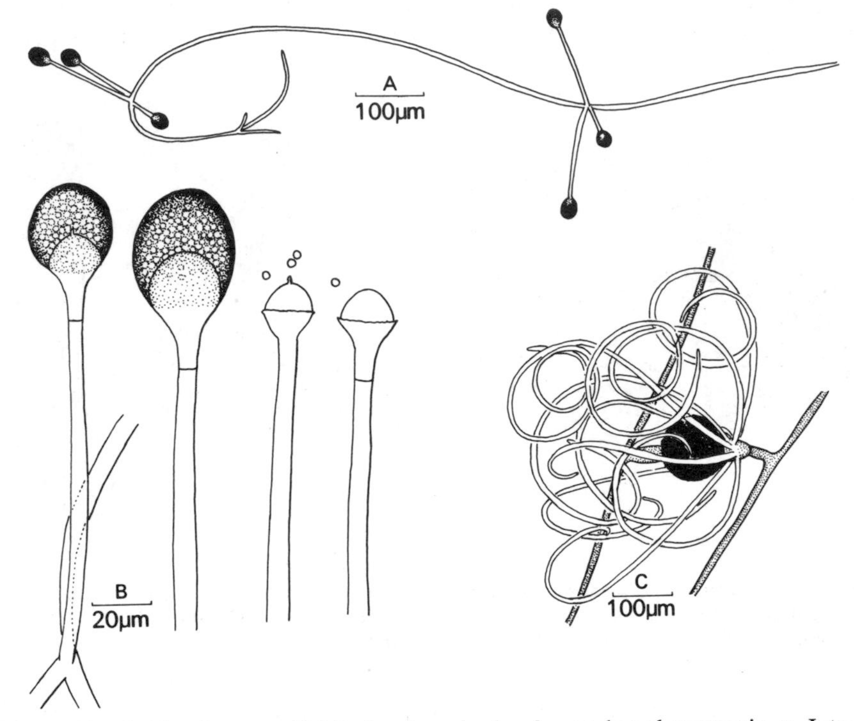

Figure 50. *Absidia glauca*. A. Habit showing whorls of pear-shaped sporangia. B. Intact and dehisced sporangia. Note the single spine-like projection on certain columellae. C. Zygospore showing the arching suspensor appendages.

ASEXUAL REPRODUCTION

Asexual reproduction is by non-motile sporangiospores contained in globose or pear-shaped sporangia. The sporangia may be borne singly at the tip of a sporangiophore or may occur on a branched sporangiophore. In some genera, e.g. *Absidia* (Fig. 50), the sporangia may be arranged in whorls on aerial branches, and in many species of *Rhizopus* the sporangiophores arise in groups from a clump of rhizoids (Fig. 51). The sporangiophores are commonly phototropic, and numerous studies on the phototropism of the large sporangiophores of *Phycomyces* have been made (Castle, 1966, Shropshire, 1963).

It has been suggested that the cylindrical sporangiophore containing sap acts as a lens focusing light on to the distal wall of the sporangiophore.

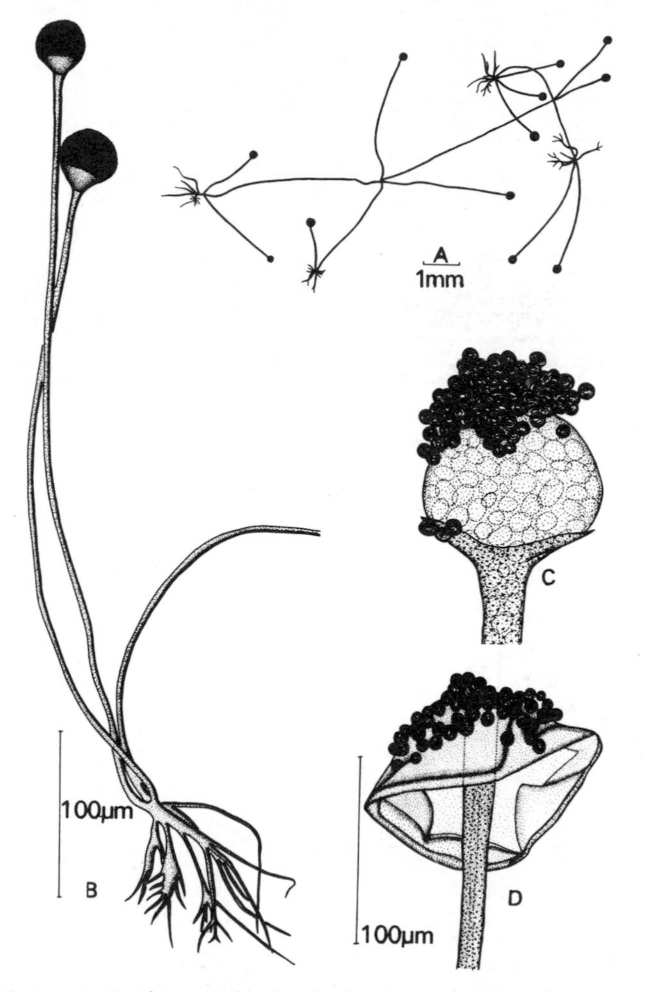

Figure 51. *Rhizopus stolonifer*. A. Habit sketch, showing stolon-like branches which develop rhizoids and tufts of sporangiophores. B. Two sporangiophores showing basal rhizoids. C. Columella and attached spores. D. Invaginated columella.

Parallel light is in fact brought to a focus just outside the distal wall (see Fig. 52A) and knowing the position of the focal point and the diameter of the sporangiophore, the refractive index of the sap can be shown to be about 1·38. It has been claimed that because the distal side is illuminated more intensely than the side proximal to the light it responds by more rapid growth resulting in curvature towards the light. This idea is supported by two experiments. If a sporangiophore is illuminated by a narrow pencil of light which grazes one edge of the sporangiophore it responds by bending away almost

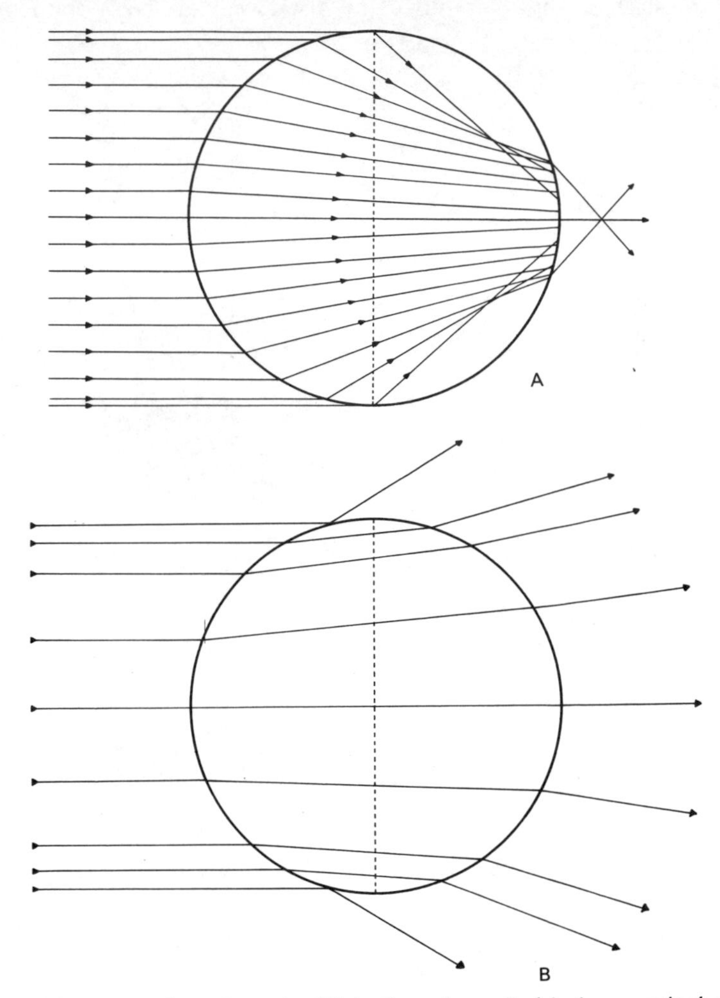

Figure 52. Diagram to show the path of light through a cylindrical sporangiophore. In A the liquid in the sporangiophore has a higher refractive index than the surrounding medium, and the sporangiophore acts as a converging lens. In B the sporangiophore is immersed in a liquid of higher refractive index than the sporangiophore sap, and the light rays are therefore divergent (after Banbury, 1959).

perpendicularly to the direction of the light beam. Secondly, if a sporangiophore is immersed in liquid paraffin with a refractive index of 1·47 it should function as a diverging lens, resulting in more intense illumination on the proximal side (see Fig. 52B) and therefore show curvature towards the light.

This has been found to be the case. The actual mechanism by which the more intensely illuminated wall responds is not clear. Since the force for curvature is derived from the turgor of the sporangiophore it is probable that the more intensely illuminated wall is either rendered more plastic or becomes a site for more rapid intussusception of wall material.

Another hypothesis concerning the phototropic response is that the site of perception is the cytoplasm of the sporangiophore, not the wall. Castle

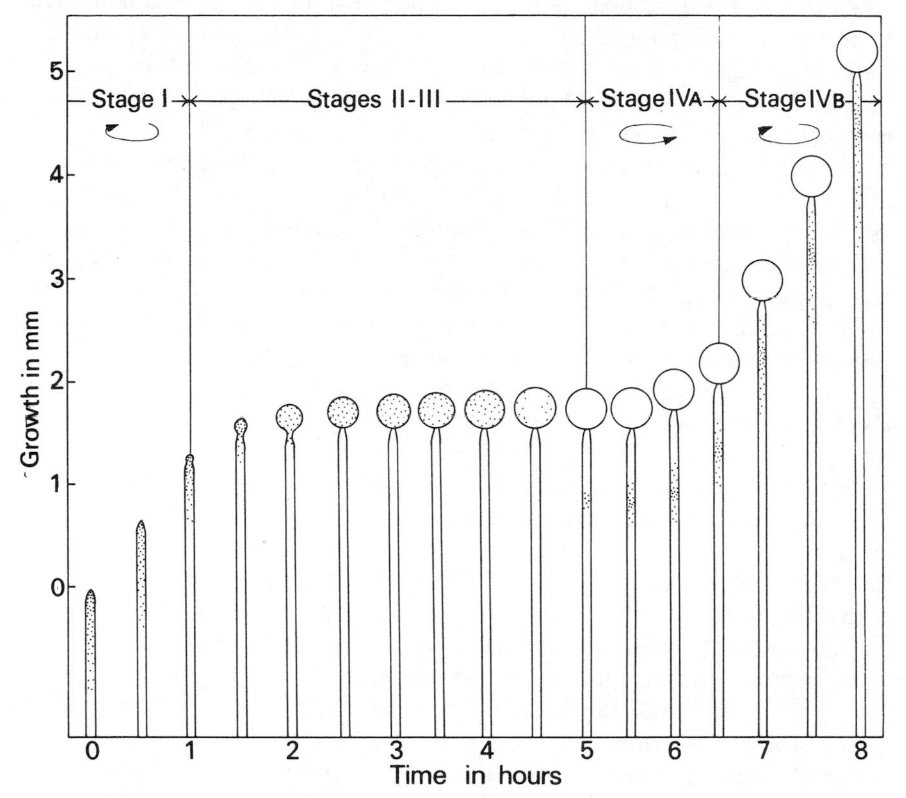

Figure 53. Diagram of developmental stages of the sporangiophore of *Phycomyces*. Regions in which growth is taking place are stippled. The rotary component of growth is indicated. During Stage I the axis of growth is directed sinistrally, in Stages II and III growth is unoriented. In Stage IVA dextral spiralling occurs and in Stage IVB sinistral spiralling again takes place (after Castle, 1942).

(1933*b*) calculated that the mean free path of light rays in a sporangiophore in air is significantly greater in the distal half of the sporangiophore, so that a larger proportion of incident light would be absorbed on this side. This idea receives support from experiments in which the action spectrum for phototropic curvature is followed over the range of visible light. A broad-action spectrum with a maximum around 430 nm has been found, which corresponds to the maximum absorption of β-carotene. It has also been suggested that a flavoprotein is involved in photoreception (Carlile, 1962; 1965). Both these substances are constituents of the cytoplasm and not the wall.

As the sporangiophore of *Phycomyces* develops it rotates. Castle (1942) followed the growth and rotation of the sporangiophore by attaching *Lycopodium* spores as markers. The displacement of the markers was then followed. His findings are illustrated in Fig. 53. After a period of apical growth of the tubular sporangiophore (Stage I) the sporangium appears as a terminal swelling and growth ceases. During this period (Stage II) growth is limited to sporangial enlargement. In the next period (Stage III) no further enlargement of the sporangium occurs, and elongation is at a standstill. During Stages IVA and IVB further elongation occurs and growth is mainly localised in a zone somewhat below the sporangium. During Stage I the tip of the sporangiophore rotates clockwise (as seen from above looking down) through a maximum angle of about 90°. There is no rotary movement during Stages II and III. However when the sporangium is completed and elongation recommences in Stage IVA, rotation is again detected so that the markers move upwards spirally. The direction of rotation is now *anti-clockwise* (as seen from above) and the direction of spiral growth is *dextral* (right-handed) instead of *sinistral* (left-handed) as in Stage I. During this stage, which lasts about one hour, markers attached to the growth zone may make up to two complete revolutions around the axis. Finally, during Stage IVB the direction of rotation reverses once more and becomes sinistral.

The reasons for the spiral growth are far from clear. It is known that the chitin microfibrils which make up the wall of the sporangiophore show a right-handed or Z spiral orientation. One possible explanation is that the laying down of the fibrils in this way is responsible for the rotation. A second is that the extension due to turgor of a cylinder whose walls are composed of spirally oriented fibrils would naturally result in a passive rotation. The phenomenon of spiral growth is not peculiar to *Phycomyces* but occurs during the extension of various cylindrical plant cells. For a fuller discussion of the problems involved see Castle (1953), and Roelofsen (1950, 1959).

Sporangiophores develop as coarser, blunt-tipped, aerial hyphae, which grow away from the substratum. The tip expands to form the sporangium initial containing numerous nuclei which continue to divide. A dome-shaped septum is laid down cutting off a distal portion which will contain the spores from a central cylindrical or subglobose spore-free core, the columella. It should be stressed that the columella is curved from its inception and does not arise by the arching of a flat septum into the sporangium. Cleavage planes separate the nuclei within the sporangium and finally the spores are cleaved out. They may be uninucleate or multinucleate according to the species, e.g. *M. hiemalis* and *Absidia glauca* have uninucleate spores whilst *Phycomyces blakesleeanus*, *Rhizopus stolonifer* and *Syzygites megalocarpus* have multinucleate spores (Robinow, 1957*a*,*b*; Hawker & Abbott, 1963*b*; Sjöwall, 1945). The number of spores formed is very variable. On nutrient-poor media minute sporangia containing very few spores may be formed, but in *Phycomyces blakesleeanus* the number may be as high as 50,000–100,000 per sporangium (Ingold & Zoberi, 1963).

The fine structure of developing sporangia of *Gilbertella* (Choanephoraceae) has been studied by Bracker (1966). The cleavage of the sporangial cytoplasm to form spores is accomplished by the fusion of cleavage vesicles associated with endoplasmic reticulum which is continuous with the nuclear envelope.

The sporangial wall often darkens and may develop a spiny surface due to the formation of crystals. Despite the apparent similarity in sporangial structure in members of the Mucoraceae at least two distinct mechanisms are involved in spore liberation. In many of the commonest species of *Mucor* (e.g. *M. hiemalis*) the sporangium becomes converted at maturity into a 'sporangial drop'. The sporangial wall dissolves and the spores absorb water so that the tip of the sporangiophore bears a drop of liquid containing spores adhering to the columella. The remnants of the sporangial wall can often be seen as a frill at the base of the columella. In large species such as *M. plasmaticus* and *M. mucedo* and in *Phycomyces* the spores are embedded in mucilage. The sporangial wall does not break open spontaneously, but only when it is touched; the slimy contents then exude. Such sticky spore masses are not readily detached by wind or by mechanical agitation, but possibly by insects or rain splash, or after drying. In other species, e.g. *M. plumbeus*, the sporangial wall breaks into pieces. Here air currents or mechanical agitation readily liberate spores. In this species the columella may terminate in one or more finger-like or spiny projections (Fig. 48B) and in *Absidia* the columella may also bear a single nipple-like projection (Fig. 52). In *Rhizopus stolonifer* the columella is large, and as the sporangium dries the columella collapses so that it appears like an inverted pudding bowl balanced on the end of a stick represented by the stiff sporangiophore (Ingold & Zoberi, 1963)—see Fig. 53. Associated with these changes in columella shape the sporangium wall breaks up into many fragments and the dry spores can escape in wind currents.

SEXUAL REPRODUCTION

The Mucoraceae reproduce sexually by a process of conjugation resulting in the formation of zygospores. Some species are homothallic, zygospores being formed in cultures derived from a single sporangiospore (e.g. *Rhizopus sexualis*, *Syzygites megalocarpus*, *Zygorhynchus molleri* and *Absidia spinosa*). However, the majority of species are heterothallic, and only form zygospores when compatible strains are mated together. If the appropriate strains are inoculated at opposite sides of a Petri dish the mycelia grow out and where they meet in the centre of the dish a line of zygospores develops (Fig. 54). The two compatible strains rarely differ in any regular way from each other, although there may be slight differences in growth rate and carotene content.

Because it was not possible to designate one strain as male and the other as female Blakeslee labelled one strain as + and the other as −. The two compatible strains can be said to differ in *mating type*. The morphological events preceding zygospore formation are sufficiently similar to allow general description of the process, although there are some morphological features peculiar to certain genera, and also differences in cytology.

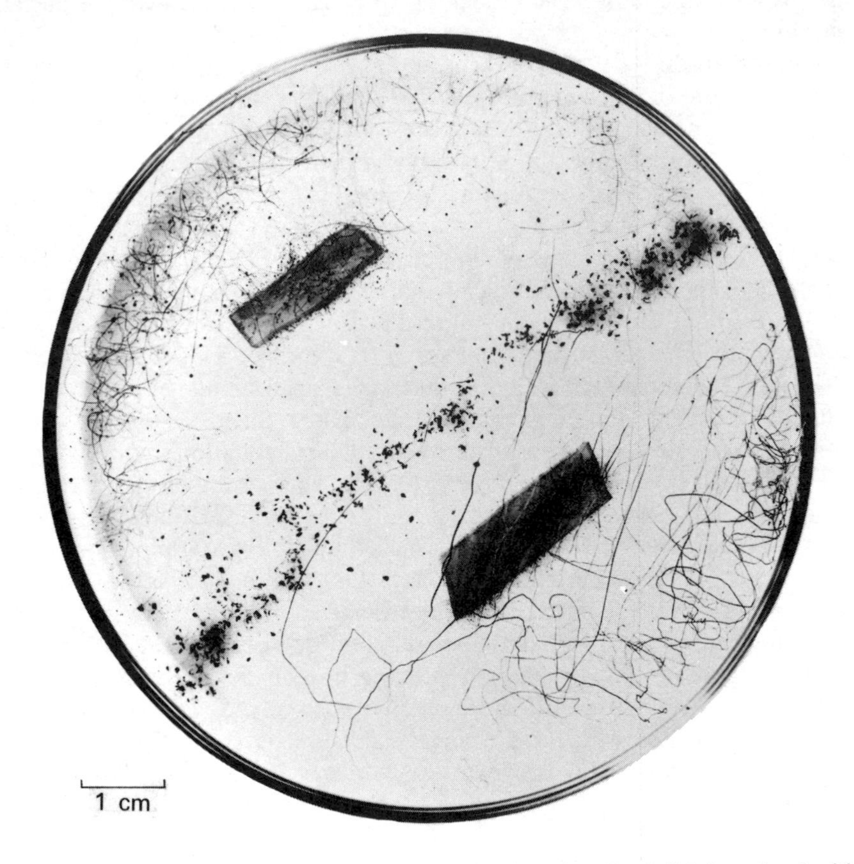

Figure 54. *Phycomyces blakesleeanus*. Ten-day-old culture in a Petri dish inoculated with two compatible strains. The black zone between the two inoculum blocks contains numerous zygospores.

When two compatible strains approach each other aerial club-shaped branches or zygophores develop which show directional growth towards zygophores of the opposite strain. Zygophores of the same strain repel each other (Banbury, 1955). The directional growth of the zygophores has been termed zygotropism. Banbury has postulated the existence of volatile growth stimulators responsible for zygotropism.

There is evidence that development of zygophores in heterothallic forms is initiated by hormones (Plempel, 1963). These hormones are not species-specific. Hormones obtained from *Mucor mucedo* can induce zygophore development in other Mucorales. The substances have been isolated in chemically pure form. When + and − strains of *Mucor mucedo* are inoculated simultaneously into aerated liquid culture (see Fig. 55) and allowed to grow together for 4 days, cell-free extracts stimulate the development of zygophores on isolated + or − mycelia. When + or − strains are inoculated

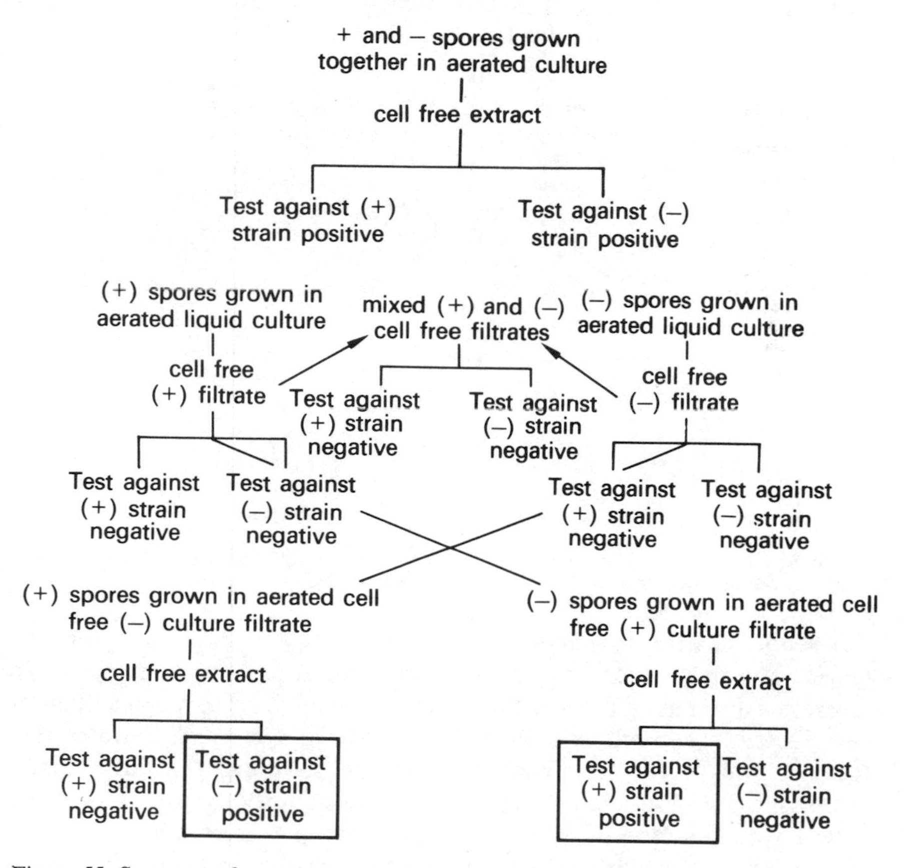

Figure 55. Sequence of experiments to demonstrate the presence of mating hormones in *Mucor mucedo* (after Plempel, 1957).

singly into aerated liquid cultures no such stimulation results when the filtrates are tested. If, however, + spores are inoculated into a cell-free filtrate from a − culture, cell-free extracts from these cultures are capable of stimulating zygophore formation on isolated − mycelia. Similar results are obtained when − spores are grown in cell free extracts of + liquid cultures and tested against + strains but mixed culture filtrates from + and − strains grown separately are ineffective.

To explain these results the existence of two hormones (gamones) has been postulated, a + gamone and − gamone secreted by the respective strains. The secretion of these gamones is elicited by the secretion of progamones formed by the vegetative hyphae of both strains, presumed to be different substances. The + gamone is only formed in the presence of the − progamone, and vice-versa. The postulated sequence is shown in Fig. 56. Although Plempel's scheme requires the existence of two substances, Van den Ende (1967) has indicated that a single substance only may be involved, probably a mixture of *cis* and *trans* isomers.

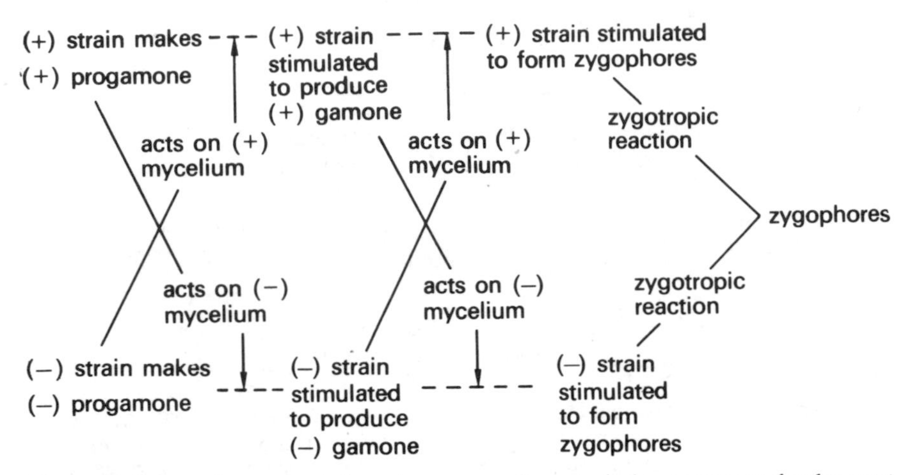

Figure 56. Schematic representation of sequence of events during zygospore development in heterothallic Mucoraceae (modified from Plempel, 1957).

When compatible zygophores make contact they develop into progametangia. The tip of each progametangium becomes cut off by a septum to separate a distal multinucleate gametangium from the subterminal suspensor (see Fig. 57). The walls separating the two gametangia break down so that the numerous nuclei from each cell become surrounded by a common cytoplasm. The fusion cell, or zygospore, swells and develops a dark warty outer layer. After a resting period the zygospore may germinate by developing a germ sporangium, usually on an unbranched sporangiophore, terminated by a columellate sporangium of the usual type. In some cases vegetative mycelium develops from the germinating zygospore. The precise conditions for zygospore germination are imperfectly known, and the development of reliable techniques for germination would greatly facilitate genetical analysis of Mucoraceae. There have been numerous accounts of the cytology of zygospore formation, and among more recent accounts the investigations of Cutter (1942*a*,*b*) and Sjöwall (1945, 1946) may be cited. Four main types of nuclear behaviour can be distinguished:

1. In *Mucor hiemalis*, *Absidia spinosa* and some other members of the

Mucorales all the nuclei fuse in pairs within a few days, then quickly undergo meiosis so that the mature zygospore contains only haploid nuclei.

2. In *Rhizopus stolonifer* and *Absidia glauca* some of the nuclei entering the zygospore do not pair, but degenerate. The remainder fuse in pairs but do not undergo meiosis immediately. Meiosis is delayed until germination of the zygospore.

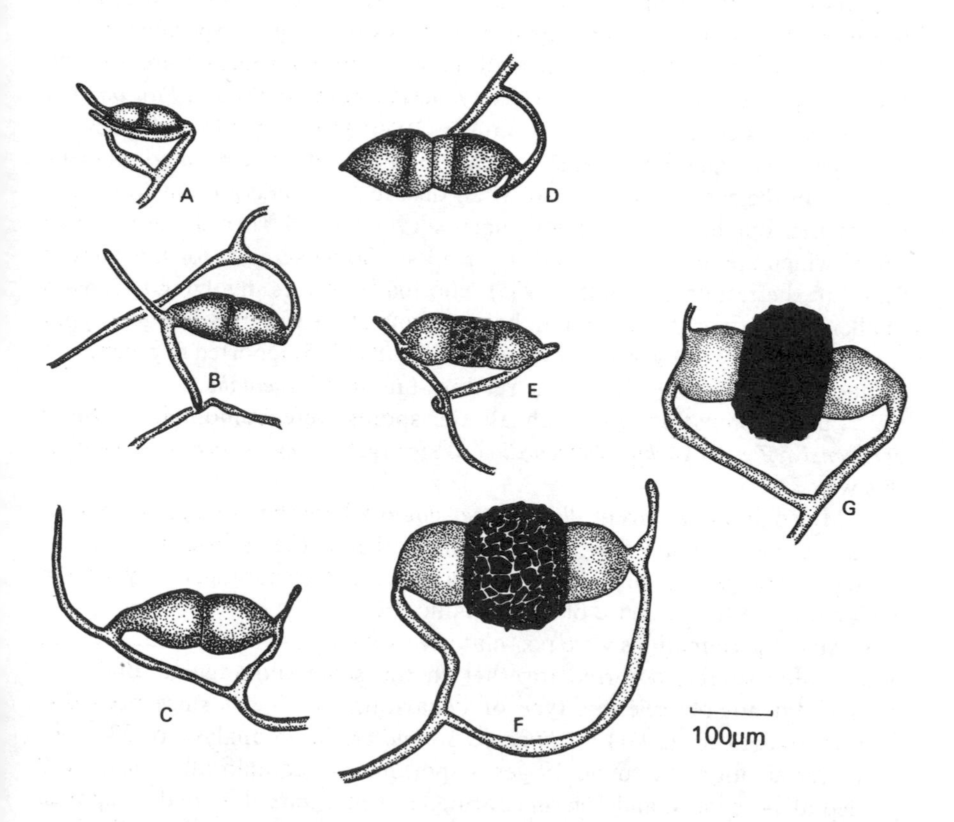

Figure 57. *Rhizopus sexualis*. A–G. Successive stages in the formation of zygospores. The fungus is homothallic.

3. In *Phycomyces blakesleeanus* the nuclei continue to divide in the young zygospore, so increasing in number. They then become associated in groups each containing several nuclei, with occasional single nuclei. Before germination some of the nuclei pair, and in the germ sporangium are found diploid nuclei and also haploid nuclei some of which are probably meiotic products; others may represent the scattered solitary nuclei which failed to pair. This assortment of nuclei in the germ sporangium is reflected in the mating-type reactions of the spores it contains (see below).

4. In *Syzygites megalocarpus* nuclear divisions continue in the young

zygospore, but nuclear fusion and meiosis apparently do not occur. This fungus can therefore be described as *amictic* (Burnett, 1956).

The results of various workers (Burgeff, 1914; Köhler, 1935) suggest that there is a single mating type locus with two alternative alleles, + and −, which segregate at meiosis. However there are a number of anomalous results for which a full cytological explanation is still awaited.

Blakeslee (1906) defined four kinds of zygospore germination in relation to the distribution of mating type in the spores of the germ sporangium.

1. Pure germinations, in which all germ sporangiospores were of one mating type, i.e. all + or all −. *Mucor mucedo*, *M. hiemalis* and *Phycomyces blakesleeanus* behave in this way (Gauger, 1956; Hocking, 1967). A possible cytological explanation of this finding is that only one diploid nucleus survives in the zygospore and that when this nucleus undergoes meiosis only one of the four haploid daughter nuclei survives which then divides to give nuclei which are all alike—i.e. all + or all −. Some support for this view is found in the results of Köhler (1935) who made crosses involving two pairs of alleles in *M. mucedo*, but found that only one of the four possible genotypes was found in any one germ sporangium. Sjöwall (1945) reported degeneration of nuclei in zygospores during the resting state in *M. hiemalis*.

2. Pure germinations in which all the spores were homothallic. *Mucor genevensis*, *Zygorhynchus dangeardi* and *Syzygites megalocarpus* behave in this way.

3. Mixed germinations in *Phycomyces nitens* where the same germ sporangia sometimes contained +, − and homothallic (i.e. self-fertile) spores. Cutter's findings that diploid nuclei enter the germ sporangia may be the explanation of the presence of homothallic spores.

4. Mixed germinations were postulated in which + and − spores, but not homothallic spores, occurred together in the same sporangium. Blakeslee himself did not observe this type of behaviour, but it has since been discovered by Gauger (1961) in *Rhizopus stolonifer*. In an analysis of 33 zygospore germinations he found 19 germ sporangia to contain all + spores, 9 yielded all − spores, and 5 germ sporangia yielded mixed + and − spores. The pure germinations would presumably be explained cytologically in a similar way to *M. mucedo*, whilst for the mixed germinations it would be necessary only to postulate the survival of more than one meiotic product so that both mating types were represented. In some germ sporangia 'neuter' spores were found, i.e. spores which on germination yielded mycelia which failed to mate both with tester + and − strains. The ability to mate was restored in time in certain isolates. Loss of ability to mate seems to be a fairly common phenomena in laboratory cultures after repeated subculturing, but an explanation of the effect is awaited.

In *Phycomyces* and *Absidia* the suspensors may bear appendages which arch over the zygospore. In *Phycomyces* the appendages are dark black and are forked, whilst in *Absidia* they are unbranched but curved inwards or coiled (see Figs. 50, 58). The function of such appendages is unknown: possibly they assist in attaching zygospores to passing animals. The forked appendage tips of *Phycomyces* bear a drop of liquid, and Burgeff (1925) regarded them as hydathodes. In the homothallic species *A. spinosa* the appendages arise on only one suspensor. The suspensors of *Zygorhynchus*, also differ in size, one being appreciably larger than the other (Fig. 59). Forms in which a distinction can be made between the two suspensors are said to be *heterogametangic*.

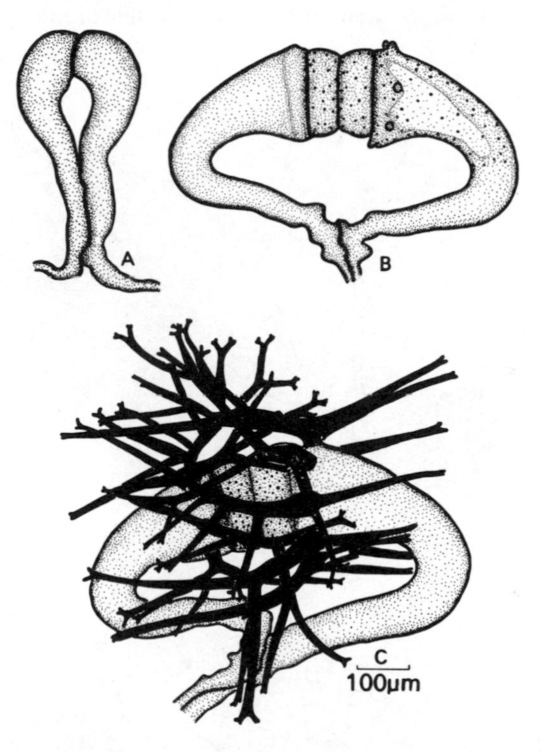

Figure 58. *Phycomyces blakesleeanus*. A–C. Stages in zygospore formation. The fungus is heterothallic. Note the dichotomous suspensor appendages.

Hybridisation experiments have been conducted between different species and genera of Mucoraceae. In some cases imperfect zygospores are formed. Attempted copulation has also been observed between homothallic and heterothallic strains (Burgeff, 1924).

Notes on the characteristic features of some common genera of the Mucoraceae are given below.

Mucor (Figs. 47, 48)

Cosmopolitan, and widespread in soil and on dung and other organic substrata. The sporangia are globose and borne on branched and unbranched sporangiophores. Most species are heterothallic and isogamous, i.e. the gametangia and suspensors are equal in size. Amongst the most common species from soil are *M. hiemalis*, *M. racemosus* and *M. spinosus*. *Mucor mucedo* is common on dung.

Zygorhynchus (Fig. 59)

Mostly reported from soil, often from considerable depth (Hesseltine, *et al.*, 1959). All species are homothallic and have heterogametangic zygospores. The sporangiophores are usually branched and the columella is often broader than high.

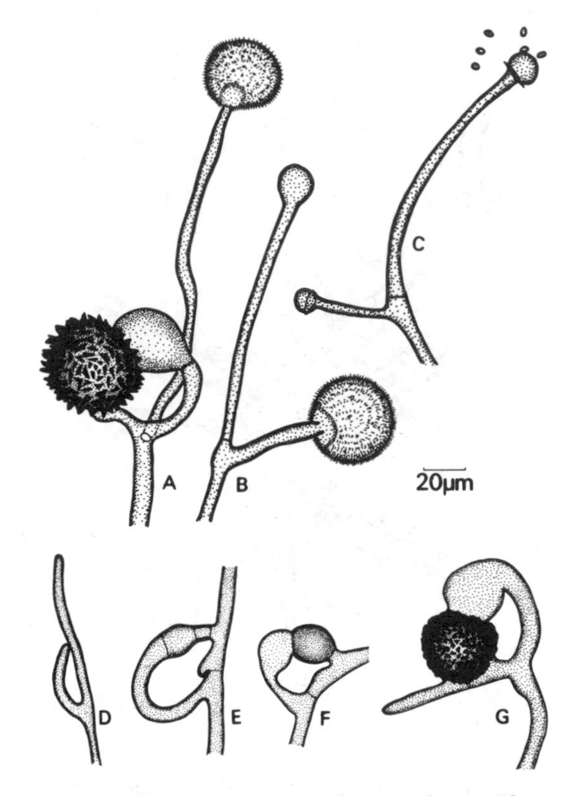

Figure 59. *Zygorhynchus molleri*. A. Zygospore and sporangium. B. Young sporangiophores. C. Dehisced sporangia. D–G. Stages in zygospore formation. Note that the fungus is homothallic and the suspensors unequal.

Rhizopus (Figs. 51, 57)

Occurs not only in soil but on fruit, other foods, all kinds of decaying materials and as a laboratory contaminant. *Rhizopus stolonifer* is often found on ripe bananas, especially if they are incubated in a moist atmosphere. The characteristic features are the presence of rhizoids at the base of the sporangiophores (which may grow in clusters), and the stoloniferous habit. An aerial hypha grows out and where it touches on the substratum it bears rhizoids and sporangiophores. Growth in this manner is repeated. Most species are heterothallic but *R. sexualis* is homothallic and forms zygospores freely within two days in the laboratory.

Absidia (Fig. 50)

The characteristic features are the pear-shaped sporangia produced in partial whorls at intervals along stolon-like branches. These branches produce rhizoids at intervals but not opposite the sporangiophores. The zygospores are surrounded by curved unbranched suspensor appendages which may arise from one or both suspensors. Most species are heterothallic but *A. spinosa* is homothallic. *Absidia glauca*, *A. orchidis* and *A. spinosa* are amongst the most commonly isolated species.

Phycomyces (Fig. 58)

Benjamin & Hesseltine (1959) recognise three species but the two best-known are *P. nitens* and *P. blakesleeanus*. The spores of *P. nitens* are larger than those of *P. blakesleeanus*, but it is likely that many workers confused the two before Burgeff (1925) pointed out the difference. Much of the literature on *P. nitens* probably refers to *P. blakesleeanus*. Neither species is particularly common in Britain, but a likely habitat is on fatty products and on empty oil casks. Bread and dung are other recorded substrata.

Syzygites (Fig. 60)

S. megalocarpus (= *Sporodinia grandis*, Hesseltine, 1957) is found on decaying sporophores of various toadstools, especially *Boletus*, *Lactarius* and *Russula*. It grows readily in culture and is homothallic (Davis, 1967). The sporangiophores are dichotomous and the sporangia thin-walled.

Pilobolaceae

There are only two genera in this family, *Pilobolus* and *Pilaira*. Both grow on the dung of herbivores occurring early in the succession of fungi which fruit regularly on such substrata. In both genera the sporangium dehisces by a transverse crack running around the base, and when this bursts a mucilaginous secretion is exuded so that the spores themselves are not released at this stage. In *Pilobolus* the sporangiophores are swollen and the sporangia are shot away violently by a jet of liquid, whilst in *Pilaira* the sporangiophore is cylindrical and the sporangium becomes converted into a sporangial drop. A monograph of the family has been given by Grove (see Buller, 1934).

Pilobolus

The generic name means literally the hat-thrower, referring to the sporangial discharge. If fresh horse-dung is brought into the laboratory and incubated in a glass dish on a window-sill, after a preliminary phase of fruiting of *Mucor* lasting about four to seven days the characteristic bulbous sporangiophores of *Pilobolus* appear (Fig. 61). Common species are *P. kleinii*, *P. longipes* and *P. crystallinus*. A full account of the development and discharge of the sporangium has been given by Buller (1934). The sporangia of *Pilobolus* become attached to the vegetation and are eaten by herbivores: in the

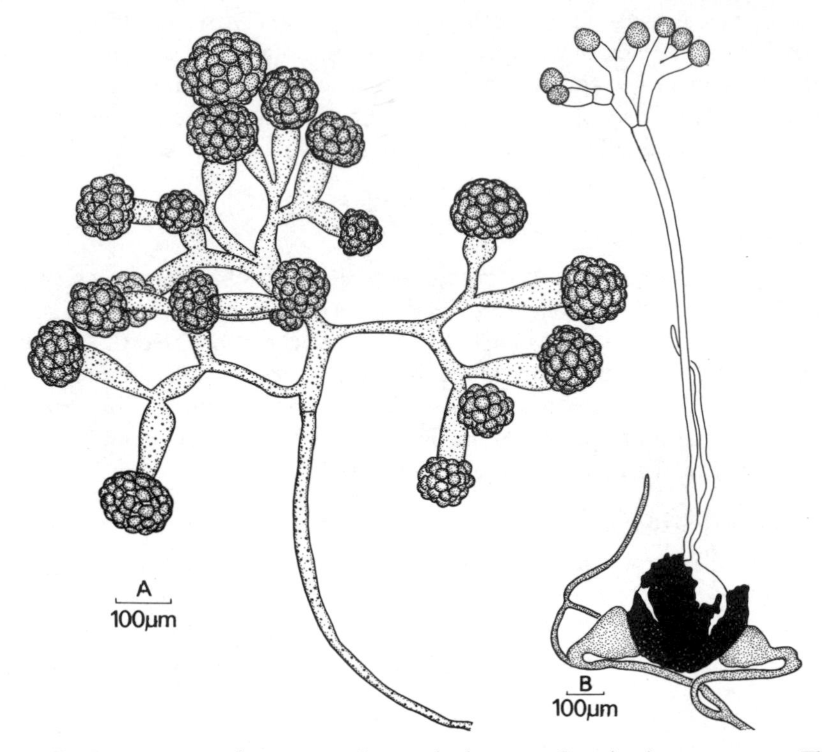

Figure 60. *Syzygites megalocarpus*. A. Sporangiophore. B. Germinating zygospore. The fungus is homothallic.

gut the spores are released and germinate. In the faeces the spores develop into mycelium which after about four days forms at the surface of the dung trophocysts, swollen segments coloured yellow by carotene (Fig. 62). Sporangiophores develop from the trophocysts in a regular daily sequence, and the stage of development can be correlated fairly closely with the time of day. During the late afternoon the sporangiophore grows away from the trophocyst towards the light and during the night its tip enlarges to become the sporangium. The swelling of the subsporangial vesicle takes place mainly between midnight and eariy morning.

The fully developed sporangiophores are highly phototropic. Light projected along the axis of the sporangiophore is brought to a focus at a point

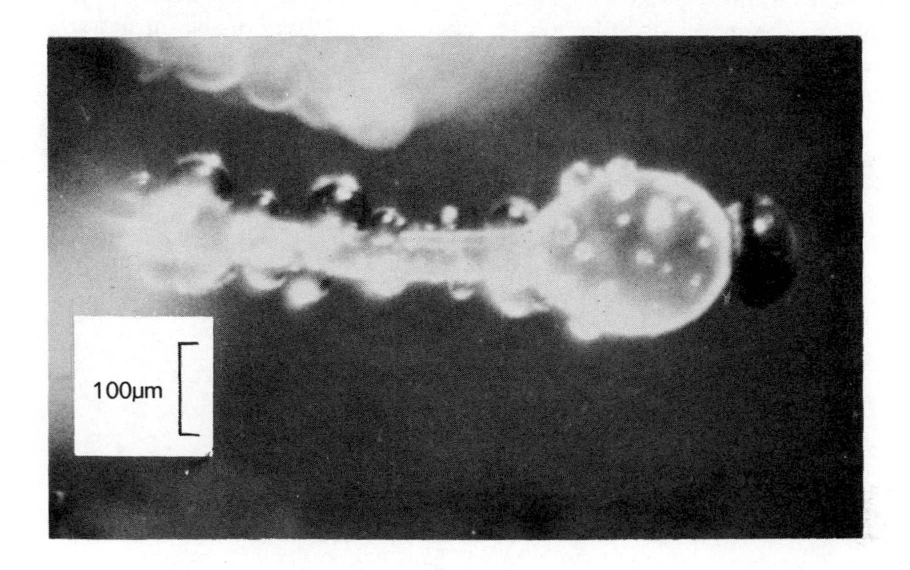

Figure 61. *Pilobolus* sp. Sporangiophore growing on dung. Note the flattened black sporangium, with a line of dehiscence around its base and the swollen subsporangial vesicle bearing drops of liquid.

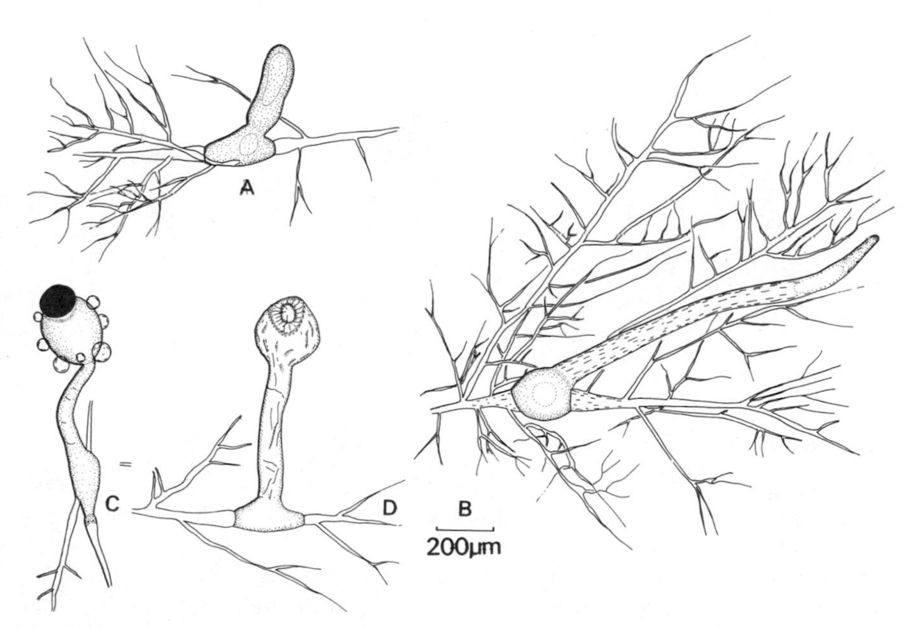

Figure 62. *Pilobolus* sp. A. Mycelium and a trophocyst from a culture. B. Sporangiophore elongating. C. Fully developed sporangiophore. D. Collapsed sporangiophore.

beneath the swollen vesicle. In this region is an accumulation of carotene-rich cytoplasm which glows orange when illuminated. When light falls asymmetrically on the sporangiophore it is focused on to the back of the subsporangial vesicle near its base, and some stimulus is probably transmitted to the cylindrical part of the sporangiophore resulting in more rapid growth on the side

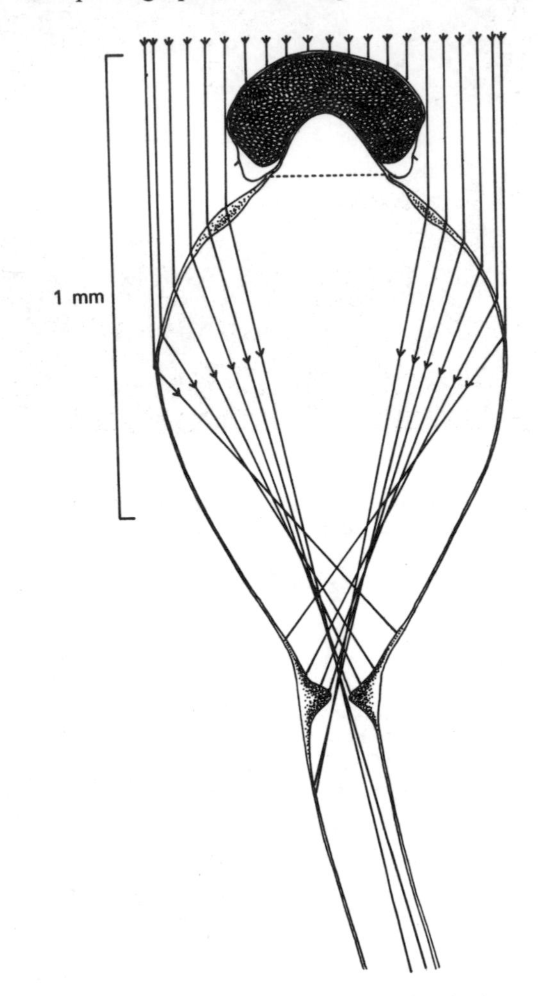

Figure 63. *Pilobolus kleinii.* Diagrammatic L.S. of sporangiophore showing the path of light rays which are brought to a focus beneath the subsporangial vesicle. The sporangiophore illustrated is orientated symmetrically with respect to the incident light. Note the mucilaginous ring extruded at the base of the sporangium (after Buller, 1934).

away from the light. Curvature of the whole sporangiophore thus occurs until it is orientated symmetrically with respect to the light (see Fig. 63). The structure of the sporangium differs in a number of ways from that of the Mucoraceae. The sporangium is flattened, its wall is dark black, shiny, tough and unwettable. At the base of the sporangium is a conical columella which is separated from the spores by a pad of mucilage. During the late morning the sporangium cracks open by a suture running around the base, just above the columella. The spores are, however, not released at this time because they are prevented from escaping by the mucilaginous pad which protrudes through the crack in the sporangium wall as a ring of mucilage.

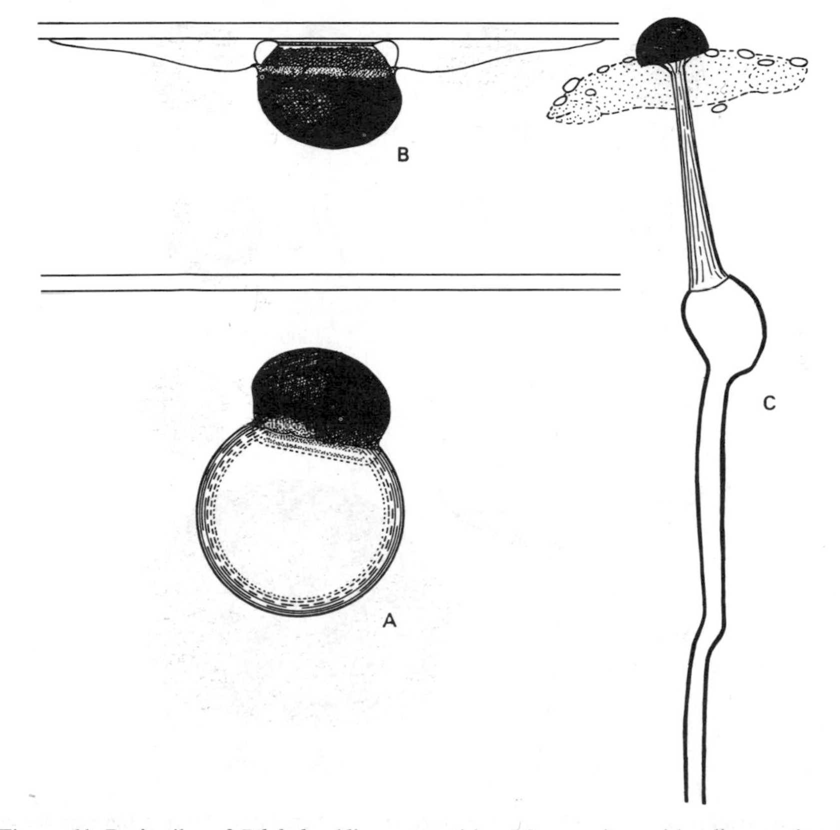

Figure 64. Projectiles of *Pilobolus* (diagrammatic). A. Sporangium with adherent drop o sporangiophore sap about to strike an obstacle. B. Sporangium after striking the obstacle. The sporangiophore sap has flowed round the sporangium which has turned outwards so that the mucilage ring adheres to the surface of the obstacle (after Buller, 1934). C. Sporangiophore projecting a sporangium. Note the jet of liquid and the bending of the narrow base of the sporangiophore under the recoil of the discharge (after Page, 1964).

The subsporangial vesicle is turgid: it contains liquid under pressure. It has been estimated that the osmotic pressure of the liquid is of the order of 5·5 atmospheres. Drops of excreted liquid commonly adhere to the sporangiophore. Eventually, usually about midday, the sporangial vesicle explodes at a line of weakness just beneath the columella. Due to the elasticity of the vesicle wall the liquid contents are squirted out, projecting the entire sporangium forwards in the direction of the light. Photographs of the jet show that it is at first cylindrical, but eventually breaks up into fine droplets (Page, 1964). The velocity of projection varies between wide limits in *P. kleinii*, 4·7–27·5 m/sec. with a mean of 10·8 m/sec. (Page & Kennedy, 1964). The sporangia can be projected vertically upwards for as much as 6 feet, and horizontally

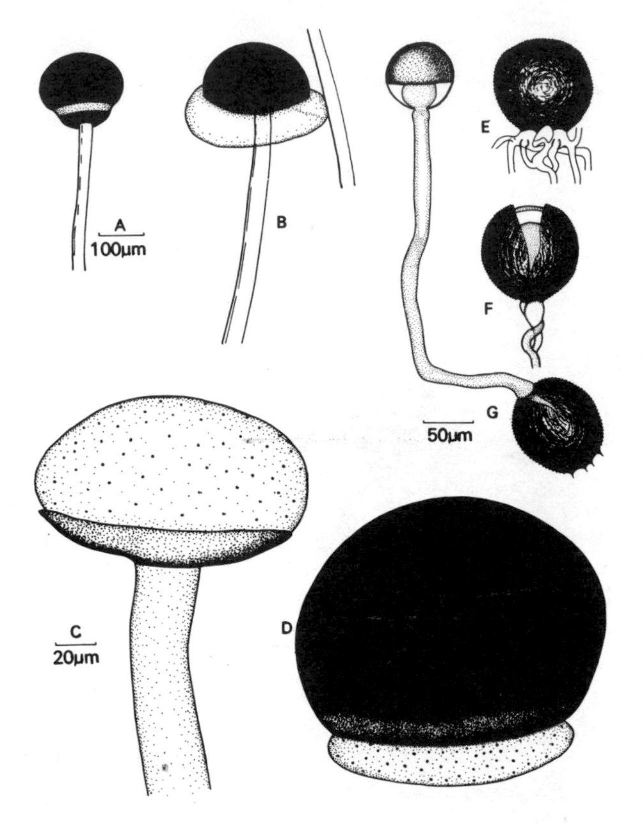

Figure 65. *Pilaira anomala*. A. Sporangiophore from rabbit dung showing the rupture at the base of the sporangium. B. Sporangium with extruded mucilage ring adhering to an adjacent hypha. C. Columella after sporangium has been detached. D. Detached sporangium showing basal mucilage ring. E. Zygospore. F, G. Stages in zygospore germination (E, F, G after Brefeld, 1881).

for up to 8 feet. On striking herbage surrounding the dung the sporangia become attached in such a way that the mucilaginous ring adheres, with the black sporangium wall facing outwards. Buller has suggested that the reason why the *Pilobolus* sporangium adheres in this way is that the projectile consists of a drop of liquid attached to the sporangium (Fig. 64A). When the projectile strikes an object the liquid flows around the sporangium, but because the sporangium wall is unwettable and because the base of the sporangium is surrounded by the wettable mucilaginous ring the sporangium turns round in the drop of liquid so that its wall faces outwards (Fig. 64B). As the drop of liquid dries the mucilage becomes more firmly attached, and dried sporangia are extremely difficult to detach from vegetation. The spores of *Pilobolus* are therefore not released at the time of sporangial discharge, but only after the sporangia have been eated by an animal, when they are released into the gut.

Most species of *Pilobolus* are heterothallic, but homothallic species are also known.

The physiology of *Pilobolus* shows a number of interesting features possibly correlated with its coprophilous habit. The spores germinate best above pH 6·5, and can be induced to germinate by treatment with alkaline pancreatin. Mycelial growth is best above pH 7·0. Growth on synthetic media is stimulated by the addition of thiazole, hemin and coprogen, an organo-iron compound produced by various fungi and bacteria. Ammonia stimulates sporangial production and in dual culture *Mucor plumbeus* can release sufficient ammonia to enhance sporulation (Page 1952, 1959, 1960; Hesseltine *et al.*, 1953; Pidacks *et al.*, 1953; Lyr, 1953).

Pilaira (Fig. 65).

P. anomala is found on dung of various herbivorous animals such as horse and rabbit The structure of the sporangium closely resembles that of *Pilobolus*, in that the spores are separated from the columella by a mucilaginous ring which extrudes from the base of the sporangium. There is, however, no subsporangial vesicle and sporangial release is non-violent. The sporangiophores are phototropic, and when they are mature they elongate quite rapidly. The mucilaginous ring on making contact with adjacent herbage becomes firmly attached to it and the sporangium is detached from the columella. In a moist atmosphere the mucilaginous ring may absorb water and swell considerably so that a large sporangial drop is formed (Ingold & Zoberi, 1963).

P. anomala is heterothallic and forms zygospores resembling those of *Pilobolus*. On germination a sporangium is produced.

Thamnidiaceae

In this family two kinds of asexual reproductive structure are found; columellate sporangia of the *Mucor* type, and smaller, usually non-columellate sporangia, termed sporangiola, often borne in whorls or at the tips of branches. The branches bearing the sporangiola may be borne laterally on the columellate sporangiophores or may arise separately. In some cases the branch system bearing the sporangiola is terminated by a spine. There are relatively few spores in the sporangiola and in *Chaetocladium* there may be a single spore only. Hesseltine (1955), recognises six genera within the family and Embree (1959) later included a new genus, *Radiomyces*, but we shall consider only *Thamnidium*, *Helicostylum* and *Chaetocladium*.

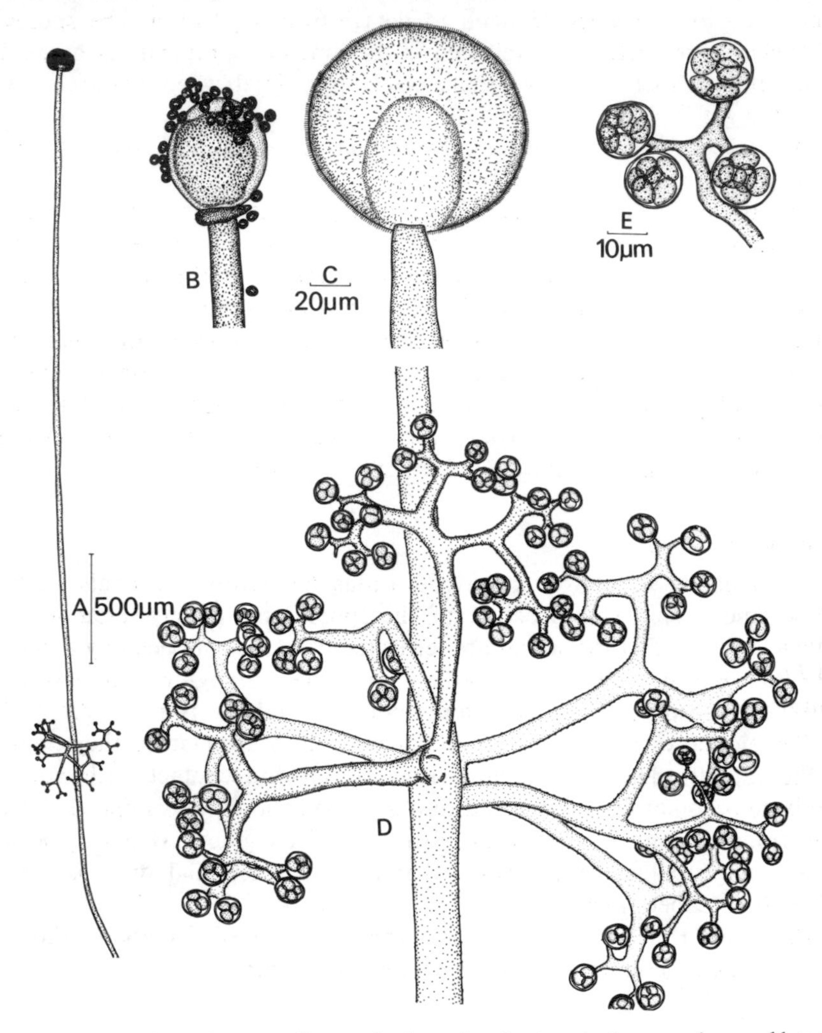

Figure 66. *Thamnidium elegans*. A. Sporangiophore showing terminal sporangium and lateral branches bearing sporangiola. B. Dehisced terminal sporangium showing columella and spores. C. Immature terminal sporangium showing columella. D. Branches bearing sporangiola. E. Sporangiola. Note the absence of a columella. B, C, D to same scale.

Thamnidium (Fig. 66)

The commonest species is *T. elegans* which grows on dung, in soil and has been reported from meat in cold storage. In culture large terminal columellate sporangia are produced with dichotomous lateral branches bearing fewer-spored non-columellate sporangiola. The sporangiola may also be borne on separate branch systems. Low temperature and light induce the formation of sporangia as opposed to sporangiola. During the development of the sporangiophores spiral growth occurs as in *Phycomyces* (Lythgoe, 1961, 1962). At maturity the columellate sporangia become converted into sticky sporangial drops not easily detached by wind currents or mechanical agitation. The sporangiola are, however, easily detached by such treatment

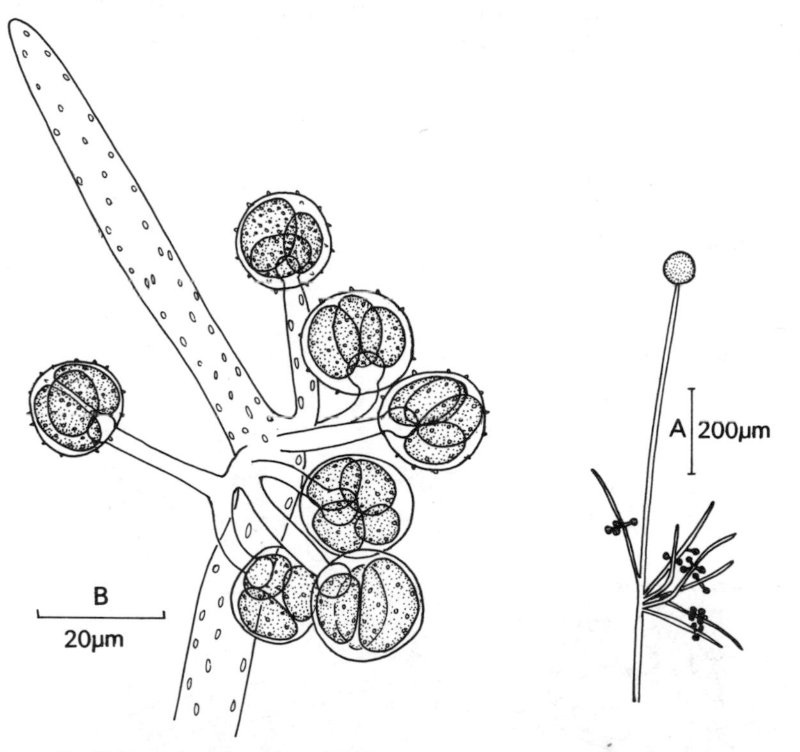

Figure 67. *Helicostylum fresenii*. A. Habit sketch to show terminal sporangium and lateral sporangioles. B. Branch ending in a terminal spine and bearing lateral sporangiola. Note that the sporangiola are columellate.

and in wind tunnel experiments become attached to slide traps, whilst the sporangia do not. Change from damp to dry air leads to an increase in sporangiole liberation (Zoberi, 1961; Ingold & Zoberi, 1963). *Thamnidium elegans* is heterothallic and forms zygospores resembling those of *Mucor* or *Rhizopus*, but they are produced best at low temperatures such as 6–7°C and not at 20°C (Hesseltine & Anderson, 1956).

Helicostylum (Fig. 67)

In this genus some of the sporangiolum-bearing branches end in terminal spines, but a *Mucor*-type sporangium is borne at the tip of the sporangiophore. *H. fresenii* illustrates this situation well. This fungus, which has been isolated from frozen meat, was earlier known as *Chaetostylum fresenii* (Lythgoe, 1958; Hesseltine & Anderson, 1957). In the wind tunnel the sporangia and sporangiola behave as in *T. elegans*.

Chaetocladium (Fig. 68)

In *Chaetocladium* there are no *Mucor*-like sporangia. Sporangiola, each containing a single spore, are borne on branches which end in spines. Such monosporous sporangiola are sometimes termed conidia. The distinction

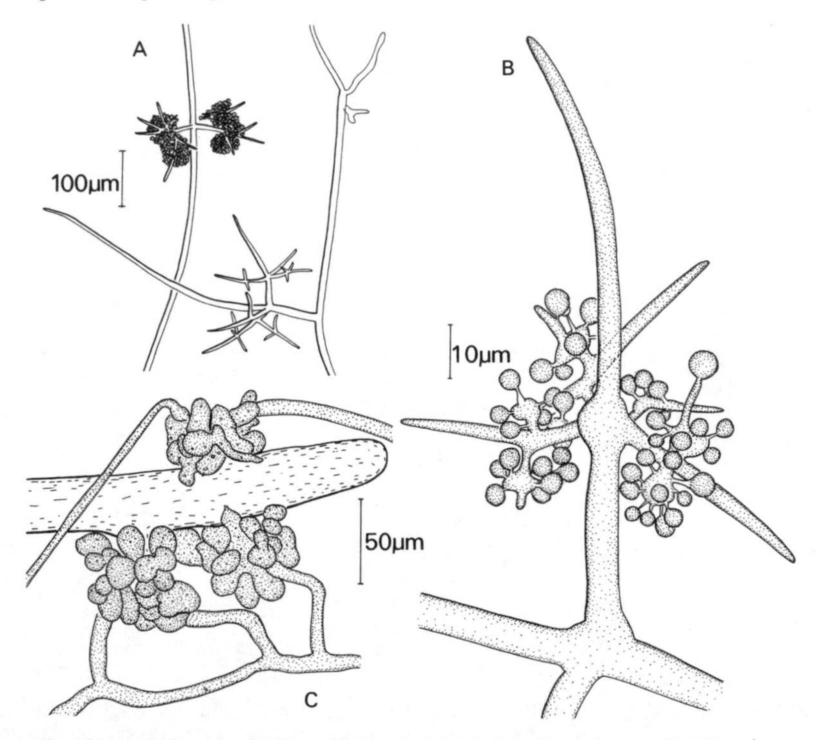

Figure 68. *Chaetocladium brefeldii*. A. Habit sketch to show branches ending in spines and bearing lateral sporangiola. B. Branch showing spine and sporangiola. C. Hypha of *Pilaira anomala* bearing bladder-like outgrowths following parasitism by *Chaetocladium*.

between the wall of the spore and the sporangiolum is sometimes visible on germination. There are two common species, *C. jonesii* and *C. brefeldii*, both parasitic on other members of the Mucorales. They often occur on *Mucor* or *Pilaira* on dung. At the point of attachment to the host there are numerous bladder-like outgrowths, which contain nuclei of both the host and the parasite (Burgeff, 1924). The parasites can, however, be cultured in the absence of a host (Hesseltine & Anderson, 1957). Both species are heterothallic. *C. brefeldii* is heterogametangic, forming zygospores resembling *Zygo-*

rhynchus. Burgeff (1920, 1924) has claimed that a given strain of *Chaetocladium* can only parasitise one of the two mating type strains of heterothallic Mucors. He has made the interesting suggestion that the parasitic habit of fungi such as *Chaetocladium* may have originated from attempted copulation with other members of the Mucorales.

Choanephoraceae

In this family, as in the Thamnidiaceae, both sporangia and sporangiola occur. The sporangia are usually columellate and often hang downwards. They contain dark brown sporangiospores with a striate epispore and bristle-like appendages. The sporangiola contain one to a few spores of similar construction. The zygospores also have striate walls. Two genera are recognised, *Choanephora* and *Gilbertella*. The genus *Blakeslea*, formerly regarded as distinct, is treated by some authorities as a synonym of *Choanephora* (Hesseltine, 1953, 1960; Hesseltine & Benjamin, 1957; Poitras, 1955).

Choanephora (Figs. 69, 70)

Species of *Choanephora* are found in warmer soils. *C. cucurbitarum* causes a rot of cucumbers and related fruits and is also commonly isolated from decaying flowers of various kinds. *C. trispora* (= *Blakeslea trispora*), which has been isolated from cowpeas, tobacco and cucumber leaves, forms in culture two kinds of asexual reproductive structure: nodding columellate or non-columellate sporangia with brown, faintly striate spores which usually bear bristle-like appendages, and non-columellate sporangioles borne in large numbers on globose vesicles. The sporangioles contain from two to five, but typically three, distinctly striate, dark brown spores also with bristle-like appendages. In *C. cucurbitarum* two similar structures can be recognised, but here the sporangiole contains only a single spore (Poitras, 1955), showing the characteristic striations. Such monosporous sporangiola are sometimes termed conidia. The sporangiola of *C. trispora* are readily detached by wind and break open in water like the two halves of a bivalve shell to release the spores. The function of the appendages is not known, but the 'conidia' are carried by insects from one plant to another. In the related genus *Gilbertella* there are no sporangioles but the sporangia split open to release the striate appendaged spores. *Gilbertella persicaria* is a parasite of peaches and tomatoes.

The Choanephoraceae have been the subject of physiological investigations (Hesseltine, 1961). In *C. cucurbitarum* it has been shown that the optimum temperature for 'conidial' production is near 25°C. A temperature of 31°C is unfavourable for 'conidial' formation but stimulates the production of large sporangia. High relative humidity inhibits conidial formation but stimulates sporangial formation at 20°C. All species studied are heterothallic, and interspecific crosses may also result in the formation of zygospores. An interesting phenomenon in connection with intra- and interspecific crosses is that the production of β-carotene is markedly enhanced when + and − strains are mated on liquid media as compared with production from either strain grown singly. This method of producing β-carotene may prove to be commercially valuable.

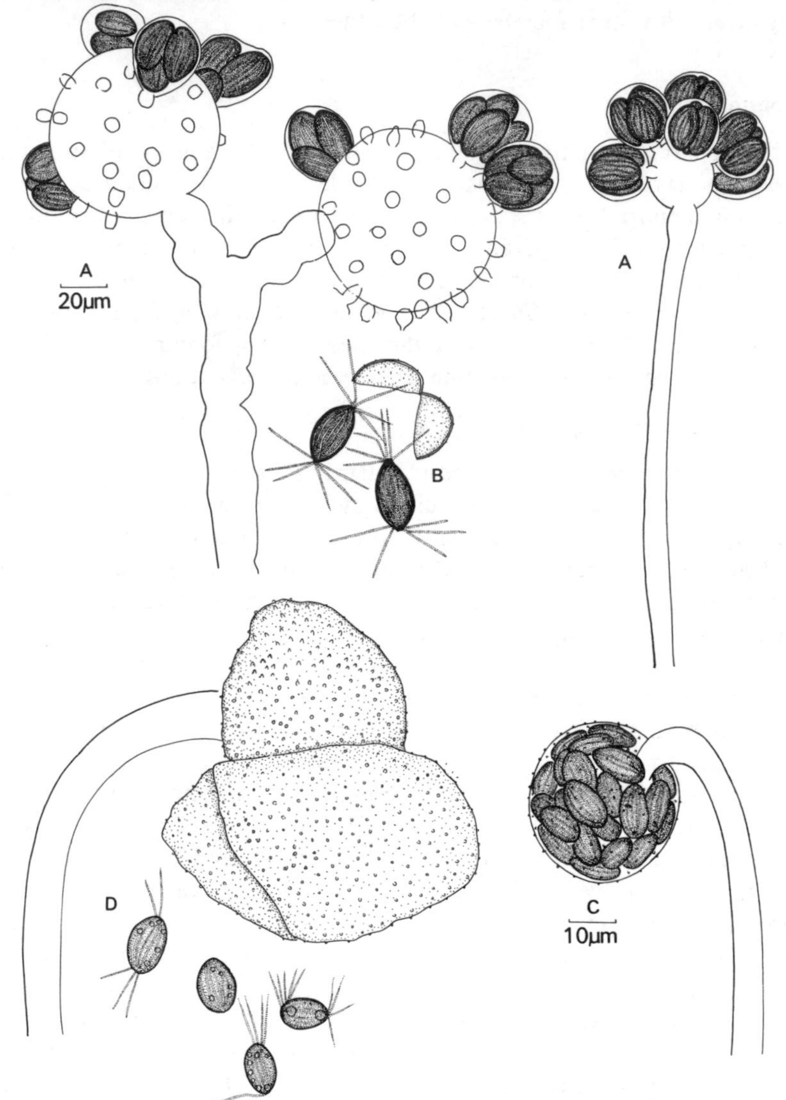

Figure 69. *Choanephora trispora*. A. Sporangiophores with globose vesicles bearing sporangiola containing three or four spores. B. A dehisced sporangiolum showing two spores. Note the striate epispore and mucilaginous appendages, and that the sporangiolum splits into two halves. C. Sporangiophore bearing a drooping sporangium. No columella was observed but columellate sporangia have been described. D. Dehisced sporangium also lacking a columella. Note the split sporangial wall and the sporangiospores with striate epispore and mucilaginous appendages.

Piptocephalidaceae

The characteristic feature of this family is that asexual reproduction occurs by means of cylindrical sporangia containing typically a single row of sporangiospores. Such sporangia are termed merosporangia and are formed in groups on inflated vesicles or at the tips of dichotomous branches (Benjamin,

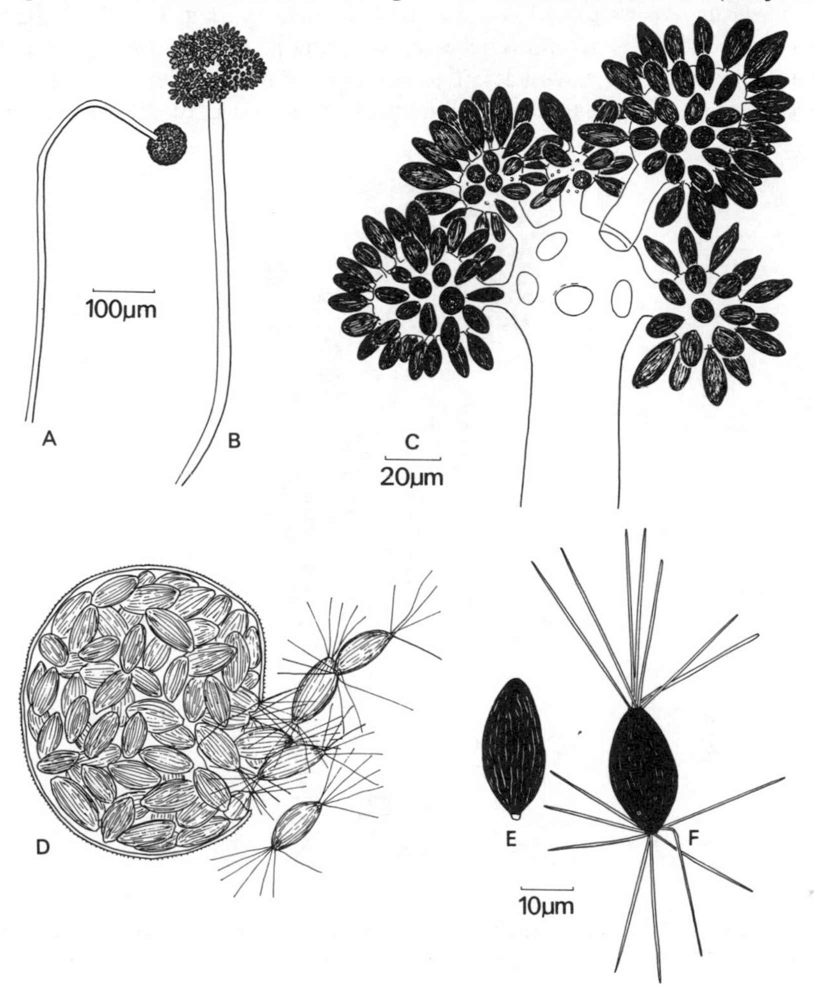

Figure 70. *Choanephora cucurbitarum.* A. Sporangiophore with drooping sporangium. B. Conidiophore and conidia. C. Apex of conidiophore showing vesicles and conidia. D. Dehisced sporangium showing striate spores with appendages. E. Conidium. F. Sporangiospore. C and D to same scale.

1966). Hesseltine (1955) recognises four genera, *Syncephalastrum*, *Syncephalis*, *Dispira* and *Piptocephalis*. *Syncephalastrum* is saprophytic and is sometimes segregated from the remaining genera into a separate family, the Syncephalastraceae. Benjamin (1959) includes only *Syncephalis* and *Piptocephalis* in the

Piptocephalidaceae. Both genera are parasitic on other fungi, usually members of the Mucorales. *Dispira* and related genera, also parasitic on Mucorales, are placed by Benjamin in a third family, the Dimargaritaceae.

Syncephalastrum (Fig. 71)

S. racemosum grows in soil and on dung. In culture its growth is similar to *Mucor*, and it forms aerial branches terminating in club-shaped or spherical enlargements known as vesicles. The vesicles are multinucleate and bud out all over their surface to form cylindrical outgrowths, the merosporangial

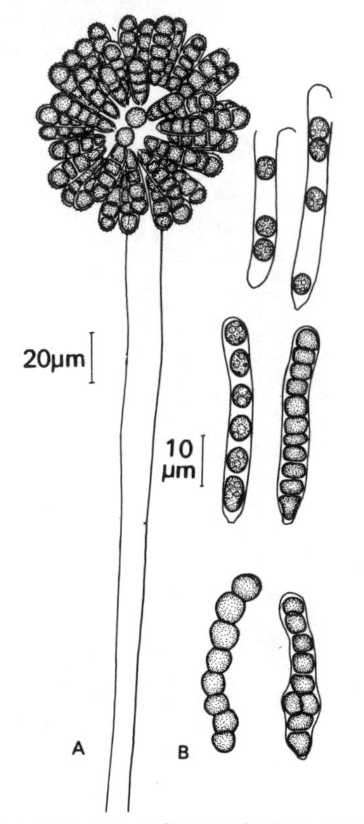

Figure 71. *Syncephalastrum racemosum*. A. Sporangiophore bearing a vesicle and numerous merosporangia. B. Merosporangia and merospores.

primordia. Into these outgrowths one or perhaps several nuclei pass, and nuclear division continues. The cytoplasm in the merosporangium cleaves into a single row of five to ten sporangiospores each with one to three nuclei. The sporangial wall shrinks at maturity so that the spores appear in chains reminiscent of an *Aspergillus*. Occasionally the merospores may lie in more than a single row. The spore heads remain dry and entire rows of spores are detached by wind (Ingold & Zoberi, 1963). *Syncephalastrum racemosum* is heterothallic and forms zygospores resembling those of the Mucoraceae.

Piptocephalis (Fig. 72)

Most species of *Piptocephalis* parasitise the mycelium of other members of the Mucorales, but *P. xenophila* attacks various Ascomycetes and Fungi-Imperfecti. A characteristic habitat for *P. freseniana* is dung at the end of the phase of fruiting of *Mucor*. If situated near to the hypha of a suitable host the cylindrical spore of *Piptocephalis* germinates laterally (not from the ends) forming one or more germ tubes which make contact with the host hypha and form enlarged appressoria (Fig. 72D). Beneath the appressorium fine haustoria develop. The mycelium of the parasite then develops externally to

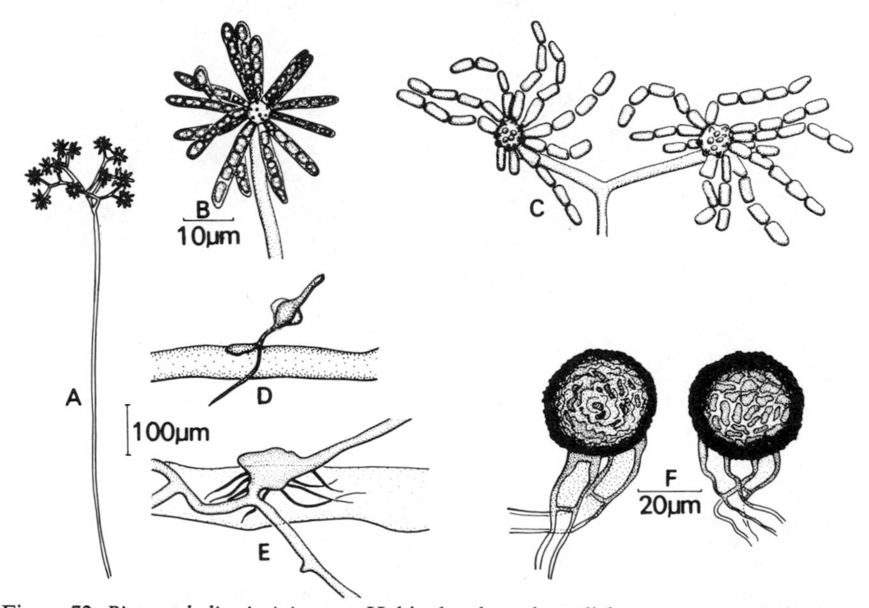

Figure 72. *Piptocephalis virginiana.* A. Habit sketch to show dichotomous sporangiophore. B. Head cell and intact merosporangia. C. Head cells showing breakdown of merosporangia to form chains of spores. D. Spore germination and formation of appressorium on a host hypha. E. Appressorium and branched haustorium on host hypha. The parasite mycelium is branched and extending to other host hyphae. F. Zygospores. The fungus is homothallic.

the host and forms stolon-like branches which develop further appressoria and haustoria. Merosporangia are formed at the tips of dichotomously branched aerial hyphae, arising from specialised swellings or head-cells. The merospores in *Piptocephalis* usually contain one or occasionally two nuclei. At maturity the merosporangia behave in two distinct ways. In *Piptocephalis virginiana* the spore chains remain dry and entire chains as well as head-cells may be dispersed by air currents, although single spores are also dispersed. In *P. freseniana* all the spore chains in any one head become involved in the formation of a sticky spore drop, and the entire spore drop is dispersed by wind (Ingold & Zoberi, 1963). Most species are homothallic (Leadbeater &

Mercer, 1956, 1957*a,b*; Benjamin, 1959). Zygospores are usually formed within the agar in cultures. The mature zygospore is a spherical dark brown sculptured globose cell held between two tong-shaped suspensors.

Cultures of *Piptocephalis* in the absence of a host have only slight success. On suitable media spores will germinate and give rise to a limited mycelium producing dwarf sporophores. The spores so formed do not germinate in the absence of a host but can infect a suitable fungus. There is evidence that spore germination and chemotropic growth of the germ tubes are stimulated by fungal secretions (Berry & Barnett, 1957).

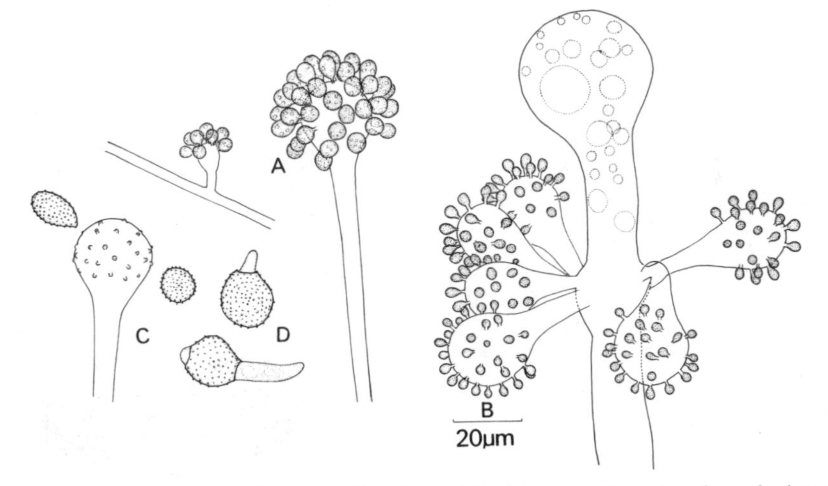

Figure 73. *Cunninghamella elegans*. A. Simple conidiophores. B. Immature branched conidiophore showing developing conidia. C. Mature conidiophore showing scars of attachment of conidia. D. Germinating conidia, 5 hr.

Cunninghamellaceae

In this family asexual reproduction is entirely by means of 'conidia': sporangia and sporangiola are not formed. For this reason the genera included here have been separated from the Choanephoraceae with which they were formerly classified. Hesseltine (1955) recognised three genera, *Mycotypha*, *Cunninghamella* and *Thamnocephalis*.

Cunninghamella (Fig. 73)

Species of *Cunninghamella* are found in soil in the warmer regions of the world. The asexual 'conidia' are hyaline and borne singly on globose vesicles on branched or unbranched conidiophores. Although the conidia are possibly to be interpreted as one-spored sporangiola there is no evidence of this from their structure. In some species, e.g. *C. echinulata* and *C. elegans*, the conidia are spiny, but in others they are smooth. Zygospores are of the *Mucor* type.

Mortierellaceae

The characteristic feature of this family is that the sporangia lack columellae. In some species stalked globose detachable one-celled spores, or stylospores, occur, possibly representing modified sporangia, In the most frequently encountered genus, *Mortierella*, zygospores may be enclosed in a weft of

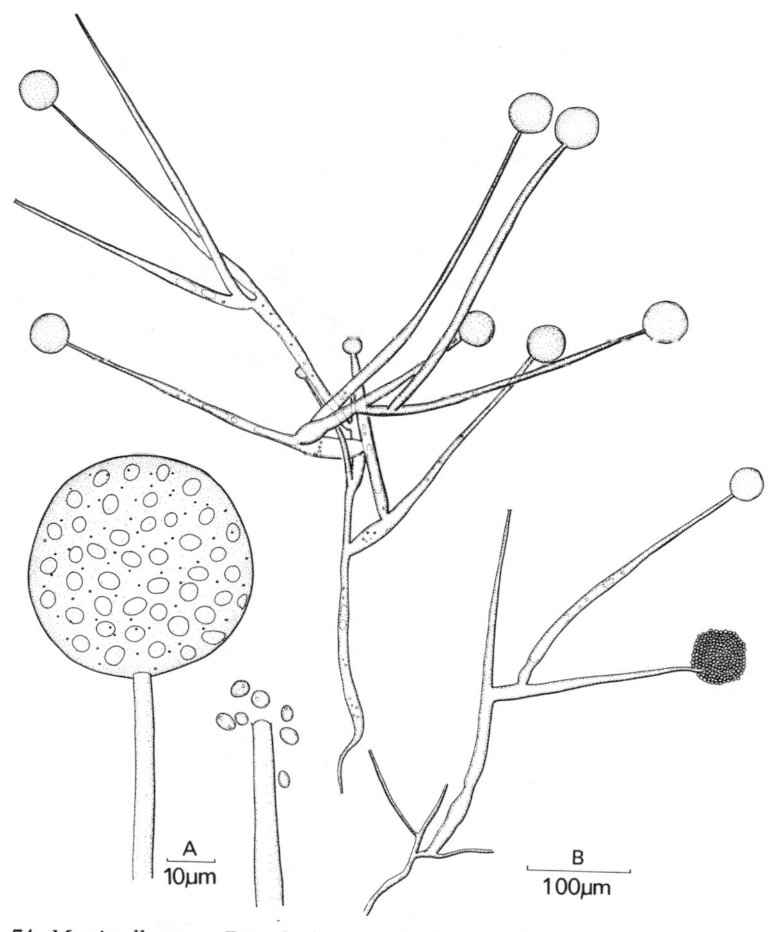

Figure 74. *Mortierella* sp. A. Branched sporangiophores. B. Intact sporangium and tip of sporangiophore after sporangium dehiscence. Note the absence of a columella.

mycelium. Hesseltine (1955) includes the genera *Mortierella*, *Haplosporangium* and *Dissophora* in the family. *Haplosporangium parvum* is associated with pulmonary mycosis of rodents.

Mortierella (Figs. 74, 75)

About 60 species of *Mortierella* are known, occurring widely in soil and on plant and animal remains in contact with soil (Linneman, 1941). These fungi can be isolated readily on nutrient-poor media which prevent the growth of more vigorous moulds. Some species can cause a mycorrhizal infection of *Vaccinium* roots (Wolf, 1954). The mycelium is fine and often shows a characteristic series of fan-like zones. The sporangia are borne on branched or

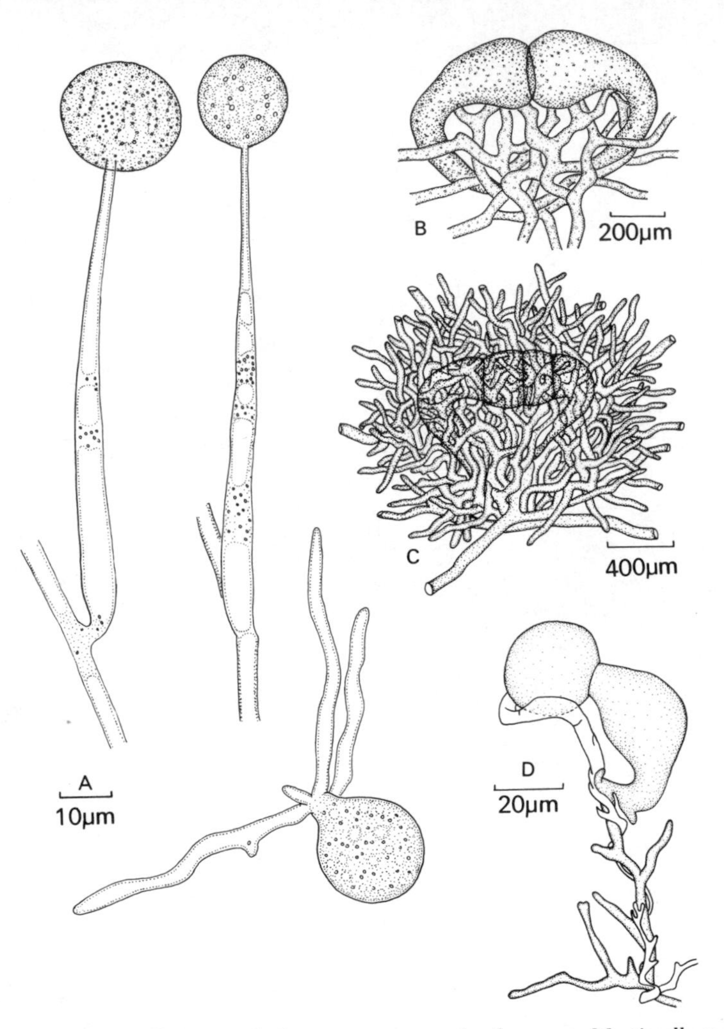

Figure 75. A. *Mortierella zonata*. Stylospores, one germinating. B, C. *Mortierella rostafinskii* (after Brefeld). B. Developing zygospore. C. Older zygospore surrounded by weft of hyphae. D. *M. parvispora* (after Gams & Williams). Zygospore formed by fusion of two compatible strains. Note the absence of investing hyphae.

unbranched, usually tapering, sporangiophores. The sporangium wall is delicate and may collapse around the spores. There is no columella. Frequently the entire sporangium is detached. In a number of species there may be only two or three spores per sporangium. Some species bear stalked globose unicellular stylospores which are easily detached and can germinate to form new mycelium, and in some species, e.g. *M. stylospora* and *M. zonata*, only stylospores are present; true sporangia are lacking.

The zygospores of some *Mortierellas* may be surrounded by an investment of sterile hyphae (see Fig. 75B,C). Whilst some species are homothallic, *M. parvispora* is heterothallic and heterogametangic, and its zygospores are not surrounded by sterile hyphae (Fig. 75D; Gams & Williams, 1963). *Mortierella marburgensis* is also heterothallic. Here the suspensors are at first equal but later one increases in size whilst the other loses most of its cytoplasm (Williams *et al.*, 1965).

Endogonaceae

This family is included by many authorities in the Mucorales because of the presence of zygospores formed after conjugation. About four genera are recognised (Zycha, 1935; Hawker, 1954, 1955).

Endogone (Fig. 76)

Species of *Endogone* form subterranean fruit-bodies (sporocarps) from 1 to 25 mm in diameter, consisting of aggregations of coarse, angular, thick-walled coenocytic hyphae enclosing several or numerous thick-walled spherical or balloon-shaped spores often up to 100 μm or more in diameter. The spores are of two kinds: chlamydospores and zygospores. In some species, e.g. *E. lactiflua*, found in summer and early autumn around twigs and leaves in conifer woods the fruit-body is entirely zygosporic. In others, e.g. *E. microcarpa*, found usually under yew or pine trees, both zygospores and chlamydospores occur in the same fruit-body or the fruit-body may be entirely chlamydosporic. A distinction between the two types of spore is that the zygospores arise by conjugation of the two progametangia and are thus subtended by two hyphae, whilst chlamydospores are usually terminal enlargements of a single hypha. Moreover the endospore of a zygospore is continuous, whilst the contents of a chlamydospore are in protoplasmic connection with the subtending hypha. In addition to zygospores and chlamydospores it has been claimed that thin-walled sporangia are also present (Thaxter, 1922; Walker, 1923). Kanouse (1936) reported *Mucor*-type sporangia in cultures of the zygosporic *E. sphagnophila*. Sporangia have not, however, been reported from British material (Hawker, 1954; Godfrey, 1957*a*).

In *E. lactiflua* the young progametangia are at first multinucleate (Bucholtz, 1912; Hawker, 1955). A single nucleus in each progametangium becomes cut off in a gametangium from the multinucleate suspensors. The wall separating the two gametangia breaks down, and the mature zygospore develops as a sac from the fused gametangia. Nuclear fusion has not been seen, but the two nuclei of the gametangia are closely associated. As the zygospore matures numerous nuclei become visible. Chlamydospores of *Endogone* will germinate fairly readily on water agar (Godfrey, 1957*b*; Mosse, 1959), especially if stimulated by secretions from other soil micro-organisms. Germination of zygospores has not been described.

Interest in *Endogone* has been aroused by the demonstration that chlamydospores, when germinated aseptically, can infect roots of various plants producing infections of the endotrophic mycorrhizal type (Mosse, 1963; Nicolson, 1967; Nicolson & Gerdemann, 1968). Within the root coarse coenocytic hyphae grow between the cells, and within the cells finely branched

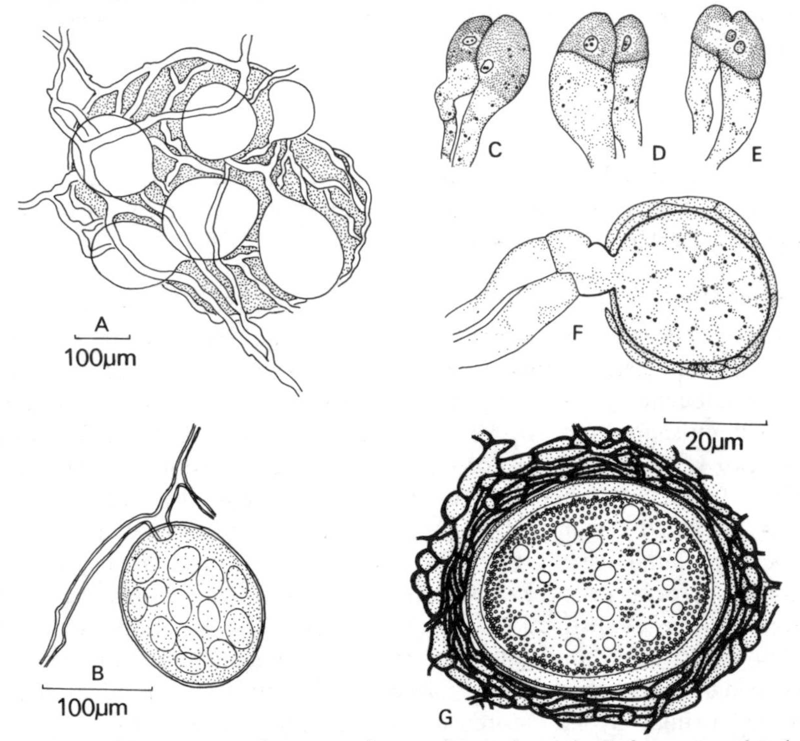

Figure 76. *Endogone* spp. A. Sporocarp from a chlamydosporic *Endogone* associated with strawberry roots. B. Single chlamydospore from A. C, D, E, F. *E. lactiflua*. Sections showing successive stages in zygospore formation (after Hawker). C. Progametangia showing enlargement of a single nucleus in each. D. Uninucleate gametangium cut off from suspensor cells. E. Nuclear migration. F. Young multinucleate zygospore with sheath hyphae beginning to develop. J. Section of mature zygospore with hyphal sheath. C–G to same scale.

haustoria (arbuscules) are found which may become digested to form globose swellings (the so-called sporangioles). Large vesicles may also occur in a terminal or intercalary position. Such mycorrhiza are often termed Phycomycetoid, or vesicular–arbuscular mycorrhiza. Although there is evidence that similar types of mycorrhizal infections may be caused by *Pythium*, the evidence for *Endogone* is very convincing. Single large spores of the *Endogone* type are apparently common in soil and can be removed by fine sieves (Gerdemann & Nicolson, 1963). The guts of rodents, especially mice, rabbits and voles, commonly contain large spores of the *Endogone* type, and it is presumed that these animals eat the subterranean fruit-bodies. Attempts to germinate spores from the caecal contents have been unsuccessful (Silver-Dowding, 1955).

ENTOMOPHTHORALES

Many but not all members of the Entomophthorales are pasasites of insects and other animals. Asexual reproduction is generally by means of violently projected propagules (conidia). Sexual reproduction is by anisogamous or isogamous copulation of gametangia and results in the formation of a zygospore.

Martin (1961) has distinguished three families as follows:

(a) Conidia borne singly or in chains, not forcibly discharged; parasitic on amoebae and nematodes **Zoopagaceae**

(a) Sporangium modified to function as a single conidium, forcibly discharged at maturity b

(b) Mycelium persistent, of uninucleate cells; gametangia unequal; saprobic on dung or in humus, occasionally parasitic in mammals **Basidiobolaceae**

(b) Mycelium usually breaking up into multinucleate segments; gametangia equal; usually parasitic on insects, occasionally on plants or saprobic **Entomophthoraceae**

Entomophthoraceae contains about six genera. Of these *Entomophthora* and *Massospora* include species parasitic on insects; *Completoria* is parasitic on fern prothalli; *Ancylistes* is a parasite of desmids. The Basidiobolaceae contain a single genus *Basidiobolus*, of which the best-known species is *Basidiobolus ranarum* which grows on dung of amphibia. The Zoopagaceae contains several genera mostly occurring as endozoic parasites of soil protozoa. Some authorities include also a fourth family, the Paracoccidioidaceae, which contains a number of fungi pathogenic to man such as *Paracoccidioides* and *Blastomyces* and *Histoplasma*. They are considered to be 'imperfect' members of the group, i.e. sexual reproduction has not been described.

Basidiobolaceae

Basidiobolus (Figs. 77–79)

If a frog is captured and placed in a jar with a little water the dung can be filtered off. When the damp filter paper is placed in the lid of a Petri dish containing a suitable medium (e.g. 1% peptone agar) conidia of *B. ranarum* will be projected on to the agar surface and within a few days the coarse septate mycelium will become visible. *Basidiobolus* is present in the gut of frogs in the form of spherical uninucleate cells up to 20 μm in diameter (Levisohn, 1927). These cells are voided with the faeces and under dry conditions are capable of surviving for several months. Under moist warm conditions

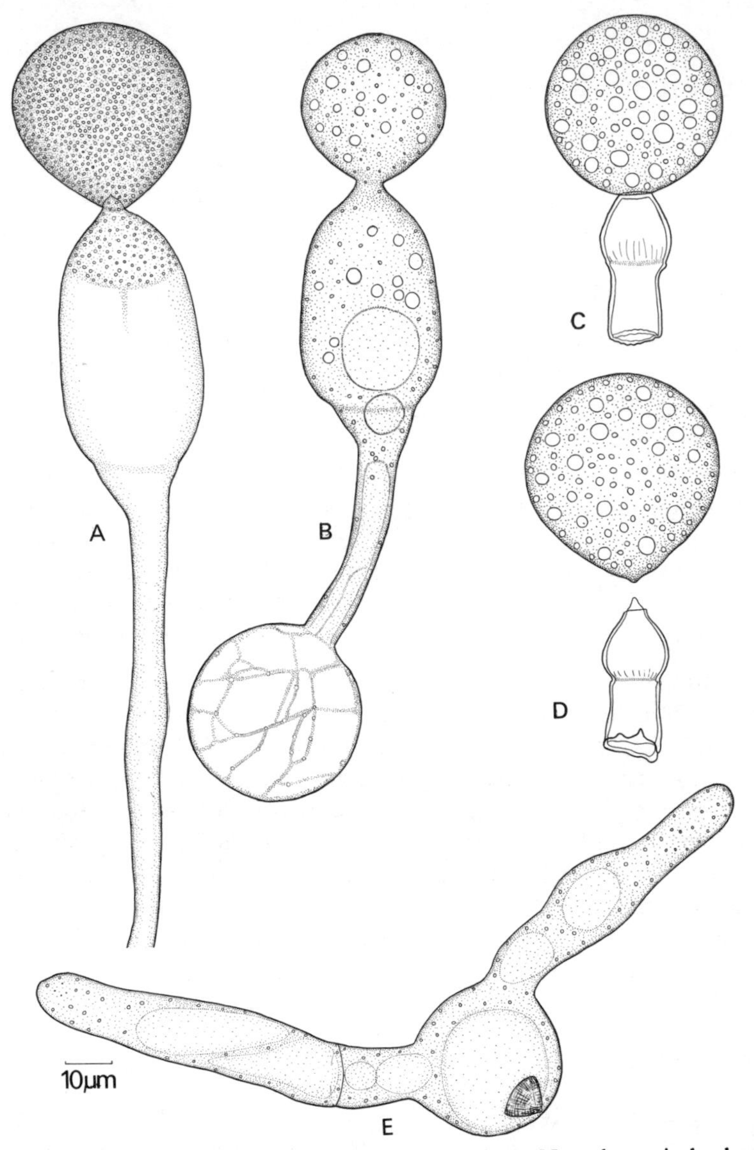

Figure 77. *Basidiobolus ranarum*. A. Conidiophore from culture. Note the conical columella and the swollen vesicle with the line of weakness around its base. B. Conidium germinating to produce a secondary conidiophore. C. Discharged conidium with remnant of vesicle attached. D. Discharged conidium separated from the remnant of the vesicle. E. Conidium germinating to form a septate mycelium.

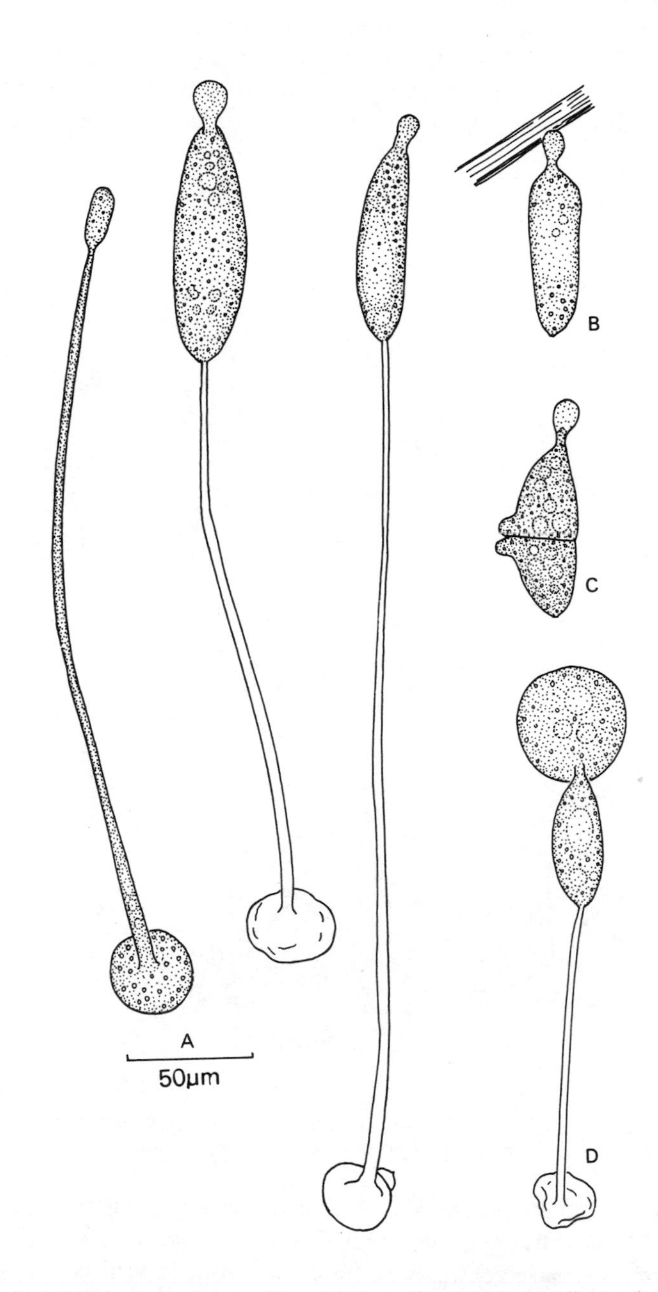

Figure 78. *Basidiobolus ranarum*. A. Germination of primary conidia to form slender conidiophores bearing secondary, non-propulsive adhesive conidia. B. Attachment of secondary conidum to a hair by a terminal adhesive pad. C. Septate secondary conidium, germinating. D. Secondary conidium enlarging to form a globose conidium.

the spherical cells germinate to give a coarse septate mycelium from which conidiophores develop. The conidiophores are phototropic and resemble the sporangiophores of *Pilobolus*, but bear a colourless pear-shaped conidium. A conical columella projects into the conidium. Beneath is a swollen sub-conidial vesicle containing liquid under pressure. A line of weakness can

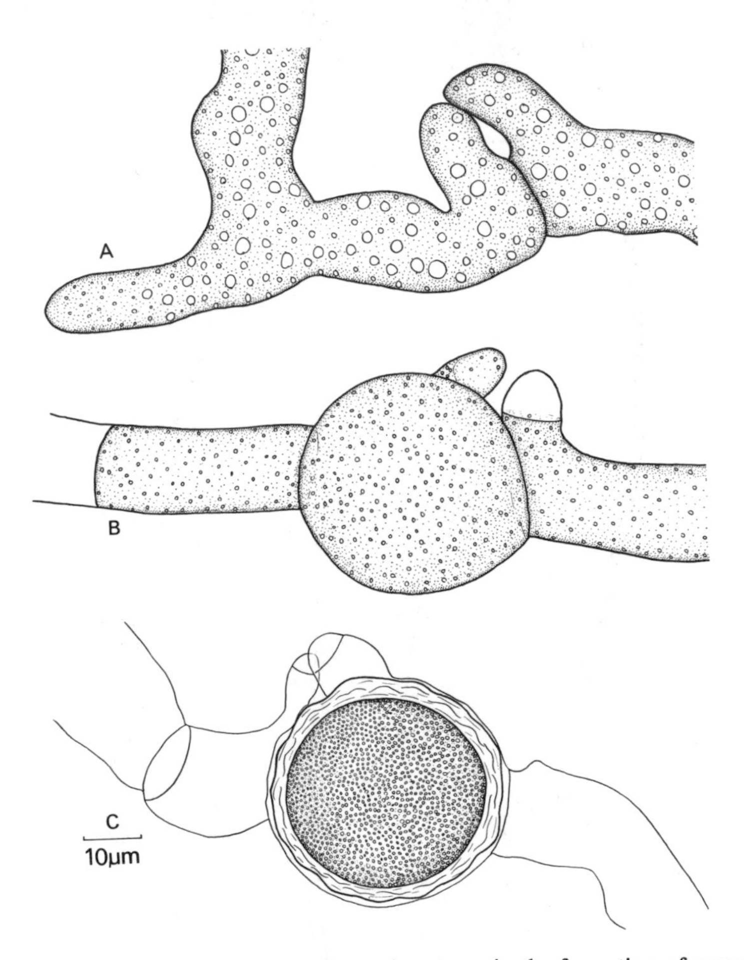

Figure 79. *Basidiobolus ranarum*. A–C. Successive stages in the formation of zygospores. A. Progametangia. B. Young zygospore. C. Mature zygospore.

be detected around the base of the vesicle, and when this ruptures the conidium and vesicle fly forwards for a distance of 1–2 cm. The elastic upper portion of the vesicle contracts and the sap within it squirts out backwards, so that the projectile behaves as a minute rocket. During its flight the conidium and the rocket motor (i.e. the vesicle) are often separated although the two parts may also remain attached. Such conidia may germinate to form secondary conidia of the same type, or may germinate directly to form a septate mycelium. The conidia are not very resistant to drying, and they are eaten by beetles. Within the gut of the beetle the conidium remains unchanged, but when beetles which have eaten *Basidiobolus* conidia are ingested by frogs the conidia are released, following digestion, into the gut of the frog. The conidial contents divide into numerous spherical cells and it is from these cells that the mycelium develops later. There is no evidence that the fungus harms either the frog or the beetle. A second type of asexual reproduction has been reported by Drechsler (1956). The globose conidia on germination may give rise to germ tubes, to further globose conidia of the same type or to elongate sausage-shaped uninucleate or binucleate secondary conidia with an adhesive pad at the distal end (Fig. 78). The contents of the secondary adhesive conidia may divide by transverse and longitudinal division to form a multicellular structure. These bodies have been found attached to the bristles of mites, and it is possible that frogs ingest secondary conidia.

Zygospores are formed following conjugation. Development can readily be followed in four to five-day-old agar cultures derived from a single conidium. On either side of a septum beak-like projections develop. The tips of these branches become cut off by septa and the subterminal cells beneath them fuse to form the zygospore which has a wrinkled thick wall when mature (Fig. 79). The cytology of zygospore formation has been described by several workers (see Woycicki, 1927). The vegetative cells of *Basidiobolus* are uninucleate and a single nucleus migrates into each beak-like projection. There each nucleus divides mitotically. One daughter nucleus is cut off by a septum in the terminal cell of the beak, and later disintegrates. The other nucleus resulting from division migrates back into the parent cell. Following this one of the parent cells enlarges to several times the volume of the adjacent cell and a pore is formed connecting the two cells through the original septum separating them. The nucleus from the smaller cell passes through the pore and lies close to the nucleus of the larger cell. Nuclear fusion may occur directly or after a further division, following which one daughter nucleus from each pair fuses, whilst the others disintegrate. Reduction division occurs within the mature zygospore to give four haploid nuclei, of which three usually degenerate. On germination the zygospore may form a germ tube, or may form a conidiophore terminated by a globose conidium (Drechsler, 1956).

There are occasional reports of the isolation of *Basidiobolus* from man and domestic animals. The isolates pathogenic to man have been identified as *B. meristosporus* (Greer & Friedman, 1966). Benjamin (1962) has given a key to the species.

Entomophthoraceae

Lakon (1963) has listed important literature on the family. An account of the

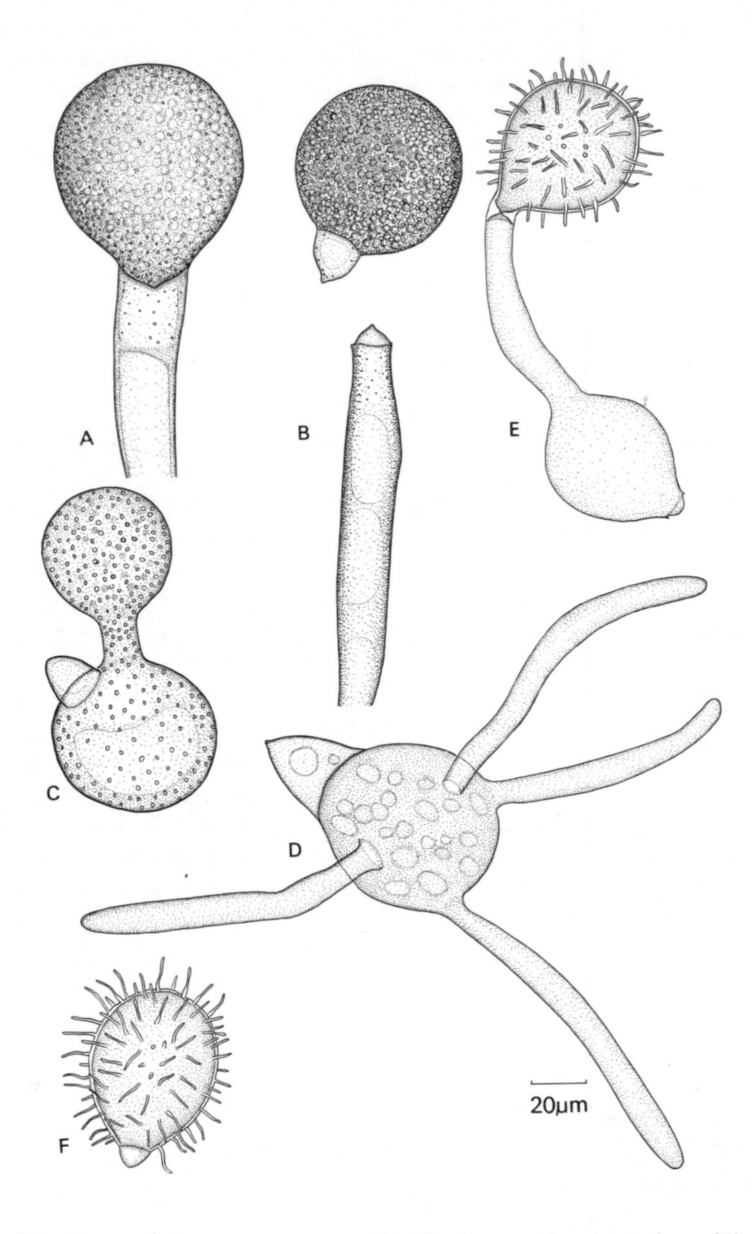

Figure 80. *Entomophthora coronata*. A. Conidiophore with attached conidium. B. Apex of conidiophore and conidium after discharge. C. Conidium germinating to produce a secondary conidium. D. Conidium germinating to form several germ tubes. E. Conidium germinating to form a spiny spore. F. Detached spiny spore.

Figure 81. A. *Entomophthora muscae*. Dead house-fly attached to a window-pane, surrounded by discharged conidia. B. *Entomophthora americana*. Dead blow-fly attached to a leaf. Note the three bands of conidiophores penetrating between the abdominal segments, and the conidia on the leaf surface.

pathology of the two insect pathogens *Entomophthora* and *Massospora* has been given by MacLeod (1963).

Entomophthora (Figs. 80–84)

The name *Entomophthora* means insect-destroyer. About 100 species are known, mostly as insect parasites. Whilst some appear to be rather specific in their host range, others are capable of attacking a wide range of hosts. A number of species have been grown in culture. One of the easiest to maintain is *E. coronata*, a fungus which has been referred to under various names, e.g. *Delacroixia coronata*, *Conidiobolus villosus*, *C. coronata* (see Srinivasan & Thirumalachar, 1964). This fungus is a parasite of aphids and termites, but is also a common saprophyte of plant debris. Infection of termites can occur by penetration of germ tubes through the exoskeleton, or via the oesophagus following ingestion of germinated conidia (Yendol & Pashke, 1965). It has also been isolated from nasal granulomatous lesions in horses (Emmons & Bridges, 1961). In culture it grows rapidly forming a septate mycelium which within two to three days forms numerous phototropic conidiophores (Fig. 80), which shoot off conidia on to the lid of the Petri dish. The terminal conidium is separated from the conidiophore by a hemispherical columella which bulges into the spore. The columella is double-walled and the spore is violently projected up to 4 cm from the tip of the conidiophore by the pushing outwards, due to turgor, of the wall of the conidium which was at first folded inwards over the columella. After discharge this part of the spore wall can be seen projecting outwards as a conical papilla. In addition to the smooth globose conidia, more pointed ciliate conidia are also formed and appear to be projected in a similar way. They may occasionally arise as secondary conidia on germination of the smooth conidia. The precise conditions under which ciliate conidia are formed are not known, nor is it known whether they are physiologically different from the smooth conidia, although Martin (1925) has called them resting spores. Structures resembling zygospores have also been described (Kevorkian, 1937; Emmons & Bridges, 1961). *Entomophthora coronata* can grow well in a purely synthetic medium containing mineral salts, arginine hydrochloride and glucose, but the introduction of peptone results in more rapid growth. It appears to be autotrophic for vitamins.

Another well-known species of *Entomophthora* is *E. muscae*. This fungus is a parasite of the house-flies and other insects, and the disease is apparently more frequent in wet weather. Diseased flies can be found occasionally attached to the glass of a window-pane surrounded by a white halo about 2 cm in diameter made up of discharged conidia (Fig. 81A). The fly shows a distended abdomen with white bands of conidiophores projecting between the segments of the exoskeleton. The conidiophores are unbranched and multinucleate and arise from the coenocytic mycelium which plugs the body of the dead fly. The conidia are also multinucleate (Fig. 82). They are pro-

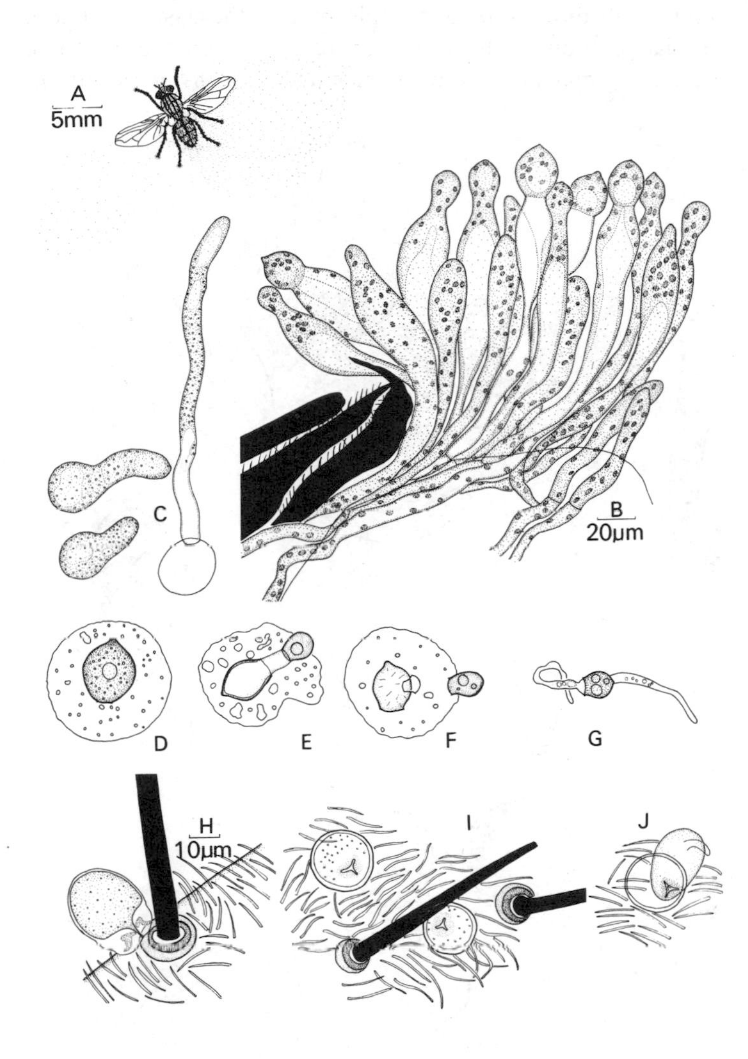

Figure 82. *Entomophthora muscae*. A. House-fly adhering to window-pane, surrounded by halo of conidia. B. L.S. house-fly showing palisade of unbranched conidiophores projecting between segments of exoskeleton. The conidiophores are multinucleate. C. Hyphal bodies from body of recently dead fly. The hyphal bodies are extending to form conidiophores. D. Conidium immediately after discharge surrounded by cytoplasm from the conidiophore. E–F. Germination of primary conidia to form secondary conidia within 12 hr of discharge. Note the septum cutting off the secondary conidium in E, and the rounding-off of the septum on discharge of the secondary conidium. G. Germination of a secondary conidium by two germ tubes. H. Attachment of primary conidium to integument of a fly. Note the thickened appressorium and the narrow point of penetration. I. Two primary conidia attached to integument, and penetrating it by a tri-radiate fissure. J. View of penetration from within the integument. Note the bladder-like expansion inside the tri-radiate fissure. B–G to same scale; H–J to same scale.

jected by a forwardly directed jet of cytoplasm from the elastic conidiophores. Recently discharged conidia have a drop of cytoplasm around them. The cytoplasmic coating may act as a protective agent against desiccation. If the

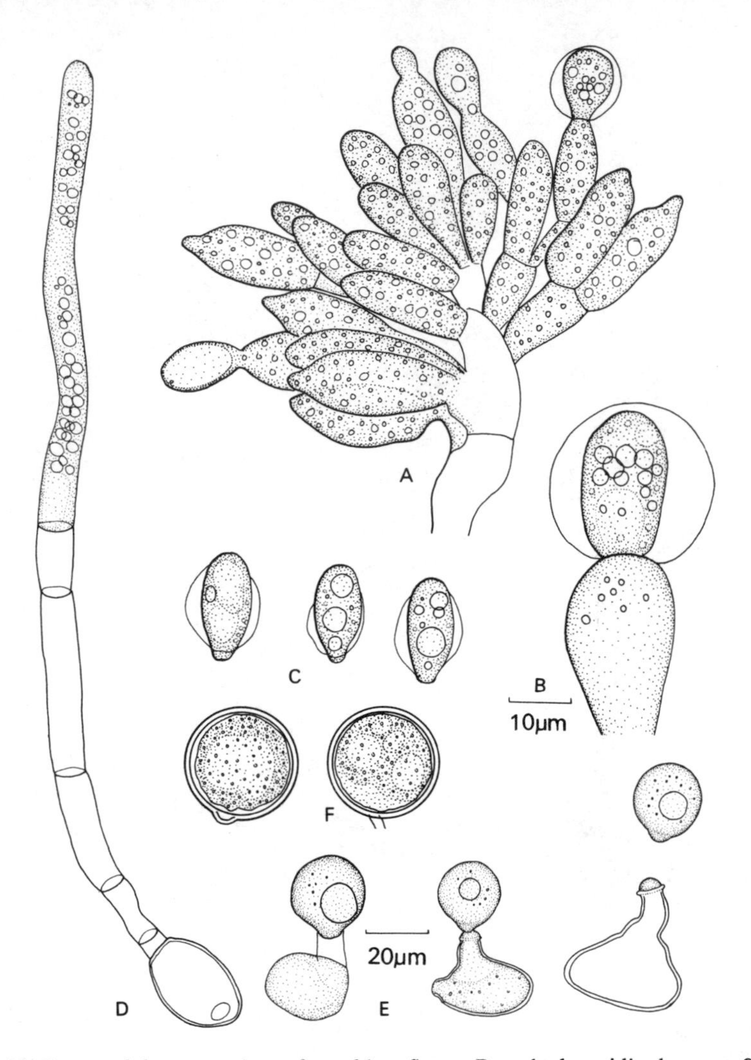

Figure 83. *Entomophthora americana* from blow-fly. A. Branched conidiophoιe. B. Single conidiophore and conidium. C. Conidia after discharge. D. Conidium germinating by germ tube. E. Conidia germinating to produce secondary sporangia. F. Spherical resting bodies from dead fly.

conidium impinges on the body of a fly it secretes an appressorium, or adhesive pad, which attacks the conidium firmly to the cuticle. Penetration of the cuticle is probably brought about by mechanical means. A few hours after infection tri-radiate fissures can be seen in the cuticle beneath attached conidia. If the cuticle in such a region is examined from the inside a thin-walled bladder-like expansion can be seen above the tri-radiate fissure. From this cell mycelial branches develop. The hyphae grow towards the fatty tissues, and as these are consumed the hyphae break up into rounded cells termed hyphal bodies, which are carried by the circulation to all parts of the body (Schweizer, 1947) (Fig. 82c). About a week after infection the flies die, often crawling to the top of a grass stem and clasping it or adhering to walls or window-panes by the proboscis. The hyphal bodies then grow out into coenocytic hyphae which penetrate between the abdominal segments and develop into conidiophores. The primary conidia remain viable for only three to five days. If they fail to penetrate a fly the primary conidia may

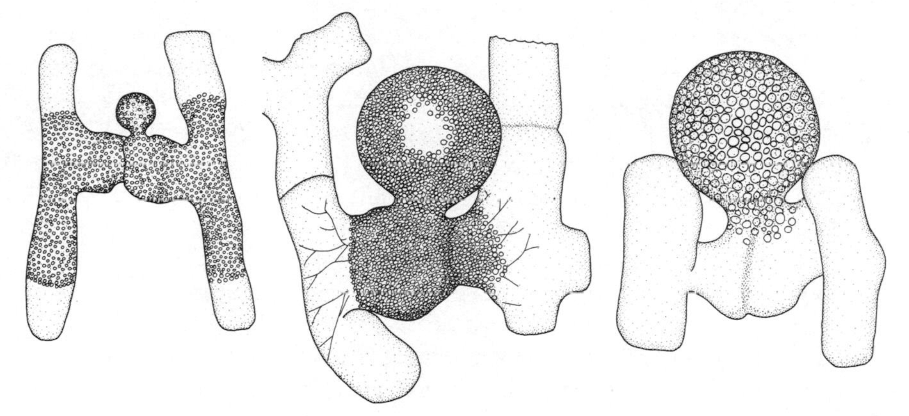

Figure 84. *Entomophthora sepulchralis.* Three stages in zygospore formation. Two hyphal bodies conjugate and the zygospore arises as a bud from the fusion cell (after Thaxter).

produce secondary conidia within 12 hours. The secondary conidia are formed at the tips of short conidiophores and are discharged by a different mechanism, by the rounding-off of a two-ply columella as in *E. coronata.* The secondary conidia may germinate by a germ tube or may produce tertiary conidia.

Within the body of the dead fly multinucleate spherical resting bodies are formed asexually, and it is presumably from such cells that infection begins each year (Goldstein, 1923). Their germination is said to be stimulated by the action of chitin-decomposing bacteria. *Entomophthora muscae* has been grown in pure culture on extracts of animal tissue but only if the media are sterilised cold (i.e. chemically). Growth is markedly stimulated by the presence of animal fat and by glucosamine, a breakdown product of chitin (Schweizer, 1947). Successful cultures have also been established on a medium containing wheat grain extract, peptone, yeast extract and glycerine (Srinivasan *et al.*, 1964).

In many other species of *Entomophthora* the conidiophores are branched, and this is well shown by *E. americana* (Figs. 81, 83), a fungus commonly found on blow-flies in the autumn, especially around corpses of dead animals or stinkhorns, where blow-flies congregate. In wet weather severe epidemics of blow-fly populations may occur, severely reducing their numbers. The dead flies are often attached to the adjacent plants by filamentous rhizoid-like hyphae. The conidiophores form yellowish pustules between the abdominal segments and the branched tips bear conidia. The wall of the conidium is double and the two layers are frequently separated from each other by liquid (Fig. 83A–C). These conidia are projected for several centimetres from the host and on germination may form germ tubes, or may produce secondary conidia which are projected by the rounding-off of a two-ply columella. Within the dry body of the dead fly numerous smooth hyaline thick-walled resting spores are formed.

In some species of *Entomophthora*, e.g. *E. sepulchralis*, resting spores (zygospores) are believed to be formed following conjugation between hyphal bodies (see Fig. 84). Riddle (1906) has claimed that the resting spores of *E. americana* also form zygospores by fusion of hyphal bodies but because doubt has been cast on these observations (Krenner, 1961) it would be valuable to have them confirmed.

The resting spores of many parasitic Entomophthoras do not germinate readily in the laboratory, but it seems likely that they remain viable for one or possibly two years. In *E. virulenta* 2–5% of the resting spores (azygospores) are ready to germinate immediately on formation. Germination is stimulated by soaking the resting spores in water, and no other special treatment, such as exposure to chitin-splitting bacteria appears to be necessary (Hall & Halfhill, 1959).

Interest in the parasitic Entomophthoras has recently been aroused by the possibility of using them in the biological control of insect pathogens. A number of them have been grown in artificial culture (for references see Müller-Kögler, 1959; Gustafsson, 1965). Many have complex requirements and will not grow easily in synthetic media, but can be grown on animal substrata such as meat, fish, coagulated egg-yolk, and milk. Unfortunately, however, attempts to use such cultures to induce epidemics in natural populations have not been successful.

3

Ascomycotina (Ascomycetes)

This is the largest class of fungi, containing some 15,000 species, although it is likely that many remain to be described. They occur in a wide variety of habitats: in soil, dung, in marine and in fresh-water; as saprophytes of plant and animal remains; as animal and plant pathogens. Their characteristic feature is that the sexually produced spores (sometimes called the 'perfect' spores) are borne in a sac or ascus, typically containing eight spores (ascospores) which are explosively ejected. The vegetative structure consists either

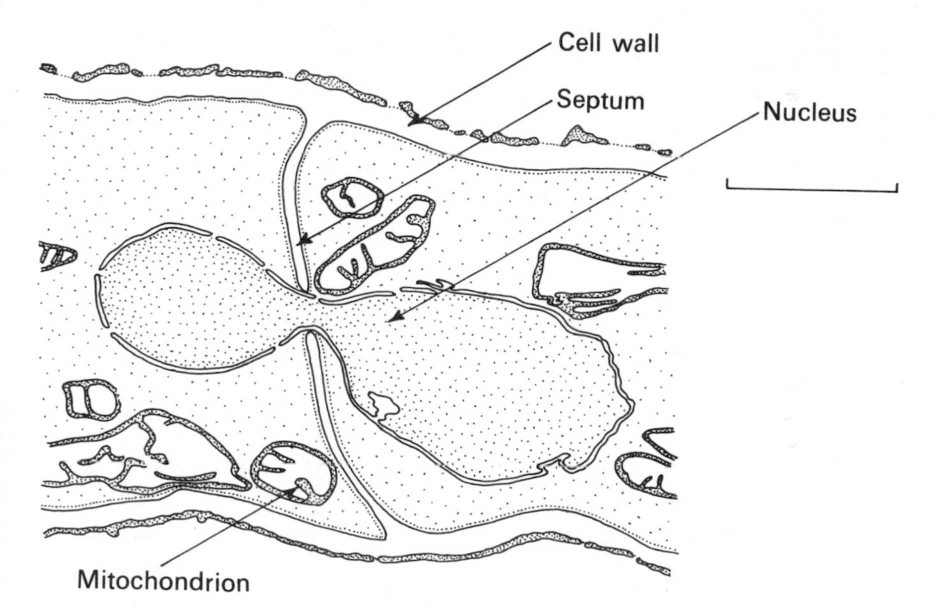

Figure 85. *Neurospora crassa.* L.S. hypha showing the perforated transverse septum through which a nucleus is passing (based on an electron micrograph by Shatkin & Tatum, 1959).

of single cells (as in yeasts) or septate filaments, each segment often containing several nuclei. The cell walls of filamentous Ascomycetes contain a microfibrillar skeleton of chitin, and in addition various other compounds have been detected such as other amino-sugars, protein, mannose and glucose. It is important to realise that the cell wall is not a functionally inert coating, but may contain surface enzymes (Sussman, 1957; Aronson, 1965). Specialised vesicular organelles, possibly derived from the endoplasmic reticulum, or from the plasma membrane, are probably concerned with the synthesis of wall material. These organelles have been termed *lomasomes* (= border body) and are also found in Oomycetes and Basidiomycetes (Moore & McAlear,

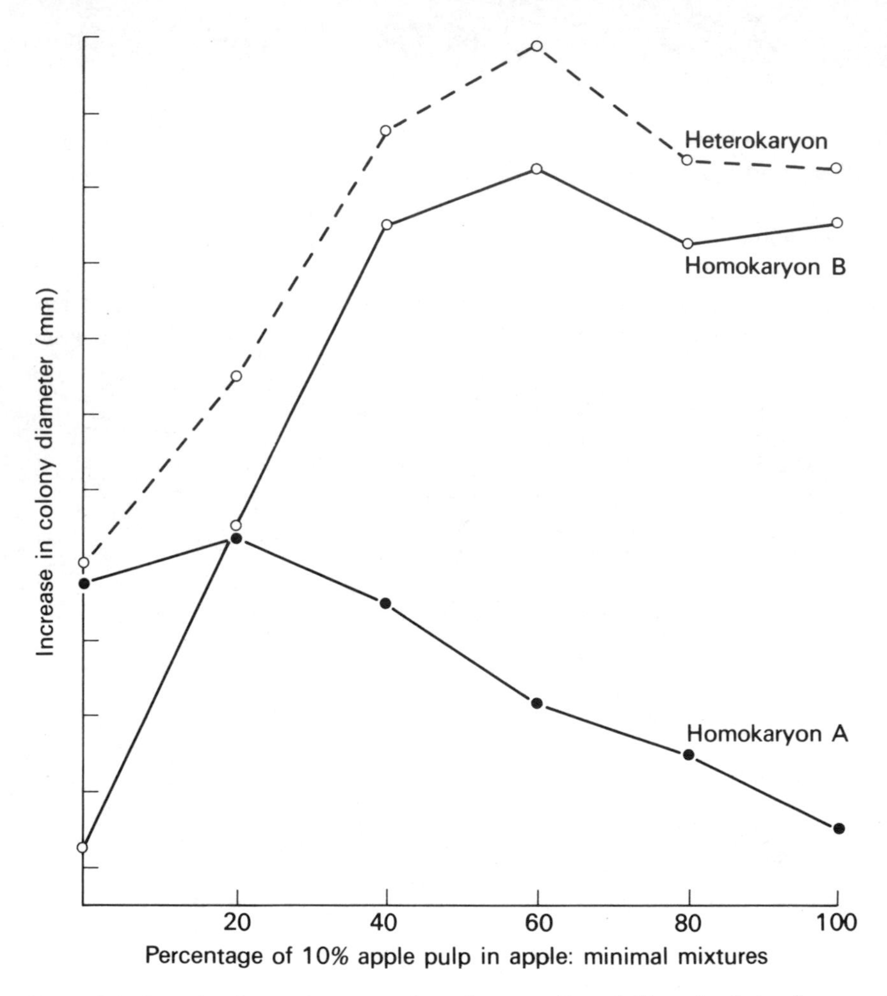

Figure 86. Growth of two component homokaryons (A and B) and a heterokaryon of *Penicillium* containing A and B nuclei on a medium of varying concentration of apple pulp. Note that the growth of the heterokaryon is superior to that of either homokaryon (after Jinks).

1961; Wilsenach & Kessel, 1965*a*; Moore, 1965). Electron micrographs show that the septum is perforated by a pore so that cytoplasmic continuity between adjacent segments occurs. The pore is also wide enough to allow mitochondria and nuclei to pass through (Fig. 85, Shatkin & Tatum, 1959; Moore, 1965). Where several nuclei occur in a single cell or mycelium they are not always genetically identical. Differences between nuclei may arise by mutation or by anastomosis of hyphae of different genotype followed by nuclear migration. A cell (or mycelium) which contains nuclei of more than one genotype is said to be a heterokaryon (or heterokaryotic) and the phenomenon of heterokaryosis has been demonstrated in numerous Ascomycetes and in Basidiomycetes and Fungi Imperfecti (Davis, 1966). Its significance is potentially profound:

1. ADAPTABILITY

The ratio of the different kinds of nuclei in a heterokaryon may vary during the growth of a colony, or in relation to the changing nutrient content of the substratum. Jinks (1952) showed that *Penicillium cyclopium* (one of the Fungi Imperfecti *not* an Ascomycete) can occur in nature as a heterokaryon. In pure culture the heterokaryon gave rise to two distinguishable homokaryotic components by virtue of the fact that individual conidia are uninucleate. The growth of the component homokaryons was inferior to that of the heterokaryon on semi-natural media over a range of concentrations (Fig. 86). Moreover, using the proportion of the different kinds of conidia as a means of estimation, it was shown that the proportion of nuclei in the heterokaryon varied with the concentration of apple pulp in the medium used (as shown in the table below).

TABLE

Nuclear ratios in a heterokaryon of *Penicillium cyclopium* in mixtures of apple pulp and minimal medium (After Jinks, 1952)

Medium		
10% apple	Minimal	% of 'A' nuclei in heterokaryon
40	60	12·82
35	65	15·15
30	70	13·70
25	75	15·87
20	80	15·38
15	85	17·86
10	90	20·41
5	95	29·41
0	100	57·47

2. COMPLEMENTATION

The observation that heterokaryons may make better growth than their component homokaryons has also been made in other fungi. Using biochemical mutants of *Neurospora crassa* (i.e. mutants produced by irradiation which differ from the wild type in their inability to synthesise from the ingredients of a minimal medium a particular substance, and which therefore need to be supplied with an exogenous supply of this substance) Beadle & Coonradt (1944) showed that two different mutants, individually incapable of growth on minimal medium were able to grow if inoculated simultaneously on to this medium. One of the mutants lacked the ability to synthesise p-amino-benzoic acid, and the other nicotinic acid. Since the genes controlling the synthesis of these two substances are non-allelic the p-amino-benzoic acid-less mutant would have an intact gene for nicotinic acid synthesis, whilst the nicotinic-acid-less mutant would have unimpaired ability to synthesise p-amino-benzoic acid. Once anastomosis of the mutant hyphae had occurred, the two kinds of nuclei would exist in a common cytoplasm which would then have the necessary biochemical facilities for synthesis of both the deficiencies in the original mutants. Each mutant therefore *complements* the deficiency of the other. In other experiments with the same fungus it has been found that growth of heterokaryons synthesised from complementary homokaryons was maximal over a wide range of nuclear ratios of the component homokaryons (Pittenger *et al.*, 1955; Pittenger & Atwood, 1956; Klein, 1960). Mutants of this type are often termed *auxotrophs* (auxotrophic) to distinguish them from wild-type strains which are *prototrophic*. Heterokaryons between auxotrophic mutants are termed forced heterokaryons (Caten & Jinks, 1966).

3. GENETICAL VARIABILITY

The occurrence within a single mycelium of nuclei of differing genotype represents a store of genetical material analogous to the heterozygous condition of diploid organisms.

4. BREAKDOWN OF HETEROKARYONS

A heterokaryotic mycelium may give rise to uninucleate or homokaryotic conidia, or to homokaryotic hyphal tips. Examples of multinucleate but homokaryotic conidia, are found in certain species of *Aspergillus*. In *A. tamarii* (non-ascocarpic) there are typically three or four nuclei per conidium. Wild-type strains have green conidia, but a mutant is known with white conidia. Heterokaryons between the wild-type and mutant strains give rise to conidial heads bearing chains of green conidia and chains of white conidia, but with no intermediate colours. Single spore cultures derived from white or green chains breed true. It is thus inferred that although the conidiophore is heterokaryotic, each phialide (mother cell at the base of the spore chain) receives only a single nucleus, either mutant or wild-type (Fig. 87). A different situation is found in *A. carbonarius*, which has two to five nuclei per spore. In this fungus heterokaryotic conidiophores, formed between the black-spored wild-type strain and a pale brown mutant, bear spores intermediate in colour between these two strains, and when single spores from the heterokaryotic heads are cultured they often give rise to mixed colonies of the wild-type and mutant strains. From this it can be inferred that the spores themselves are heterokaryotic and arise from phialides which contained both kinds of nuclei. The inference that the phialides are multinucleate can be confirmed cytologically (Yuill, 1950).

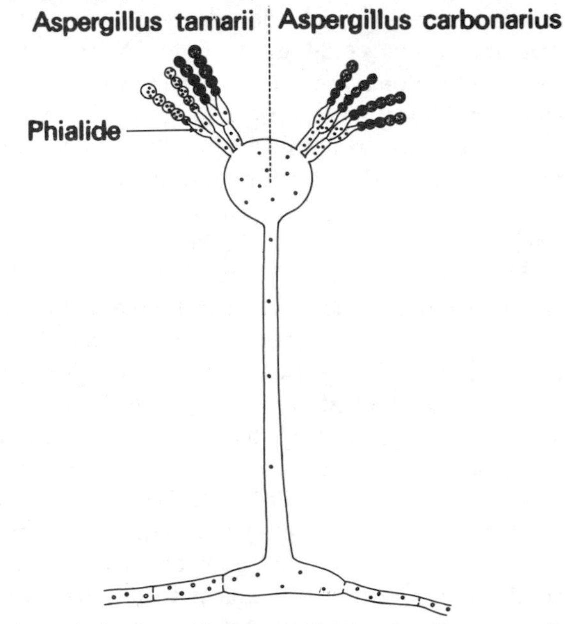

Figure 87. Heterokaryosis in *Aspergillus* spp. Left: Situation in *A. tamarii* where a heterokaryotic conidiophore formed between a dark-spored wild type and a pale-spored mutant strain gives rise to separate kinds of spore chain, because each phialide receives a single nucleus from which the nuclei of the spore chain are all derived. The spores are thus homokaryotic. Right: Situation in *A. carbonarius* where a similar heterokaryotic conidiophore gives rise to conidia which are heterokaryotic because the phialides contain several nuclei.

5. RECOMBINATION

An important consequence of heterokaryosis is that recombination can occur even in homothallic fungi and in Fungi Imperfecti. In homothallic fungi such as *Aspergillus nidulans* or *Sordaria fimicola* heterokaryon formation may result in the presence of genetically distinct nuclei in the ascus initial. Nuclear fusion and meiosis associated with sexual reproduction of a conventional kind may thus give rise to recombinants between the two genotypes, so that a degree of outbreeding is possible even in homothallic organisms (Olive, 1963). In Fungi Imperfecti, that is in fungi which are not known to reproduce sexually, recombination has been discovered in the progeny derived from heterokaryons. The recombination process is of a novel kind, not involving meiotic segregation, and has been termed parasexual recombination (Pontecorvo, 1956; Roper, 1966).

In *A. nidulans*, in which the conidia are uninucleate, the phenomenon was detected in heterokaryons between two strains marked by having spore colours different from the wild type, i.e. white or yellow as distinct from the green wild-type conidia, and in addition biochemical markers such as a requirement for adenine or lysine, which are not required by wild-type strains. The genes for white and yellow spores are non-allelic and are recessive. On minimal medium on which the wild-type strain can grow, neither of the two resultant strains could grow singly. When inoculated simultaneously, however, a heterokaryon was found and was able to grow because each mutant strain could complement the nutritional deficiency of the other, i.e. a strain requiring adenine would not require lysine and vice-versa. When the heterokaryon formed conidia, in addition to the yellow and white conidia of the two parental strains, green conidia were also occasionally formed. The green conidia were diploid and were slightly larger than haploid wild-type conidia. The diploid condition is believed to have arisen by rare fusion between two different mutant nuclei. If such diploid conidia are inoculated into cultures, the resulting colony may give rise to conidia of four types: the two original mutant strains, further diploids and, rarely, conidia which when plated out show recombination of the properties of the two parental mutants. For example when one parent had white conidia and a requirement for lysine, and the other had yellow conidia and a requirement of adenine, recombinant strains isolated from diploid colonies had yellow conidia and a requirement for lysine, or white conidia and a requirement for adenine. It is believed that recombination is not the result of meiosis, but occurs during mitosis in diploid nuclei and that subsequent to mitotic recombination the diploid nuclei become haploid by progressive occasional loss of a chromosome. Pontecorvo (1956) believes that the events in parasexual recombination form part of a cycle which he has termed the parasexual cycle. The steps in the cycle are:

(a) Fusion of two unlike nuclei in a heterokaryon.

(b) Multiplication of the resulting diploid heterozygous nucleus side by side with the parent haploid nuclei in a heterokaryotic condition.

(c) Eventual sorting-out of a homokaryotic diploid mycelium which may become established as a strain.

(d) Mitotic crossing-over occurring during multiplication of the diploid nuclei.

(e) Vegetative haploidisation of the diploid nuclei.

Parasexual recombination has been demonstrated in several Ascomycetes, Basidiomycetes and Fungi Imperfecti. The implications of parasexual recombination are great:

1. It is possible to study the genetics of asexual organisms. In the case of *A. nidulans* it has proved possible to compare genetical maps based on conventional sexual reproduction and on parasexual recombination, and there is a close degree of correspondence.

2. It is possible to 'breed' asexual organisms such as members of the Fungi Imperfecti used in fermentations and so to combine, in one strain, desirable properties from different strains.

3. The occurrence of parasexual recombination explains how variation can occur in Fungi Imperfecti, e.g. how new strains of pathogenic fungi can appear (Parmeter *et al.*, 1963).

In assessing the likely significance of heterokaryosis and parasexual recombination in nature it should be borne in mind, however, that many laboratory studies have been based on forced heterokaryons (complementary auxotrophs) which can only grow if they can form heterokaryons. Secondly, numerous studies have shown that heterokaryons are only formed readily between strains which have a closely similar genetical background (e.g. mutants derived from a common wild-type strain). This restricts the possibility of gene flow between strains which differ appreciably in genotype (Caten & Jinks, 1966).

NUCLEAR MIGRATION IN ASCOMYCETES

As shown in Fig. 85 there is evidence that nuclei can migrate through septal pores. Indirect evidence of nuclear migration has also been obtained in

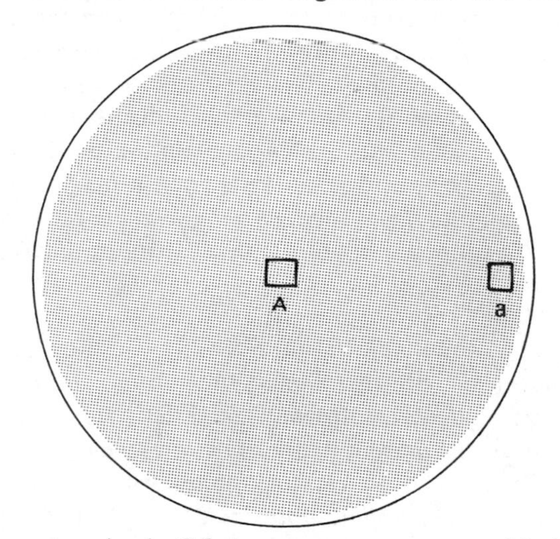

Figure 88. Nuclear streaming in *Gelasinospora tetrasperma*. A. original inoculum of 'A' mating type. a. point at which a block of mycelium of the compatible 'a' mating type is later added after the 'A' mycelium has colonised the whole plate. The migration of 'a' nuclei can be followed by removing plugs at intervals of time at various distances from the point of inoculation.

experiments with mycelia of opposite mating type in heterothallic ascomycetes. *Gelasinospora tetrasperma* forms flask-shaped ascocarps (perithecia) containing asci which have ascospores of two sizes. Cultures started from the larger ascospores are self-fertile, but those derived from small ascospores are of two distinct mating types, 'A' and 'a', and perithecia are formed only when mycelia of the two kinds are mated together. If an 'A' mycelium is allowed to fill a Petri-dish culture a small block of 'a' mycelium can be added and within a few days perithecia appear at some distance from the point of inoculation of the 'a' block. If small plugs of agar are removed from the dish at intervals after addition of the 'a' block, and transferred to fresh agar plates, those plugs which have received 'a' nuclei will develop perithecia (Fig. 88). In this way estimates of the rate of nuclear migration of the order of 10·5 mm/hr have been made, about two to three times the rate of growth of the mycelium (Dowding & Bakerspigel, 1954). This rate is probably an underestimate and direct observation and photography give rates of 40 mm/hr (Dowding, 1958). The actual mechanism of nuclear migration probably depends on cytoplasmic streaming.

SOMATIC NUCLEAR DIVISION

At present there is no general agreement about the behaviour of the nuclei of Ascomycetes during mitotic division. Whilst some authors claim that a mitosis of the conventional type occurs (e.g. Somers *et al.*, 1960; Ward & Ciurysek, 1962) others claim that nuclear division does not follow such a pattern, and have described constriction of nuclei, and the arrangement of chromosomes upon a filament to form a threadlike nucleus (Weijer *et al.*, 1965; Weijer & Weisberg, 1966; Dowding, 1966; see also Olive, 1953; Robinow & Bakerspigel, 1965). Most workers agree that there is no spindle apparatus visible, and that the nuclear membrane remains intact during somatic cell division. The constriction of the nucleus may be accompanied by invagination of the nuclear membrane, and the term *karyochorisis* (= nuclear sundrance) has been proposed for this process (Moore, 1965).

MATING BEHAVIOUR

Ascomycetes may be homothallic or heterothallic. The basis for heterothallism is typically a single gene with two alleles, 'A' and 'a', and because of segregation during the meiotic division which precedes ascospore formation, the eight ascospores normally present in an ascus will include four of one mating type and four of the other. In certain species with four-spored asci (e.g. *Neurospora tetrasperma*, *Podospora anserina*), the ascospores are binucleate, and commonly contain nuclei of both mating types. Such ascospores on germination would give rise to fully fertile mycelia, and it would appear that the fungus is homothallic. However, occasionally uninucleate ascospores are formed, and on germination the resulting mycelium is not fertile: ascocarps are only formed in 50% of matings between such mycelia. Since basically these fungi are heterothallic in their mating behaviour, the term *secondary homothallism* is used to describe the behaviour of their binucleate ascospores (Whitehouse, 1949*a*,*b*).

Sex organs are formed in some Ascomycetes. The female organ or *asco-*

gonium is commonly a coiled, multinucleate cell, sometimes surmounted by a receptive trichogyne. The male organs may take the form of slender branches, antheridia, or minute unicellular *spermatia* incapable of germination, or *microconidia* (oidia) which, whilst capable of fulfilling a sexual role are nevertheless capable of germination. In some heterothallic species, e.g. *Neurospora crassa*, ascogonia and microconidia are formed on a single strain, but they are self-incompatible—i.e. the microconidia are unable to fertilise the ascogonia on the same mycelium. Only when a fertilising element from the opposite mating type is brought into contact with the ascogonium does fertilisation occur. This paradoxical phenomenon is found in numerous Ascomycetes. Each mating type possesses both kinds of sex organs, and is morphologically indistinguishable from the opposite mating type. Whitehouse (1949) has used the term *physiological heterothallism* to describe this type of behaviour, and since there are two alleles involved it has also been termed *two-allelomorph physiological heterothallism*.

In many Ascomycetes there are no recognisable sex organs; fusion takes place between ordinary hyphae. In heterothallic forms this can only happen where heterokaryosis is possible between hyphae of opposite mating type.

DEVELOPMENT OF ASCI

In yeasts and related fungi the ascus may arise directly from a single cell,

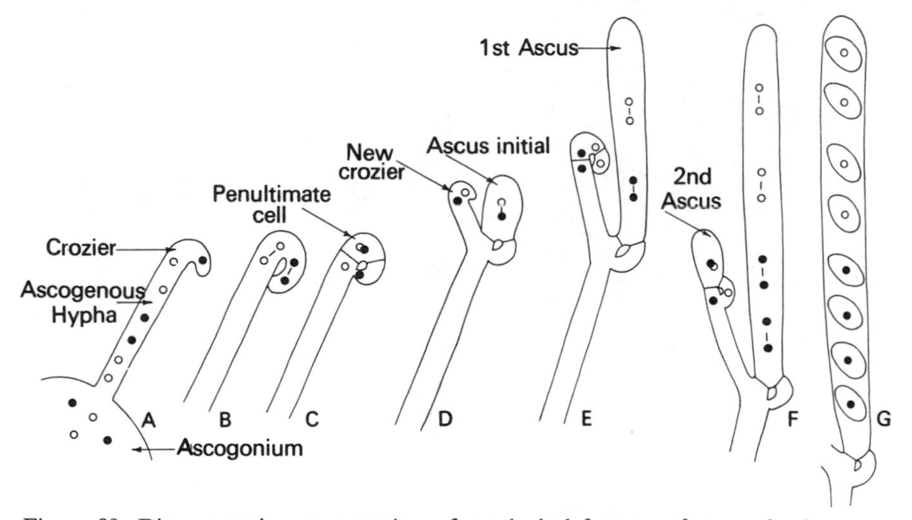

Figure 89. Diagrammatic representation of cytological features of ascus development. A. Ascogenous hypha with a crozier at its tip developing from an ascogonium. B. Conjugate division of the two nuclei in the crozier. C. Two septa have cut off a binucleate penultimate cell. These two nuclei fuse to form a diploid fusion nucleus. The uninucleate terminal segment has fused with the ascogenous hypha. D. Penultimate cell enlarges to become the ascus within which the fusion nucleus begins to divide meiotically. A new crozier is developing from beneath the ascus. E. Second division of meiosis has occurred in the young ascus. The behaviour of the second crozier repeats that of the first. F. Mitotic division of the four haploid nuclei in the ascus. G. Ascospores formed.

but in most other Ascomycetes the ascus develops from a specialised hypha, the ascogenous hypha, which in turn develops from an ascogonium. The ascogenous hypha is multinucleate, and its tip is recurved to form a *crozier* (shepherd's crook). Within the ascogenous hypha nuclear division occurs simultaneously. Two septa at the tip of the crozier cut off a penultimate binucleate cell (Fig. 89C) destined to become an ascus. The terminal cell of the crozier curves round and fuses with the ascogenous hypha behind the

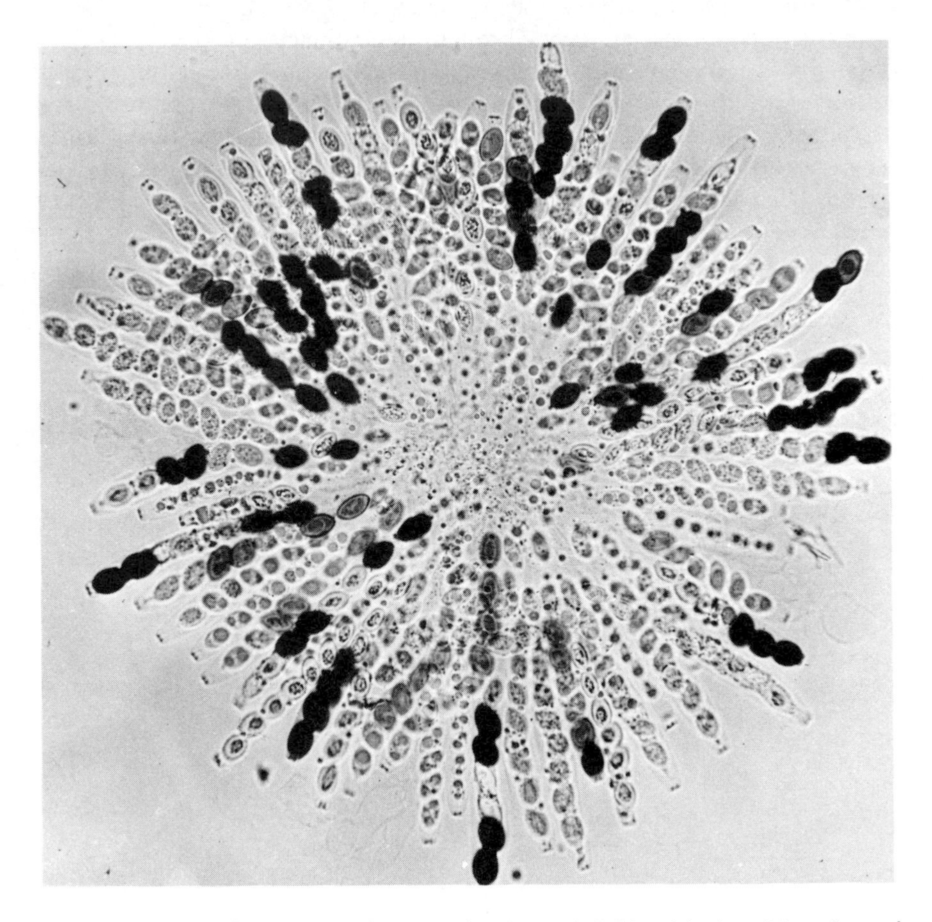

Figure 90. *Sordaria fimicola*. Squash preparation from a hybrid perithecium. Most ripe asci contain four black and four white ascospores.

penultimate cell, and this region of the ascogenous hypha may grow on to form a new crozier in which the same sequence of events is repeated. Repeated proliferation of the tip of the crozier can result in a cluster of asci. In the ascus initial the two nuclei fuse and the fusion nucleus undergoes meiosis to form four haploid daughter nuclei (Fig. 89D,E) (Olive, 1965). These nuclei then undergo a mitotic division so that eight haploid nuclei result. During these nuclear divisions the ascus is elongating, and the plane of the division is parallel to the length of the ascus. Cytoplasm is cleaved out around each nucleus to form an ascospore. In some forms the eight nuclei divide further so that each ascospore is binucleate. Where the ascospores are multicellular there are repeated nuclear divisions. In some forms more than eight ascospores are formed, or the eight ascospores may break up into part-spores. Studies of the fine structure of asci during cleavage of the ascospores have shown that a system of double membranes continuous with the endoplasmic reticulum extends from the envelope of the fusion nucleus in developing asci. The double membrane forms a cylindrical envelope lining the young ascus. The ascospores are cut out from the cytoplasm within the ascus by invagination of the double membrane. Between the two layers forming the membrane the spore wall is secreted, and the inner membrane forms the plasma membrane of the ascospore (Carroll, 1967; Delay, 1966; Reeves, 1967).

It is important to note that because the division which follows the four-nucleate stage is mitotic and because the division plane is usually parallel to the length of the ascus, adjacent pairs of spores starting from the tip of an ascus are normally sister spores which are thus genetically identical. Rare exceptions to this situation are occasionally found where the division planes are oblique, or for other reasons. A neat illustration of the normal arrangement of sister spores is provided by crossing experiments between black-spored Ascomycetes and mutant derivatives with colourless spores. Hybrid asci with four black and four white ascospores are formed and, as shown in Fig. 90, the majority of asci show either alternating pairs of black and pairs of white, or groups of four black then four white spores. Alternating pairs result from second-division segregation of the gene for spore colour, whilst the four–four patterns result from first-division segregation.

The actual form of the mature ascus is very variable. In forms with non-explosive ascospore release the ascus is often a globose sac, but in the majority of Ascomycetes the ascus is cylindrical, and the spores are expelled from the ascus explosively. It is thought that the explosive release follows increased turgor, caused by water uptake. In the young ascus, after the spores have been cut out, epiplasm (residual cytoplasm rich in polysaccharides such as glycogen) remains lining the ascus wall surrounding a large central vacuole containing ascus sap, within which the ascospores are suspended. Conversion of the polysaccharide to sugars of smaller molecular weight is believed to bring about an increased osmotic concentration of the ascus sap, which is followed by increased water uptake by the ascus, and the resulting increase in turgor causes the ascus to stretch. Ingold (1939) has estimated that the osmotic pressure of the ascus sap in *Ascobolus furfuraceus* (= *A. stercorarius*) is about 10–13 atmospheres and for *Sordaria fimicola* 10–30 atmospheres (Ingold, 1966). In the latter fungus glucose is the most abundant sugar present in ascus sap, but the amount present represents only a negligible fraction of the total solutes; the balance being made up by salts. In many cases the asci are surrounded by packing tissue in the form of paraphyses, pseudoparaphyses and other asci, so that they cannot expand laterally but are forced to elongate. In the cup-fungi or Discomycetes the elongation of the asci raises their tips above the general level of the hymenium. The ascus tips are often phototropic and when the increased pressure causes the ascus tip to burst the spores are shot out in a drop of liquid, the ascus sap. In this group large numbers of asci may be discharged simultaneously, so that a cloud of ascospores is visible. This phenomenon is known as puffing. In some Discomycetes (e.g. the Pezizales) the ascus tip is surmounted by a cap or operculum which is blown aside or actually blown off the tip of the ascus by the force of the explosion. However in other Discomycetes (e.g. the Helotiales) the ascus tip is perforated by a pore, and there is no operculum. These two types of asci are respectively termed operculate and inoperculate, and the

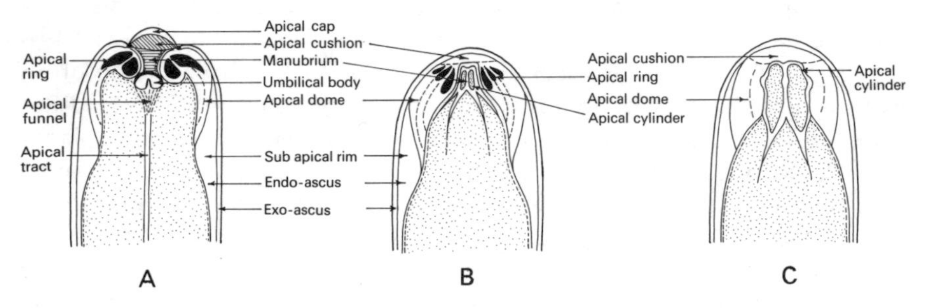

Figure 91. Diagrammatic interpretation of apical apparatus of asci (after Chadefaud, 1960). A. Annellascous type of ascus as found in certain Pyrenomycetes, e.g. Sphaeriales. B. Archetypal ascus as found in some lichens, e.g. Lecanorales. C. Nassascous type of ascus as found in Loculoascomycetes, e.g. Pleosporales.

presence or absence of an operculum is an important feature of classification.

In the flask-fungi (often loosely termed the Pyrenomycetes) the asci are enclosed in a cavity which opens to the exterior through a narrow pore, the ostiole. As an ascus ripens it elongates and takes up a position inside the ostiole, often gripped in position by a lining layer of hairs, periphyses. In this case the asci discharge their spores singly and puffing does not happen.

The asci of Pyrenomycetes are never operculate. In many groups the ascus tip has a distinctive apical apparatus, the detailed structure of which may be of importance in classification. The apical apparatus is seen to advantage if stained by iodine, by writing ink, or stains such as Congo Red or Janus Green. Chadefaud (1960) has summarised his interpretation of the structural details of the apical apparatus (see Fig. 91). According to Chadefaud the walls of asci are double, composed of an external layer, the exoascus (exotunica), and an inner layer, the endoascus (endotunica). However, electron micrographs of a number of asci have not revealed this double structure. Where the two layers *are* clearly visible as in the Loculoascomycetes the asci are said to be *bitunicate* (Luttrell, 1951). Where the wall appears single asci are described as *unitunicate*. Most Ascomycetes have asci of this type.

The apical apparatus may include a number of distinctive organelles which are not necessarily all present in any one ascus:

1. *An apical cap* formed from the apical part of the exoascus. The apical cap may be replaced by an apical pore with a thickened rim (Fig. 91A).
2. *An apical dome* developed from the endo-ascus. The tip of the apical dome is usually perforated by opening surrounded by a thickening (Fig. 91A,B).
3. *An apical cushion* situated between the cap and the dome. Sometimes the apical cusion is poorly developed or indistinct (Fig. 91A,B,C).
4. *An apical ring* formed by differentiation of the material of the dome. The apical ring is usually refringent and may or may not be stainable by the stains mentioned above. It is often double (Fig. 91B).
5. Within the perforation of the apical dome two further structures may be present:

   (a) An apical cylinder (Chadefaud uses the French word *nasse* = a basket-work eel trap somewhat resembling a lobster-pot with an inner funnel).

   (b) A manubrium suspended from the apical cushion. The manubrium is tubular. In the most simple cases it is a tube open at the bottom and encircled by the apical ring. In more complex structures (Fig. 91A) it includes an umbilical body and an apical funnel continued below into a narrow thread of epiplasm continuous with the epiplasm of the ascospores. This narrow thread is termed the apical tract. The ascospores may thus be attached to the apical apparatus by this cytoplasmic filament.

In respect of the varied apical apparatus Chadefaud has distinguished three main kinds of ascus:

1. *The archetypal ascus* (les Archaéascés) in which there are present simultaneously an apical ring and an apical cylinder (nasse). Asci of this type are found in some lichens, e.g. Lecanorales (Fig. 91B).
2. *The nassascous type of ascus*, which possess an apical cylinder (nasse) only. Neither manubrium nor apical ring are distinguishable. This type of ascus is found in the Loculoascomycetes (Fig. 91C).
3. *The annellascous type of ascus* in which there is a manubrium, an apical ring and no apical cylinder. In this type of ascus the wall is unitunicate.

There are numerous variations on these patterns of apical apparatus. The function of the various organelles is unknown. The dimensions of some of the organelles lie close to the limits of resolving power of the light microscope and it will be of interest to see interpretations of them through the electron microscope.

The behaviour of the bitunicate type of ascus during discharge has been described as the Jack-in-a-box mechanism. The outer wall is relatively rigid and inextensible. As the ascus expands the outer wall ruptures laterally or apically (see Figs. 171, 174) and the inner wall then stretches before the ascus explodes. In some forms, e.g. *Cochliobolus sativus*, the endo-ascus may break down and apparently plays no part in spore release (Shoemaker, 1955).

The explosive release of the ascospores appears to throw all the spores out simultaneously. In fact it is likely that in most cases the spores are spatially separated from each other as they are constricted on passing through the ascus pore. This has been neatly demonstrated by spinning a transparent disc over the surface of a culture of *Sordaria* discharging spores (Ingold & Hadland, 1959*a*). The ascus contents are laid out on the disc in the order in which they are released. Various patterns of spore clumping and separation are visible, and although in many asci the eight spores are well separated from each other there is also a tendency for spores to stick together. From measurements of the length of the ascospore deposit and the velocity of rotation of the disc calculations of the velocity of ascospore discharge have been made. A value for the minimum initial velocity of 366 cm/sec. was obtained by this method. By a separate method a value of 1078 cm/sec. was obtained, about three times the first. The actual time taken for an ascus to discharge was estimated by the rotating disc method to be 0·000048 sec.

When ascospores stick together they are discharged further than single-spored projectiles. In many coprophilous Ascomycetes (e.g. *Ascobolus*, *Saccobolus*, *Podospora*) the spores may be attached together by mucilaginous secretions so that the spores may be projected for distances of 30 cm in *Ascobolus immersus* and *Podospora fimicola*. The distances to which individual ascospores are discharged vary. Single ascospores are commonly discharged for about 1–2 cm. Where puffing of the asci occurs the distance to which the spores are projected may be increased.

In some Ascomycetes the asci do not discharge their spores violently, and in such cases the asci are often globose instead of cylindrical. The Endomycetales and Eurotiales have asci of this type. In *Ceratocystis* the asci break down to produce a mass of sticky spores which ooze out as a drop from the tip of a cylindrical neck which surmounts the ascocarp. Breakdown of asci within the fruit-body is also found in *Chaetomium*. In Ascomycetes with subterranean fruit-bodies, e.g. the Tuberales, the ascospores are again usually not discharged violently, but are dispersed when the fruit-bodies are eaten by rodents.

TYPES OF FRUIT-BODY

In yeasts and related fungi (Endomycetales) the asci are not enclosed by hyphae, but in most Ascomycetes they are surrounded by hyphae to form an ascocarp. The form of the ascocarp is very varied. In *Gymnoascus* there is a loose open network of hyphae. In *Aspergillus* and *Penicillium* and in the

Erysiphales the asci are enclosed in a globose fructification with no special opening to the outside. Such ascocarps are termed *cleistocarps* (closed fruits) or *cleistothecia*. In the cup fungi, e.g. Pezizales and Helotiales, the asci are borne in open saucer-shaped ascocarps, and at maturity the tips of the asci are freely exposed. Such fruit-bodies are termed *apothecia*. The Pyrenomycetes (Sphaeriales and Hypocreales) have *perithecia*, flask-shaped fruit-bodies opening by a pore or ostiole. The perithecial wall is formed from sterile cells derived from hyphae which surrounded the ascogonium during development. The ascus in these groups is unitunicate. The Loculoascomycetes (e.g. the Pleosporales) have ascocarps which bear a superficial resemblance to perithecia, but differ in details of development, and contain bitunicate asci. Such fruit-bodies are properly termed *pseudothecia*, but the term perithecium is sometimes applied loosely to include them.

Ascocarps may arise singly or are often clustered together. In many Pyrenomycetes the perithecia are seated on a mass of tissue termed a *stroma* (e.g. *Nectria cinnabarina*) or may be embedded in it with only the ostioles visible at the surface (see for example the Xylariaceae, and the Clavicipitaceae).

CONIDIA OF ASCOMYCETES

Whilst some Ascomycetes reproduce by means of ascospores only, many have one or more conidial states. The recognition that Ascomycetes are pleomorphic we owe to the brothers Tulasne who in their monumental *Selecta Fungorum Carpologia* (1861–5; translated into English in 1931) described the association of perfect (i.e. ascosporic) and imperfect (conidial) stages of many Ascomycetes, and illustrated their findings by exquisite figures unrivalled in their accuracy and detail. However, evidence of association can be misleading and the pure culture techniques exploited by Brefeld and later mycologists have provided conclusive evidence linking perfect and imperfect states. The type of conidial apparatus is sometimes a guide to relationships (Tubaki, 1958). For example the conidia of many Eurotiales and Hypocreales are phialospores, whilst there is a general similarity between the conidia of Erysiphales. However, such generalisations are dangerous because it is known that morphologically similar conidia may belong to quite distinct groups of fungi (compare the *Monilia* conidia of *Neurospora crassa*, a member of the Sphaeriales (Fig. 127D) with those of *Sclerotinia fructigena*, a member of the Helotiales (Fig. 150D)). It is often found that conidia have different means of dispersal from ascospores. For example the conidia of *Nectria cinnabarina* are dispersed by rain-splash, whilst the ascospores are wind-dispersed. The conditions under which ascospores and conidia are produced may also be quite different. Conidia tend to be produced fairly soon after a new host or substratum has become infected, whilst ascospores may develop rather later. The viability of ascospores and conidia is usually different: conidia are relatively short-lived.

CLASSIFICATION OF ASCOMYCOTINA (ASCOMYCETES)

Different authorities hold widely differing views about the classification of Ascomycetes. The problem has been discussed by Miller, (1949); Luttrell, (1951); von Arx & Müller, (1954); Müller & von Arx, (1962). For the purpose of this account the system proposed by Ainsworth, (1966) will be followed, with slight modification. An outline classification including only those fungal groups described here follows.

ASCOMYCOTINA

1. **Hemiascomycetes** Endomycetales Taphrinales
2. **Plectomycetes** Erysiphales Eurotiales
3. **Pyrenomycetes** Sphaeriales Hypocreales
4. **Discomycetes**
   Inoperculate discomycetes Helotiales Phacidiales Lecanorales*
   Operculate discomycetes Pezizales
   Hypogaeous discomycetes Tuberales
5. **Loculoascomycetes** (bitunicate ascomycetes) Pleosporales.

* This group is not included in Ainsworth's scheme.

1 HEMIASCOMYCETES

The features which distinguishes the Hemiascomycetes from other Ascomycetes is the absence of an ascocarp, i.e. an investment of sterile cells surrounding the asci. The asci are formed singly, usually following karyogamy, and are not borne on ascogenous hyphae. Three orders are included (Martin, 1961): the Protomycetales, Endomycetales and Taphrinales. The Protomycetales are parasitic on higher plants and produce spores in a spore-sac or synascus which is regarded as the equivalent of several asci. This group will not be considered further. The Endomycetales include a number of yeasts and related mycelial forms and are mostly saprophytic. The Taphrinales are parasitic on vascular plants, causing a variety of diseases such as leaf curl, witches' broom diseases, and diseases of fruits. The distinction between these two orders is set out below.

(a) Zygote a single cell transformed directly into an ascus; mycelium sometimes lacking. Mostly saprobic ENDOMYCETALES

(b) Hyphae bearing terminal chlamydospores or ascogenous cells, each of which produces a single ascus, usually forming a continuous hymenium-like layer on often modified tissues of host. Parasitic on vascular plants TAPHRINALES

ENDOMYCETALES

This group of fungi, sometimes also known as the Saccharomycetales, is important economically because it includes ascospore-forming yeasts such as *Saccharomyces* and *Schizosaccharomyces*, used in alcoholic fermentations and in bread-making. These two genera usually do not form true mycelium but exist as single cells which reproduce by budding or by division. There are, however, related forms which possess a true mycelium, e.g. *Eremascus*. In numerous yeast-like organisms ascospores have not been discovered. Some of these forms are possibly 'imperfect' yeasts, i.e. yeasts which have lost the power to reproduce sexually, or may represent haploid strains of heterothallic yeasts. These asporogenous yeasts include such genera as *Cryptococcus*, *Torulopsis*, *Pityrosporum* and *Candida* and some of them are important human and animal pathogens. A taxonomic account of the ascosporogenous and asporogenous yeasts has been given by Lodder & Kreger-van Rij (1952). More general accounts of the biology of yeasts will be found in Ingram (1955), Cook (1958), Reiff *et al.* (1960). The details of classification vary from one author to another. Martin (1961) recognises four families but we shall only refer to two, the Endomycetaceae and Saccharomycetaceae.

Endomycetaceae

This group includes *Schizosaccharomyces*, *Endomyces* and *Eremascus*.

Schizosaccharomyces

The best-known species is *S. octosporus*, which has been isolated from currants and honey. *S. pombe* is the fermenting agent of African millet beer (pombe), of arak in Java, and it has also been isolated from sugar molasses and from grape juice.

S. versatilis isolated from canned grape juice is of interest because as well as growing like a yeast it can also form a true mycelium (Wickerham & Duprat, 1945).

S. octosporus grows readily in liquid or on solid media such as malt extract agar forming ripe asci within three days at 25°C. Individual cells are globose to cylindrical, uninucleate and haploid. Cell division is preceded by nuclear division, and the nuclei are separated by a septum which develops centripetally from the parent wall, eventually cutting the cytoplasm into two. For this reason *Schizosaccharomyces* is sometimes termed a fission yeast. The two sister cells may separate or remain together, and if this process is repeated clumps of cells attached to each other are found (Fig. 92A). Electron micrographs of thin sections of dividing *S. octosporus* cells show that the nucleus constricts and becomes dumbbell-shaped during division (Conti & Naylor, 1959). The cell is comparatively free from vacuoles which form such a characteristic feature of the cells of *Saccharomyces cerevisiae*. Ascus formation is preceded by copulation. The cytology of this process has been reinvestigated by Widra & Delamater (1955) with the light microscope and by Conti & Naylor (1960) using the electron microscope. Two cells come into contact by a portion of the cell wall. Often the fusing cells are sister cells formed by division. A pore is formed in the centre of the attachment area and this widens and elongates to form a conjugation canal (Fig. 92B). During this process the nuclei, one from each cell, migrate towards each other and fuse. Vacuoles may appear in the young ascus following nuclear fusion. The fused nucleus elongates and may reach half the length of the ascus, and then divides by constriction, the nuclear membrane remaining intact during division. The two daughter nuclei migrate to opposite ends of the ascus and then divide further. These divisions constitute meiosis, and a single mitosis follows so

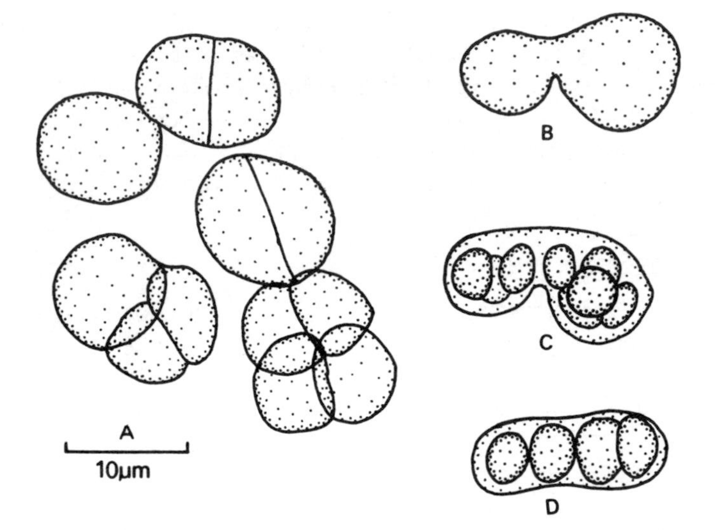

Figure 92. *Schizosaccharomyces octosporus*. A. Vegetative cells showing transverse division. B. Copulation. C. Eight-spored ascus. D. Four-spored ascus.

that eight haploid nuclei result, and eight ascospores are finally differentiated. The ascospores are released by breakdown of the ascus wall. Four-spored asci are also common. The life cycle of *Schizosaccharomyces* is thus interpreted as being based on haploid vegetative cells, which fuse to form asci, the only diploid cells. Meiosis in the ascus restores the haploid condition (Fig. 93A). Some variation in this pattern may occur. For example in *S. versatilis* and *S. pombe* limited division of the zygote may occur in the diploid state before ascospore formation takes place (Suminoe & Dukmo, 1963). All three species are homothallic. Heterothallic strains of *S. pombe* are also known.

Physical and chemical analyses of the cell walls of *Schizosaccharomyces* show an absence of mannan and dextrin in contrast with other yeasts (Kreger,

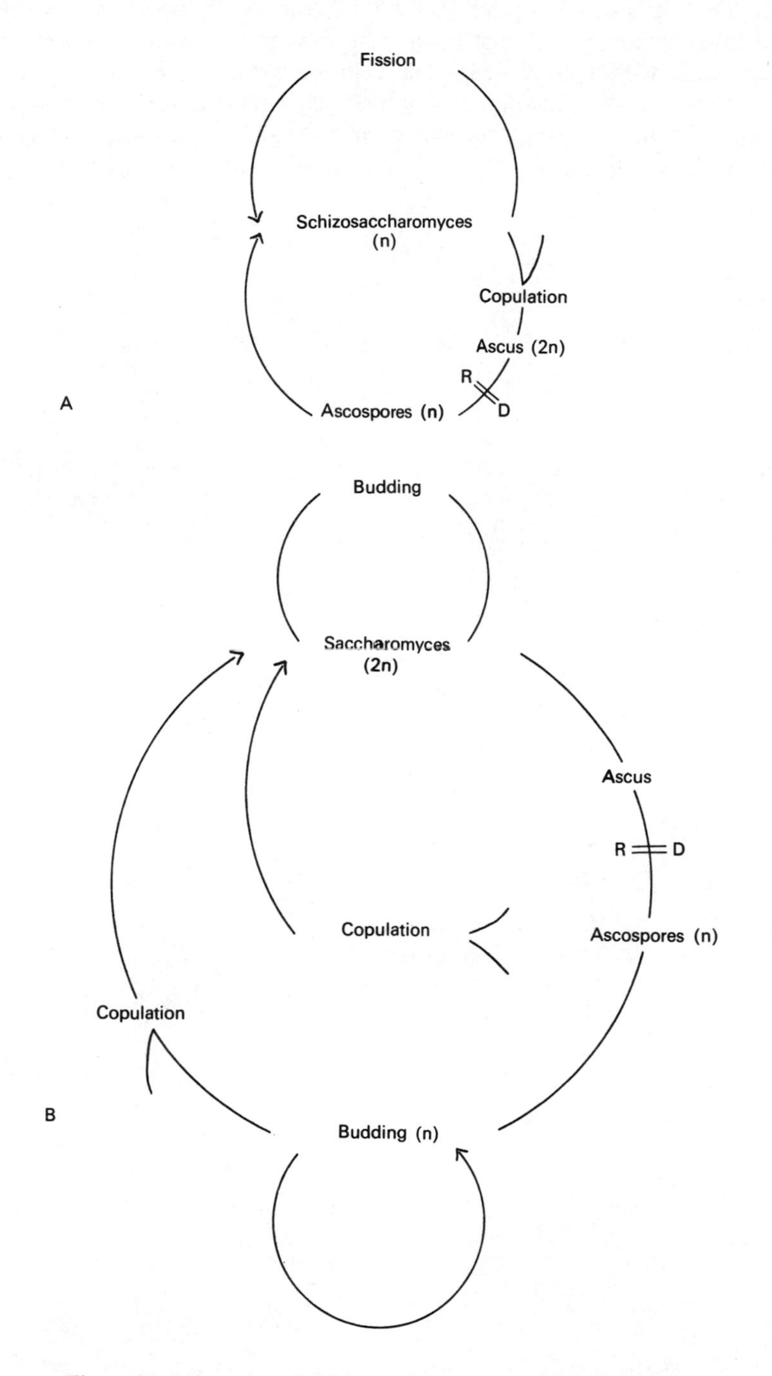

Figure 93. Life cycles of *Schizosaccharomyces* and *Saccharomyces*.

1954). The walls of the spores give a blue reaction with iodine. Growth of the cell wall occurs at one end of the cell only, at the end opposite from the division scar (MacLean, 1964). The cells of *Schizosaccharomyces* show a moderate tolerance of alcohol (about 5–7%) an important property of yeasts used in fermentations (Gray, 1941; Ingram, 1955). In contrast many wine yeasts can tolerate much higher concentrations of alcohol.

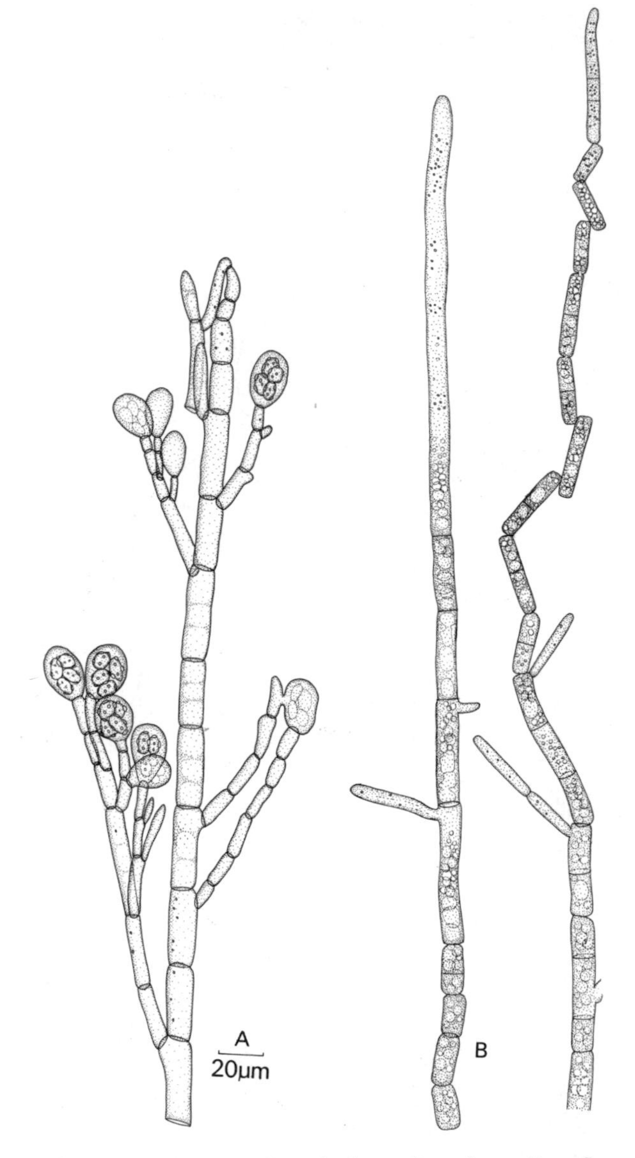

Figure 94. *Endomyces magnusii.* A. Branched mycelium from slime flux of oak showing ascus formation (after Brefeld). B. Mycelium showing arthrospore formation.

Endomyces

This genus represents a mycelial parallel to *Schizosaccharomyces* forming a mycelium which fragments into segments termed arthrospores (see Fig. 94). In *E. magnusii*, originally isolated from a slimy exudation from an oak tree, the mycelial segments are multinucleate, but the tips of the hyphae are narrower and usually uninucleate. Asci are formed following anisogamous fusion of uninucleate hyphal tips or by transformation of a swollen hyphal tip not preceded by fusion and contain four ascospores (see Fig. 94). *Geotrichum*, a common soil fungus, also reproduces by arthrospores. Since it does not form asci it is classified in the Fungi Imperfecti, but is is possibly closely related to *Endomyces*.

Eremascus

Two species of *Eremascus* are known. *E. albus* and *E. fertilis* (Fig. 95) are associated with sugary substrata such as mouldy jam, but several collections

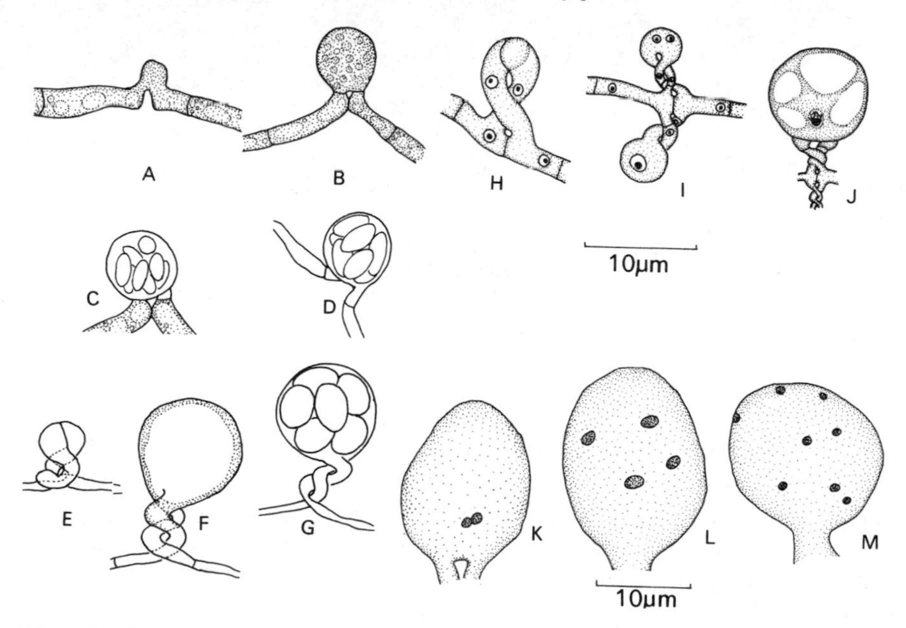

Figure 95. *Eremascus*. A–D. *E. fertilis*, stages in development of asci. E–G. *E. albus*, stages in development of asci. Note the coiling of the gametangia and the globose ascospores of *E. albus*. H–M. *E. albus*, nuclear behaviour during ascus formation (after Harrold, 1950). H. Uninucleate gametangia. I. Plasmogamy and karyogamy. K–M. Nuclear divisions preceding ascospore formation.

of *E. albus* have been made from powdered mustard. Harrold (1950) has shown that both fungi grow best on media with a high sugar content (e.g. 40% sucrose), and do not grow well in a saturated atmosphere. The mature mycelium consists of uninucleate segments. Both species are homothallic.

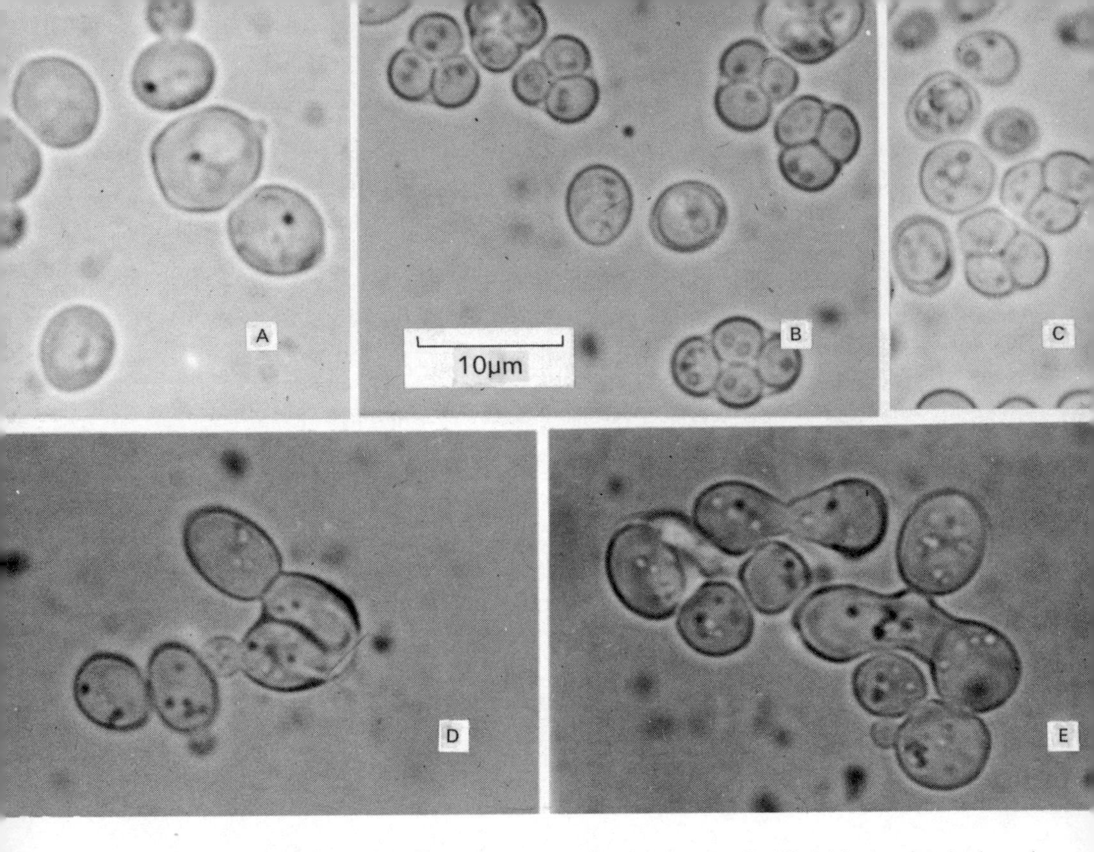

Figure 96. *Saccharomyces cerevisiae*. A. Vegetative yeast cells (diploid) showing buds and bud scars. Note the prominent vacuole. B. Yeast asci mostly containing four spores, sometimes with only three spores in focus. C. Ascus showing a budding ascospore. D. Ascus in which two spores have fused together and are budding. E. Two ascospores fusing (top left), and two fused ascospores forming a diploid bud (right).

On either side of a septum short gametangial branches arise, swollen at their tips, and coiling around each other in *E. albus*. The gametangial tips in *E. albus* are usually uninucleate and, following breakdown of the wall separating the tips of adjacent gametangia, nuclear fusion occurs. This is followed by meiosis and mitosis (Delamater *et al.*, 1953), so that eight nuclei result, each one being surrounded by cytoplasm to form the uninucleate ascospores (Fig. 95M). The ascospores are dispersed passively following breakdown of the ascus wall. On germination a multinucleate germ tube first forms, but as septa appear the uninucleate condition is established.

Saccharomycetaceae

Saccharomyces

About thirty species of *Saccharomyces* have been distinguished but the best known is *S. cerevisiae*, strains of which are used in the fermentation of certain beers and wines, and in baking. *Saccharomyces cerevisiae* var. *ellipsoideus* (sometimes known as *S. ellipsoideus*) is also used in wine-making. *Saccharomyces cerevisiae* is found in nature on ripe fruit. Grape wines are often made by spontaneous fermentation by yeasts growing on the surface of grapes. Because of the economic significance of yeasts there is an extensive literature on their cytology, genetics, ecology, nutrition and physiology and on the technology of yeast (see Cook, 1958; Ingram, 1955; Reiff *et al.*, 1960). The

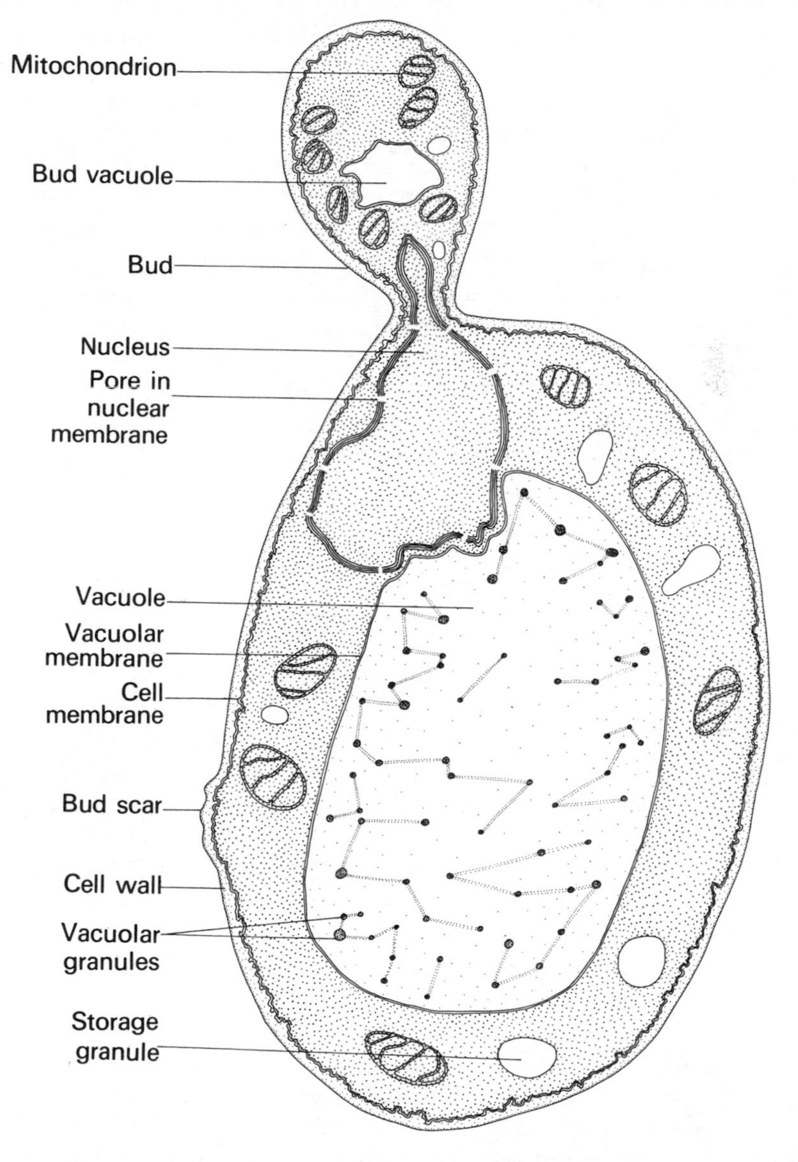

Figure 97. *Saccharomyces cerevisiae*. Diagrammatic representation of a section of a budding yeast cell as seen under an electron microscope.

cells of *S. cerevisiae* are elliptical and about 6–8 × 5–6 μm. In suitable conditions they multiply by budding (see Fig. 96A). The small size of the cell renders observation of its contents under the light microscope difficult, and there has previously been controversy about the interpretation of its structure (McClary, 1964). Studies of thin sections of yeast cells under the electron microscope (Agar & Douglas, 1955, 1957; Hashimoto *et al.*, 1959, 1960; Windisch & Bautz, 1960; Vitols *et al.*, 1961; Moor & Mühlethaler, 1963; Hagedorn, 1964; Robinow & Marak, 1966), have clarified our knowledge and the findings are represented in Fig. 97. The cell wall is about 70 nm thick but may be thinner in young cells. Chemical analyses and X-ray diffraction studies have shown that it contains protein, lipid, and at least two polysaccharides (a mannan and a glucan). Chitin has also been reported (Roelofsen & Hoette, 1951; Northcote & Horne, 1952; Houwink & Kreger, 1954; Eddy, 1958*a,b*; Phaff, 1963). At points where buds have arisen the cell wall bears circular raised bud scars, and as many as 23 have been seen on a single cell. Calculations based on the average size of a yeast cell show that the maximum number of bud scars which could be accommodated on its surface would be about 100 (Barton, 1950; Bartholomew & Mittwer, 1953). This suggests that yeast cells are not capable of unlimited budding. The cytoplasmic membrane, about 88 nm thick, underlying the cell wall is a typical unit membrane, but is invaginated at certain points (see Fig. 97). In surface view the invaginations appear as elongated folds. The membrane itself is made up of a hexagonal array of particles penetrated by fibrils which probably correspond to the glucan fibrils of the cell wall. Within the cytoplasm are mitochondria, endoplasmic reticulum and reserves of fat and glycogen. A large vacuole surrounded by a single unit membrane contains strands and granules of dense material sometimes linked into a network. The larger granules are probably volutin, i.e. polymetaphosphate. The nucleus is surrounded by a nuclear envelope composed of a double unit membrane perforated by pores. The nucleus envelope is distinct from that which surrounds the vacuole. The electron microscope provides little information about the structure of the nucleus. In dividing nuclei a fibre-apparatus consisting of parallel microtubules has been demonstrated. During nuclear division the fibre elongates (Robinow & Marak, 1966). There are conflicting reports on the chromosome number of the diploid yeast cell. Cytological studies suggest that the diploid number is 8 (McClary *et al.*, 1957; Ganesan, 1959) but this does not correspond with the fourteen linkage groups postulated (Mortimer & Hawthorne, 1966). Polyploid yeasts are also known.

When the yeast cell buds its nucleus appears to divide by constriction and the nuclear envelope does not break down. A portion of the constricted nucleus enters the bud along with other organelles. The cytoplasmic connection is closed by the laying down of wall material. Eventually the bud separates from the parent cell, but its previous point of attachment is distinguishable as a birth scar.

S. cerevisiae can be induced to form asci by growing yeast cells on a nutrient rich medium (e.g. molasses agar, or a complete medium containing peptone, casein hydrolysate, yeast extract, glucose and vitamins) and then transferring them to a sporulation medium containing sodium acetate and glucose or raffinose at 30°C (Adams, 1949; Fowell, 1952). The cells then develop directly into asci within 12–24 hr. The cytoplasm differentiates usually into four thick-walled spherical spores, although the number of spores may be fewer (see Fig. 96). The cells from which asci develop are diploid and the nuclear divisions which precede spore formation are meiotic, (a fact which has been confirmed in genetical studies). The electron microscope shows that, as in vegetative nuclear division, the nuclear membrane remains intact during division preceding sporulation (Hashimoto *et al.*, 1960).

If ascospores are dissected from the asci using a micromanipulator and allowed to germinate in a nutrient medium they form haploid buds which are often smaller and rounder than the diploid yeast cell, and single spore cultures can be maintained indefinitely in the haploid state. The diploid state may be re-established in several ways:

1. By fusion of ascospores: this may occur inside or outside the ascus. The walls separating ascospores break down, or short conjugation tubes develop which bring the cytoplasm of the two spores into contact. Nuclear fusion follows and the zygote develops diploid buds (Fig. 96A).
2. Haploid cells fuse to form diploid cells.
3. Fusion may take place between an ascospore and a haploid cell.

Many strains of *S. cerevisiae* are heterothallic, and the ascospores are of two mating types. Mating type specifically is controlled by a single gene which exists in two allelic states *a* and α, and segregation at the meiosis preceding ascospore formation results in two *a* and two α ascospores. Fusion normally occurs only between cells of differing mating type, and this has been termed legitimate copulation. Such fusions result in diploid cells which readily form asci with viable ascospores (Lindegren & Lindegren, 1943). There are, however, exceptions to the fusion of cells or nuclei of opposite mating type:

1. In haploid colonies devised from single ascospores mutation from *a* to α and from α to *a* may occur, followed by copulation (Ahmad, 1965).
2. In haploid colonies derived from single ascospores two-spored asci have been reported: the ascospores are relatively non-viable. It is believed that such asci arise from diploid cells derived from fusion of cells of the same mating type (illegimate copulation; Lindegren, 1949).
3. Spontaneous diploidisation may occur following the first mitotic division of a germinating spore by fusion of the two sister nuclei (Winge & Laustsen, 1937).
4. In *S. chevalieri* and in hybrids between this yeast and *S. cerevisiae* a gene *D* for diploidisation may be present. The presence of this gene permits

diploidisation to occur in haploid progeny of either mating type. Diploid yeast cells may be heterozygous for the *D* gene (i.e. *Dd*) and then may give rise to asci from which two ascospores will give rise to diploid colonies directly and two will not (Winge & Roberts, 1949).

Numerous studies of the genetics of yeast have been made. By means of hybridisation it has been possible to develop strains of yeast for baking, brewing and other types of fermentation which possess a number of desirable properties (see Winge & Roberts, 1958; Mortimer & Hawthorne, 1966).

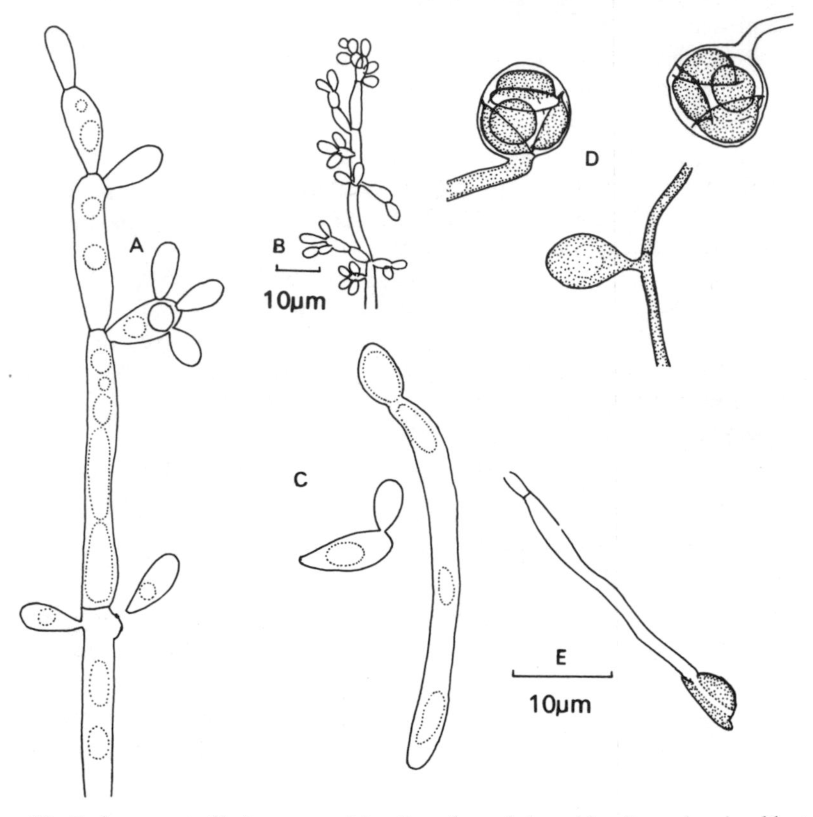

Figure 98. *Endomycopsis fibuliger*. A, B. Mycelium from 3-day-old culture showing blastospore formation. C. Blastospores germinating by germ tube, or budding to form a further blastospore. D. A young ascus and two mature asci containing four hat-shaped ascospores. E. Germinating ascospore. A, C, D, E to same scale.

Endomycopsis

Endomycopsis is a mycelial form which reproduces by buds (blastospores) and also forms asci parthenogenetically or following isogamous fusion (Kreger van Rij, 1966). *Endomycopsis fibuliger* grows in flour, bread and macaroni and produces an active extracellular diastase, an unusual property of yeasts (Wickerham *et al.*, 1944). In culture it may form budding yeast cells and septate branched hyphae which produce blastospores laterally and terminally (Fig. 98). Arthrospore formation has also been demonstrated (Müller, 1964). Ascus formation can be induced by growing the yeast for a few days on malt extract agar and transferring to distilled water. The asci are mostly four-spored, and the spores are hat-shaped (Fig. 98D), having a flange-like extension of the wall. *Endomycopsis fasciculata* is an ambrosia fungus lining the galleries of ambrosia beetles and serving as a source of food for the larvae (Batra, 1963). The fungus is homothallic and forms two-spored asci. *Encomycopsis scolyti*, associated with bark beetles of the genus *Scolytus*, parasitic on conifers, is heterothallic, and conjugation only occurs when yeast cells of both types are mixed. Conjugation results in the formation of diploid cells which give rise to asci containing one to four hat-shaped ascospores (Phaff & Yoneyama, 1961). There are a number of mycelial forms which form blastospores resembling *Endomycopsis*, but in which asci have not been seen. One of the best-known is *Candida albicans*, the cause of various diseases in man such as thrush. Other species of *Candida* have been identified as strains of heterothallic species of *Endomycopsis* (Wickerham & Burton, 1954).

TAPHRINALES

Taphrina

Taphrina (In older literature the generic name *Exoascus* was used.)
About 100 species are known, mostly parasitic on Amentiferae and Rosaceae (Mix, 1949), causing diseases of three main kinds:

1. Leaf curl or leaf blister diseases, e.g. *T. deformans* the cause of peach leaf curl; *T. tosquinetii* the cause of leaf blister of alder and *T. populina* the cause of yellow leaf blister of poplar.
2. Diseases of branches in which the infected plant undergoes repeated branching to form dense tufts of twigs called witches' brooms, e.g. *T. betulina* causing witches' brooms of birch. Similar twig proliferation is also caused by mites. *T. insititiae* causes witches' broom of plum and damsons and *T. cerasi* witches' brooms and leaf curl of cherry.
3. Diseases of fruits, e.g. *T. pruni* which causes the condition known as pocket plums in which the fruit is wrinkled and shrivelled and has a cavity in the centre in place of a stone.

T. deformans: Peach leaf curl is common on leaves and twigs on peach and almond, especially after a cool moist spring. Towards the end of May infected peach leaves show raised reddish puckered blisters which eventually acquire a waxy bloom (Fig. 99). Sections of leaves in this condition show an extensive septate mycelium growing between the cells of the mesophyll and between the cuticle and the epidermis, where the hyphae end in swollen chlamydospores (see Fig. 100). Cytological studies (Martin, 1940; Kramer, 1961; Caporali, 1964) show that the segments of the mycelium and the young chlamydospores are binucleate. In the chlamydospore the two nuclei fuse and the diploid nucleus divides mitotically. The upper of the two daughter nuclei then undergoes meiosis followed by a mitosis, so that eight nuclei result,

Figure 99. *Taphrina deformans*. Peach leaf showing leaf curl.

which form the nuclei of the eight ascospores. The lower daughter nucleus remains in the lower part of the chlamydospore and is often separated from the other nucleus by a cross-wall. During these nuclear divisions the wall of the chlamydospore has stretched to form an ascus. Within the ascus the

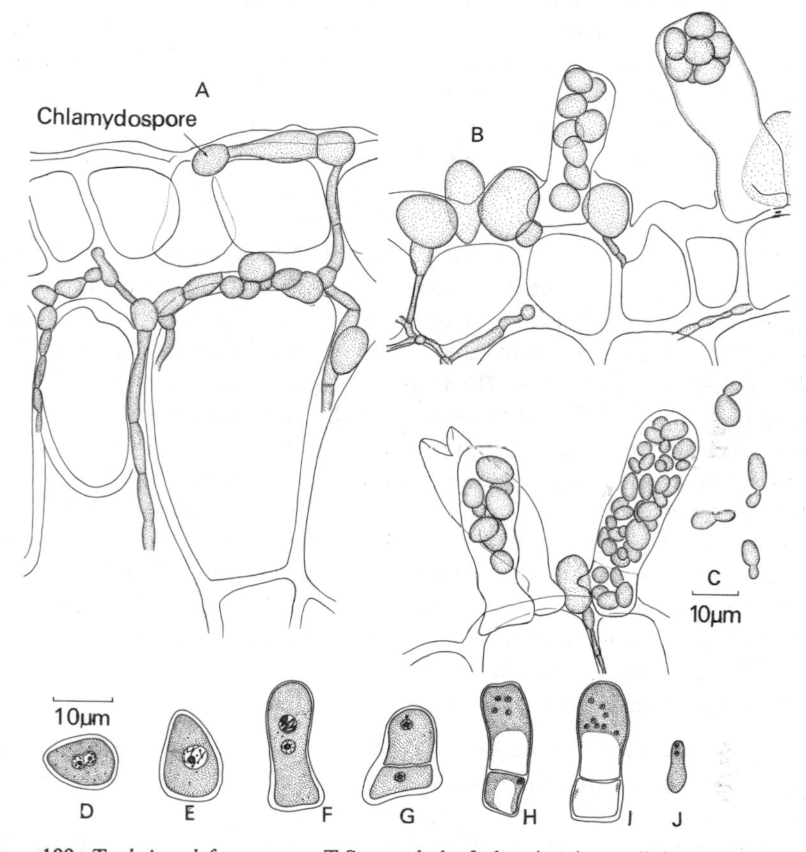

Figure 100. *Taphrina deformans*. A. T.S. peach leaf showing intercellular mycelium and subcuticular chlamydospores. B. T.S. peach leaf showing chlamydospores and asci, containing eight ascospores. C. T.S. leaf showing a dehisced ascus, and eight-spored ascus and an ascus in which the ascospores are budding. Ascospores budding outside the ascus are also shown. D–J. Cytology of ascus formation (after Martin, 1940). D, E. Fusion of nuclei in chlamydospore. F. Elongating ascogenous cell containing two nuclei formed by mitosis from the fusion nucleus. The upper nucleus has begun to divide meiotically. G. Uninucleate ascus with uninucleate basal cell. H, I. Four- and eight-nucleate asci. J. Binucleate germ tube in germinating ascospore.

ascospores may bud so that ripe asci may contain numerous buds (see Fig. 100C). The asci form a palisade-like layer above the epidermis and it is their presence which gives the leaf its waxy bloom. The ascospores or buds are projected from the asci which often opens by a characteristic slit (Fig. 100C). Following treatment with potassium hydroxide the ascus wall of *T. populina* appears double and Schneider (1956) has compared it with the bitunicate asci of other Ascomycetes. Yarwood (1941) has shown that there is a diurnal cycle of ascus development and discharge in *T. deformans*. Nuclear fusion takes place during the afternoon or evening; the nuclear divisions are complete by about 5 a.m. and the spores appear mature by 8 a.m. However, maximum spore discharge does not occur until about 8 p.m. Outside the ascus the spores or conidia may continue budding and the fungus can be grown saprophytically in a yeast-like manner in agar or in liquid culture. Young leaves can be infected from such budding cells, and it has been shown that a culture derived from a single spore can cause infection resulting in the formation of asci, so that *T. deformans* is homothallic. In this respect it differs from some other species, e.g. *T. epiphylla* where fusion of buds, presumably of different mating type, is necessary before infection can occur (Wieben, 1927). In *T. deformans* the binucleate condition is established at the first nuclear division of a bud placed on a peach leaf; the two daughter nuclei remain associated in the germ tube which penetrates the cuticle (Fig. 100J).

The distortion of the host tissue is associated with division and hypertrophy of the cells of the palisade mesophyll. In liquid cultures of *T. deformans*, especially on media containing tryptophane, considerable quantities of indole acetic acid have been demonstrated (Crady & Wolf, 1959; Sommer, 1961). Following chromatographic separation of culture extracts of *T. deformans* a substance was found which stimulated elongation of pea internodes, and another which stimulated mitosis of pea cells. If such substances are formed by the fungus in peach leaves they could explain the hypertrophy of the tissues. Infection is followed by increased respiration of the host tissues and an increase in the free amino-acid pool (Raggi, 1967).

The fungus overwinters in two ways:

1. Mycelium persists throughout the winter in the cortex of infected twigs. Towards the end of winter the mycelium penetrates buds beneath the infected area.

2. Ascospores and conidia survive on the surface of twigs and between bud scales during autumn and winter. Between November and March the spores and conidia develop thick walls, and in spring as the peach buds open the conidia produce germ tubes which penetrate the young leaves (Caporali, 1964). Control of peach leaf curl is achieved by spraying with a suitable fungicide before bud-break in spring.

2: PLECTOMYCETES

In this group are included Ascomycetes with ascocarps which are rudimentary, or consist of a loose investment of hyphae or are globose cleistocarps, i.e.

closed ascocarps not opening by an ostiole. There are two orders, the Erysiphales and the Eurotiales. It is doubtful if the two groups are closely related. The Erysiphales are obligate parasites of higher plants. The ascocarps contain one to several oval to club-shaped asci which discharge their ascospores violently. The Eurotiales are mostly saprophytic. The ascocarps are very variable in form, but the asci are small and globose. The ascospores are not violently discharged.

ERYSIPHALES

The Erysiphales are known as powdery mildews (distinct from the downy mildews which are Peronosporaceae), because of the powdery appearance of conidia on infected plants. All are obligate parasites of Angiosperms often showing physiological specialisation, i.e. having strains or races confined to a

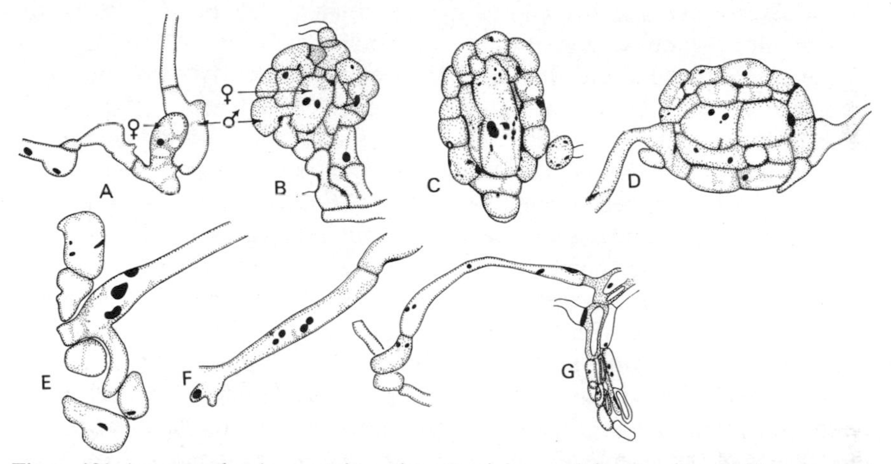

Figure 101. Ascocarp development in various Erysiphaceae (after Gordon, 1966). A. Contact between pseudoascogonium (♀) and pseudoantheridium (♂) in *Erysiphe cichoraceacum*. B. Conjugation between a cell of the pseudoantheridum (♂) and the pseudoascogonium (♀) in *E. cichoraceacum*. Note that the pseudoascogonium is binucleate. C. Binucleate pseudoascogonium of *Microsphaera diffusa*. The pseudoascogonium is surrounded by two layers of pseudoantheridial cells. D. Extension of a mother cell of the pseudoantheridium to form a receptive hypha in *E. cichoraceacum*. E. Tip of receptive hypha in *E. cichoraceacum* making contact and encircling a cell of the superficial mycelium. F. Pairs of nuclei in the receptive hyphae of *M. diffusa*. G. Receptive hypha of *E. cichoraceacum* showing its connection with a cell of the peridium, derived from the pseudoascogonium.

narrow range of host plants. Diseases caused by the Erysiphales are of economic significance—e.g. cereal and grass mildew caused by *Erysiphe graminis*, apple mildew caused by *Podosphaera leucotricha*, American gooseberry mildew caused by *Sphaerotheca mors-uvae*. The mycelium of the Erysiphales is usually superficial, with haustoria often confined to the epidermal cells. Chains of conidia arise in basipetal succession from a mother cell on the mycelium. Later ascocarps (cleistothecia) may be formed. The cleistothecia are brown globose bodies and have no ostiole. They may contain one to several asci which discharge their spores explosively.

There is a voluminous literature on ascocarp development (Luttrell, 1951), but interpretation of the ontogeny has recently been re-appraised (Gordon, 1966). Development probably follows the same pattern in most species (Fig. 101). Two branches of the superficial mycelium come into contact. The terminal cells are at first uninucleate. One cell encircles the other, and the encircled cell enlarges. Following breakdown of the walls separating them, or by the development of a conjugation tube a nucleus is transferred to the larger central cell. Fusion of the two nuclei has been reported, but accounts differ in this respect. Many early authors have interpreted the central enlarged cell as an ascogonium, and the encircling cell as an antheridium, and have claimed that the ascogonium initiates the development of asci, but Gordon, although claiming nuclear fusion in the central cell, finds no evidence that it plays a functional role in ascus development. He therefore terms the central cell a pseudoascogonium and the encircling cell (and cells derived from it) the pseudoantheridium. In Gordon's account (which is unfortunately a composite account based on the development of four species), further development is as follows. The pseudoantheridial cells which completely encircle the pseudoascogonium divide to form the peripheral cells of the ascocarp. The outer cells (which we may call mother cells) develop two- to five-celled hyphae, receptive hyphae, with uninucleate segments. The tips of the receptive hyphae make contact with vegetative hyphae on the surface of the host. A nucleus from the vegetative hypha apparently migrates into the receptive hypha and pairs with the nucleus already present. One of the paired nuclei, thought to be the nucleus from the vegetative hypha, divides, and a daughter nucleus migrates through the septal pore into the next cell of the receptive hypha. This process is repeated until a daughter nucleus pairs with the nucleus of the peripheral cell which gave rise to the receptive hypha (i.e. mother cell). This is followed by deliquescence of the terminal part of the receptive hypha. The binucleate mother cell now enlarges and becomes multinucleate and divides. Inner cells formed by division from the enlarged mother cell become binucleate whilst the outer cells remain uninucleate. Many of the outer uninucleate cells produce new receptive hyphae and the whole process of anastomosis and nuclear migration is repeated.

During the events described above the two nuclei in the pseudoascogonium fuse and the enlarged nucleus divides. Following cell wall formation between the daughter nuclei the pseudoascogonium becomes a series of three to five uninucleate cells which at first contain denser cytoplasm than the surrounding pseudoparenchymatous cells, but later this distinction is no longer obvious. Occasional binucleate cells have been found in the three- to five-celled pseudoascogonium, but it is possible that such cells were dividing. The immature ascocarp consists of a pseudoparenchymatous centrum composed largely of binucleate cells derived from the mother cells of the pseudoantheridium intermixed with some uninucleate cells, and surrounded by a peridium, some four to six cell layers thick, which become darkly pigmented. Uninucleate and binucleate cells above the middle part of the centrum lyse (break down). Karyogamy (i.e. nuclear fusion) occurs within certain of the binucleate cells, more or less isolated from surrounding cells by lysis. These cells then enlarge to form asci. The asci appear to have the ability to absorb the uninucleate and binucleate cells of the centrum. Eventually the asci (or in some cases a single ascus) occupy almost the entire centrum. Meiosis of the fusion nucleis in developing asci is usually delayed until the centrum cells have all been absorbed.

In addition to the 'receptive hyphae' the cleistothecia of all Erysiphaceae bear thick-walled hyphae (appendages) which may be branched or unbranched, and often have a highly distinctive appearance. The type of appendage has proved a useful aid to classification. Another criterion is whether there is only one ascus in each ascocarp or several, and using such criteria about seven European genera have been distinguished (Blumer, 1967).

Erysiphe

Diseases caused by *Erysiphe* include grass and corn mildew (*E. graminis*), mildew of cucurbits and other plants (*E. communis* = *E. cichoracearum*) and pea and clover mildew (*E. polygoni*). The last-named is especially common in Britain on clover, *Polygonum* and on *Heracleum sphondylium* but is has an exceptionally wide range of other hosts. *E. graminis* can be collected throughout the year in the conidial state on numerous grasses and cereals. A number of physiological races (or formae speciales) of *E. graminis* exist, e.g. *E. graminis* f. sp. *tritici* infects wheat (*Triticum*) but not barley, whilst *E. graminis* f. sp. *hordei* infects barley (*Hordeum*) but not wheat. The cleistothecia are commonly found on cereals in the summer and autumn. The appearance of conidial pustules on wheat leaves is shown in Fig. 102. The pustules are white to pale brown in colour. Conidia on wheat leaves germinate within one to two days to form short germ tubes (Fig. 103A). The germ tube attaches itself to the epidermis by a pad or appressorium and beneath the point of attachment a fine infection tube penetrates the host cell wall. Penetration is probably brought about by a combination of mechanical pressure and enzymatic activity, because around the narrow infection hole is a wider 'halo' in which the cell wall is stained by such stains as cotton-blue. The host cell wall may be thickened beneath and around the infection tube (Fig. 103C). Where infection of a suitable host takes place the infection tube enlarges within the epidermal cell to form an elongate uninucleate haustorium with finger-like projections developing from opposite ends (Fig. 103C). Such haustoria are typical of *E. graminis*, but in most other members of the Erysiphales the haustoria are simple globose bodies (Fig. 106B). The whole of the haustorium, i.e. the body and the finger-like lobes, is enclosed in a *sheath* (Fig. 104A). The lobes may be individually enclosed, or several lobes may be enclosed in a common extension of the sheath. The body of the haustorium contains a single nucleus, mitochondria and unidentified vesicles, whilst the most prominent features of the haustorial lobes are their mitochondria and branched system of tubules forming the endoplasmic reticulum. Below its point of penetration of the host cell wall the neck of the haustorium is surrounded by a collar which differs in structure from the host wall and may represent a deposit of material on it. The sheath surrounding the haustorium is probably

Figure 102. *Erysiphe graminis*. A. Conidial pustules on wheat leaf. B. Cleistothecia on wheat leaf sheath. About twice natural size.

a modified invagination of the host plasmalemma (ectoplast), so that the haustorium does not lie freely in the host cytoplasm but is bounded by a membrane continuous with the plasmalemma. Immediately outside the wall of the haustorium there is a wide *sheath matrix*, itself bounded by a *sheath membrane* contiguous with the tonoplast of the host cell. Thus the key interface between fungus and the host is the sheath membrane. At various points

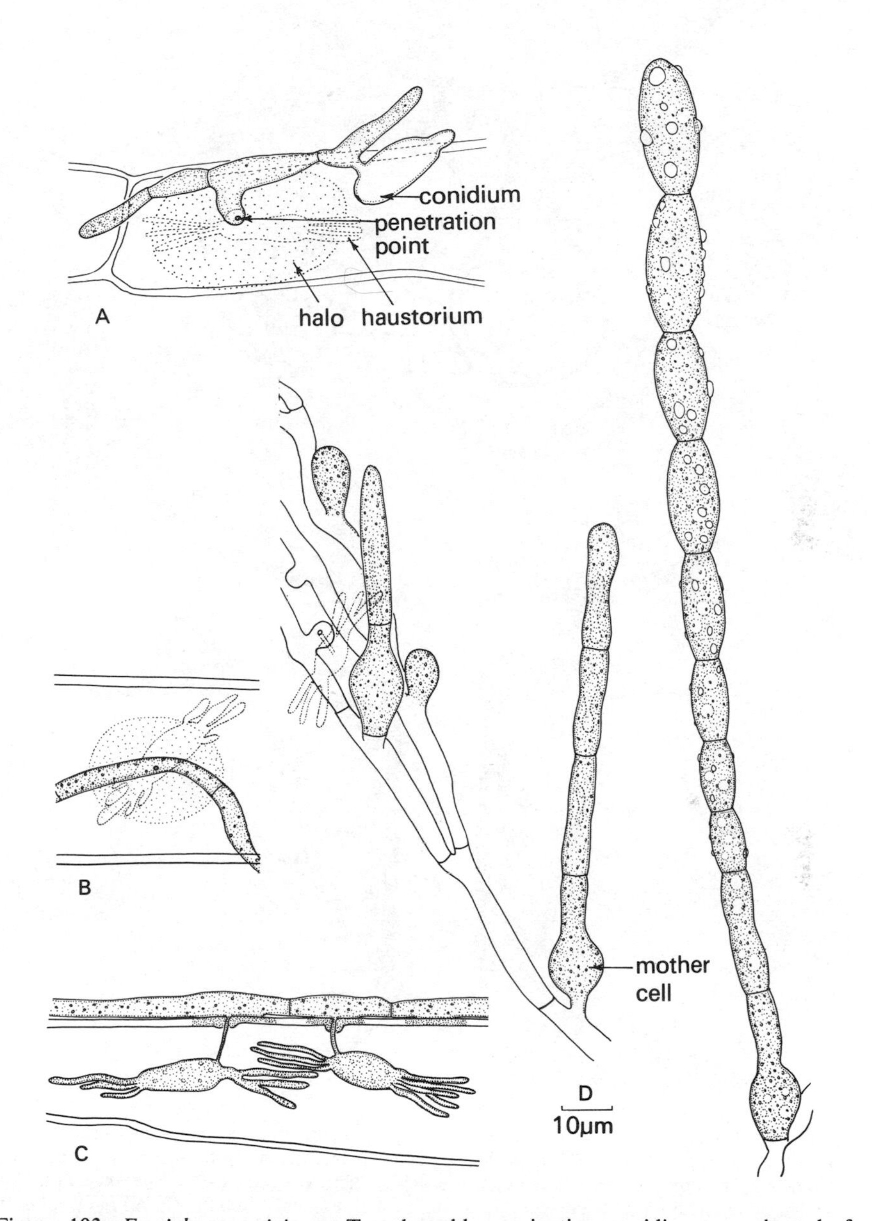

Figure 103. *Erysiphe graminis*. A. Two-day-old germinating conidium on wheat leaf, showing penetration point, surrounded by a 'halo' (stippled). A haustorium has developed beneath the penetration point. B. Penetration from an established mycelium. C. Section of an epidermal cell showing two penetration points and two haustoria. Note the thickening of the epidermal cell beneath the penetration point. D. Mycelium and conidiophores, showing the swollen flask-shaped mother cell.

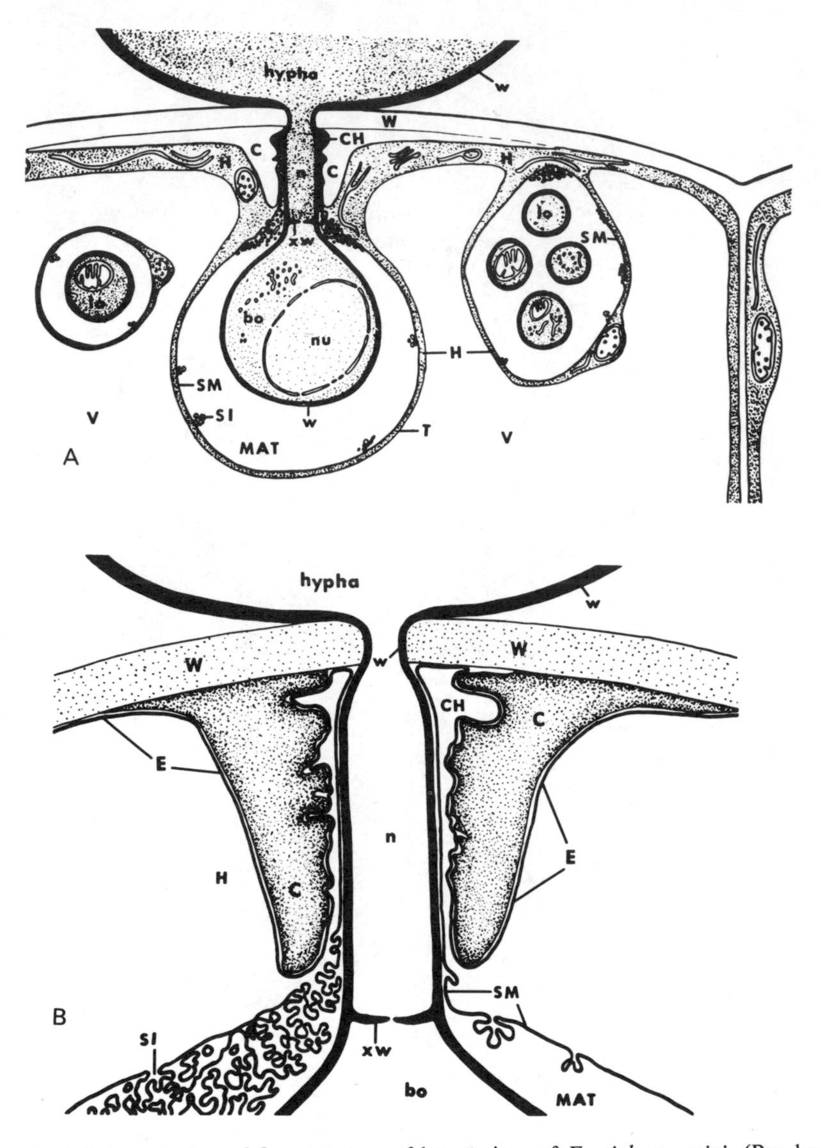

Figure 104. Interpretation of fine structure of haustorium of *Erysiphe graminis* (Bracker, 1968). A. Section of host leaf at point of penetration. The body of the haustorium (bo) containing a single nucleus (nu) lies immediately beneath the point of penetration. The body of the haustorium is enclosed in a sheath with extensive matrix (MAT). The sheath membrane (SM) is in contact with the host tonoplast (T). The sheath membrane bears invaginations (SI). A single lobe of another haustorium enclosed in an extension of the sheath is shown to the left of the diagram, and four lobes enclosed in a common sheath to the right. B. Enlargement of neck of haustorium. Note the thickened collar (C) deposited on the host cell wall (W). The sheath membrane (SM) is continuous with the host ectoplast (E). (XW), crosswall or septum; (CH), channel; (H), host cytoplasm.

the sheath membrane shows tubular invaginations, and these presumably represent the points at which interchange of material between the host cell and the pathogen are taking place (Bracker, 1968). Many of the features of the haustorium of *E. graminis* have also been found in *E. cichoracearum* (McKeen *et al.*, 1966).

In a susceptible host the infected cell remains alive. Following successful establishment in an epidermal cell the superficial mycelium develops branches, further appressoria and haustoria (Fig. 103B,D) and within 7–10 days begins to develop conidia. If, however, a non-susceptible grass host is infected, the host cell undergoes rapid necrosis and adjacent cells may also die, thus restricting further development of the pathogen. Infection of races of wheat and barley bred for resistance to mildew differs in several respects from infection of susceptible hosts. Resistance genes confer resistance in several ways:

1. The proportion of conidia which germinate and produce haustoria is reduced.
2. Haustorial development is delayed and the size of the haustoria is reduced.
3. Sporulation, i.e. development of conidia, is suppressed (Masri & Ellingboe, 1966).

The conidia develop from a flask-shaped mother cell within which nuclear division occurs. The mother cell elongates away from the host leaf and a cross-wall cuts off the hyphal tip. Further cross-walls develop so that a chain of cells is formed, increasing in length at its base—i.e. by further divisions of the mother cell. Each conidium is uninucleate. The segments become swollen and barrel-shaped, and are detached by wind. This type of conidial apparatus belongs to the form-genus *Oidium* of the Fungi Imperfecti, and is found in most members of the Erysiphales. Interpretation of the mother cell as a rudimentary phialide is not generally accepted. Hughes (1953) refers to these conidia as meristem arthrospores and not phialospores. The fine structure of conidia of *E. cichoracearum* has been studied by McKeen *et al.* (1967).

The conidia of *Erysiphe* are unusual in their ability to germinate at low humidities, even at zero relative humidity. This has been attributed to the fact that they have a water content as high as 70% compared with about 10% for other representative air-borne fungus spores (Somers & Horsfall, 1966). Possibly the ability to germinate at low humidity is related to the fact that powdery mildews are abundant in hot dry seasons (Schnathorst, 1965).

The cleistothecia of *E. graminis* are dark brown and globose, and grow on the basal leaves and leaf-sheaths of cereals nestling in a dense mass of mycelium (Figs. 102B; 105). In many species of *Erysiphe* the perithecial wall also bears unbranched dark appendages with free ends (see Fig. 106 of *E. polygoni*). Each cleistothecium has a wall made up of several layers of cells, surrounding a number of asci. There is no ostiole, and it is for this reason that the ascocarps are termed cleistothecia although the term perithecium is

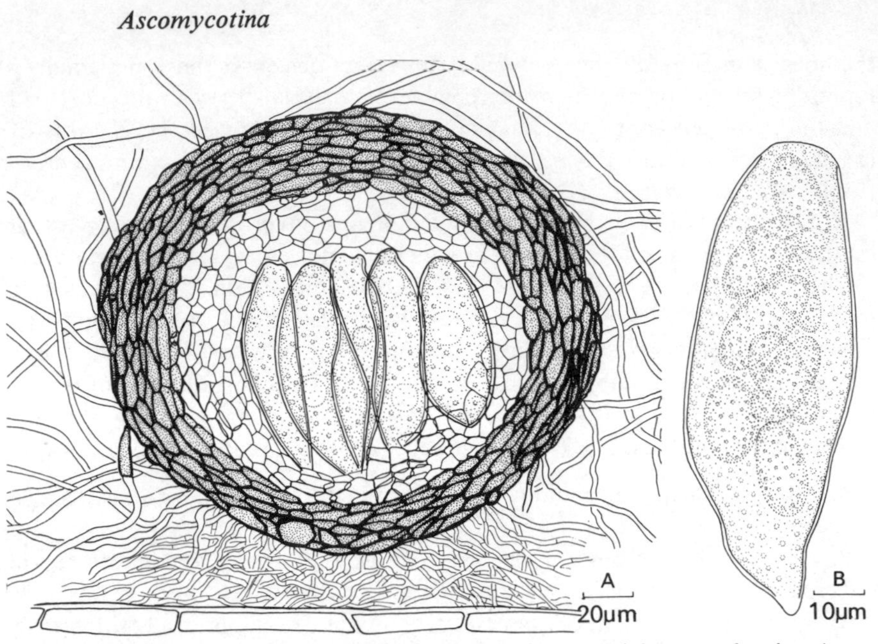

Figure 105. *Erysiphe graminis*. A. Section of cleistothecium containing several asci. B. Ascus.

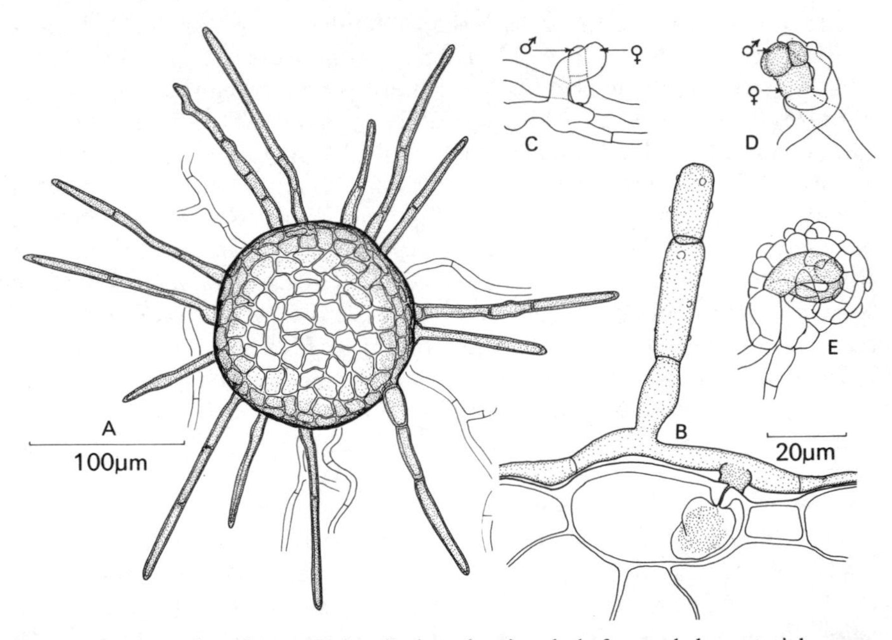

Figure 106. *Erysiphe polygoni*. Cleistothecium showing dark, free-ended equatorial appendages, and the superficial mycelium anchoring the cleistothecium to the host leaf. B. T.S. host leaf showing simple haustorium, superficial mycelium and conidial chain. C. Contact between pseudoascogonium (♀) and pseudoantheridium (♂). D. Pseudoascogonium becoming surrounded by cells of pseudoantheridium. E. Pseudoascogonium almost completely enclosed.

also used. They crack open by swelling of the contents and the asci discharge their spores. Ascospores of *E. graminis* formed in the current season are capable of infection, but they can also survive for up to 13 years (Moseman &

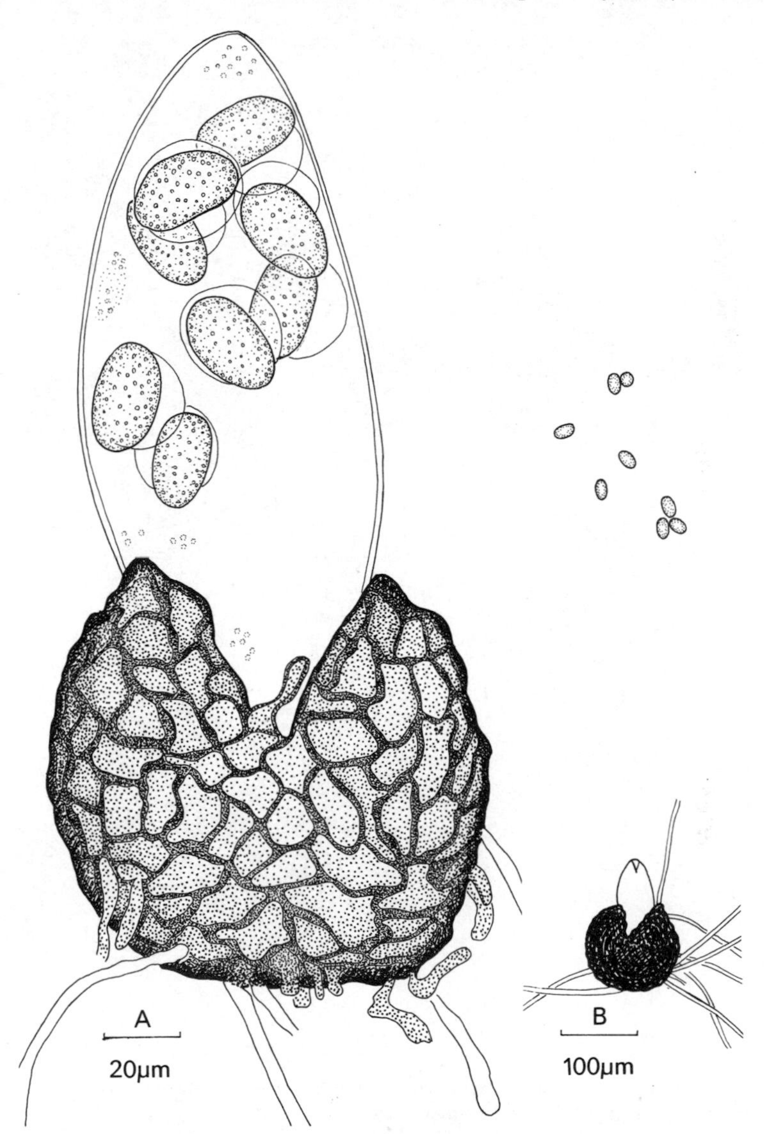

Figure 107. *Sphaerotheca pannosa*. A. Cleistocarp crushed to show the single ascus. B. Cleistocarp showing discharged ascospore.

Powers, 1957). *E. graminis* is heterothallic and this is probably true of most Erysiphaceae.

Infection of cereals by *E. graminis* results in increased respiration of the

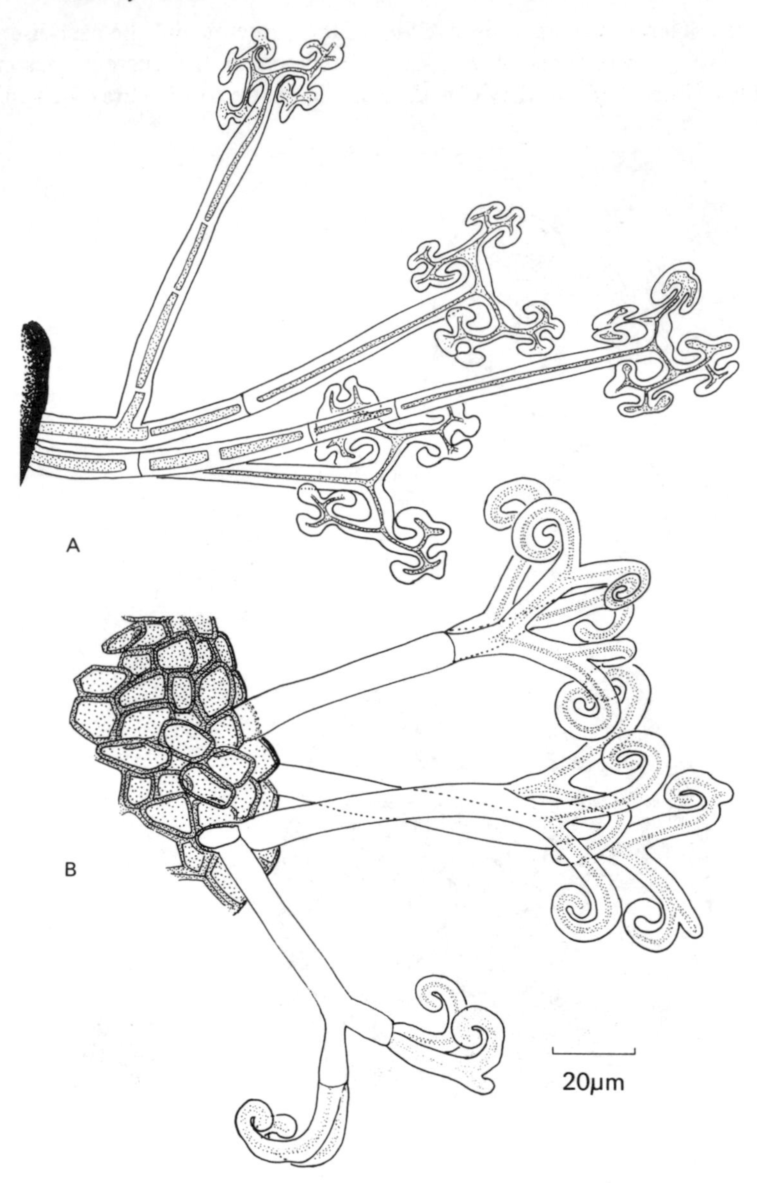

Figure 108. A. Cleistothecial appendages of *Podosphaera clandestina* from *Crataegus*. B. Cleistothecial appendages of *Uncinula bicornis* from *Acer*.

host, and a decrease in photosynthesis, and this may mean that an infected leaf is unable to export carbohydrate. This effect is ultimately reflected in the reduced weight of shoots, roots and ears (Allen, 1942; Last, 1962).

Chemical control of diseases caused by Erysiphales can be achieved by dusting or spraying host plants with a fungicide containing sulphur. Some systemic fungicides can be used. Control can also be achieved by selecting and breeding resistant host varieties (Moseman, 1966).

Sphaerotheca

Common British species are *S. macularis* var. *fuliginea* (= *S. fuliginea*) a common mildew of dandelions and other Compositae, *S. macularis* (= *S.*

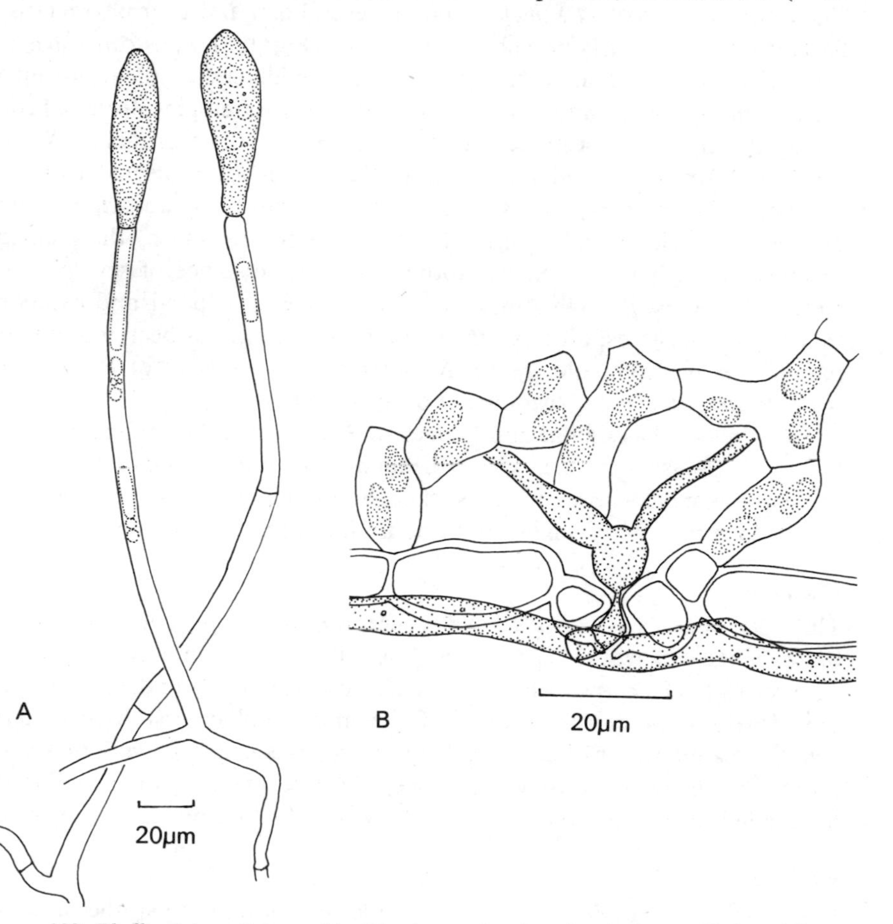

Figure 109. *Phyllactinia guttata*. A. Conidiophores showing the single terminal conidium. B. T.S. leaf of *Corylus avellana* showing penetration of stoma in lower epidermis, and extension of the mycelium into the mesophyll.

humuli) hop mildew, *S. mors-uvae*, American gooseberry mildew and *S. pannosa*, rose mildew.

The cleistothecial structure of *Sphaerotheca* closely resembles that of *Erysiphe*. The appendages are simple, but instead of containing several asci each ascocarp contains only one (Fig. 107). *S. macularis* var. *fuliginea* is homothallic.

The mycelium and conidia of *S. pannosa* are common on leaves and shoots of cultivated and wild roses. Cleistothecia are formed on twigs, embedded in a dense mycelial felt. Overwintering is not only by means of ascospores, but as mycelium within dormant buds.

Podosphaera

The cleistothecia of *Podosphaera* contain several asci and bear characteristic flattened dichotomously branched appendages (Fig. 108). *P. leucotricha* is the cause of apple mildew, and the mycelium and conidia are visible in spring on the expanding foliage and young shoots probably developing from a perennating mycelium. Cleistothecia are formed on the young branches. Woodward (1927) has reported that the asci themselves are projected from the perithecia. The ascocarp gapes open as the asci expand by absorbing water, but the wall is elastic and eventually the asci are thrown out by the snapping together of the 'jaws' of the cleistothecium, for a distance of several centimetres. If the ascus alights in water it continues to expand and explodes, shooting out its ascospores. A similar double discharge has been reported for *Erysiphe graminis* (Ingold, 1939). Another common species of *Podosphaera* is *P. clandestina* (= *P. oxyacanthae*) on hawthorn.

Microsphaera has similar appendages to *Podosphaera*, but differs in having only a single ascus in each cleistothecium. *Microsphaera alphitoides* is the cause of oak mildew, common on sucker shoots and seedlings. The cleistothecial stage is rare, and seems only to be found during hot summers.

Uncinula

The cleistothecia of *Uncinula* have several asci and appendages which may be branched but with recurved tips (Fig. 108B). The commonest species is *U. bicornis* (= *U. aceris*) which forms cleistothecia on the underside of sycamore leaves. *U. necator* is the cause of vine mildew. When first discovered the conidial state only was known and the fungus was called *Oidium tuckeri*. It threatened the wine industry of France, but experiments on controlling the disease led to the discovery of the efficacy of sulphur-containing fungicides.

Phyllactinia

P. guttata (= *P. corylea*) grows on hazel leaves forming cleistothecia in late summer and autumn. As well as the superficial mycelium there is penetration of the mesophyll (Fig. 109B). The conidia are unusual in being formed singly, and are club-shaped (Fig. 109A). Conidia of this type are classified in the form-genus *Ovulariopsis*. In a damp chamber chains of up to four conidia formed in basipetal succession as in other Erysiphales have been observed. The cleistothecia bear two types of appendage, an equatorial group of radiating unbranched appendages with bulbous bases, and a crown of repeatedly branched appendages which secrete mucilage (Fig. 110). The base of the bulbous appendage is thick-walled above and thin-walled below. On drying the appendage bends downwards as the thin part buckles inwards, and the downward pressure of the appendage-tips levers the perithecium free from the superficial mycelium. The blob of mucilage between the apical crown of branched appendages helps to stick the perithecium on to twigs and leaves. The asci usually contain only two spores (Fig. 110E).

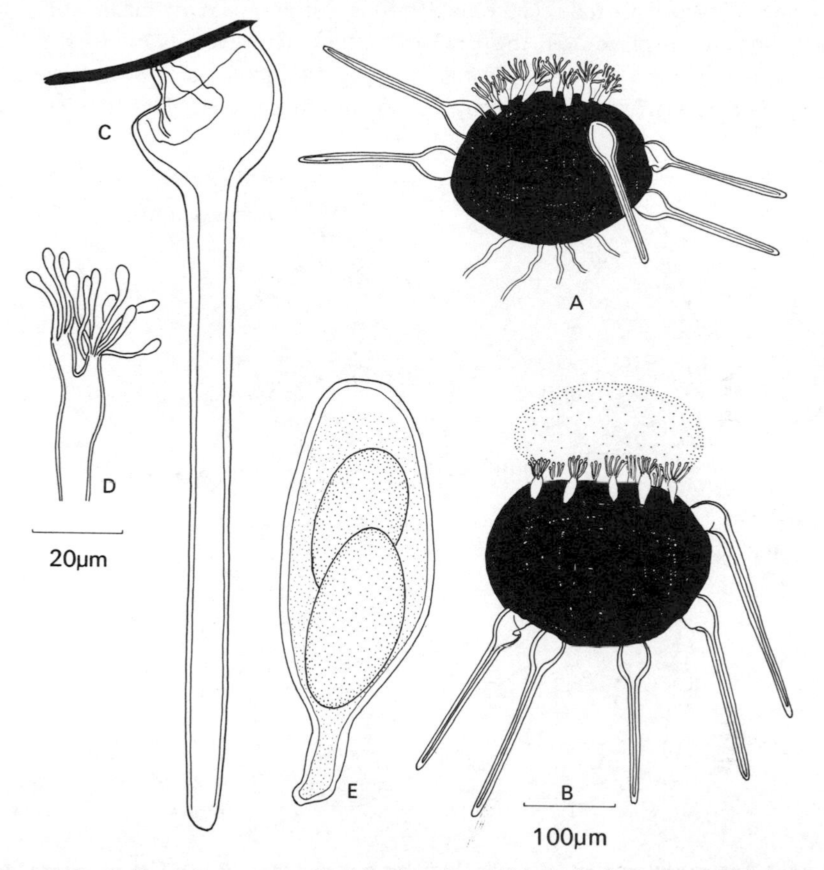

Figure 110. *Phyllactinia guttata*. A. Perithecium from *Corylus* leaf showing bulbous appendages and attaching hyphae. B. Perithecium raised from the leaf surface by the downwards pushing of the bulbous appendages. A mass of mucilage has been secreted by the secretory appendages. C. Enlarged view of a bulbous appendage showing the invagination of the thinner walled lower part of the bulb. D. Enlarged view of the secretory appendages. E. Ascus, containing two ascospores. A, B, to the same scale; C, D, E, to the same scale.

EUROTIALES

The Eurotiales include fungi of great economic importance, such as *Aspergillus* and *Penicillium*, some species of which cause spoilage of food and textiles, whilst others are used in fermentation. The Gymnoascaceae have aroused interest because of their relationship to skin pathogens of animals and man. Whilst the asci in *Aspergillus* are completely enclosed by a well-defined envelope of sterile hyphae or peridium, those of *Gymnoascus* are only partially enclosed by loose hyphae. In *Byssochlamys* the asci, although arising in clumps from ascogenous hyphae, are not enclosed. The inclusion of this genus in the Eurotiales (Kuehn, 1958) is not accepted by all mycologists and some would place it in the Endomycetales. However its conidial structures are similar to those of *Penicillium* and the genus is in some respects intermediate between the Endomycetales and Eurotiales. The inclusion of

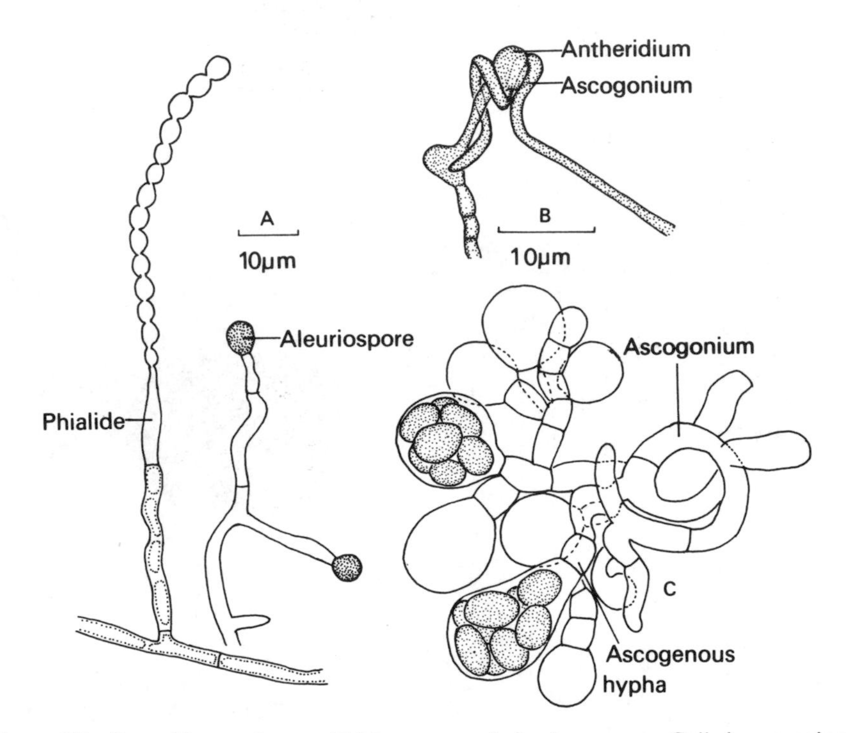

Figure 111. *Byssochlamys nivea*. A. Phialospores and aleuriospores. B. Coiled ascogonium surrounding an antheridium. C. Ascogonium bearing ascogenous hyphae which in turn bear asci. Note the absence of sterile investing hyphae.

Ceratocystis is justified in terms of its ascocarp development, despite the fact that the ascocarp has a long neck terminated by an ostiole. Three families of the Eurotiales may be distinguished as follows:

Ascocarps not possessing an ostiole — a
Ascocarps ostiolate — **Ophiostomataceae**
a Peridium of ascocarp composed of loosely interwoven hyphae or absent — **Gymnoascaceae**
Peridium pseudoparenchymatous — **Eurotiaceae**

Gymnoascaceae

About 15 genera have been grouped here (Benjamin, 1956; Kuehn, 1958; Apinis, 1964). Most species occur in soil, and fruit on animal substrata such as dung, feathers, wool, skin and bones. Some are skin pathogens (dermatophytes) of animals including man (Kuehn *et al.*, 1964). The ascocarp usually consists of a loose mesh of hyphae surrounding the asci, or, rarely, the asci may be naked.

Byssochlamys

Two species have been distinguished: *B. nivea* and *B. fulva* (Brown & Smith, 1957). *Byssochlamys fulva* is of economic interest because it may cause spoilage

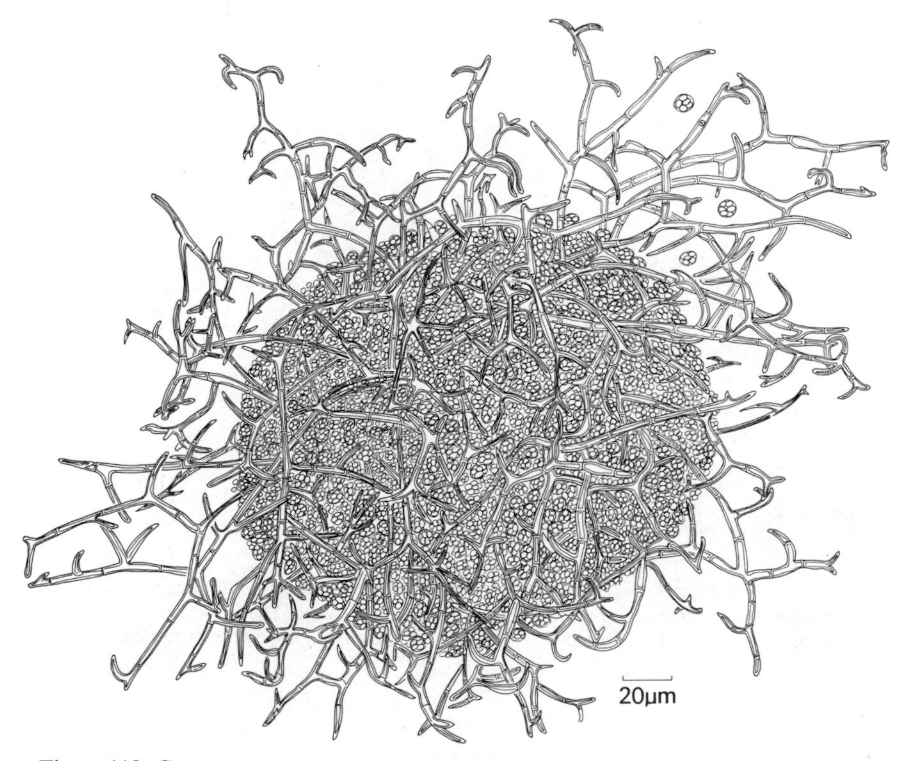

Figure 112. *Gymnoascus reessii.* Ascocarp showing branched peridial hyphae and asci.

of canned and bottled fruit. It is fairly common in the soil of orchards and so may be splashed on to fruit. Its ascospores can survive a temperature of 84–87°C for 30 min. and have been known to retain viability up to 98°C. They are also resistant to high concentrations of SO_2 and alcohol. Moreover the fungus can grow at low oxygen tension. These combined properties explain its survival and growth in canned or bottled fruit. *Byssochlamys nivea* has been isolated from soil. In culture both species reproduce asexually by the formation of chains of conidia derived from tapering open-ended phialides (see Fig. 111A) which occur singly or in groups on the aerial mycelium. Conidial apparatus of this type belongs to the form-genus *Paecilomyces* of the Fungi Imperfecti. Terminal, thick-walled unicellular aleuriospores are also found. The asci of *Byssochlamys* develop best in cultures incubated around 30°C. In *B. nivea* a club-shaped antheridium becomes encoiled by an ascogonium (Fig. 111B). Later the coiled ascogonium develops short branches, or ascogenous hyphae, which bear globose, eight-spored asci either terminally or laterally, so that eventually clusters of asci can be found, but there is no sign of sterile hyphae enclosing them (Fig. 111C).

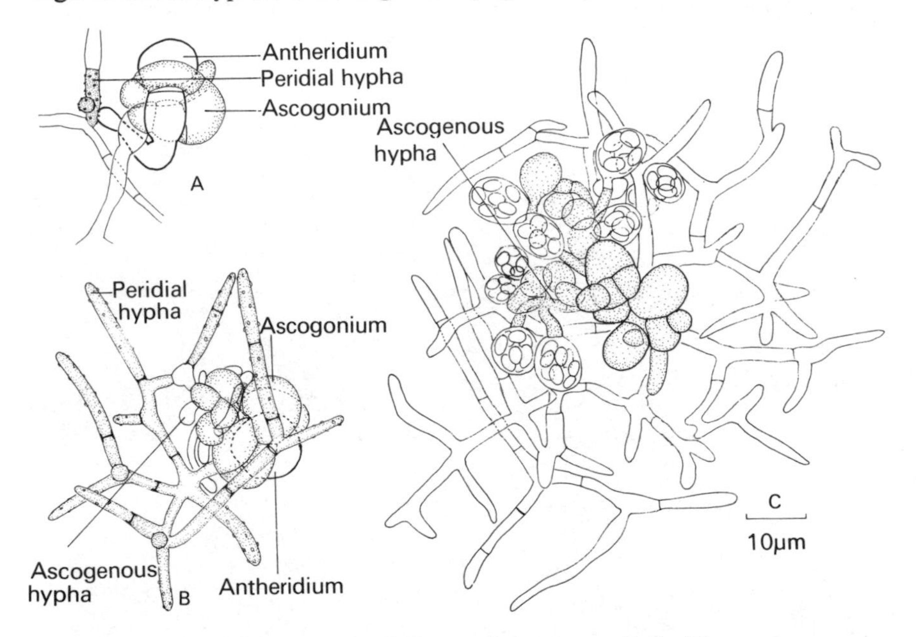

Figure 113. *Gymnoascus reessii*: development of ascocarp. A. Antheridium and ascogonium. B. Ascogonium showing development of ascogenous hyphae. The peridial envelope is also developing. C. Young ascocarp showing asci at the tips of ascogenous hyphae.

Gymnoascus

Accounts of this genus have been given by Kuehn (1969), Orr *et al.* (1963) and Apinis (1964). *Gymnoascus reessii* forms minute reddish-brown globose ascocarps on animal substrata, sacking etc., consisting of branched, recurved, thick-walled hyphae loosely enclosing a mass of asci (Fig. 112). Ascocarp development in culture begins from paired gametangia which arise from the same or from different hyphae. The antheridium is club-shaped and the ascogonium coils around it (Fig. 113A). The ascogonium becomes septate and its cells give rise to ascogenous hyphae (see Fig. 113B) whose tips develop into croziers. Asci develop from the penultimate cells of the croziers (Kuehn,

1956). The branched peridial hyphae arise from vegetative hyphae in the region of the gametangium (Fig. 113B,C). The asci do not discharge violently; the ascus wall disappears and the spores escape through the loose envelope. There is no conidial stage.

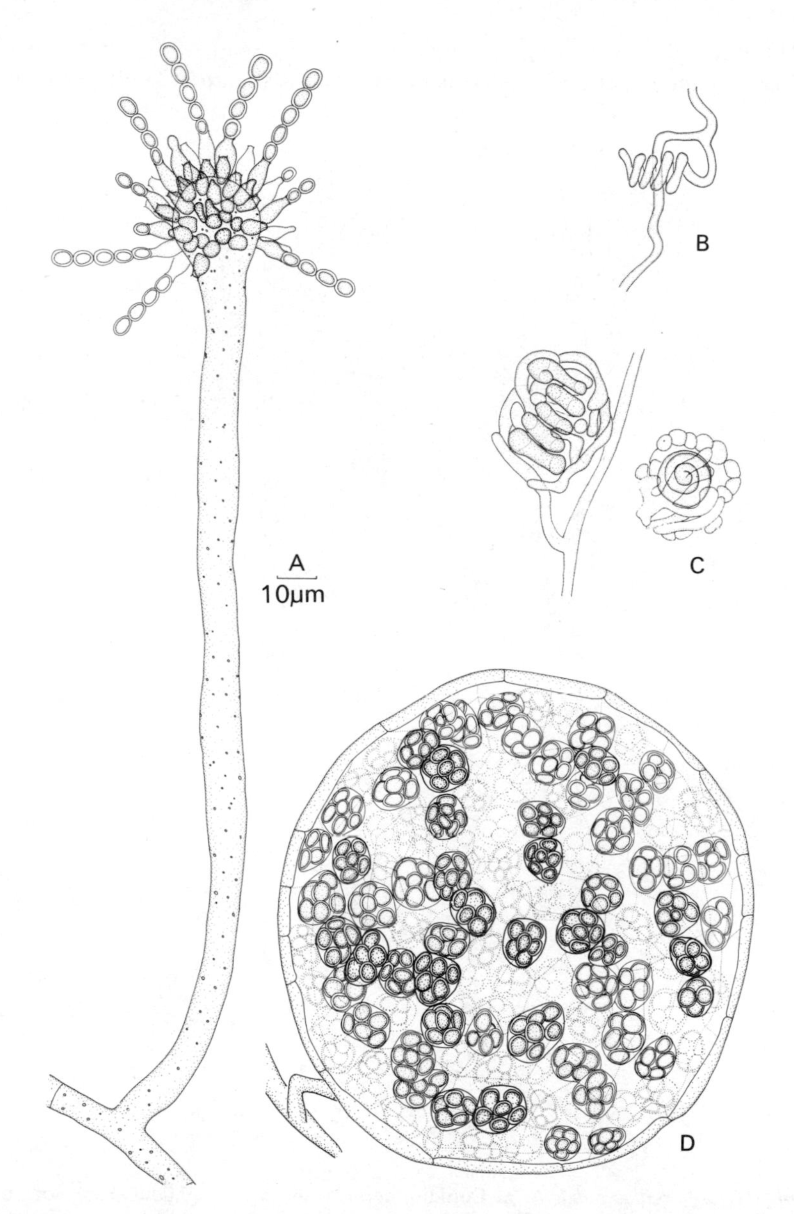

Figure 114. *Aspergillus repens*. A. Conidiophore. B. Ascogonium. C. Ascogonium surrounded by sterile hyphae. D. Cleistocarp showing mature and immature asci.

Eurotiaceae

The two best-known genera are *Aspergillus* and *Penicillium*. Both genera include species which have ascocarps and some which do not, i.e., in these species only conidia have been described.

Aspergillus

Species of *Aspergillus* are abundant in soil, and also cause spoilage of food,

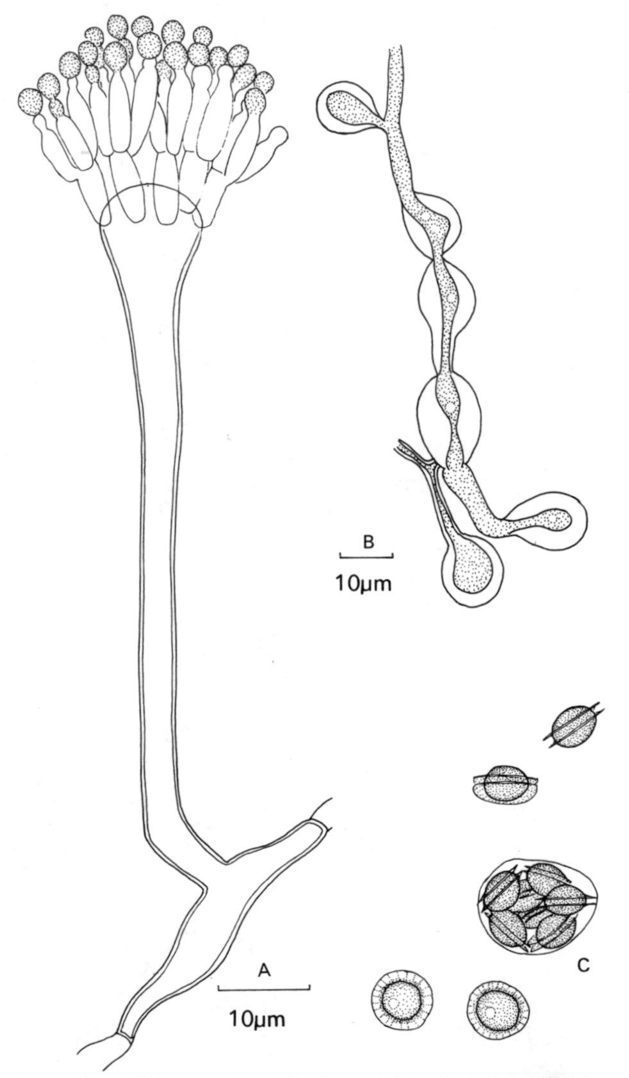

Figure 115. *Aspergillus nidulans*. A. Conidiophore. Note that the phialides are not borne directly on the vesicle. B. Hülle cells, thick-walled cells surrounding the ascocarp. C. Ascus and ascospores. Note that the ascospores bear a double flange. A and C to same scale.

textiles, etc. Some, e.g. *A. fumigatus*, cause pulmonary disease in man and animals, whilst forms of *A. nidulans* have been isolated from diseased finger-nails. Ground-nut meal infected with *A. flavus* and its allies may be carcinogenic to animals. A number of Aspergilli are used commercially; e.g. *A. niger*, used in citric acid fermentation, and *A. oryzae* used in the manufacture of diastase. Other strains are known to produce antibiotics. About 160 species of *Aspergillus* have been recognised (Raper & Fennell, 1965). *Aspergillus repens* produces both ascocarps and conidia and is common on mouldy jam. In culture, on media low in sugar content, only conidia are formed. The segment of the mycelium from which the conidiophore arises persists as a

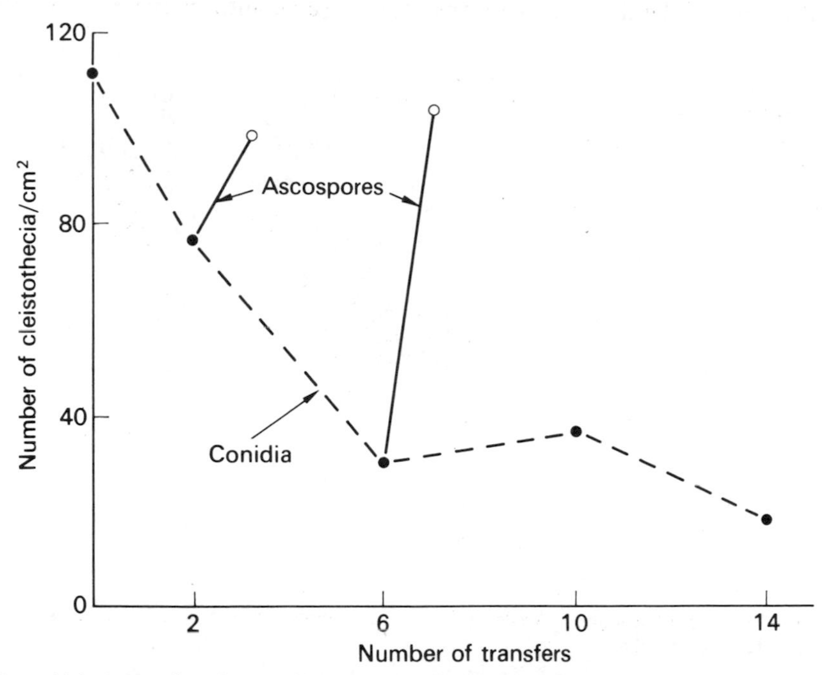

Figure 116. *Aspergillus glaucus*. Change in density of cleistothecia in subcultures transferred either by means of conidia or ascospores (after Mather & Jinks, 1958).

swollen foot cell (Fig. 114A). The tip of the conidiophore swells to form a club-shaped vesicle bearing directly on its surface a cluster of bottle-shaped phialides which give rise to chains of green conidia in basipetal succession. On media of high sugar content (e.g. 20% sucrose) yellow spherical ascocarps develop also. Aerial hyphae develop coiled ascogonia (Fig. 114B) and although there are reports of associated antheridia (Benjamin, 1955) they are not always visible. The ascogonium becomes invested by sterile hyphae which grow up from the stalk of the ascogonium. The ascogonium becomes septate, and from its segments ascogenous hyphae develop which penetrate and dissolve the surrounding pseudoparenchyma derived from the investing hyphae.

Globose asci develop from croziers at the tips of the ascogenous hyphae and when ripe the ascocarp consists of clusters of asci surrounded by a single-layered yellow-coloured peridium. The peridium breaks open irregularly. The asci do not discharge violently: but the spores escape as the ascus walls break down. In some species, e.g. *A. nidulans*, the ascospores have a prominent flanged groove running round the equator of the spore (Fig. 115C) but in *A. repens* such a groove is absent or very inconspicuous.

When first discovered the ascocarps and conidia were not recognised as belonging to the same species and the name *Aspergillus* was used for the conidia and *Eurotium* for the ascocarps. Later it was shown that the two forms formed part of the life history of one fungus. Some authors prefer to use the

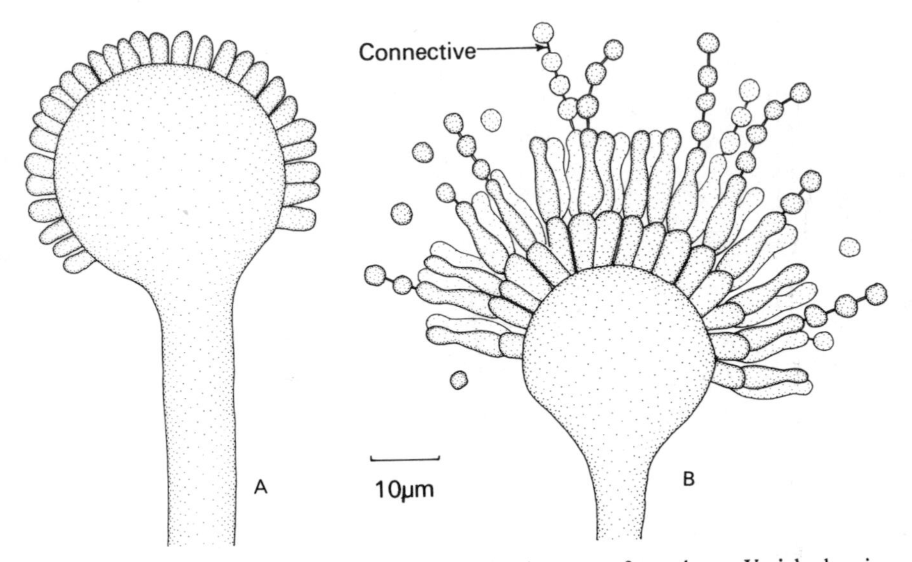

Figure 117. *Aspergillus niger*. A. Veslcle showing development of metulae. B. Vesicle showing phialides and metulae.

generic name *Aspergillus* for both ascocarpic and non-ascocarpic species (but see Benjamin, 1955; Raper, 1957).

Aspergillus repens like nearly all Aspergilli is homothallic, and cultures can be transferred by means of conidia, ascospores or hyphal tips. If repeated transfers are made by conidia only, the ability to form ascocarps declines but can be restored by making subcultures from ascospores (see Fig. 116, Mather & Jinks, 1958). This suggests that the formation of ascocarps and conidiophores is partially controlled by cytoplasmic determinants, and it also suggests a way in which the non-ascocarpic forms might have evolved. Ascocarps are lacking in many species, such as *A. niger* and *A. oryzae*. The demonstration that one species, *A. heterothallicus*, is heterothallic should stimulate investigation of other species in which ascocarps have not been described to see if any of these are also heterothallic (Kwon & Raper, 1967). In *A. niger* and some other Aspergilli the phialides are not borne directly on the vesicle, but are borne on intermediate cells (see Fig. 117).

Penicillium

This is one of the most cosmopolitan genera of fungi, occurring whenever substrata and conditions are suitable for growth. It is abundant in soil and on

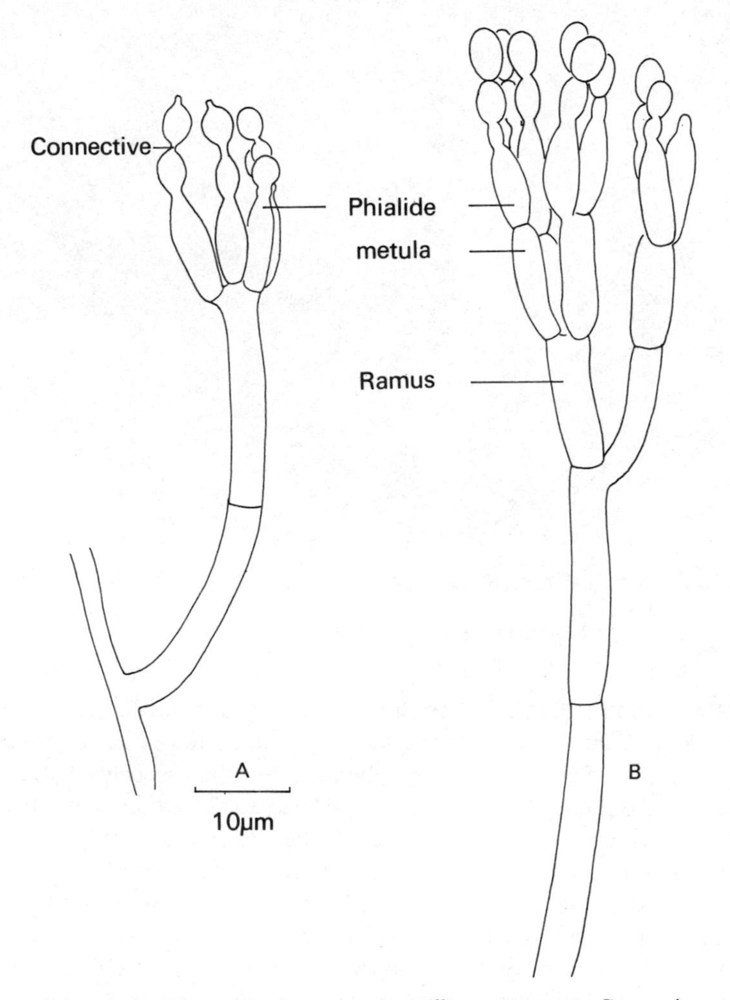

Figure 118. A. *Penicillium spinulosum*. B. *Penicillium expansum*. Comparison of conidiophore structure.

all kinds of decaying materials, and the spores are almost universally present in air so that it is a frequent contaminant of cultures. The contamination of a bacterial culture by *P. notatum* resulted in inhibition of bacterial growth,

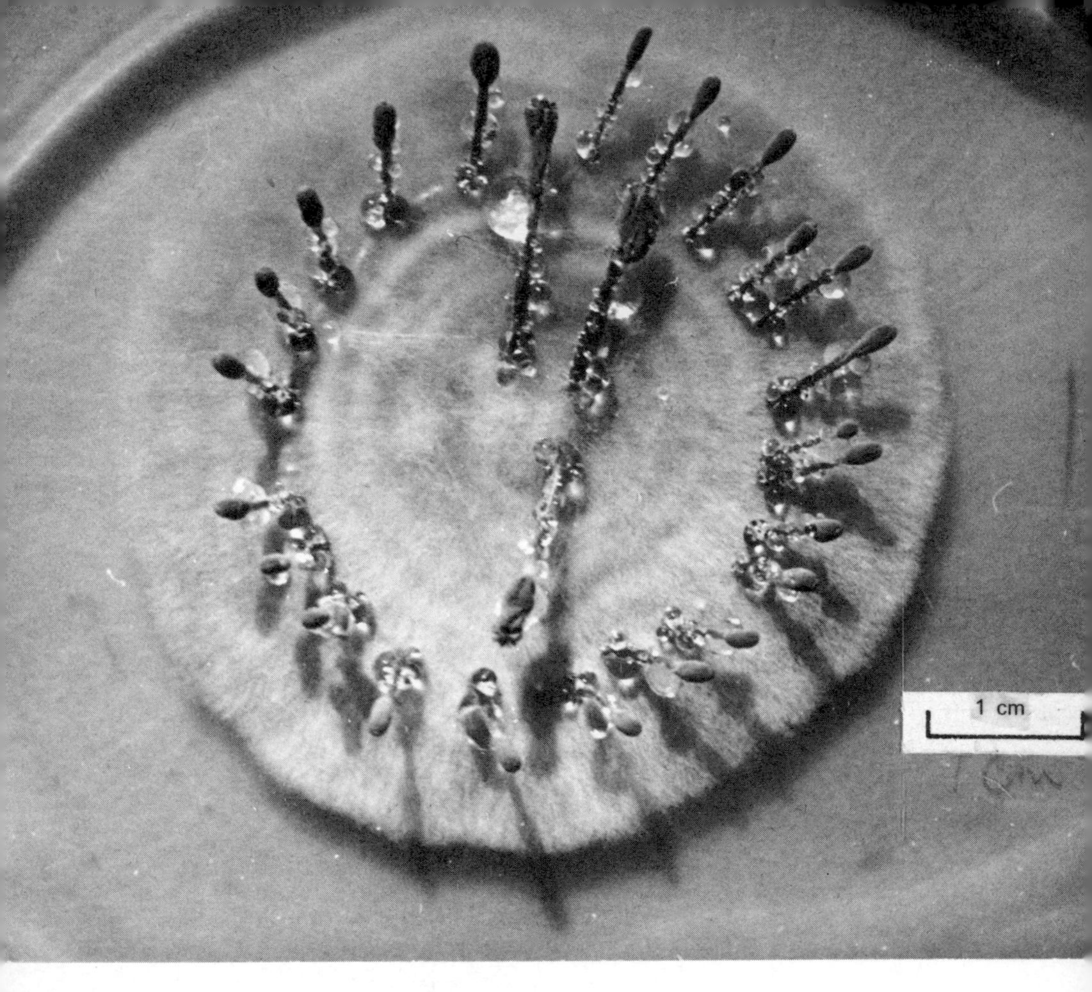

Figure 119. *Penicillium claviforme.*

and this observation led Fleming to the discovery of the antibiotic penicillin (Fleming, 1944). Although *P. notatum* was first used in antibiotic production, an intensive search for antibiotic production by other species led to its replacement by the related *P. chrysogenum* (Raper, 1952). Other economically important species are *P. camemberti* and *P. roqueforti* which play a part in cheese fermentation, and *P. griseo-fulvum*, a source of the antibiotic griseofulvin which causes distortion of fungal hyphae and has proved useful clinically in the treatment of skin and nail infections (Brian 1960). *Penicillium italicum* and *P. digitatum* cause rotting of citrus fruits, whilst *P. expansum* causes a brown-rot of apples.

The classification of the genus is difficult and over 100 species have been recognised (Raper & Thom, 1949). Whilst some form ascocarps, the majority do not. The characteristic conidial apparatus is a branched conidiophore, bearing successive whorls of branches terminating in clusters of phialides. In some species (belonging to the section Monoverticillata), e.g. *P. spinulosum*, the phialides are borne directly on the conidiophore. More commonly they are borne on a further whorl of branches or metulae, and these may in turn arise on a further verticil of branches, the rami, e.g. *P. expansum* (Fig. 118B). In some species, e.g. *P. claviforme*, the individual conidiophores may be aggregated together into club-shaped fructifications or *coremia* (see Fig. 119).

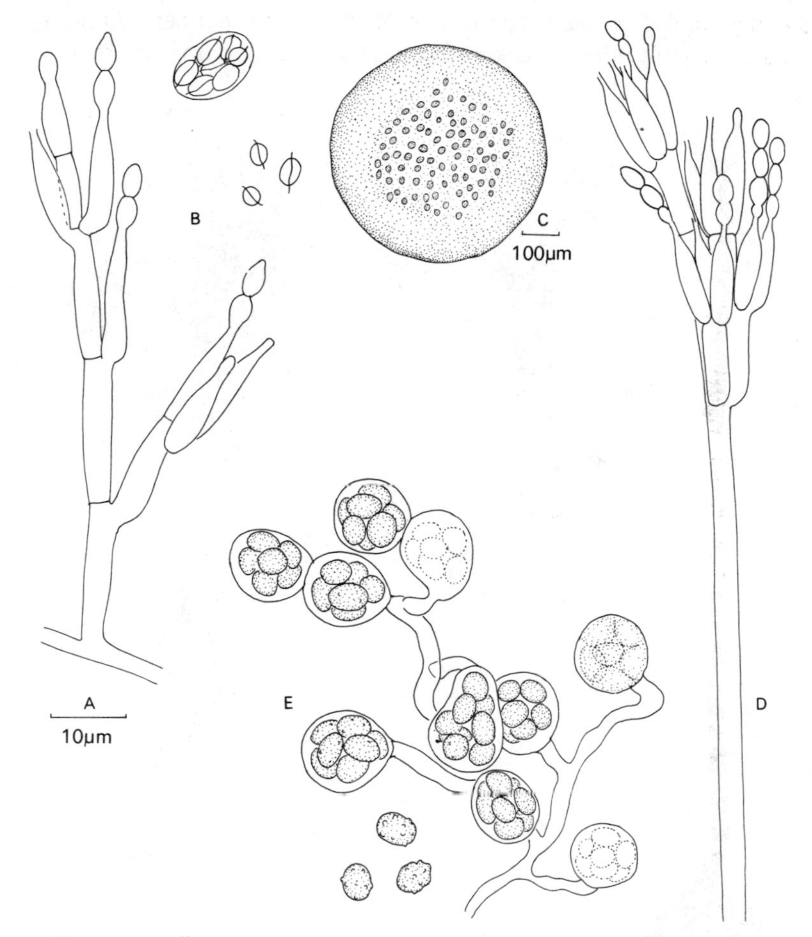

Figure 120. A. *Penicillium stipitatum*, conidiophore. B. Ascus and ascospores: note the equatorial frill. C. *Penicillium vermiculatum*, ascocarp. D. Conidiophore: note the long tapering phialides characteristic of the Biverticillata–Symmetrica. E. Ascogenous hyphae and asci: note that some asci arise in chains.

The ascocarps of *Penicillium* (Fig. 120) are usually yellow and are of two main types: loose cottony wefts enclosing the asci, as in *P. wortmanni*, or more compact pseudoparenchymatous fructifications as in *P. brefeldianum*. All species in which ascocarps have been found are homothallic (Emmons, 1935, Raper & Fennell, 1952).

Ophiostomataceae

Ceratocystis

Over 40 species are known (Hunt, 1956), some causing serious plant diseases —e.g. *C. ulmi* the cause of Dutch elm disease; *C. fagacearum* the cause of oak wilt; *C. fimbriata* the cause of a rot of sweet potatoes, wilt diseases of rubber and coffee; and *C. adiposa* the cause of black rot of sugar cane. Other species cause staining of timber, attacking mainly the medullary rays of the sapwood,

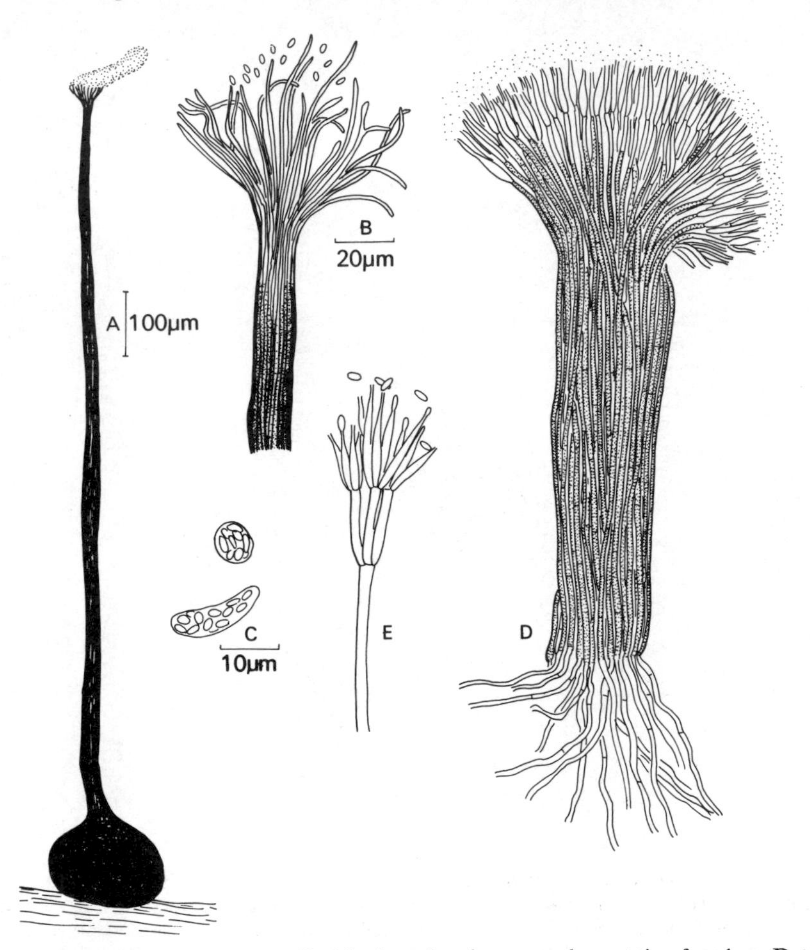

Figure 121. *Ceratocystis piceae*. A. Perithecium showing spore drop at tip of neck. B. Details of ostiole showing ring of setae. C. Asci. D. Conidial fructification bearing sticky mass of spores. E. Details of apex of a conidiophore. B and D to same scale; C and E to same scale.

e.g. *C. piceae* and *C. coerulescens*. The perithecia have a swollen base and a long slender neck often bearing a fringe of hairs at its tip (Fig. 121). The ascus wall breaks down early during the development of the ascospores so that intact asci are difficult to find. The ascospores are forced along the narrow

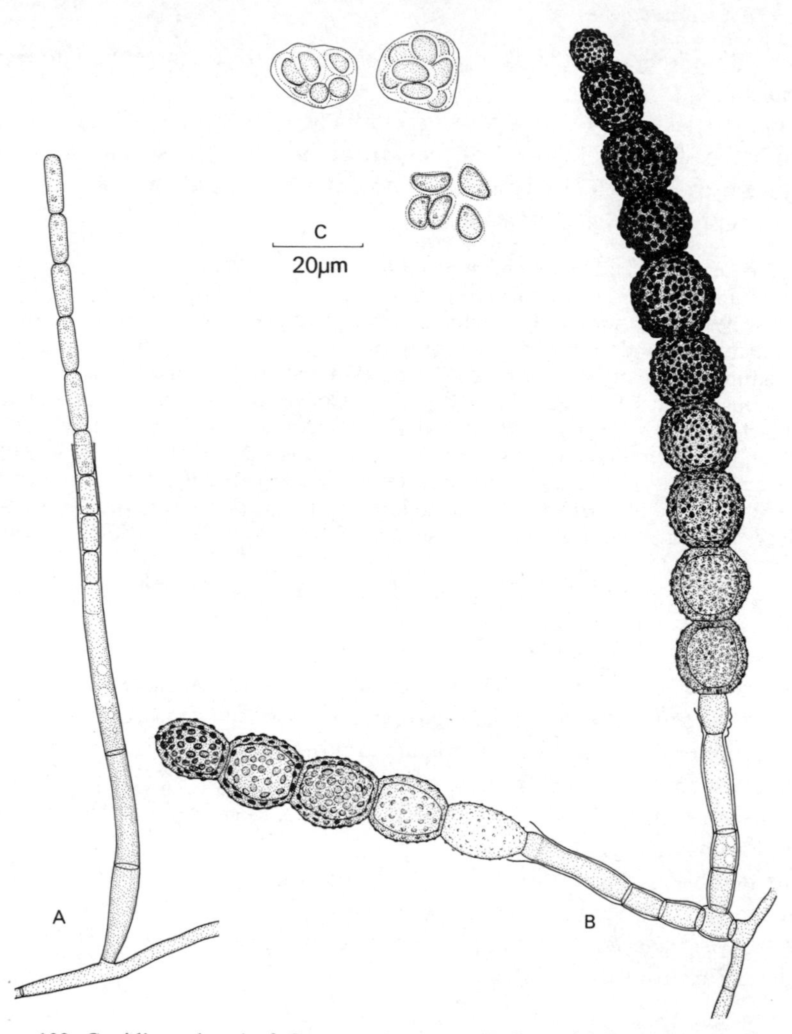

Figure 122. Conidia and asci of *Ceratocystis* spp. A. *Endoconidiophora* type conidia of *C. coerulescens*, B. Conidia of *C. adiposa*. C. Asci and ascospores of *C. adiposa*.

neck and accumulate as sticky spore drops held in place by the ostiolar fringe (Fig. 121) probably as an adaptation to insect dispersal. The ascospores are small and hyaline and vary in shape from ellipsoidal to bean-shaped, hat-shaped, quadrangular or needle-shaped. Various kinds of conidial apparatus are known (Fig. 122). In *C. ulmi* and *C. piceae* there are two distinct kinds of conidia. Coremia, consisting of a parallel bundle of dark hyphae branching at their tips give rise to a mass of sticky conidia also forming a spore drop. Such conidia were earlier described as *Graphium* spp. A second dry-spored conidial type on short sterigmata is also known in *C. ulmi*. In other species of *Ceratocystis* conidia are borne on a single stalk with a much-branched head, belonging to the form-genus *Leptographium*, or are formed in chains within long tapering phialides, belonging to the form-genus *Endoconidiophora*.

Some species, e.g. *C. ulmi*, are heterothallic; others, e.g. *C. piceae*, are homothallic. The development of the perithecia follows a distinctive course designated by Luttrell (1951) as the *Ophiostoma* type, but the name *Ceratocystis* has priority over *Ophiostoma*.

The ascogonia are free upon the mycelium. Branches from the stalk cell of the ascogonium or from neighbouring vegetative hyphae envelop the ascogonium to form the perithecial initial. The outer layers of hyphae develop into a perithecial wall which surrounds a centrum composed of pseudoparenchymatous cells. The asci mature progressively from apex to base of the perithecium along chains of ascogenous cells derived from the ascogonium. Consequently, the asci are produced irregularly throughout the centrum and never form a definite wall layer. As the asci mature the sterile cells of the centrum collapse and disintegrate to form the perithecial cavity. The walls of the asci deliquesce and free the ascospores within the perithecial cavity. Growth of hyphae in the apical region of the perithecial wall produces an elongated neck which is penetrated by a lysigenous ostiole.

The essential feature of this kind of development is that here the asci are produced irregularly throughout the centrum instead of forming a lining to the base of the perithecium.

This kind of development has been found in various species of *Ceratocystis*. Whilst some accounts claim that asci do not arise following crozier formation, later accounts show that typical croziers are formed at the tips of ascogenous hyphae which may arise from a cushion in the base of the perithecium (Andrus & Harter, 1937; Gwynne-Vaughan & Broadhead, 1936; Bakshi, 1951; Rosinski, 1961).

The taxonomic disposition of *Ceratocystis* is not generally agreed. Whilst some authors include it in the Plectomycetes (Nannfeldt, 1932; Hunt, 1956), others have suggested that it might be placed in a separate order, the Ophiostomatales (Rosinski, 1961).

DUTCH ELM DISEASE

Dutch elm disease is at present not severe in Britain, but in North America it is the most destructive disease of shade trees. It was probably introduced about 1930 and now occurs in many of the eastern and central states. It is particularly severe on *Ulmus americana*, but other species such as *U. pumila* and *U. parvifolia* are fairly resistant. The disease shows itself as a wilting and yellowing or drying of the foliage, followed by defoliation and death of branches. Infected trees may die within a few weeks. These symptoms are the result of occlusion of the vessels, especially those of the current season with gum and tyloses derived from the wood parenchyma. Affected wood shows patches of brown discoloration.

The disease is spread by bark-boring beetles, especially *Scolytus multistriatus* which tunnel into wood and bark to feed or to lay eggs. The sticky conidia and ascospores adhere to the bodies of the insects and may be introduced into the vessels of the spring wood where they can be carried upwards with the sap. Control measures are centred largely on insecticide treatments, but attempts to breed resistant elm varieties are also being made (Whitten & Swingle, 1964).

3: PYRENOMYCETES

The term is used here in the narrow sense of Ainsworth (1966) for ascomycetes with true perithecia containing unitunicate asci. The definition of a true perithecium is given by Miller (1928, 1949):

> that is, walls arising from an archicarp, soon forming a globose hollow ball with a hymenium of asci and apically free paraphyses developing from a wall layer and an ostiole formed by extension of the wall tissue and lined with periphyses. The mature asci have thin lateral and thick apical walls penetrated by a pore. The spores are typically liberated through the pore.

The term Pyrenomycetes is often used in a much wider sense to include several orders of Loculoascomycetes, forms with bitunicate asci, but since the forms with unitunicate and bitunicate asci are possibly not closely related it will avoid confusion if the use of the term is restricted. Perithecia of the Pyrenomycetes can develop in various ways. Following Luttrell (1951) we may distinguish the following patterns of development.

1. THE *Xylaria* TYPE

(Note: Dennis, 1958*a,b*, has claimed that the name *Xylosphaera* should have priority over *Xylaria*, but this proposal has not been welcomed by other mycologists, e.g. Petrak, 1962; Holm & Müller, 1965.)

The ascogonia are produced free upon the mycelium or more commonly within a stroma. Branches from the stalk cells of the ascogonium or from neighbouring vegetative hyphae surround the ascogonium and form the perithecial wall. Hyphal branches with free tips (paraphyses) grow upward and inward from the inner surface of the wall over the base and sides of the perithecium. Pressure exerted by the growth of opposed paraphyses expands the perithecium and creates a central cavity. The perithecium becomes pyriform as a result of growth of hyphae in the apical region of the wall to form a neck. The layer of inward-growing hyphae is continuous up the sides and into the perithecial neck. Growth of these hyphae within the neck produces a schizogenous ostiole lined with free hyphal tips (periphyses). The ascogonium produces ascogenous hyphae which typically grow out along the inner wall over the base and sides of the perithecium. Asci derived from the ascogenous hyphae grow among the paraphyses to form a continuous hymenium of asci and more or less persistent paraphyses lining the perithecial cavity. In some forms the paraphyses are evanescent, and the ascogenous hyphae form a plexus in the base of the perithecium. The asci then arise in a single aparaphysate cluster.

This type of development occurs in *Xylaria polymorpha*, *Cordyceps militaris* (Varitchak, 1931), and *Claviceps purpurea* (Killian, 1919). Whilst some authors (van der Weyen, 1954; Whiteside, 1961) interpret the development of *Chaetomium* as belonging to this type, others (Chadefaud & Avellanas, 1967) believe that it shows greater resemblance to the *Diaporthe* type (see below). In *Xylaria* and its allies the ascogenous hyphae spread out along the base and sides of the perithecial cavity, so that the asci are arranged in a continuous wall layer and are interspersed by paraphyses which are usually persistent. In *Cordyceps*, *Claviceps* and *Chaetomium* the ascogenous hyphae are restricted to the base of the perithecial cavity and the asci arise in a single basal cluster. Although in *Chaetomium globosum* evanescent paraphyses arise amongst the asci (Whiteside, 1961), in the other genera they are limited to the sides of the perithecial wall and are evanescent (Luttrell, 1951).

2. THE *Nectria* TYPE

The ascogonia, which are formed within a stroma, become surrounded by concentric layers of vegetative hyphae which form a true perithecial wall. The cells of the inner layer of the wall in the apical region of the young perithecium produce a palisade of inward-growing hyphal branches. These hyphal branches grow downward to form a vertically arranged mass of hyphae with free ends termed apical paraphyses (Luttrell, 1956*b*). Pressure exerted by the elongation of the apical paraphyses, accompanied by expansion of the wall, creates a central cavity within the perithecium. The free tips of the apical paraphyses ultimately push into the lower portion of the wall so that they become attached at both the top and bottom of the perithecial cavity. Ascogenous hyphae arising from the ascogonium spread out across the floor and sides of the cavity and produce asci by means of croziers. The asci grow upward among the apical paraphyses and form a concave layer lining the inner surface of the wall in the basal region of the perithecium. A schizogenous ostiole lined with periphyses (hyphae with free apices, attached at their bases to the inner wall of the neck) develops in the wall of the perithecium. At maturity the perithecia may protrude from the stroma and appear to be seated on its surface (Luttrell, 1951). This type of development has been found in *Nectria* (Hanlin, 1961, Gilles, 1947; Strickmann & Chadefaud, 1961), *Hypocrea* (Hanlin, 1965; Doguet, 1957), and in a few other genera of Hypocreales.

3. THE *Diaporthe* TYPE

The ascogonia are formed within a stroma or free upon the mycelium. Branches from the stalk cell of the ascogonium or from neighbouring vegetative hyphae envelop the ascogonium to form a spherical mass of tissue, the perithecial initial. The outer layers of this mass become differentiated into a perithecial wall. The central portion develops into a centrum composed of pseudoparenchymatous cells. Expansion and ultimate disintegration of these pseudoparenchyma cells produces the perithecial cavity. The asci expand as a group into the disintegrating centrum pseudoparenchyma and ultimately form a layer lining the base of the perithecial cavity. Growth of hyphae in the apical region of the perithecial wall produces a more or less elongated perithecial neck. The neck is penetrated by a schizogenous periphysate ostiole.

Development of this type has been described in *Sordaria fimicola* (Piehl, 1929; Greis, 1927; Dengler, 1937; Ritchie, 1937) and *Podospora minuta* (Wicker, 1929).

SPHAERIALES

The simpler members of this group have separate perithecia, but in many the perithecia are embedded in a stroma. Perithecia and stromata are typically dark-coloured and carbonaceous, and this characteristic distinguishes them from the Hypocreales which have brightly coloured fleshy perithecia or stromata. Another distinction is that in the Sphaeriales the paraphyses are true paraphyses, free at their apex, whilst in the Hypocreales there are apical paraphyses which grow downwards from the roof of the perithecium. However, many authorities do not regard these criteria sufficient to warrant separation. Many of these fungi grow as saprophytes or as weak parasites on woody hosts: some are common in soil and on dung. There are several families (for details see von Arx & Müller 1954, Müller & von Arx, 1962), but we shall consider examples from only the Sordariaceae, Melanosporaceae and Xylariaceae.

Sordariaceae

The perithecia are usually single and dark-walled, with an ostiole lined by periphyses. The asci are thin-walled, and the apical apparatus of the ascus is in the form of a thickened annulus or apical plate which does not stain blue with iodine. Free-ended paraphyses are often present but may dissolve at maturity. The ascospores are black and sometimes surrounded by a mucilaginous epispore or have mucilaginous appendages. They are mostly unicellular and germinate by means of a germ pore.

Sordaria

Perithecia of *Sordaria* are common on the dung of herbivores and occasionally on other substrata (Moreau, 1953). The best-known species is *S. fimicola*, which has been used in experiments on nutrition, the physiology of fruiting, spore liberation and genetics. Perithecial development occurs within nine days on a wide range of media. A longitudinal section of a perithecium (Fig. 123) shows a basal tuft of asci at different stages of development. The asci elongate in turn and only one ascus can occupy the ostiole at a time. The spores are flung out for distances of up to 8 cm. Because each spore is about 13 μm wide and the apical apparatus of the ascus is only about 4 μm wide the spores are gripped as they leave the ascus, and projectiles which vary in size from one to eight spores may be formed. The larger the number of spores in the projectile the greater the distance of discharge, doubtless due to the fact that the surface/volume ratio of single spores is greater than that of multiple-spored projectiles, so that the effects of wind resistance are disproportionately high (Ingold & Hadland, 1959*a*) (Fig. 124).

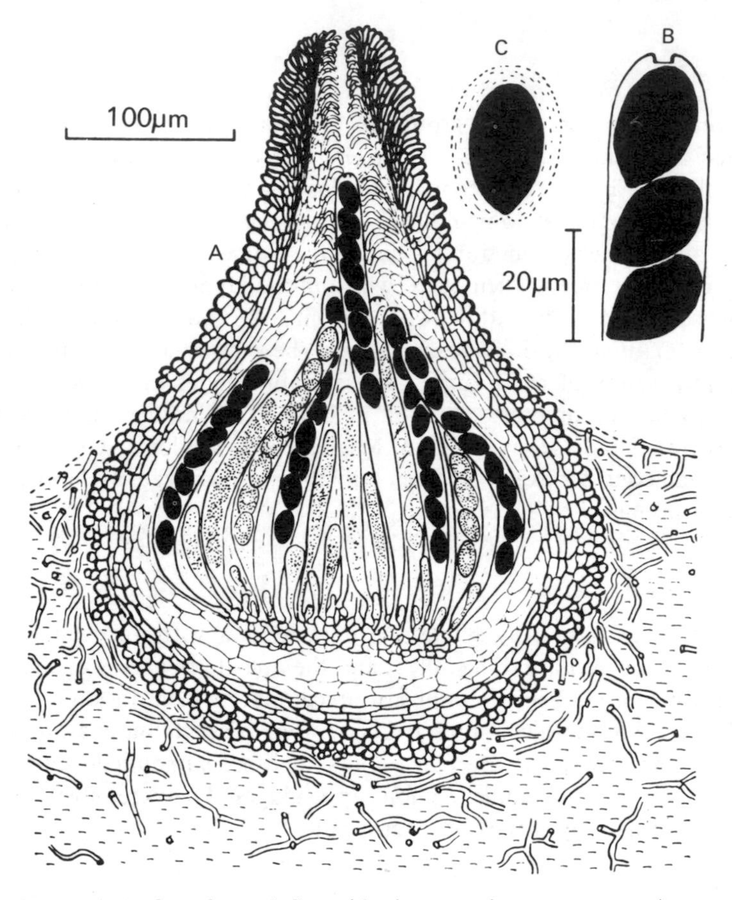

Figure 123. *Sordaria fimicola*. A. L.S. perithecium growing on agar. B. Ascus apex. C. Ascospore showing mucilaginous epispore (after Ingold).

The ascospores of *S. fimicola* have a distinct mucilage envelope which enables them to adhere to herbage. They can survive for long periods and, on drying, gas vacuoles may appear in the spores, but they retain their viability (Ingold, 1956). Spore germination does not readily occur unless the spores are ingested by a herbivore, but chemical treatment with sodium acetate or pancreatin can induce germination. In pure culture growth and fruiting is stimulated by the addition of biotin (Hawker, 1957). Although fruiting occurs equally well in light or dark the ascospore discharge is stimulated by light (Ingold & Dring, 1957; Ingold & Hadland, 1959*b*). The necks of the perithecia are phototropic, and as in many other coprophilous fungi this is probably an adaptation to ensure that spores are projected into the air away from the dung substratum.

The cytological details of perithecial and ascus development in *S. fimicola* do not present unusual features. The cells of vegetative hyphae may contain over 100 nuclei, and several nuclei enter the young ascogonium. Nuclear fusion occurs in the penultimate cell of typical croziers, and this is followed by four nuclear divisions. The first two are meiotic, and the last two mitotic, so that 16 nuclei are present in the ascus. Each spore is binucleate (Carr & Olive, 1958; Heslot, 1958).

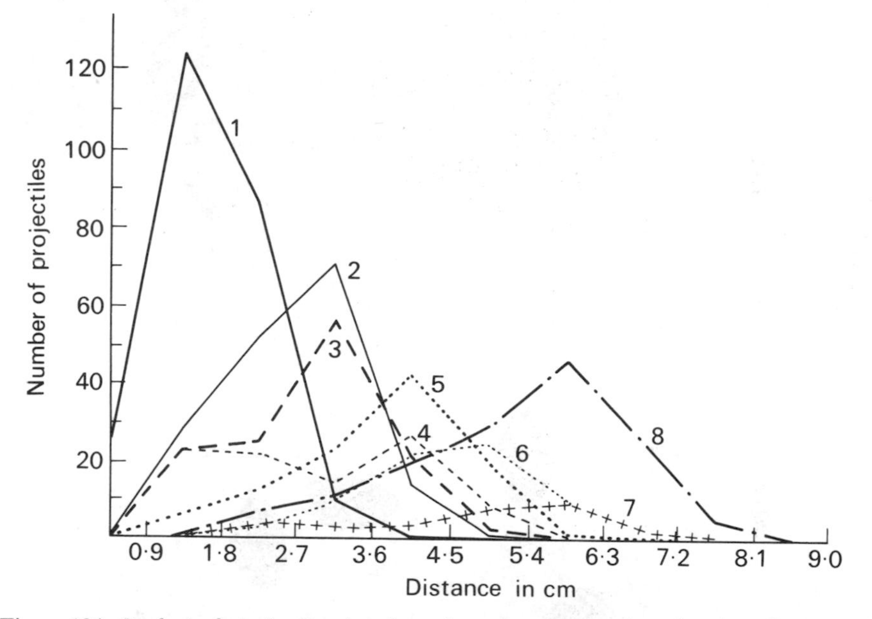

Figure 124. *Sordania fimicola*. Graphs of number of projectiles plotted against distance of projection for projectiles containing one to eight ascospores. The figure associated with each graph shows the number of spores per projectile (from Ingold & Hadland, 1959*a*).

Although *S. fimicola* is homothallic it has the capacity to hybridise. Wild-type strains have black ascospores, but mutants are known with colourless or pale-coloured spores. If a wild-type strain and a white-spored mutant are inoculated together into a culture, a proportion of hybrid perithecia may develop from heterokaryotic segments of the mycelium. Most of the asci from hybrid perithecia contain four black and four white ascospores, and six arrangements of the two kinds of ascospores are found (Fig. 90). Asci which have four black or four white ascospores at the tip of the ascus are those in which the gene for spore colour segregated at the first meiotic division of the fusion nucleus in the young ascus. In those with two black or two white ascospores at the tip of the ascus segregation of the gene for spore colour occurred at the second meiotic division. First-division segregation results from the absence of a cross-over between gene and centromere, whilst second-

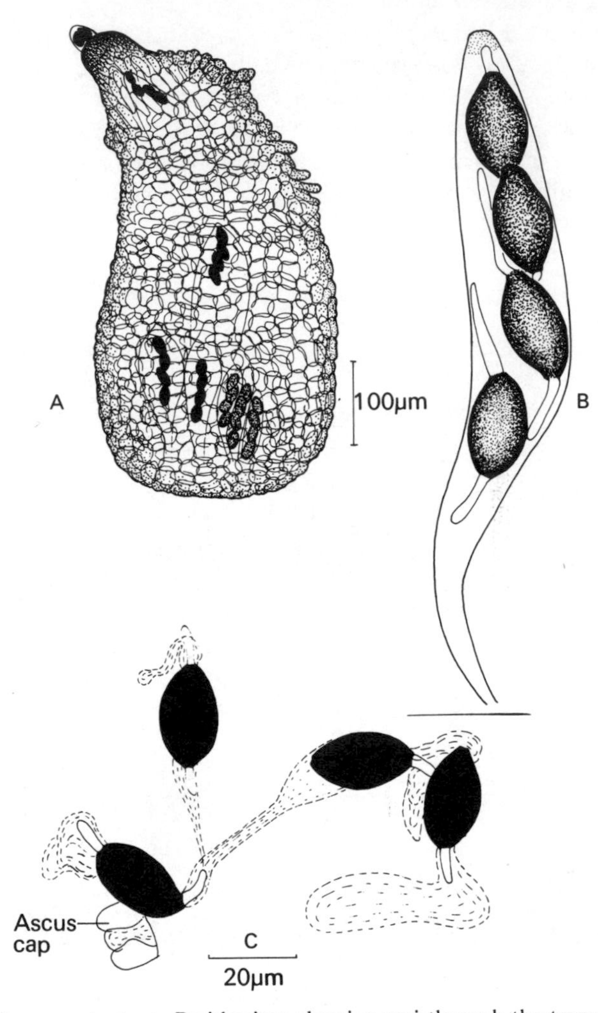

Figure 125. *Podospora minuta*. A. Perithecium showing asci through the transparent wall. B. Ascus. C. Projectile consisting of four spores attached to the ascus cap and to each other by means of mucilaginous appendages.

division segregation results from a single cross-over between gene and centromere. Since the likelihood of crossing over depends on the distance between gene and centromere the frequency of the two kinds of segregation can be used in mapping the position of the gene for spore colour (Catcheside, 1951; Olive, 1956, 1963). In a low proportion of hybrid asci spores showing 5 : 3,

6 : 2, or very rarely 7 : 1 segregations are found. These findings have been largely explained in terms of an eight-strand model of homologous chromosome pairs in meiotic prophase (Kitani *et al.*, 1962; Fincham & Day, 1965). Similar findings have been reported for other Ascomycetes.

Podospora

Perithecia of *Podospora* are also to be found on herbivore dung. Unfortunately there is disagreement about the correct name, and species discussed here have at times been placed in genera such as *Sordaria*, *Pleurage* and *Philocopra* (Cain, 1962; Moreau, 1953). Most species have semi-transparent perithecia within which the outline of the asci can be seen (Fig. 125A). The number of spores in the ascus varies from 4 to 512, and although in the past, spore number has been used as a taxonomic criterion, the species concept has been widened to include forms with a range of spore numbers so that, for example, 8-, 16-, 32- and 64-spored forms of *Podospora decipiens* are recognised (Moreau, 1953; Remacle & Moreau, 1962). The name *Podospora* (podos = foot, spora = seed) refers to the mucilaginous appendage attached to one or both ends of the black ascospore (Fig. 125B). In some of the commonest species, *P. curvula* and *P. minuta*, the spore appendages are attached to the cap of the ascus, and when the ascus explodes, the spores, roped together by their appendages, are shot out as a single projectile (Fig. 125C). As in *Sordaria* it has been shown that multi-spored projectiles are discharged further than single spores (Walkey & Harvey, 1966*a*).

Podospora anserina normally has four-spored asci, each ascospore being binucleate. Cultures derived from single binucleate ascospores usually form perithecia readily. Occasionally, however, smaller uninucleate ascospores may occur in some asci and when such spores are grown fruiting does not occur. Instead, perithecia only develop when certain of the uninucleate ascospore cultures are paired together. On each strain derived from a uninucleate ascospore, ascogonia bearing trichogynes, and spermatia develop, but these are self-incompatible: perithecia only develop if trichogynes of one strain are spermatised by spermatia of a genetically distinct strain (Ames, 1934). Thus although the behaviour of the large ascospores suggests that *P. anserina* is homothallic, it is clear that the underlying mechanism controlling perithecial development is a heterothallic one, of the usual one-gene, two-allele type (i.e. with '+' and '−' strains). Whitehouse (1949) has termed this type of mating behaviour secondary homothallism. The majority of large ascospores (about 97%) contain nuclei of the two distinct mating types (Esser, 1965). In this fungus a novel kind of incompatibility system has also been discovered. If 'uninucleate' strains of different geographical origin are mated together a phenomenon termed 'barrage' can often be observed as a white zone between the two strains (see upper and lower right-hand sectors in Fig. 126). Instead of the expected reciprocal behaviour of '+' and '−' mating types it is commonly found that deviations, occur, e.g.

1. If reciprocal crosses are made, only one of the two reciprocal crosses is compatible; the other is incompatible; this non-reciprocal incompatibility has been termed semi-incompatibility.
2. Reciprocal incompatibility also occurs: between + and − strains of different races no perithecia are formed.

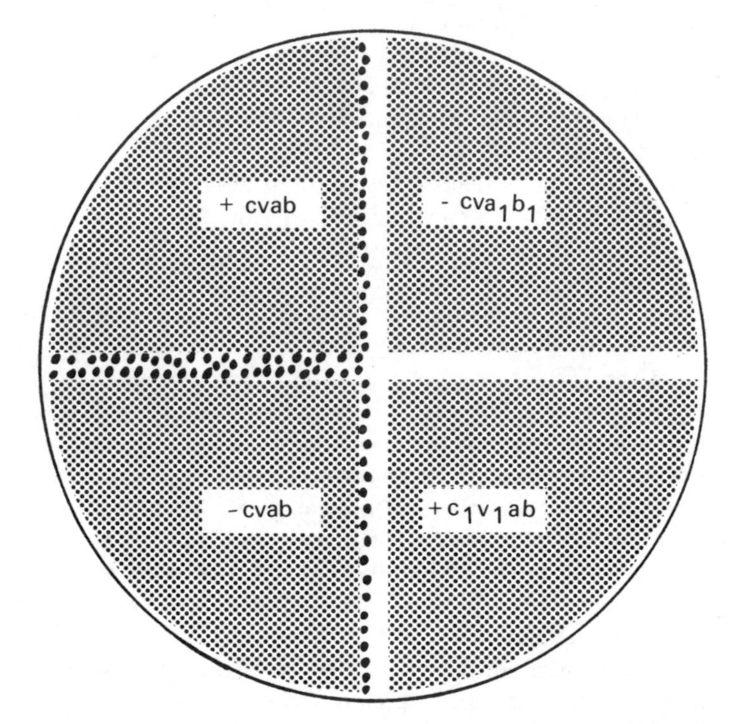

Figure 126. *Podospora anserina*. Genetic basis for heterogenic incompatibility. A Petri dish has been inoculated with four different mycelia. The presence of perithecia is indicated by heavy stippling. In these crosses the upper and lower left-hand sectors contain fully compatible mycelia homozygous at all loci except the mating type locus (+ and −). The upper sectors show semi-incompatibility because the a and b loci are heterozygous, whilst the lower sectors are semi-incompatible because the c and v loci are heterozygous. Semi-incompatibility is demonstrated by the presence of perithecia on only one of the sectors. The two right-hand sectors are fully incompatible because all four loci a, b, c, v are heterozygous (after Esser, 1965).

Esser (1965) has explained the behaviour of this fungus on the basis of four unlinked loci, *a*, *b*, *c*, *v*, each with two alleles. Semi-incompatibility occurs in crosses of the type $ab \times a_1b_1$ and $cv \times c_1v_1$, i.e. where two different alleles are involved. Thus, in Fig. 126 the cross $+\ cvab \times -cvab$ is fully compatible since the two strains differ only in mating type: all the other genes are homozygous. The cross $+\ cvab\ \times -cva_1b_1$ is semi-incompatible since only one pair of genes differ (ab vs a_1b_1). The cross $-\ cvab \times +c_1v_1ab$ is also semi-incompatible but the genes *c* and *v* differ (cv vs c_1v_1). Total incompatibility results from differences between both pairs of genes simultaneously (e.g. cross $-cva_1b_1 \times +c_1v_1ab$). In other words even if two strains differ in mating type they are not fully compatible unless the four genes are identical in each strain.

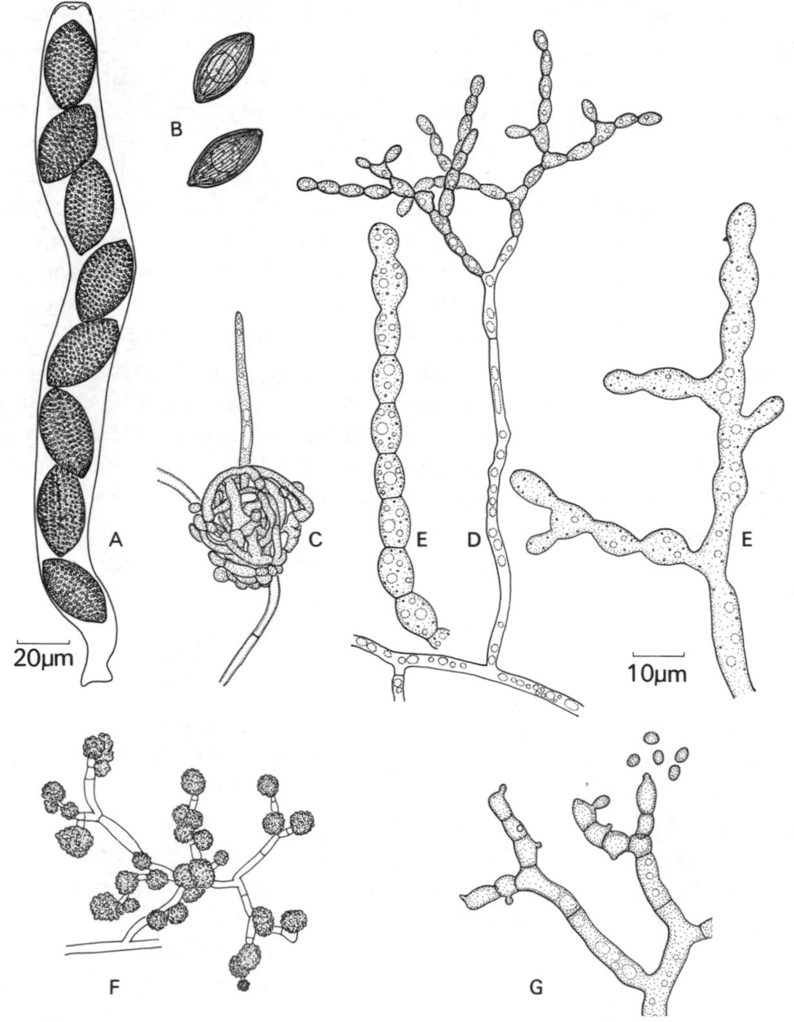

Figure 127. *Neurospora crassa*. A. Ascus. B. Ascospores showing ribbed surface. C. Protoperithecium showing projecting trichogyne. D. Macroconidia from one-day-old culture. E. Enlarged view of developing macroconidia. F. Microconidia forming sticky clusters. G. Enlarged view showing origin of microconidia. A, B, C, D, F to same scale; E, G to same scale.

Incompatibility is here brought about by *differences* in genetic make-up, and Esser has coined the term *heterogenic incompatibility* for this kind of behaviour in contrast with the more common type of incompatibility where genetic similarity precludes mating. This type of incompatibility is not confined to *P. anserina* but is also found in *Sordaria* (Olive, 1956).

Neurospora

Species of *Neurospora* have been widely used in genetical and biochemical studies. The best-known are *N. crassa* and *N. sitophila* both of which are eight-spored and heterothallic. *N. tetrasperma* is four-spored and secondarily homothallic. The reasons why *Neurospora* has proved so useful as a tool in biochemical and genetical research are that wild-type strains have simple nutritional requirements (a carbohydrate source; simple mineral salts; and one vitamin, biotin); mutations can be induced readily by irradiation of conidia; growth and sexual reproduction is rapid; and tetrad analysis by means of ascus dissection is relatively easy (Tatum, 1946; Beadle, 1959). The extensive literature has been compiled by Bachmann & Strickland (1965) as a bibliography containing 2310 references. In nature these species of *Neurospora* colonise burnt ground and charred vegetation, and are also found in warm humid environments such as wood-drying kilns and bakeries where they can cause serious trouble because of their rapid growth and sporulation. For this reason *N. sitophila* is sometimes called the red bread mould (Shear & Dodge, 1927; Ramsbottom & Stephens, 1935). The generic name is derived from the characteristically ribbed ascospores (Lowry & Sussman, 1958; Sussman & Halvorson, 1966) (Fig. 127). The ascospores of *N. crassa* are viable for many years and do not germinate readily unless treated chemically (e.g. by chemicals such as furfural) or by heat shock (e.g. 60°C for 20 min.). Following such treatment the spores germinate and produce a coarse septate rapidly growing mycelium, each segment of which is multinucleate. In contrast the conidia are killed by such heat treatment (Sussman, 1966).

Within 24 hr the mycelium can begin asexual reproduction. Upright branches develop from which branched chains of multinucleate pink conidia arise. Further conidia develop by budding of the terminal conidium on a chain, and when the terminal conidium gives rise to two buds the chain branches (Fig. 127D,E). Conidia of this type belong to the form-genus *Monilia*. The individual segments of the spore chain break apart and are readily dispersed by wind. The spores are formed in vast numbers and if released into a laboratory can cause serious contamination of other cultures.

Cultures derived from a single ascospore also develop two other types of reproductive structure. In contrast to the large dry wind-dispersed macroconidia, clumps of smaller oval sticky microconidia develop laterally (Fig. 127F,G). Ascogonia, terminated by long tapering trichogynes, and surrounded at the base by hyphae also develop (Fig. 127C). Such structures are termed protoperithecia or bulbils.

In *N. crassa* and *N. sitophila* no further development occurs in single ascospore cultures, i.e. each strain is self incompatible. Incompatibility is controlled by a pair of alleles A and a, and if two compatible strains are grown together in a culture vessel for a few days microconidia of one strain can be transferred to the trichogynes of the opposite strain by flooding with sterile water. Transfer of macroconidia of the opposite strain to a trichogyne can also affect fertilisation. Fusion between the trichogyne and the fertilising cell is followed by migration of one or more nuclei from the fertilising cell down the trichogyne into the ascogonium (Backus, 1939). Development of ripe perithecia occurs within 7–10 days, and follows a pattern typical for Ascomycetes generally (Colson, 1934; McClintock, 1945; Singleton, 1953) although Mitchell (1965) has suggested that there may be some anomalous features.

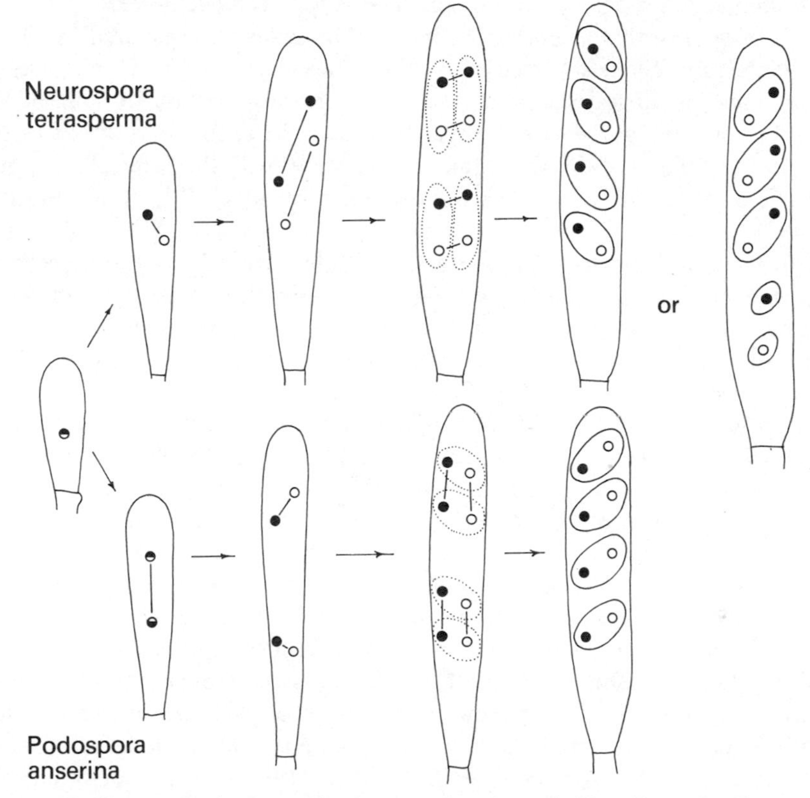

Figure 128. Two mechanisms resulting in secondary homothallism. Top row, *Neurospora tetrasperma* showing first-division segregation of mating-type alleles and overlapping spindles. Bottom row, *Podospora anserina* showing second-division segregation. At the second meiotic division the spindles lie transversely in the ascus. Third-division spindles are parallel to the ascus (after Fincham & Day, 1965).

In *N. tetrasperma* the asci normally contain four binucleate ascospores, and single-spore cultures derived from four-spored asci produce perithecia. As in *Podospora anserina* occasional asci have five or six ascospores, some smaller than the rest and uninucleate. Cultures derived from single small ascospores do not develop perithecia, but when certain of such cultures are paired perithecia develop. The genetical basis for incompatibility in uninucleate ascospores is the same as in *N. crassa*, and the reason why single large ascospores give rise to perithecial cultures is that the two nuclei present usually include one of each mating type. The segregation of the two mating-type alleles almost invariably takes place at the first meiotic division in developing asci, i.e. crossing-over occurs very infrequently between the locus for incompatibility and the centromere. During the second and third nuclear divisions in the ascus the spindles overlap and when the four ascospores are cut out they normally contain two nuclei of distinct mating type, or, occasionally, uninucleate ascospores are formed (Fig. 128) (Dodge, 1927).

It is of interest that in *Podospora anserina* which is also secondarily homothallic the same end is achieved by different means. In this fungus segregation of mating-type alleles occurs at the second meiotic division in 98% of the asci. The spindles at the second division lie transverse to the long axis of the ascus, whilst the third-division spindles lie parallel with the ascus. Ascospores are cut out around pairs of genetically different nuclei (Fig. 128) (Franke, 1957).

In *N. crassa* heterokaryons are not normally formed between mycelia of opposite mating type (Sansome, 1946) and this implies that plasmogamy normally occurs only between a trichogyne of one strain and a fertilising agent (e.g. a microconidium or macroconidium) of the opposite strain. This condition is termed *restricted heterokaryosis*. Later work has shown that even within one mating type the ability to form heterokaryons is under genetical control. Two genes involved are designated *C* and *D*. Stable heterokaryons are normally only formed between strains which have a like genotype (e.g. *CD*+*CD*, *Cd*+*Cd* or *cd*+*cd*). When strains of unlike genotype anastomose there is evidence of cytoplasmic incompatibility resulting in vacuolation and disorganisation of cell contents in the region of the anastomosis. Similar cytoplasmic reactions are visible when anastomosis occurs between hyphae of wild-type strains differing in mating type (Garnjobst, 1955; Garnjobst & Wilson, 1956; Wilson *et al.*, 1961). In contrast heterokaryons are readily formed between different mating-type strains of *N. tetrasperma* (Dodge, 1942), which thus exhibits *unrestricted heterokaryosis*. Whitehouse (1949*a*) has made the interesting suggestion that whilst restricted heterokaryosis necessitates the intervention of sexual organs in plasmogamy, where unrestricted heterokaryosis occurs, the need for sex organs as a prerequisite of plasmogamy no longer holds, although they may still occur and function. This view is supported by evidence that in many Ascomycetes with unrestricted heterokaryosis there are no differentiated sex organs: plasmogamy occurs between vegetative hyphae.

Melanosporaceae

This is a small family with simple perithecia, often with long projecting necks, but sometimes (as in *Chaetomium*) lacking necks entirely. The asci are club-shaped and thin-walled, and break down within the perithecia so that the ascospores ooze out in tendrils and are not violently discharged. The ascospores are unicellular and black in colour. The only genus we shall consider is *Chaetomium* which some authorities place in a separate family (Chaetomiaceae) or order, the Chaetomiales (Ames, 1961).

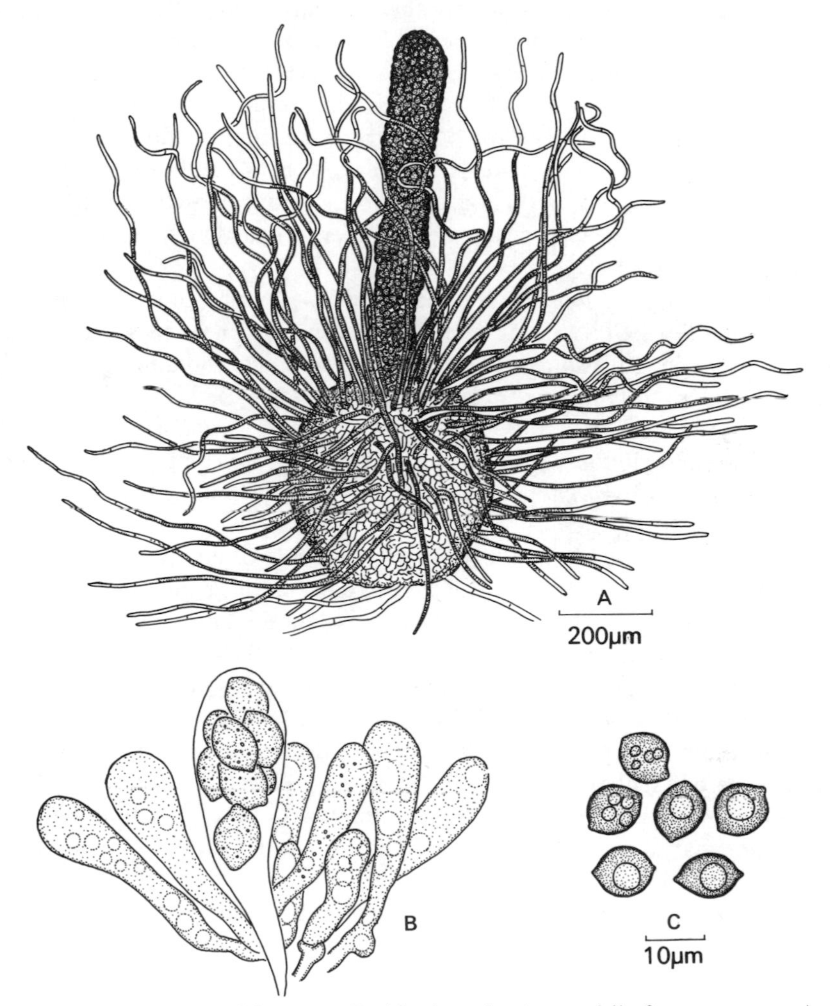

Figure 129. *Chaetomium globosum*. A. Perithecium showing tendril of ascospores. B. Asci. C. Ascospores.

Chaetomium

Over 80 species are known (Ames, 1961) many of them causing decay of cellulose-rich substrata such as textiles in contact with soil, straw, sacking, dung (Lodha, 1964*a*,*b*) and wood, which may undergo a superficial decay known as soft-rot (Savory, 1954). The ability to degrade cellulose is related to the fact that many species possess a powerful cellulase.

The perithecia of *Chaetomium* are superficial and barrel-shaped, and they are clothed with dark stiff hairs. In some species, e.g. *C. elatum*, one of our commonest species, growing on damp rotting straw, the hairs are dichotomously branched. In others, e.g. *C. cochliodes*, the body of the perithecium bears straight or slightly wavy unbranched hairs, whilst the apex bears a group of spirally coiled hairs. When the perithecia are ripe a column-like mass of black ascospores arises from the apex (Fig. 129A). In most species the spores are lemon-shaped. The spore column results from the breakdown of the asci within the body of the perithecium, i.e. the asci do not discharge their spores violently. The young asci are cylindrical to club-shaped, but this stage is very evanescent, and is only found in young perithecia (Fig. 129B). Some species form typical croziers, but in others croziers are lacking (Corlett, 1966; Berkson, 1966).

Most species of *Chaetomium* are homothallic, but a few, e.g. *C. cochliodes*, are heterothallic (Seth, 1967). *Chaetomium elatum* is reported by some workers to be homothallic, but some isolates appear to be heterothallic (Fox, 1953). Conidial states are rare in *Chaetomium* but simple phialides and phialospores occur in *C. elatum* and *C. globosum* whilst *C. piluliferum* forms both phialospores and globose aleuriospores of *Botryotrichum* type (Daniels, 1961). *Chaetomium trigonosporum* has conidia belonging to the form-genus *Scopulariopsis* (Corlett, 1966).

The fruiting of *Chaetomium* species in culture is often stimulated by the inclusion of cellulose in the form of filter-paper, cloth or jute-fibre. Extracts of jute stimulate sporulation of *C. globosum* in pure culture, and stimulation has also been found when *C. globosum* is grown with *Aspergillus fumigatus*. Attempts to analyse the chemical nature of the stimulus have shown that the effect of jute extract can be partially substituted by the addition of calcium to the medium (Basu, 1951). Sporulation is also stimulated by presence in the medium of sugar phosphates and phospho-glyceric acid, and it has been shown that *A. fumigatus* excretes such compounds into the medium and that sugar phosphates are also present in jute extract (Buston & Khan, 1956; Buston *et al.*, 1953).

The breakdown of cellulose by *Chaetomium* is brought about by a powerful cellulase (Abrams, 1950; Agarwal *et al.*, 1963*a*,*b*). When fed with sucrose *C. globosum* breaks down the disaccharide into its two hexose moieties, fructose and glucose, but the glucose moiety is absorbed preferentially. Little sucrose is absorbed directly (Walsh & Harley, 1962). Attack of wood by *C. globosum* is associated with depletion of cellulose and pentosan, but a considerable proportion of the lignin remains (Savory & Pinion, 1958; Levi & Preston, 1965). Some species of *Chaetomium*, e.g. *C. cochliodes*, are antagonistic to soil-borne and seed-borne fungal pathogens, and the possibility of using them to control plant diseases has been investigated (Tveit, 1953, 1956; Tveit & Moore, 1954; Tveit & Wood, 1955). An antibiotic, chaetomin, has been isolated from *C. cochliodes* (Waksman & Bugie, 1944).

Figure 130. A. *Xylaria polymorpha*. Perithecial stromata at base of a stump of *Fagus sylvatica*. B. *Daldinia concentrica*. Perithecial stromata at base of *Fraxinus excelsior*. One stroma has been cut open to show the concentric zonation, and the perithecia in the outer layers.

Xylariaceae

Typical members of this group have perithecia embedded in stromata. Most members are saprophytic on wood, and occasionally on other substrata such as dung, but a few are parasitic on woody hosts, e.g. species of *Rosellinia*. *Ustulina deusta* causes a butt-rot of beech. General accounts of British Xylariaceae have been given by Miller (1930, 1932*a*,*b*) and Dennis (1968).

The ascus tip contains a blue-staining apical apparatus pierced by a narrow pore (Greenhalgh & Evans, 1967). Ascospores are unicellular, black, inequilateral (i.e. with one side more strongly curved than the other) and often with a hyaline equatorial germ pore.

Daldinia

Daldinia concentrica grows parasitically on ash but can continue fruiting on dead trunks and branches. It is rare on other hosts but grows on wood of birch or gorse that has been burnt. On ash it forms large (5–10 cm diameter) hemispherical brown stromata annually which contain ripe asci between May and October. In section (Fig. 130) the stromata show a concentric zonation of alternating light and dark bands. The surface of young stromata may be covered with a pale fawn powdery mass of conidia. The conidia are dry and

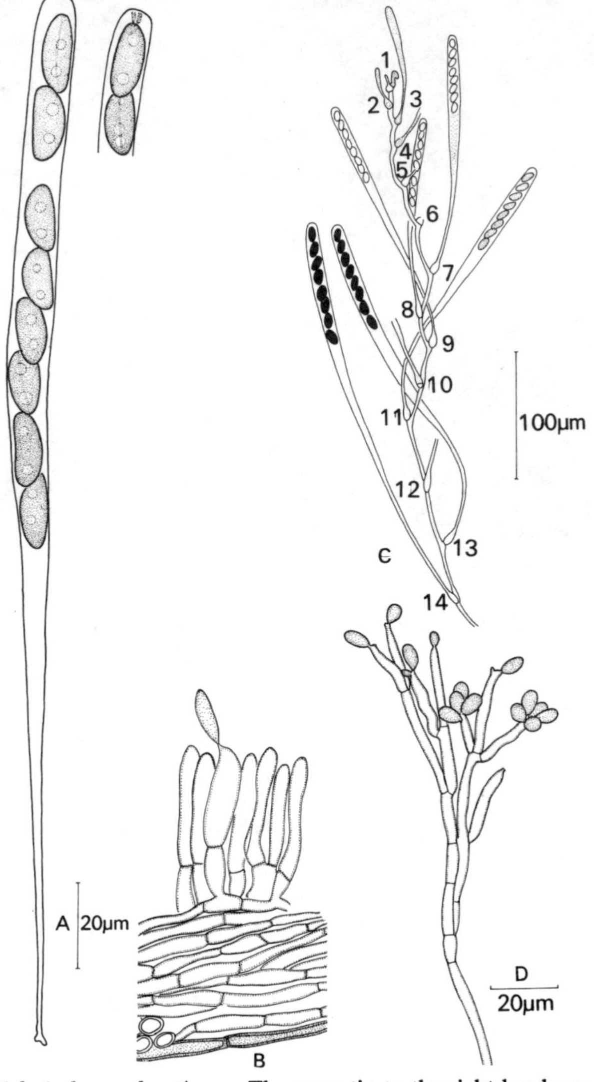

Figure 131. A. *Xylaria hypoxylon*. Ascus. The ascus tip to the right has been stained with iodine to reveal the apical apparatus. B. *X. hypoxylon* conidiophores. C. *Daldinia concentrica*. Ascogenous hypha. The numbers represent successive asci working backwards from the apex. D. *D. concentrica: Nodulisporium*-type conidia. C after Ingold (1954*a*).

oval in shape, developing successively at the tips of branched conidiophores by the outgrowth of the wall, and when detached leave a small scar (Fig. 131D). Conidia of this type have been named *Nodulisporium tulasnei*.

Figure 132. *Xylaria hypoxylon*. A. Conidial stroma. Conidia are borne on the white tips of the branches. B. Perithecial stroma. Perithecia develop at the base of the old conidial stroma.

Perithecia develop in the outer layers of the stroma, each arising from a coiled archicarp. The perithecial wall is lined by ascogenous hyphae which are unusual in that there is often a considerable distance separating successive asci (Ingold, 1954*a*) (Fig. 131C). The stroma of *Daldinia* apparently functions as a water reserve and detached stromata will continue to discharge ascospores for about three weeks even if placed in a desiccator (Ingold, 1946*a*). Spore discharge is nocturnal and the rhythm of spore discharge is maintained for several days if detached stromata are maintained in continuous dark. In continuous light periodic spore discharges ceases after about three days but is restored on return to alternating light and dark (Ingold & Cox, 1955). The output of spores from a single stroma of average size is about 10 million a night.

Xylaria

Stromata of *Xylaria hypoxylon*, the candle-snuff fungus, are common on stumps and fallen branches of deciduous trees. As in most Xylariaceae growing on wood the limits of the mycelium within infected tissues are visible as conspicuous black zone lines. The stromata are branched and cylindrical.

At the upper end the stroma is covered by a white powdery mass of conidia (Fig. 132A, 131B). Perithecia develop later at the base of the stroma and are visible externally as swellings at the surface (Fig. 132B). The apical apparatus

Figure 133. *Hypoxylon fragiforme* and *H. multiforme*. A. *H. fragiforme*: perithecial stromata on *Fagus sylvatica*. One stroma has been broken open to show the perithecia embedded in the outer layers. B. *H. multiforme*: perithecial stromata on *Betula verrucosa*.

of the ascus is visible after staining in iodine as a bright blue cylindrical collar pierced by a narrow pore, even in immature asci (Fig. 131A). *Xylaria polymorpha* (dead men's fingers) grows in late summer and autumn at the base of

Figure 134. ***Hypocrea pulvinata.*** Perithecial stromata on fruit-body of *Piptoporus betulinus*. The darker dots are the perithecial ostioles. Tendrils of white ascospores can be seen issuing from some of the ostioles.

old tree stumps. The stromata are finger-like and clustered (Fig. 130). The surface is at first covered by an inconspicuous conidial layer, but eventually perithecia develop beneath the surface of the whole stroma and are not restricted to the basal region as in *X. hypoxylon*.

Hypoxylon

This is a large genus of over 100 species (Miller, 1961) forming stromata which are often hemispherical of sometimes flattened on the surface of wood and bark. Different species often show a preference for a particular host. In

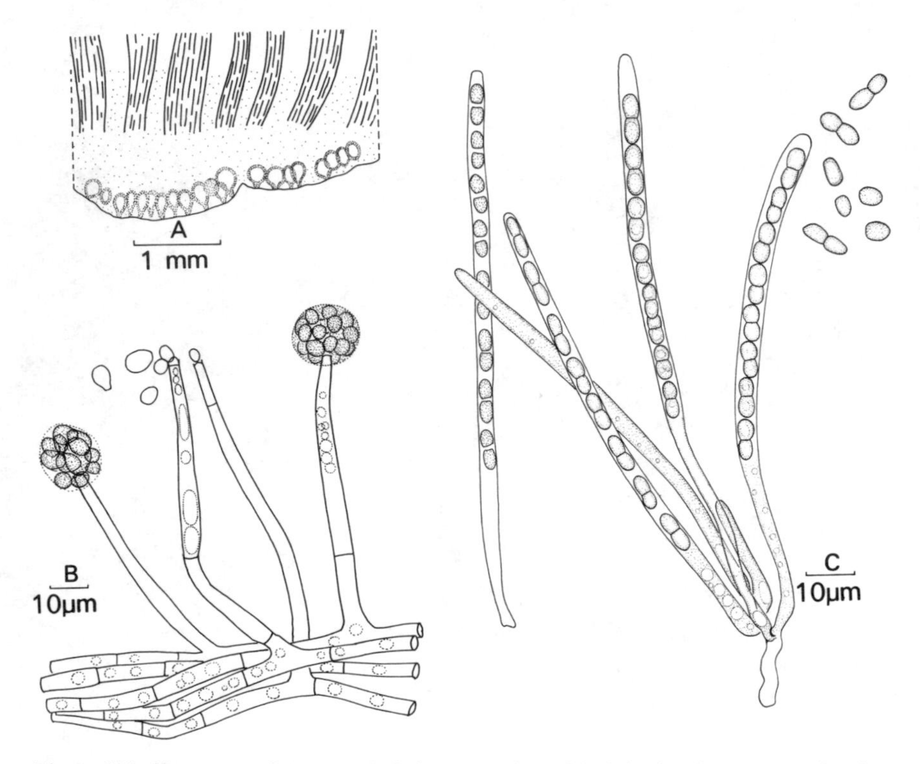

Figure 135. *Hypocrea pulvinata*. A. L.S. lower portion of fruit-body of *Piptoporus betulinus* showing perithecial stroma of *Hypocrea* in section. B. Conidia produced from upright phialides. C. Asci and ascospores. Note how the two-celled ascospores break up into part-spores.

Britain common species are *H. fragiforme* (= *H. coccineum*) almost confined to freshly dead branches and trunks of *Fagus* (Fig. 133A), *H. multiforme* on *Betula* (Fig. 133B), and *H. rubiginosum* which forms flat stromata on decorticated wood of *Fraxinus*. The young stroma of all these species bears a conidial felt of the *Nodulisporium* type (Chesters & Greenhalgh, 1964). Most species show nocturnal spore discharge (Walkey & Harvey, 1966*b*).

HYPOCREALES

The distinctions between this group and the Sphaeriales have already been discussed, but they are not great. We shall study examples from two families the Hypocreaceae and Clavicipitaceae. The latter have a clearly defined apical cap to the ascus, not found in the Hypocreaceae.

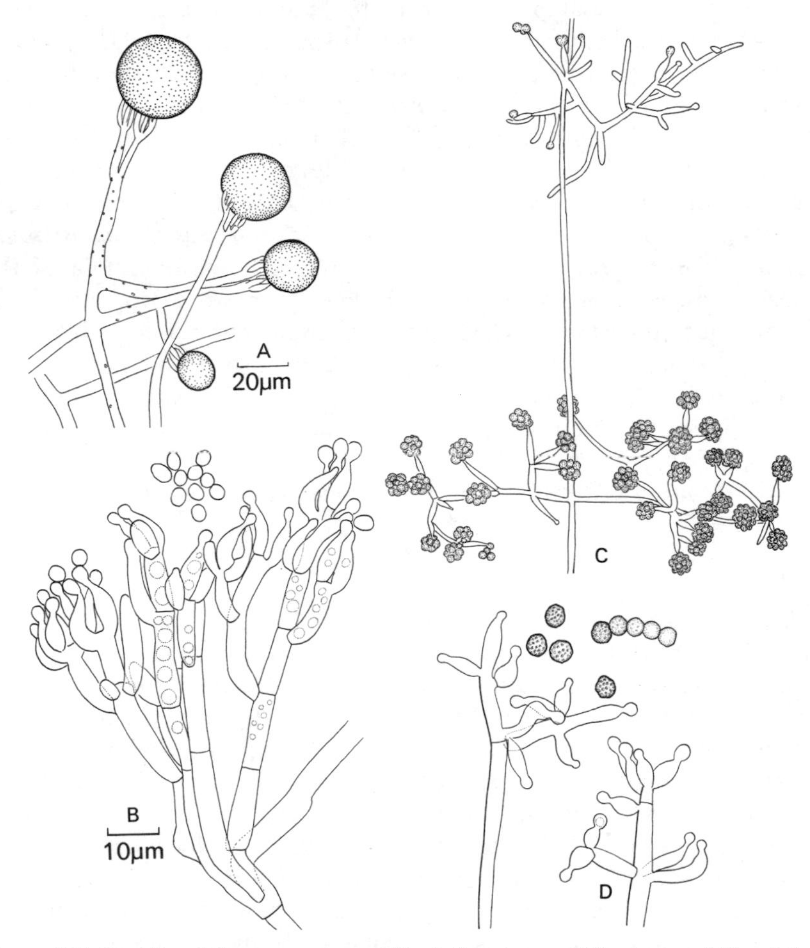

Figure 136. Conidia of *Hypocrea* spp. A. *Gliocladium*-type conidiophores of *H. gelatinosa*. B. Detail of phialides of *H. gelatinosa*. C. *Trichoderma viride* conidial state of *H. rufa*. Arrangement of conidiophores. D. Detail of phialides of *H. rufa*. A, C to same scale; B, D to same scale.

Hypocreaceae

Perithecia brightly coloured (white, yellowish, red, violet), fleshy, single or seated on a stroma, containing an ostiole lined by periphyses. Asci thin-walled. The ascospores are often two-or-more-celled, and may break up within the ascus to form part-spores. The conidial states of Hypocreaceae are typically phialospores.

Hypocrea

Species of *Hypocrea* form brightly coloured fleshy perithecial stromata, with perithecia embedded in the outer layers. The thin-walled asci contain eight two-celled ascospores and in many species the spores separate into two part-spores before ascus discharge, so that 16 part-spores are released (Fig. 135C). Our commonest species is *H. pulvinata* which forms bright yellow stromata on dead fruit-bodies of *Piptoporus betulinus*, the birch polypore. It is possible that this fungus grows parasitically on the polypore. The ascospores are often visible as white tendrils issuing from the ostioles of the perithecia (Fig. 134). In culture, conidia are formed in sticky masses at the tip of single phialides (Fig. 135B). Conidia of this type belong to the form-genus *Cephalosporium* (Rifai & Webster, 1966). Some species of *Hypocrea* have conidia of the *Trichoderma* type, in which whorls of phialides give rise to separate sticky, green or white spore masses. *Hypocrea rufa* forms conidia referable to *T. viride* (Fig. 136C,D). *Trichoderma* species are important soil saprophytes, and several are antagonistic to other fungi, including some pathogens. Conidia of the *Gliocladium* type are found in cultures of *H. gelatinosa* (Webster, 1964) (Fig. 136A,B).

The development of perithecia in *Hypocrea* is of the *Nectria* type (Hanlin, 1965).

Nectria

Perithecia of *Nectria* are common on twigs and branches of woody hosts. Many are saprophytic but some cause economically important diseases, e.g. *N. galligena* causes apple and pear canker. *Nectria cinnabarina* or coral spot is common on freshly cut twigs, but may occasionally be a wound parasite. The name coral spot refers to the pale pink conidial pustules, about 1–2 mm in diameter, which burst through the bark (Fig. 137). Before the connection with *Nectria* was understood these conidial pustules had been named *Tubercularia vulgaris*. They consist of a column of pseudoparenchyma bearing a dense tuft of conidiophores, long slender hyphae producing phialides at intervals along their length (Fig. 138B,D). Conidial pustules of this type are termed sporodochia. The conidia are sticky and form a slimy mass at the surface of the sporodochium. They are dispersed very effectively by rain-splash (Gregory *et al.*, 1952). Around the base of the old conidial pustule perithecia arise (Fig. 137), and eventually the pustule may bear perithecia over its entire surface. Perithecial pustules develop in damp conditions in late summer and autumn and are readily distinguished from conidial pustules by their bright red colour and their granular appearance. The pustules are regarded as perithecial stromata, and bear as many as 30 perithecia. Ripe perithecia contain numerous club-shaped asci each with

Figure 137. *Nectria cinnabarina*. Conidial and perithecial stromata on a twig of *Acer pseudoplatanus*. The smooth pale structures are conidial pustules and the rough clustered bodies are perithecia which develop around the base of conidial pustules.

eight two-celled hyaline ascospores (Fig. 138C). There is no obvious apical apparatus to the ascus (Strickmann & Chadefaud, 1961).

Perithecial development begins by the development of ascogonial primordia beneath the surface of the conidial pustule. The details of development have already been described. An important feature is the development of apical paraphyses which grow downwards from the upper part of the perithecial cavity. As the asci develop from ascogenous hyphae lining the base of the perithecial cavity, they grow upwards through the mass of apical paraphyses, which are difficult to find in mature perithecia (Strickmann & Chadefaud, 1961).

Other species of *Nectria* differ from *N. cinnabarina* in a number of ways. In many, the perithecia are not grouped together on a stroma, but occur singly (e.g. in *N. galligena*). The asci of some species, e.g. *N. mammoidea*, have a well-defined apical apparatus (Fig. 139A). The conidial states of some *Nectria* species are classified in several form-genera of the Fungi Imperfecti, including *Cephalosporium*, *Cylindrocarpon*, *Fusarium* and *Verticillium* (Booth, 1959, 1960, 1966) (Figs. 140, 141).

Clavicipitaceae

The fungi in this group have several characteristics. The perithecia are developed on a fleshy stroma; the asci have a well-defined thick apical cap, and the ascospores are long and narrow, often breaking up into short segments. They escape singly and successively through a narrow pore in the ascus cap.

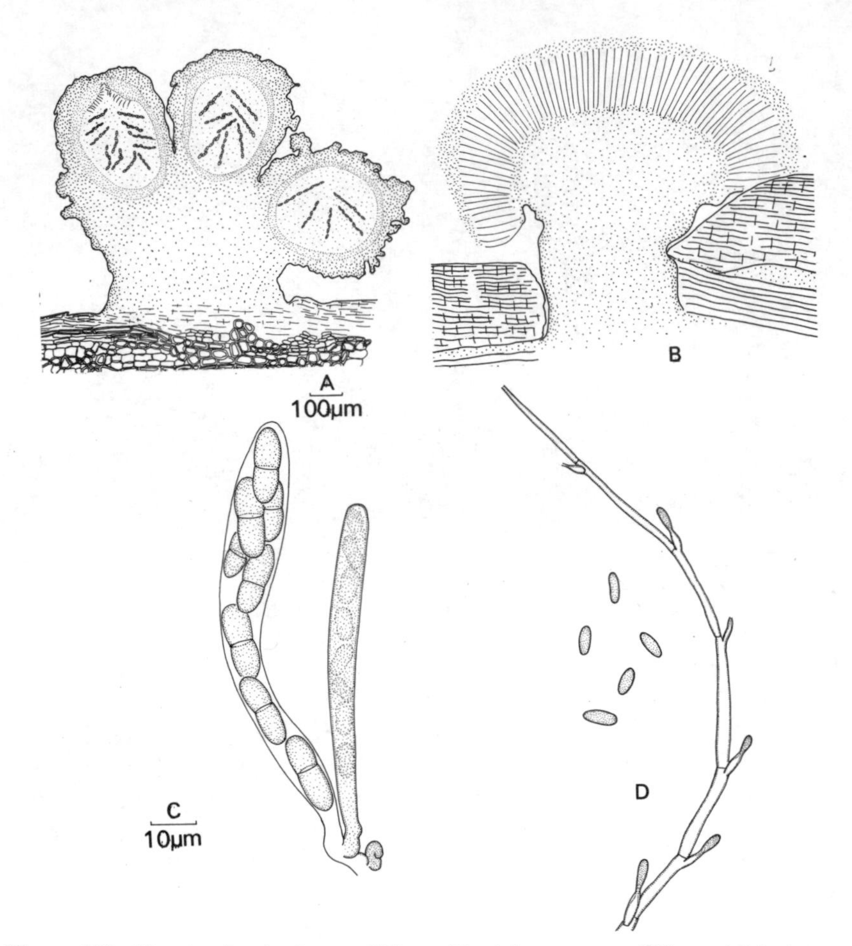

Figure 138. *Nectria cinnabarina*. A. V.S. perithecial stroma. B. V.S. conidial stroma or sporodochium. C. Asci. D. Conidiophore, phialides and conidia.

Most members are parasitic on grasses (e.g. *Claviceps*, *Epichloe*) or on insects (*Cordyceps*). Whilst some mycologists accord the group ordinal rank (Nannfeldt, 1932; Gäumann, 1964; von Arx & Müller, 1954), others regard it as a family of the Hypocreales (Bessey, 1950) or Sphaeriales (Miller, 1949; Luttrell, 1951).

Claviceps

Claviceps purpurea, the cause of ergot of grasses and cereals, grows on a wide range of grasses in late summer and autumn. Although common on rye and other cereals in Europe and North America it is not usually troublesome in Britain. In the occasional years in which its incidence is high there is an apparent correlation with high relative humidity and low maximum temperature in June (Marshall, 1960*a*). This is probably the result of a prolongation of the period during which the cereal hosts are susceptible to infection. A form with small sclerotia on *Molinia*, *Phragmites* and *Nardus* is regarded by some authorities as a separate species, *C. microcephala*. Grasses and cereals infected with *C. purpurea* develop purple curved sclerotia (ergots) in the place of healthy grain (Fig. 142A). The sclerotia contain a number of toxic alkaloids and if they are eaten they can cause severe illness and sometimes death. One effect of the toxins is to constrict the blood vessels, and the

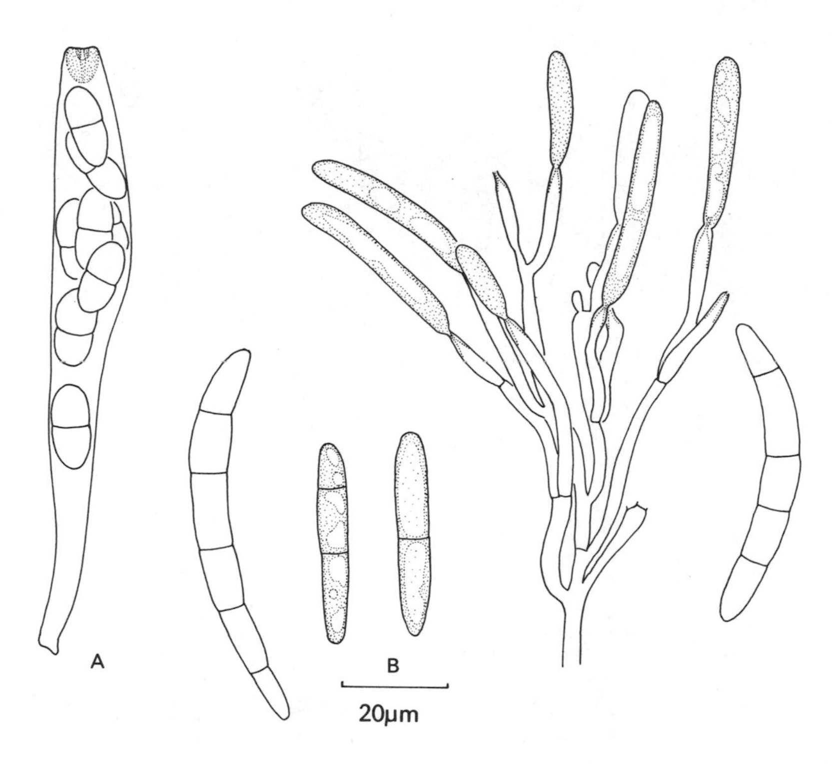

Figure 139. *Nectria mammoidea*. A. Ascus: note the apical apparatus. B. *Cylindrocarpon*-type conidia.

impaired circulation may result in gangrene or loss of limbs. Another effect is on the nervous system, resulting in convulsions and hallucinations. In the Middle Ages the symptoms of ergotism were called 'St Antony's fire' and there are numerous records of outbreaks of the disease (see Barger, 1931; Ramsbottom, 1953). With improved grain-cleaning techniques the disease is now rare in humans, and the last recorded outbreak in Britain was in the Jewish community in Manchester in 1928 as a result of eating rye-bread (Robertson & Ashby, 1928; Morgan, 1929). Cattle and sheep which have eaten sclerotia from pasture grasses may also be affected and if pregnant animals are involved there is a risk of abortion (Ainsworth & Austwick, 1959).

The sclerotia of *C. purpurea* are used medicinally to hasten uterine contraction during childbirth. The ergot of commerce is produced by cultivating the fungus on rye (*Secale cereale*), and crops are produced commercially in Eastern Europe, Spain and Portugal. Attempts are being made to extract the medically important alkaloids from pure cultures of the fungus (Abe & Yamatodani, 1964).

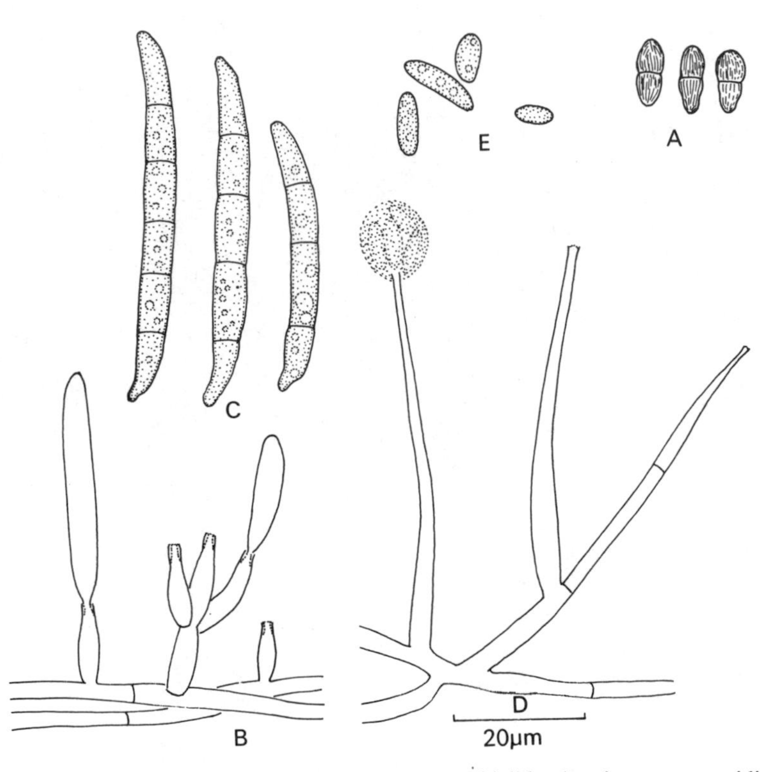

Figure 140. *Nectria haematococca*. A. Ascospores. B. Phialides bearing macroconidia. C. Macroconidia of the *Fusarium* type. D. Phialides bearing microconidia. E. Microconidia.

The sclerotia fall to the ground and overwinter near the surface of the soil, and probably need a period of low temperature before they can develop further (Cooke & Mitchell, 1966). The following summer they develop one or more perithecial stromata about 1–2 cm high, shaped like miniature drumsticks (Fig. 142B). The enlarged spherical head contains a number of perithecia, embedded in the stroma, each surrounded by a distinct perithecial wall (Fig. 143A). The cytological details of perithecial development have been studied in *C. purpurea* by Killian (1919) and in *C. microcephala* by Kulkarni (1963). In

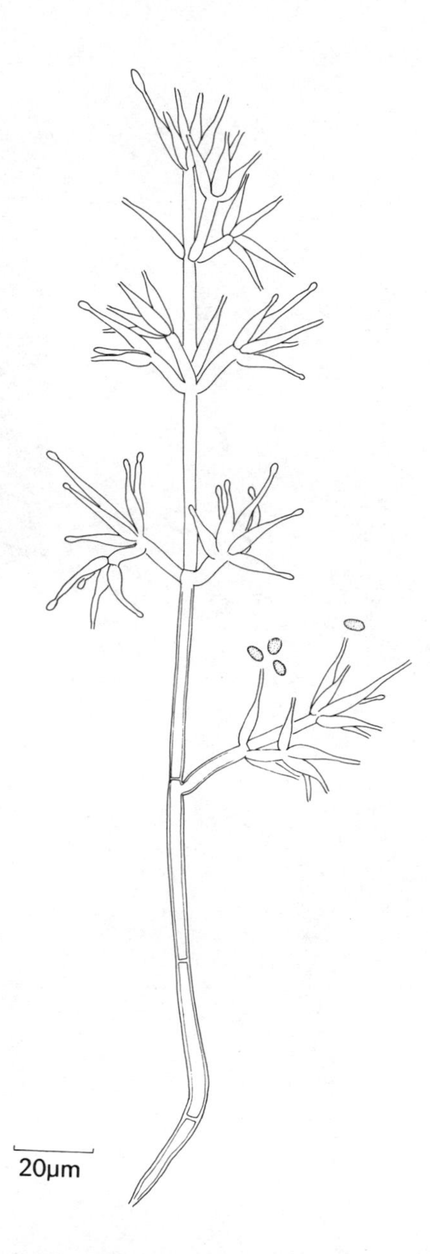

Figure. 141. *Nectria inventa. Verticillium cinnabarinum* conidial state. The whorled phialides bear globose masses of sticky phialospores.

Figure 142. *Claviceps* and *Epichloe*. A. Sclerotia of *C. purpurea* on *Lolium perenne*. B. Germinating sclerotium showing perithecial stroma. C. *Epichloe typhina*, perithecial stroma on *Agrostis tenuis*.

the outer layers of the head of the perithecial stroma club-shaped multinucleate antheridia and ascogonia undergo plasmogamy. Ascogenous hyphae made up of predominantly binucleate segments develop from the base of the ascogonium and the tips of the ascogenous hyphae form croziers

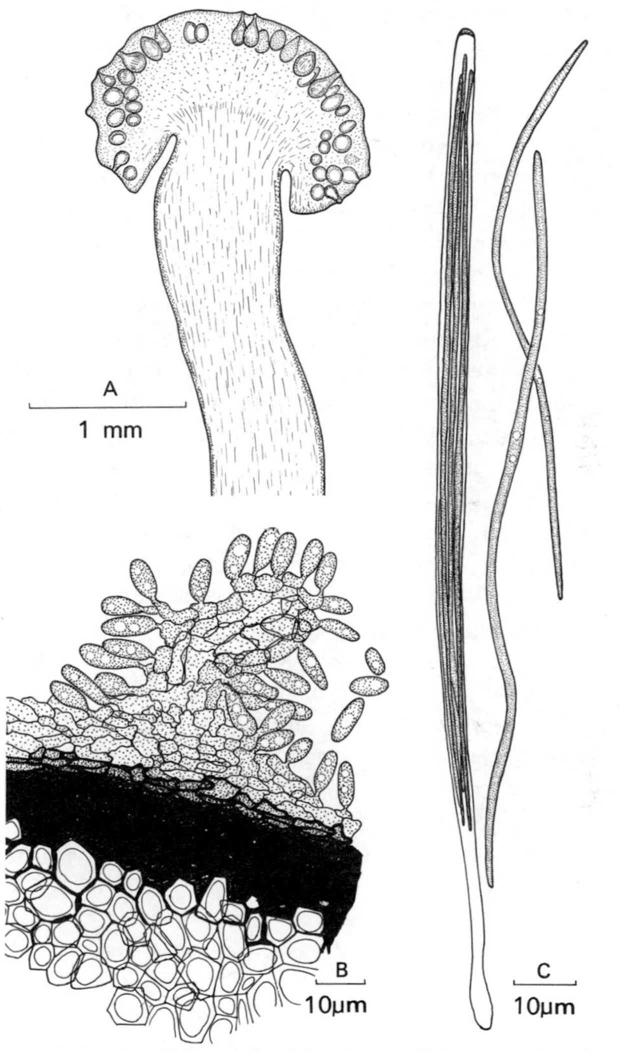

Figure 143. *Claviceps purpurea*. A. L.S. perithecial stroma. B. T.S. young sclerotium showing the formation of the conidia on the surface. C. Ascus and ascospores. Note the cap of the ascus.

with binucleate penultimate segments. The penultimate cell elongates to form the ascus and fusion between the two nuclei occurs. There are numerous asci in each perithecium, each containing a bundle of eight filiform ascospores. The ascus bears at its tip a conspicuous cap (Fig. 143C). Ascospore release coincides approximately with anthesis of the grass or cereal host and infection of the developing ovary occurs. Whether ascospore infection occurs via the stigmata or the meristematic tissue at the base of the ovary is not certain, but within a few days of infection a crop of conidia are formed, which are also capable of infecting young ovaries (Fig. 143B). The conidia are unicellular and are budded off from the surface of the infected ovary. They are enveloped in a sticky, sweet liquid called honeydew, which is attractive to insects. The honeydew contains glucose, fructose, sucrose and other sugars. These sugars are formed by the host following infection by the fungus. Cereal grains normally store starch, but the sclerotium which replaces the seed contains none. *Claviceps purpurea* probably produces an inhibitor of starch phosphorylase which inhibits starch formation by its host. The fungus is unable to utilise starch and it has been suggested that the inhibitor prevents the conversion of sugars to starch and preserves the sugars of the host in forms available to the fungus (Campbell, 1959). Insects ingest conidia and honeydew and, on reaching uninfected flowers, may bring about new infections. The conidia do not germinate in the concentrated honeydew but only after it has been diluted. Before it was realised that these conidia belonged to *C. purpurea* they had been described under the name *Sphacelia segetum*. Conidia are formed freely in cultures derived from ascospores or from sclerotial tissue, and can be used experimentally to bring about infection. Although earlier studies had indicated that there was considerable degree of physiological specialisation it now appears that an isolate from one grass host may be capable of infecting a wide range of other cereals and grasses (Campbell, 1957). The period when the host is susceptible to infection varies, and corresponds with the time when the glumes are open (Campbell & Tyner, 1959). The course of infection following experimental inoculation with conidia is from the base of the young ovary, not from the stigmata (Campbell, 1958), and within about five days of infection a further crop of conidia may develop. Eventually the whole of the ovary may become converted into a sclerotium which is often considerably larger than the normal grain.

Epichloe

Epichloe typhina causes 'choke' of pasture grasses, and is common on grasses such as *Dactylis*, *Holcus* and *Agrostis*. At flowering time the uppermost leaf sheath becomes surrounded by a white mass of mycelium 2 cm or more in length, and at the surface small unicellular conidia are produced (Fig. 144B). Later the conidial stroma becomes thicker and turns orange in colour as perithecia are formed (Fig. 142C). The perithecia contain numerous asci, each with a well-defined apical cap, and eight long narrow ascospores which break up within the ascus to form part spores (Fig. 144D) (Ingold, 1948). The mycelium is systemic, for the most part intercellular, unbranched and mainly located in the pith, but in the region of the inflorescence primordium intracellular penetration of the vascular bundles is found. Because of the

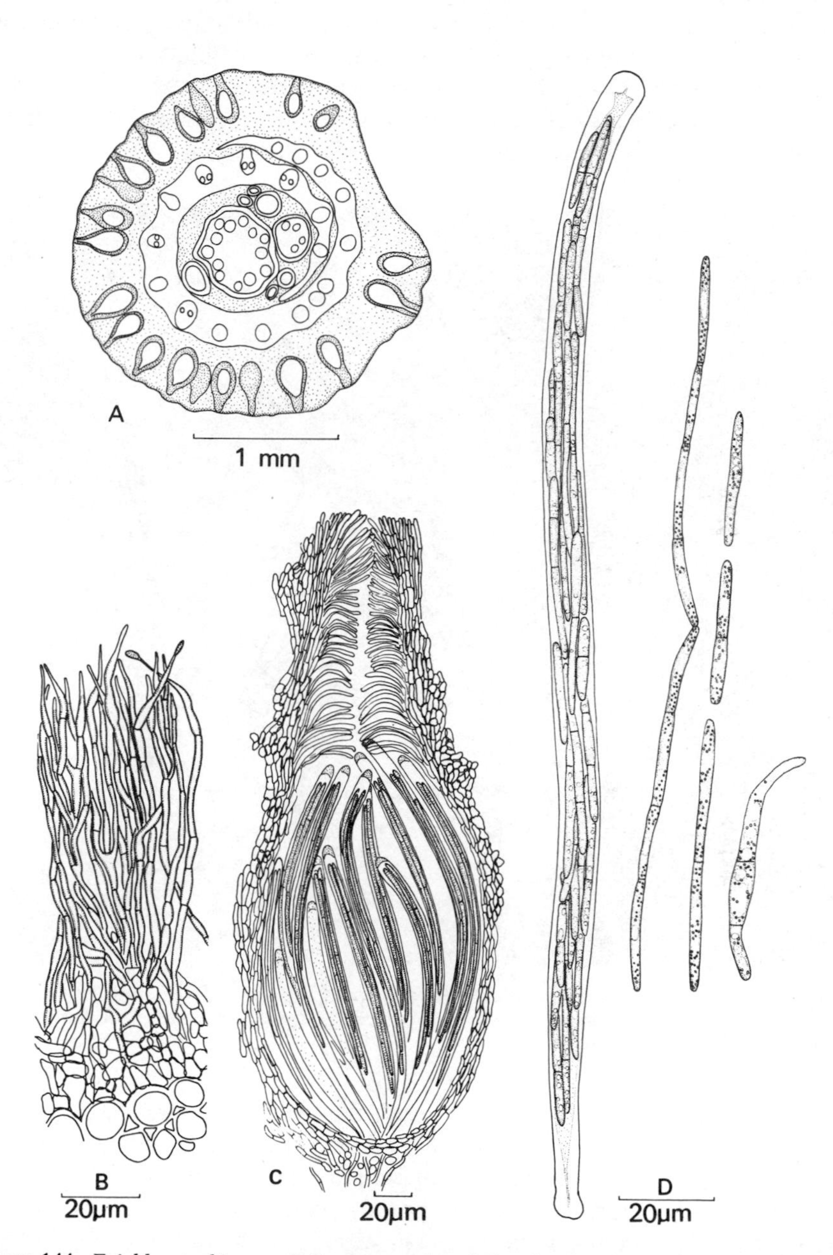

Figure 144. *Epichloe typhina*. A. T.S. stem and leaf sheath of *Agrostis* surrounded by a perithecial stroma. Note the axillary shoots between the leaf sheath and stem. B. Part of conidial stroma. C. A single perithecium. Note the periphyses lining the ostiole. D. Ascus and ascospores. Note the apical apparatus of the ascus.

systemic condition grasses infected with *Epichloe* can be propagated clonally and cultivated to yield plants almost all of which are infected. Perithecial stromata are formed only on tillers containing inflorescence primordia, and by manipulation of daylength or auxin treatment in such a way as to stimulate induction of inflorescence primordia it has been shown that the formation of stromata is correlated directly with the presence of an inflorescence primordium rather than with external conditions (Kirby, 1961).

Figure 145. A. *Cordyceps militaris*: perithecial stroma arising from lepidopterous pupa. B. Enlargement of tip of stroma to show perithecia. C. *Cordyceps ophioglossoides* attached to ascocarp of *Elaphomyces*.

In most infected grass hosts, although flowering is suppressed, vegetative growth is only slightly reduced, and the number of vegetative tillers may be increased. Thus choke is not a serious disease in pasture crops, but can be serious in crops grown for seed, especially in cocksfoot (*Dactylis*) (Large, 1952). Infection in *Festuca rubra* can be carried over by mycelium in the seeds produced by occasional flowering heads formed on infected plants (Sampson, 1933), but flowering of infected *Dactylis* is rare, and seed derived from infected plants does not produce infected offspring. Experimentally it has not

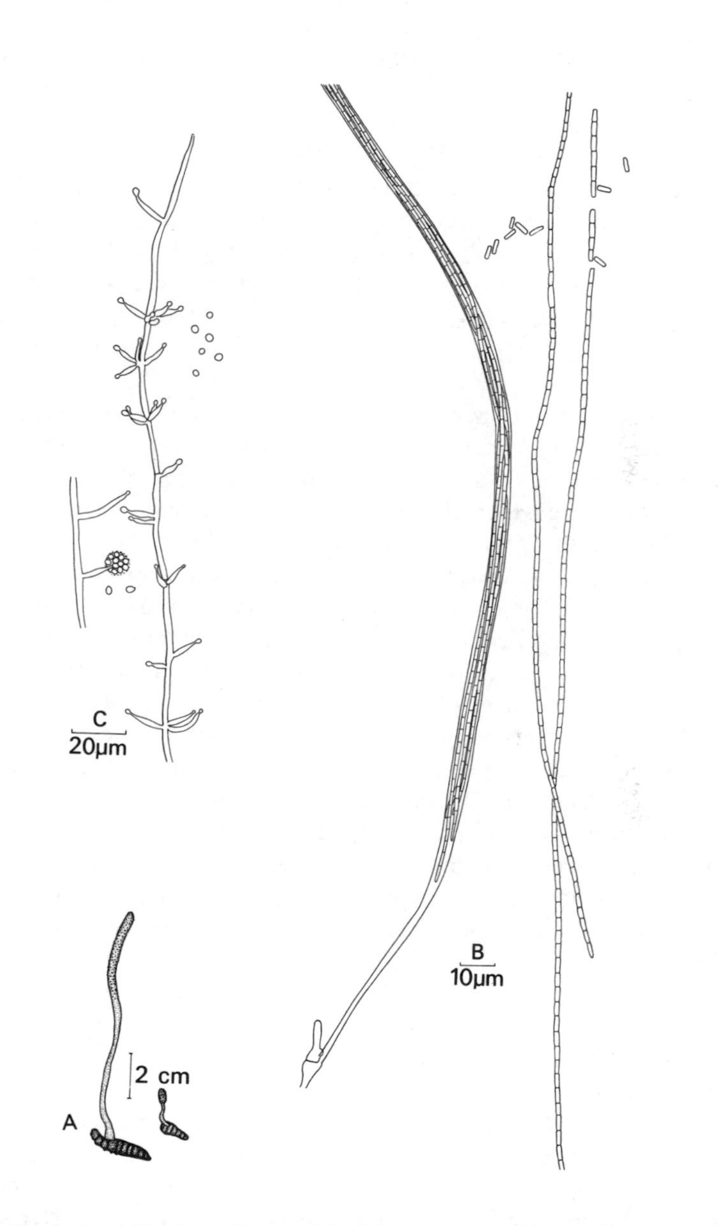

Figure 146. *Cordyceps militaris*. A. Two perithecial stromata attached to pupae. B. Ascus and ascospores. The ascospore to the right contains 82 segments. Note the ascus cap. C. Conidiophores and conidia.

been possible to infect seeds of *Dactylis*, and the only effective method of infection is by application of ascospores or conidia to the cut ends of stubble (Western & Cavett, 1959). If this is the natural route of infection in other grasses it may explain the greater incidence of the disease in *Agrostis* in heavily grazed pastures (Bradshaw, 1959).

Perithecia develop on the surface of the conidial stroma and possess true perithecial walls. Paraphyses arise from the inside of the upper part of the perithecial wall and as the asci develop most of these paraphyses disappear, but some may persist to line the ostiole (Fig. 144C). Most authorities interpret the ascus wall as single but Doguet (1960) has suggested that it might be bitunicate, and that the spores may be released by a mechanism similar to the 'Jack-in-a-box' type found in some other Ascomycetes. The apical apparatus consists of a thickened ring pierced by a narrow canal continuous with the cytoplasm of the ascus (Fig. 144D).

Cordyceps

About 150 species of *Cordyceps* are known, most of them parasitic on insects. Others are parasitic on spiders or on the subterranean fruit-bodies of *Elaphomyces*. Some eight species occur in Britain (Petch, 1938). *Cordyceps militaris* forms club-shaped orange-coloured stromata which project above the surface of the ground in autumn from buried Lepidopterous pupae (Fig. 145A). Several genera of Lepidoptera and some Hymenoptera are susceptible. The stromata bear numerous perithecia. The asci have conspicuous apical caps and contain eight long narrow ascospores which break up after discharge into numerous short segments (Fig. 146B). If the ascospores alight on the integument of a susceptible pupa germ tubes may penetrate, possibly aided by their ability to hydrolyse chitin (Huber, 1958; McEwen, 1963). After infection cylindrical hyphal bodies appear in the haemocoel of the pupa. The hyphal bodies increase by budding and the buds are distributed within the insect's body. Following death, mycelial growth follows and the body of the insect becomes transformed into a sclerotium, from which the perithecial stromata later develop. In pure culture a conidial state of the *Cephalosporium* type is formed (Fig. 146C). Earlier observations that coremia of the *Isaria* type were also conidia of *Cordyceps* are erroneous (Petch, 1936). When Lepidopteran pupae are inoculated by introducing conidia into the body cavity, death follows within five days. Mature fruit bodies can develop within 45–60 days of infection (Shanor, 1936).

Development of perithecia has been studied in *C. agariciformia* (probably synonymous with *C. canadensis*) (Jenkins, 1934) and in *C. militaris* (Varitchak, 1931). Coiled septate ascogonia arise in the peripheral layers of the perithecial stroma. The segments of the ascogonium become multinucleate and give rise to ascogenous hyphae from which asci develop in a single cluster at the base of the perithecium. The perithecial wall arises from hyphae which develop from the stalk of the ascogonium or from surrounding hyphae. Paraphysis-

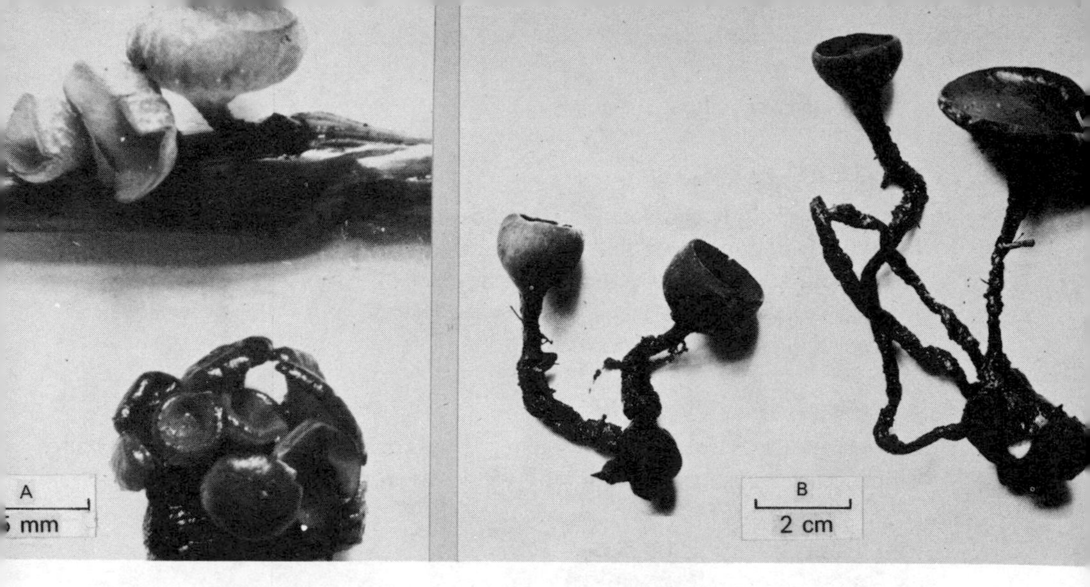

Figure 147. A. *Sclerotinia curreyana*. Above: stem base of *Juncus effusus* splitting to show a black sclerotium bearing three apothecia. Below: sclerotium removed from the host bearing eight apothecia. B. *Sclerotinia tuberosa*. Two groups of apothecia rising from sclerotia formed on rhizomes of *Anemone nemorosa*. The apothecia are formed at the surface of the soil.

like hyphae grow inwards from the perithecial wall, but at maturity these hyphae are dissolved and disappear.

Cordyceps ophioglossoides grows on the subterranean fruit-bodies of *Elaphomyces*, forming bright yellow mycelial strands over the surface. Brown club-shaped perithecial stromata grow above the soil surface in autumn (Fig. 145c).

4: DISCOMYCETES

This is an assemblage of Ascomycetes in which the ascocarp is generally cup-shaped or saucer-shaped, with a hymenium which is usually freely exposed when ripe. An exception is the hypogaeous discomycetes (Tuberales) which form subterranean fruit-bodies with an enclosed hymenium. An important feature in classification is the presence or absence of an operculum at the ascus apex. The inoperculate forms (Helotiales, Phacidiales and Lecanorales) often form rather smaller and less conspicuous fruit-bodies than the operculate forms (Pezizales). The Pezizales are mostly terrestrial or coprophilous whilst the Helotiales usually grow saprophytically or parasitically on plants. The Phacidiales include a number of important leaf pathogens.

HELOTIALES

We need not study the detailed classification (see Dennis, 1968; Nannfeldt, 1932) but will consider a few representatives. Whilst most of the Helotiales are saprophytic, the group includes a number of important plant pathogens such as species of *Sclerotinia* and its segregates, and *Trichoscyphella willkommii*, the cause of larch canker.

Sclerotinia

The characteristic feature of this genus is the formation of stalked apothecia which grow from stromata developed within the infected host tissue. The

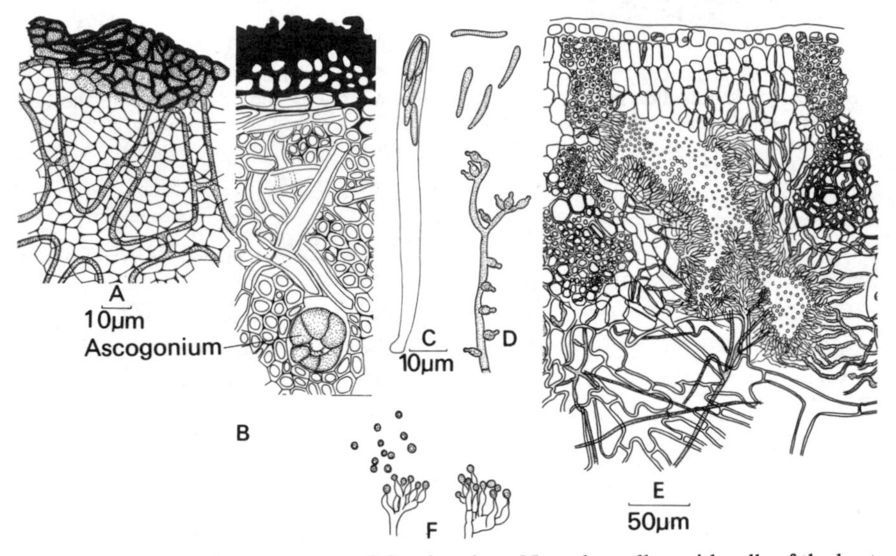

Figure 148. *Sclerotinia curreyana*. A. T.S. sclerotium. Note the stellate pith cells of the host, *Juncus effusus*. B. T.S. sclerotium showing an ascogonium. C. Ascus and ascospores. D. Microconidia in culture. E. T.S. spermodochidium on *Juncus effusus*. Note the cavity lined by phialides. F. Microconidia from host.

apothecia usually develop in spring from over-wintered stromata. The stroma is a food storage organ and is differentiated into two parts, a medulla of hyaline cells and a rind of dark thick-walled cells. Two generalised types of stroma have been distinguished. To quote from Whetzel (1945):

> The *sclerotial stroma* (commonly called the *sclerotium*) has a more or less characteristic form and a strictly hyphal structure under the natural conditions of its development. While elements of the substrate may be embedded in its medulla they occur there only incidentally and do not constitute part of the reserve food supply. The *substratal stroma* is of a diffuse or indefinite form, its medulla being composed of a loose hyphal weft or network permeating and preserving as a food supply a portion of the suscept or other substrate (e.g. culture media).

Various types of macroconidia are formed and some authors (e.g. Whetzel, 1945; Dennis, 1956) have separated species with different macroconidia into distinct genera or subgenera. For example *Sclerotinia fuckeliana* (= *Botryotinia fuckeliana*) has *Botrytis cinerea* as its conidial state (Groves & Loveland, 1953) whilst *Sclerotinia fructigena* (= *Monilinia fructigena*) has *Monilia fructigena* as its conidial state. Other species may have no macroconidial state, e.g. *S. curreyana* a parasite of *Juncus effusus* and *S. tuberosa* a parasite of *Anemone nemorosa*. Apothecia of the last two species are common in May. In *S. curreyana* the apothecia arise from black sclerotia in the pith at the base of the *Juncus* stem (see Fig. 147A). Infected stems look paler than healthy stems, and by feeling down to the base of an infected stem the sclerotium can be felt as a swelling, between finger and thumb. The sclerotium has an outer layer of dark cells and a pink interior which includes some of the

Figure. 149. *Sclerotinia fructigena*. Apple showing brown rot caused by this fungus, and bearing conidial pustules.

stellate pith cells of the host (Fig. 148A). One to several apothecia may grow from a single sclerotium. The ascospores are released in late spring and infect the current season's stems. In culture germinated ascospores form a mycelium which develops microconidia, from small phialides (Fig. 148D). Similar clusters of microconidia can be found on infected *Juncus* later in the season and line cavities beneath the epidermis in the upper part of infected culms. Whetzel (1946) has used the term *spermodochidium* for these microconidial fructifications (Fig. 148E). It is probable that the microconidia play a role in fertilization.

The apothecia of *S. tuberosa* (Fig. 147B) are about 2 cm in diameter and arise from sclerotia within rhizomes of *Anemone nemorosa* (Spaeth, 1957;

Siegel, 1958). They may also occur on garden *Anemone* associated with black-rot disease. Microconidia are formed in culture.

Sclerotinia fructigena is the cause of brown-rot of apples, pears and some other fruits, and although the apothecial state has not been found in Britain the disease is common and is transmitted by means of conidia. Apples and pears showing brown-rot bear buff-coloured pustules of conidia often in concentric zones (Fig. 149). Sporulation is stimulated by light, and adjacent zones correspond to daily periods of illumination. The conidia are formed in

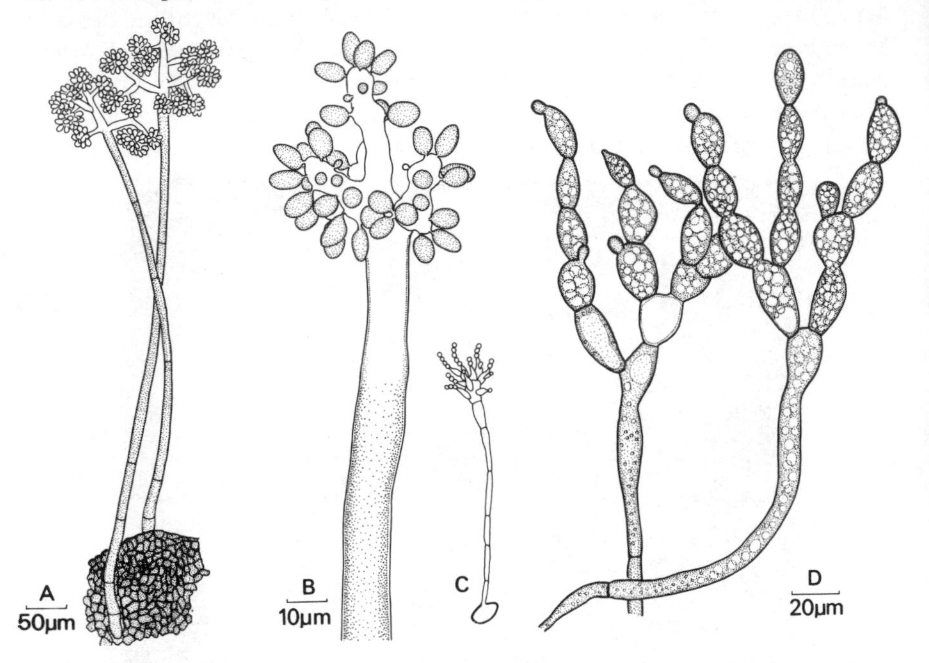

Figure 150. A–C. *Botrytis cinerea.* A. Conidiophores developing from a sclerotium. B. Apex of conidiophores showing origin of conidia as blastospores. C. Conidium germinating to produce phialides and microconidia (after Brierley, 1918). D. *Sclerotinia fructigena.* Conidia of the *Monilia* type.

chains which extend in length at their apices by budding of the terminal conidium. Occasionally more than a single bud is formed, and this results in branched chains (see Fig. 150D). Conidial formation of this type is characteristic of the form-genus *Monilia* of the Fungi Imperfecti. Infection of the fruit is commonly through wounds, caused mechanically or by insects such as codling moth, wasps and earwigs (Croxall *et al.*, 1951). Fruit left lying on the ground is the source of infection in the following season. Infected fruit becomes mummified during the winter, but in the following year such fruit may develop conidial pustules. The disease may develop in stored apples, and in some varieties a twig infection (spur canker) may occur. A similar group of diseases of apple and plum is caused by *Sclerotinia laxa* which also has a *Monilia* conidial state. The apothecial state has been found once in Britain (Wormald, 1921, 1954).

Although the apothecia of *Sclerotinia fuckeliana* are not commonly collected, the conidial state, *Botrytis cinerea*, is abundant on all kinds of moribund plant material, and the fungus is also associated with a wide range of disease often referred to as grey mould. It is probable that the name *Botrytis cinerea* is a collective name used to describe a number of closely similar, but genetically distinct, species, possibly with distinct apothecial states. For this reason many authors prefer to write of a *Botrytis* of the *cinerea* type. Serious diseases caused by this fungus are grey mould of lettuce, tomato, strawberry and raspberry, die-back of gooseberry and damping-off of conifer seedlings. The macroconidia are formed on infected host tissue as dark-coloured branched conidiophores. The tips of the branches are thin-walled and bud out to form numerous elliptical multinucleate conidia (blastospores) (Hughes, 1953) which are easily detached by wind, or are thrown off as the conidiophores twist hygroscopically (Fig. 150A,B). Microconidia are also formed from clusters of phialides (Fig. 150C) and the microconidia are capable of germination (Brierley, 1918) but they are also involved in sexual reproduction. Sclerotia are also formed at the surface of infected tissue and the fungus overwinters in this form. In spring the sclerotia may develop to give rise to tufts of macroconidia or, much less commonly, to apothecia. *Sclerotinia fuckeliana* is heterothallic with ascospores of two kinds. In a single ascospore culture macroconidia, microconidia and sclerotia develop, but not apothecia. Apothecia will develop if microconidia of one mating type are applied to sclerotia of opposite mating type (Groves & Drayton, 1939; Groves & Loveland, 1953). In Whitehouse's (1949) terminology *S. fuckeliana* shows physiological heterothallism. Another example of the same kind is *Sclerotinia gladioli* (= *Stromatinia gladioli*) (Drayton, 1934). In this species both sclerotial and substratal stromata are formed. The stromata bear columnar receptive bodies within which are ascogonial coils. When microconidia of opposite mating type are applied to the receptive bodies apothecia develop. A second kind of behaviour is found in *Sclerotinia narcissi* (Drayton & Groves, 1952). Of the eight spores formed in the asci of this fungus four produce mycelia bearing microconidia but no sclerotia, whilst the other four produce mycelia bearing sclerotia and stromata. Apothecia develop on the strains forming stromata if microconidia are transferred to them. Thus the mating behaviour of *S. narcissi* differs from that of *S. gladioli*, and we can say that *S. narcissi* is sexually dimorphic. This type of behaviour is not common in Ascomycetes and it is possible that an incompatibility system of this has been derived from the more usual system exemplified by *Sclerotinia gladioli* by aberrations which prevent the normal sequence of development of sexual organs in basically hermaphroditic forms (Raper, 1959). A third kind of mating behaviour is seen in *S. sclerotiorum* which is homothallic. A single ascospore culture produces microconidia and sclerotia which bear ascogonial coils beneath the rind. Transfer of microconidia to the sclerotia on the same mycelium results in the formation of apothecia (Drayton & Groves, 1952). A similar process of self-fertilisation also occurs in *Sclerotinia porri* (= *Botryotinia porri*) (Elliott, 1964).

Microconidia of some species of *Sclerotinia* fail to germinate when transferred to nutrient media. In the related fungus *Gloeotinia temulenta*, the cause of blind seed disease of ryegrass, it has been shown that the cytoplasm of the microconidia is deficient in RNA. In contrast, the microconidia of *Neurospora crassa* which do germinate, are heavily charged with RNA (Griffiths, 1959). It has therefore been suggested that, since RNA is essential for protein synthesis, the microconidia are incapable of growth.

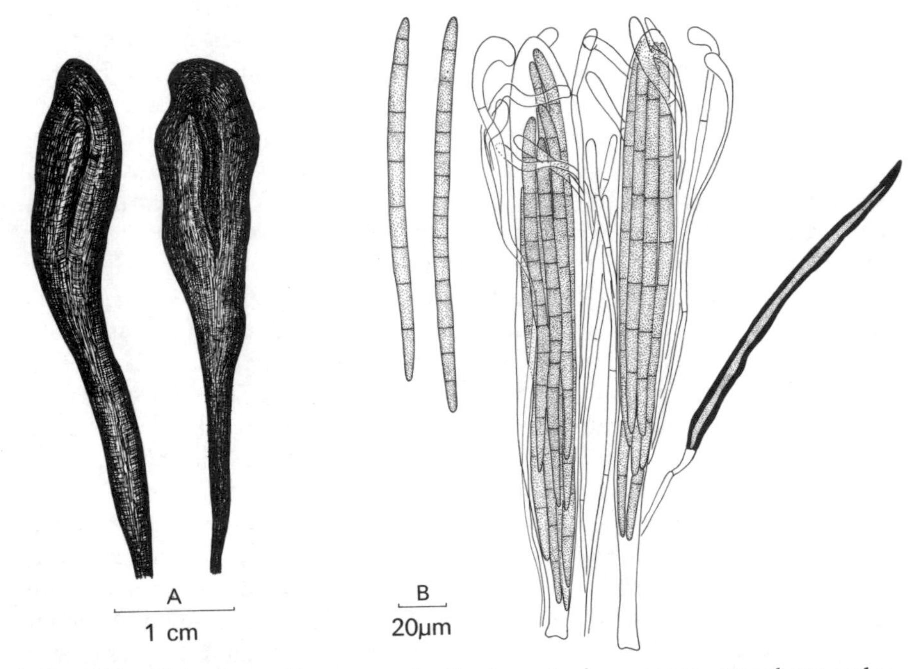

Figure 151. *Trichoglossum hirsutum.* A. Apothecia. B. Asci, ascospores, paraphyses and a hymenial seta.

Trichoglossum

Trichoglossum is a representative of the Geoglossaceae, or earth-tongues, which form club-shaped stalked fruit-bodies, growing usually on the ground, but sometimes on dead leaves or amongst *Sphagnum* (e.g. *Mitrula*). Accounts of the family have been given by Nannfeldt (1942) and Dennis (1968). *Trichoglossum hirsutum* has black, somewhat flattened fruit-bodies up to 8 cm high, and grows in pastures and lawns. The ascospores are long, dark and septate, and the asci are interspersed by black, thick-walled, pointed hymenial setae whose function is not known (Fig. 151). The presence of hymenial

Figure 152. *Rhytisma acerinum*. A. Leaf of sycamore, *Acer pseudoplatanus* with tar spot. B. Tar spot from an overwintered leaf showing the cracking of the surface to reveal the hymenia.

setae separates *Trichoglossum* from *Geoglossum* which grows in similar habitats. The elongate ascospores of *Trichoglossum* and *Geoglossum* are discharged singly through a minute pore at the tip of the ascus. When the ascus is ripe the pore bursts and one ascospore is squeezed into it, blocking

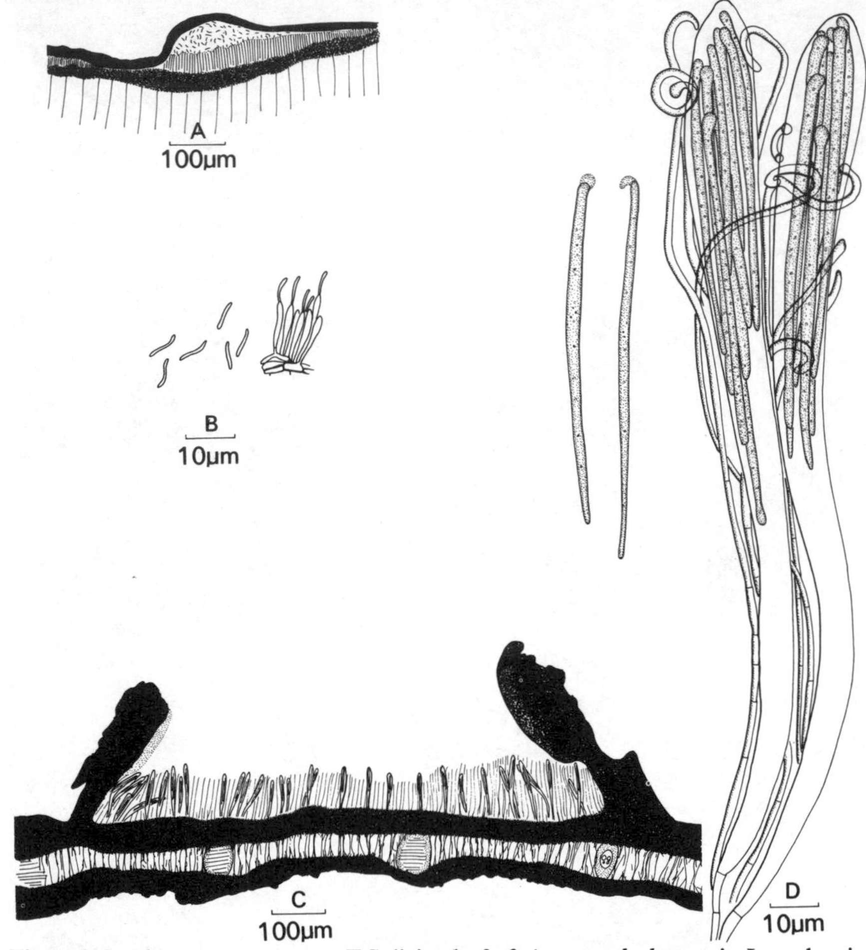

Figure 153. *Rhytisma acerinum*. A. T.S. living leaf of *Acer pseudoplatanus* in June showing spermogonium. B. Details of cells forming spermatia. C. T.S. overwintered leaf of *Acer* showing the opening of the lips of the epithecium to reveal the hymenium. D. Asci, paraphyses and ascospores. Note the mucilaginous appendage at the upper end of the ascospore.

it. The pressure of the ascus sap behind the spore causes the spore to protrude, at first slowly, but when about half the spore is projecting the spore gathers velocity and is rapidly discharged. Another ascospore immediately takes the place of the first and the process of discharge is continued until all eight ascospores are released (Ingold, 1953).

PHACIDIALES

The best-known example is *Rhytisma acerinum*, the cause of tar-spot diseases of sycamore. Some other plant pathogens also belong here, e.g. *Lophodermium pinastri*, the cause of leaf-cast of pine. For details of the group see Dennis (1968).

Rhytisma

Rhytisma acerinum is common on leaves of sycamore, *Acer pseudoplatanus*, forming black shining lesions (tarspots) about 1–2 cm wide (Fig. 152). The lesions arise from infection by ascospores released from apothecia on overwintered leaves. Lesions become visible to the naked eye in June or July, some two months after infection, as yellowish spots which eventually turn black. Sections of the leaf at this stage show an extensive mycelium filling the cells of the mesophyll, and especially the cells of the upper epidermis. Within the epidermal cells spermogonia develop. These are flask-shaped cavities which give rise to uninucleate curved club-shaped spermatia measuring about 6×1 μm (Fig. 153A,B). The spermatia exude from the upper surface of the centre of the lesion through ostioles in the spermogonial wall. The spermatia do not germinate, even on sycamore leaves, and it is believed that they play a sexual role (Jones, 1925). Apothecia begin development in the portion previously occupied by spermogonia and the hymenium is roofed over by several layers of dark cells formed within the upper epidermis. The asci complete their development on the fallen leaves and are ripe about May when sycamore leaves of the current season have unfolded (Aragno, 1968). The hymenium is exposed by means of cracks in the surface layer of the fungal stroma (Fig. 152B) and the asci discharge their spores, sometimes by puffing. Although the spores are only discharged to a height of about 1 mm above the surface of the stroma they are carried by air-currents to leaves several metres above the ground. The ascospores are needle-shaped and have a mucilaginous epispore especially well developed at the upper end (Fig. 153D), and this probably helps in attaching them to leaves. Infection occurs by penetration of the germ tubes through stomata on the lower epidermis.

LECANORALES

This is a group of inoperculate Discomycetes which live symbiotically with algae in lichen thalli. Other fungi also form lichens, e.g. the Pleosporales, Hysteriales, Sphaeriales, Basidiomycetes and some Fungi Imperfecti (Santesson, 1953; Ciferri & Tomaselli, 1955), but the majority of lichens belong in this group. Lichens inhabit the surfaces of rocks and trees, and some grow on infertile soil. Although they contain two organisms lichens are classified into species, genera and families. Several hundred British species are known (Smith, 1918, 1921, 1926; Duncan, 1959, 1963; Kershaw & Alvin, 1963;

James, 1965). General accounts of lichens have been written by des Abbayes (1951), Ahmadjian (1967*a*) and Hale (1967). Members of the Chlorophyceae (green algae) and Myxophyceae (blue-green algae) may be the algal components of lichens (Ahmadjian, 1967*b*). It is possible to grow the algal and fungal partners of many lichens separately in pure culture, and physiological studies of the separate components and of intact lichen thalli have been made (Quispel, 1959; Smith, 1962, 1963; Ahmadjian, 1965). Attempts have also been made to synthesise algal thalli *in vitro* from cultures of the two components, but typical lichen thalli have not been formed (Ahmadjian, 1962).

In most lichen thalli the algae are confined to a special region, the algal zone, interspersed by fungal hyphae (see Figs. 156 and 157). Above the algal zone there is often a cortex made up entirely of closely packed fungal cells and below the algal zone there is the medulla made up of loosely woven thick-walled hyphae. Although algal cells are occasionally penetrated by fungal haustoria, it is probable that digestion of the algal cells is not a normal feature of lichen symbiosis. When $[^{14}C]O_2$ is supplied to intact lichen thalli in the light, organic compounds containing $[^{14}C]$ accumulate first in the algal zone, and later in the medulla (Smith, 1961). In experiments in which discs of lichen tissue are allowed to absorb $[^{14}C]$-labelled sugars, much of the labelling can later be detected in the fungal tissue in the form of sugar alcohols such as mannitol and arabitol (Smith *et al.*, 1969). There is also evidence that in lichens which contain blue-green algae such as *Nostoc*, nitrogen is fixed by the alga and the products of fixation are passed on to the fungus (Scott, 1956). No such fixation has been demonstrated in lichens with green algae, although in some of these, pockets of blue-green algae within the thallus termed cephalodia may be fixing nitrogen. Experiments with pure cultures of lichen algae and lichen fungi show that growth of either may be stimulated with culture filtrates of the other component (Quispel, 1959).

The reproduction of these lichen fungi is by means of ascospores which are violently discharged from the apothecia. If the spores germinate close to a suitable algal partner a new lichen thallus may be initiated. Many species also reproduce by means of detachable fragments of the thallus which contain both fungal and algal cells, and sometimes the propagules form a powdery mass of granules (soredia) on the surface of special upright branches of the thallus, e.g. in *Cladonia* (Fig. 158).

Xanthoria

The most abundant species is *X. aureola* (= *X. parietina* var *aureola*) which forms bright yellow crusts on the surface of rocks, roofs, trees and farm buildings, especially near the sea (Fig. 154A). The thallus is lobed and attached to the substratum by rhizoids. The algal component is the green alga *Trebouxia*, which forms single globose cells. The apothecia are saucer-shaped, and about 2–3 mm in diameter, on the upper surface of the thallus, and the algal zone extends into the apothecial margin (Fig. 156). The ascospores are

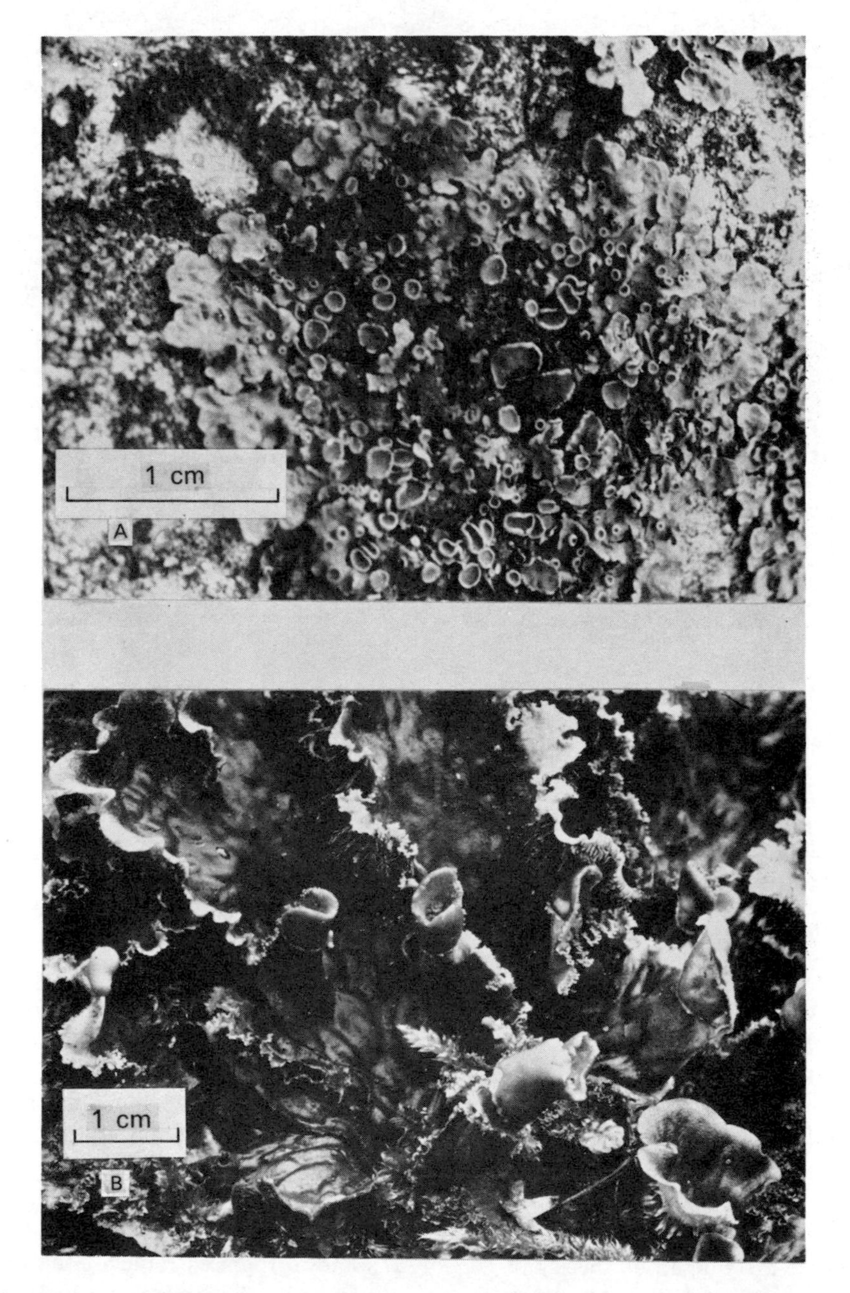

Figure 154. A. *Xanthoria aureola*, thallus with apothecia. B. *Peltigera polydactyla*. Note the rhizinae growing from the lower side of the thallus and the folded marginal apothecia.

Figure 155. *Cladonia pyxidata*. Primary foliose thallus bearing funnel-shaped podetia growing at the base of a Millstone grit wall. Note the granular soredia outside and inside the podetia.

at first one-celled, but ingrowth from the wall of the ascospore eventually divides the contents of the spore into two. The yellow colour of the thallus is due to the presence of the quinone parietin.

Peltigera

Species of *Peltigera* form large lobed leaf-like thalli attached to the ground or to rocks by groups of white rhizinae. The commonest species are *P. polydactyla* (Fig. 154B) and *P. canina* which grow amongst grass on heaths, and on sand dunes and on rocks amongst moss. The algal component is the blue-green alga *Nostoc*. The apothecia are reddish-brown, folded extensions of the thallus which do not contain algal cells. The red colour of the apothecia is due to pigments in the tips of the paraphyses (Fig. 157B).

Cladonia

There are numerous species of *Cladonia*, some of them extremely common, growing on heaths and moors, on rocks and walls. There are two kinds of thallus. The primary thallus is prostrate and lobed, and the secondary thallus is upright and cylindrical, often consisting of a hollow podetium which may open out into a cup, around the margins of which apothecia develop as in *C. pyxidata* (Fig. 155). The podetia frequently bear the granular soredia, containing algal and fungal cells (Fig. 158). In wind-tunnel experiments, using *C. pyxidata*, Brodie & Gregory (1953) showed that soredia were removed from the funnel-shaped podetia at winds of 1·5–2·0 m/sec. although they were not removed from horizontal glass slides at the same wind-speeds. They suggested that funnel-shaped structures generate eddy currents when placed in a wind-stream and that the eddy currents effectively remove soredia.

In some species of *Cladonia*, e.g. *C. sylvatica*, the primary basal thallus

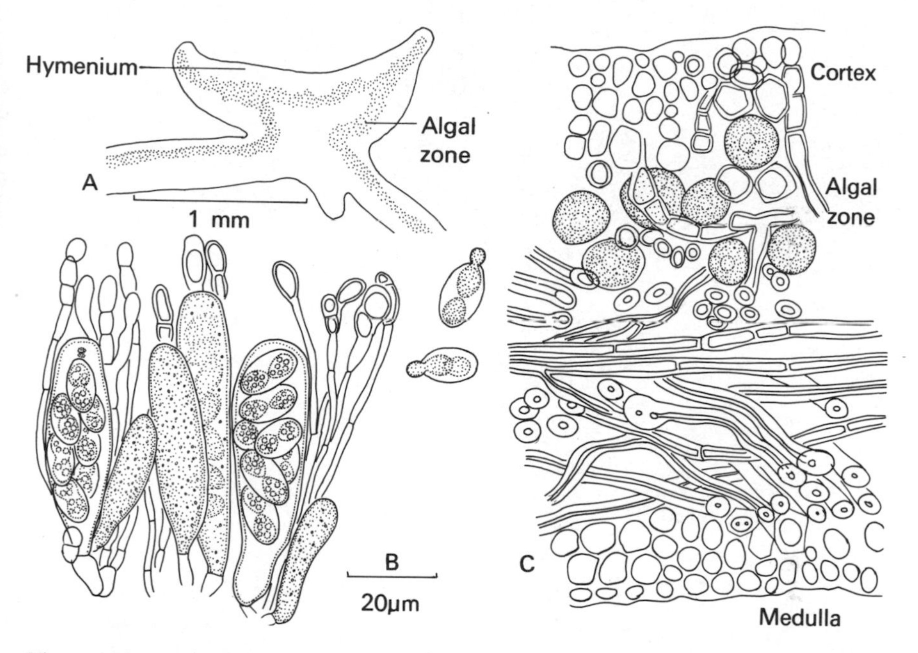

Figure 156. *Xanthoria aureola*. A. V.S. thallus and apothecium showing the extension of the algal zone into the apothecium. B. Asci, paraphyses and two germinating ascospores. C. V.S. thallus.

quickly disappears and a secondary thallus made up of much-branched cylindrical axes persists.

PEZIZALES

In the Pezizales the asci are operculate, opening by a lid or operculum, and in this respect the group differs from the Helotiales in which the asci are inoperculate. The detailed classification will not be discussed (see Dennis, 1968) but a number of common examples illustrating some of the major groups will be described.

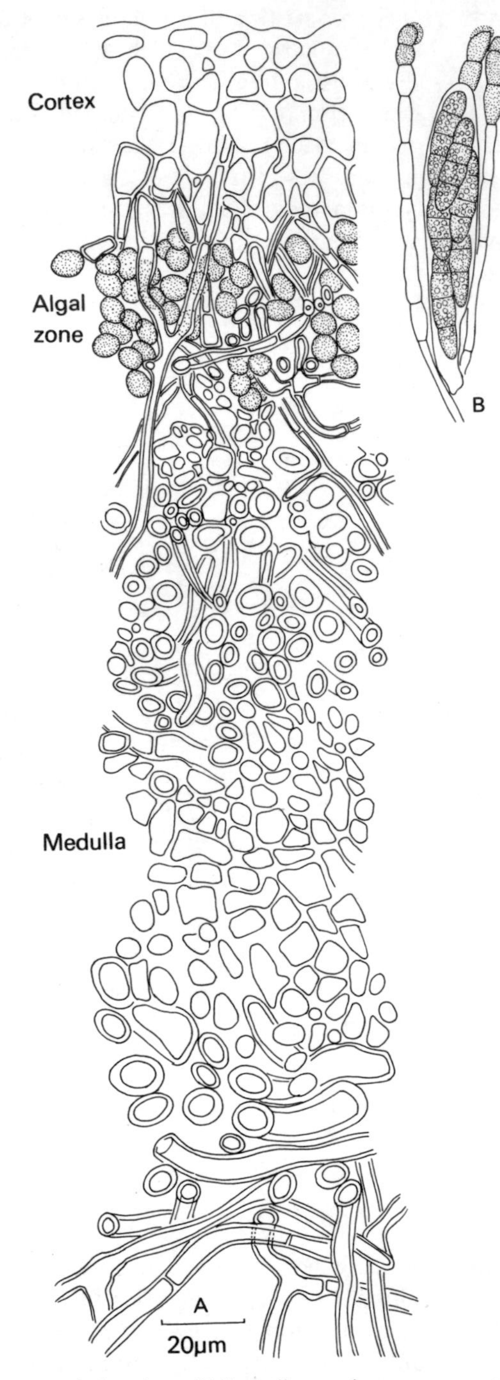

Figure 157. *Peltigera polydactyla*. A. V.S. thallus. B. Ascus, ascospores and paraphyses.

Pyronema

Pyronema is found on bonfire sites and on sterilised soil. There are two common species, *P. omphalodes* (= *P. confluens*) and *P. domesticum*. In *P. omphalodes* the apothecia are often confluent and lack marginal hairs, whilst in *P. domesticum* the apothecia are discrete, and surrounded by tapering hairs (see Fig. 159A). *Pyronema domesticum* forms sclerotia in culture whilst *P. omphalodes* does not (Moore & Korf, 1963). In earlier studies the distinction between the two species was often not appreciated and some of the work on '*P. confluens*' may well have been done on *P. domesticum*. Both species are homothallic and grow well in agar culture or on sterilised soil and form their pink apothecia 1–2 mm in diameter, within four to five days. There have been numerous accounts of the cytology of apothecial development (for references see Moore, 1963), but earlier claims for a double nuclear fusion and reduction

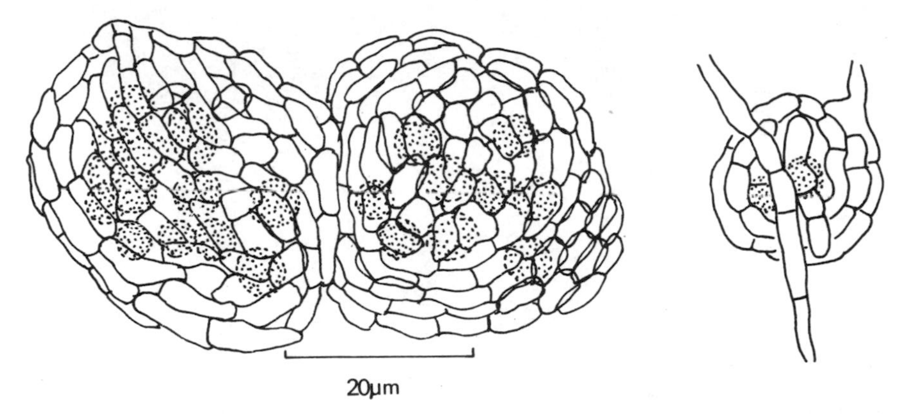

Figure 158. ***Cladonia pyxidata.*** **Soredia. Note the algal cells surrounded by fungal hyphae.**

(brachymeiosis) are no longer accepted (Hirsch, 1950; Wilson, 1952; McIntosh, 1954). In *P. domesticum* the apothecia arise from clusters of paired ascogonia and antheridia formed by repeated dichotomy of a single hypha. The ascogonia are fatter than the antheridia and each ascogonium is surmounted by a tubular recurved trichogyne which makes contact with the tip of the antheridium (Fig. 159B). The ascogonia and antheridia are multinucleate and following fusion of the antheridium with the trichogyne numerous antheridial nuclei enter the ascogonium. Contrary to earlier reports there is no nuclear fusion at this stage. Multinucleate ascogenous hyphae develop from the ascogonium (Fig. 159D). They are septate and often branched and their tips are recurved to form croziers. The tip of the crozier is at first binucleate (Fig. 159E) and each nucleus divides mitotically. Septa develop between the nuclei to cut off a uninucleate terminal cell, a binucleate penultimate cell containing two non-sister nuclei, and a uninucleate antepenultimate cell or stalk cell (Fig. 159FG). The binucleate penultimate cell usually develops into an ascus. The two nuclei fuse and the diploid fusion nucleus undergoes meiosis, followed by mitosis so that eight haploid nuclei are formed around which the ascospores are cleaved. The fine structure of developing asci of *Pyronema* has been studied by Reeves (1967). Occasionally, instead of forming an ascus, the binucleate cell may grow out to form a

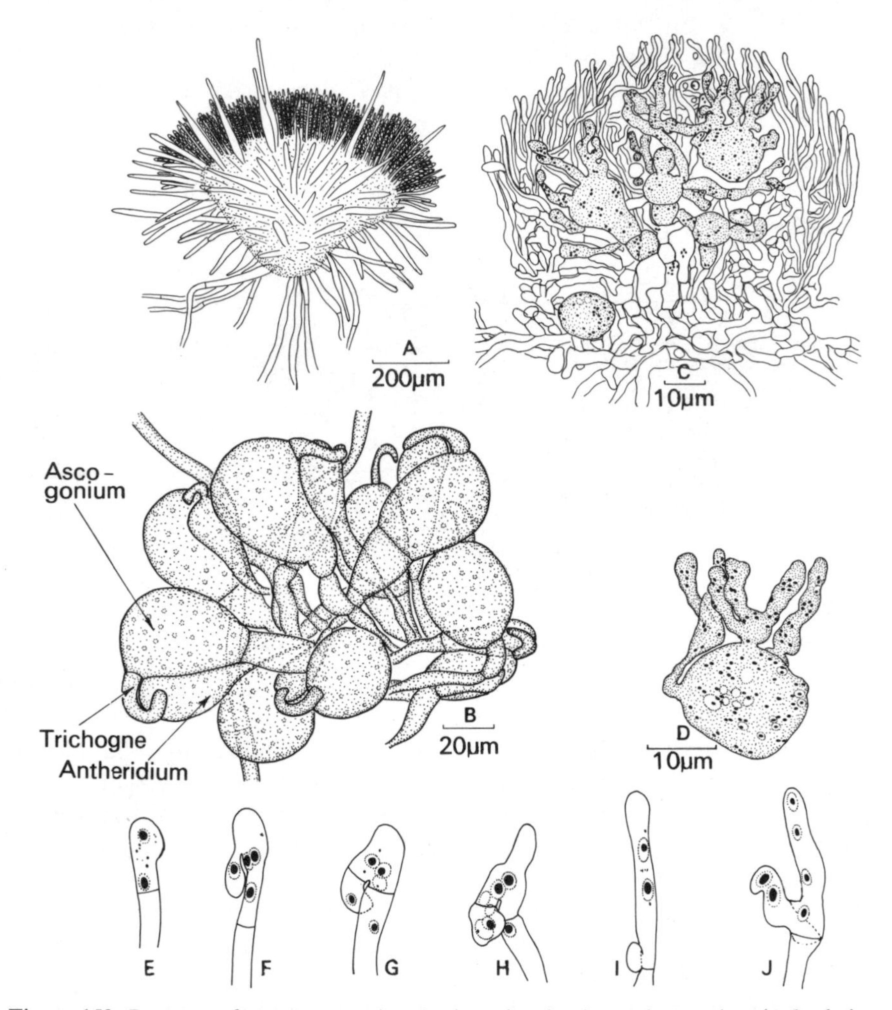

Figure 159. *Pyronema domesticum*. A. Apothecium showing hymenium and excipular hairs. B. Group of ascogonia and antheridia. C. V.S. through developing apothecium showing several ascogonia producing ascogenous hyphae and the development of paraphyses and excipulum from the ascogonial stalks. D. Enlarged view of the ascogonium and developing ascogenous hyphae. E–J. Stages in development of asci. E. Binucleate tip of ascogenous hyphae beginning to form a crozier. F. Quadrinucleate stage. G. Septation of crozier to form a binucleate penultimate cell. H. Development of ascus from binucleate cell. I. First meiotic division complete. J. Second meiotic division complete. Note the proliferation of a new ascogenous hypha from the stalk cell.

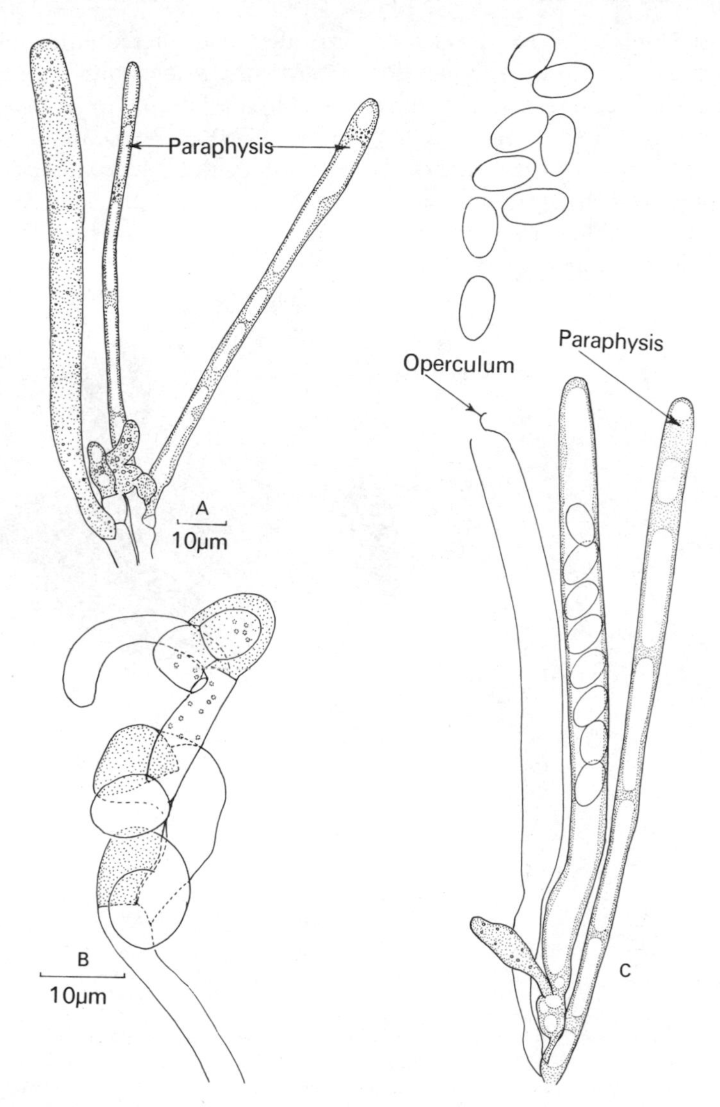

Figure 160. *Pyronema domesticum*. A. An immature ascus (left) showing the ascogenous hyphae from which it developed continuing to proliferate. B. More magnified view of tip of an ascogenous hypha showing repeated proliferation. The three stippled cells represent penultimate cells of croziers probably destined to develop into asci. C. Mature asci, one discharged and showing an operculum. A paraphysis is also shown apparently arising from the ascogenous hypha.

new crozier. Proliferation of the original crozier can also occur by fusion of the uninucleate terminal cell with the stalk cell, which then grows out to form a new crozier. In this way a complex branched system bearing several asci may arise from the apex of a single ascogenous hypha (Fig. 160B).

The remaining tissues of the apothecium develop from the hyphae bearing the ascogonia. The paraphyses develop after the differentiation of the sex organs and are mostly fully developed before the ascogenous hyphae appear, so that the asci develop between the paraphyses. Paraphyses apparently also develop from the ascogenous hyphae (Fig. 160C). Pseudoparenchymatous cells develop around the base of the ascogonia and fill space not occupied by paraphyses or ascogenous hyphae. To the outside of the apothecium they form a specialised layer, or excipulum, which in *P. domesticum* bears the tapering excipular hairs (Fig. 159A).

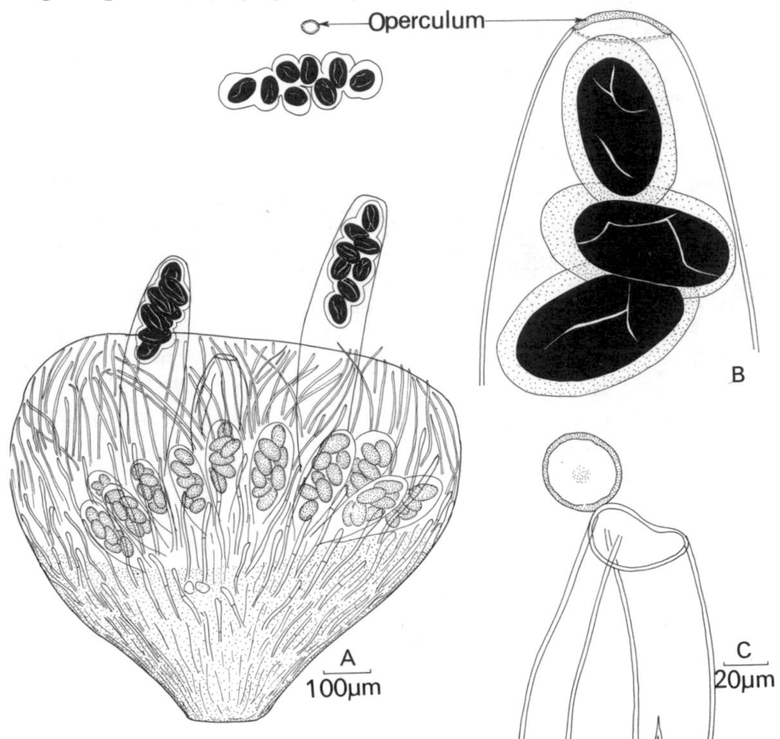

Figure 161. *Ascobolus immersus*. A. Apothecium showing two projecting asci. Immature asci can be seen below the general level of the surface. A single projectile consisting of eight adherent ascospores is shown above the apothecium. Note the operculum which has also been projected. B. Tip of ripe ascus showing operculum. C. Tip of discharged ascus. In this case the operculum has remained attached to the ascus tip.

Pyronema is one of a group of Discomycetes which are associated with burnt ground. The reasons for their preference for this habitat are not fully understood, but it is known that *Pyronema* has a very rapid growth rate and appears to be relatively intolerant of fungal competitors (El-Abyad & Webster, 1968*a*,*b*).

Ascobolus

Most species of *Ascobolus* are coprophilous, growing on the dung of herbivorous animals, but *A. carbonarius* grows on old bonfire sites (van Brummelen, 1967). Common coprophilous species are *A. furfuraceus* (= *A. stercorarius*) which is to be found on old cow dung, often along with *A. immersus* (Fig. 161). Whilst these species are heterothallic others, e.g. *A. crenulatus* (= *A. viridulus*), are homothallic. Characteristic features of all species are the purple colour of the ascospore and the operculate photo-

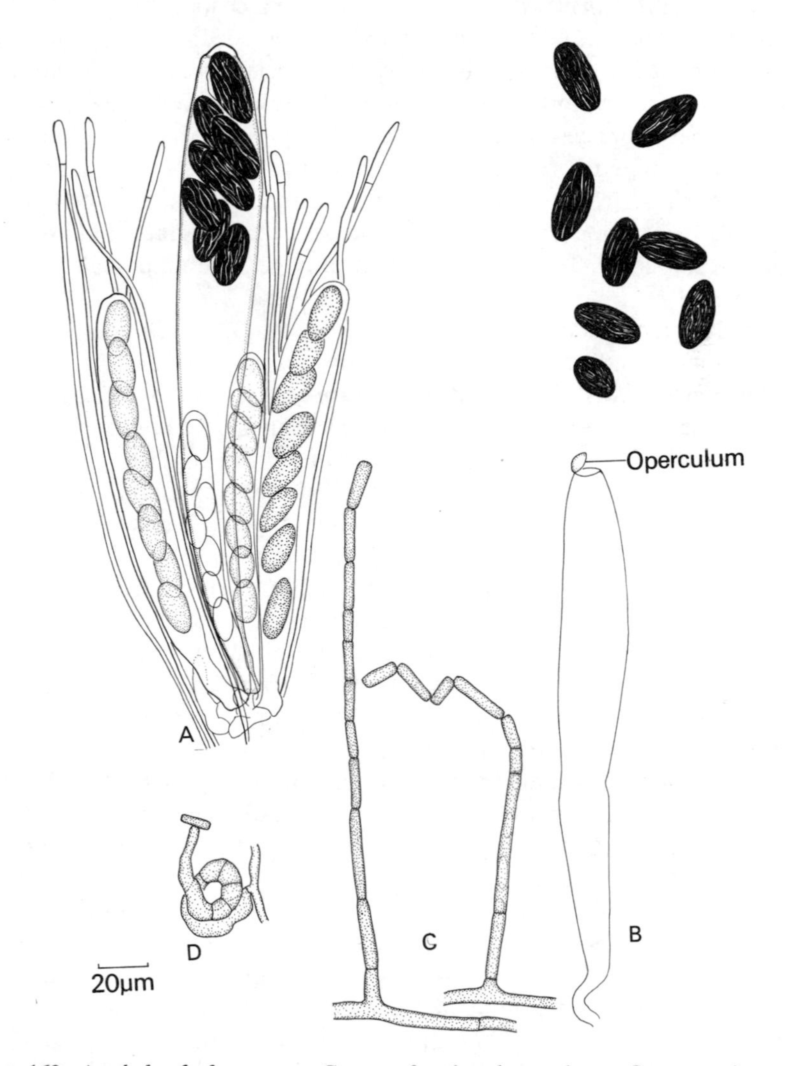

Figure 162. *Ascobolus furfuraceus*. A. Group of asci and paraphyses. One ascus is mature and contains purple-pigmented ascospores. B. The same ascus as shown in A after discharge. The ascus decreases in size after discharge. Note the operculum. C. Arthrospores (or oidia) developed in 5-day-old culture. D. Coiled ascogonium formed in a single ascospore culture within 48 hr of adding oidia of the opposite mating type. The trichogyne of the ascogonium has grown towards the oidium and has fused with it.

tropic asci. *Ascobolus furfuraceus* forms yellowish saucer-shaped apothecia up to 5 mm in diameter, and when it is mature the surface of the apothecium is studded with purple dots which mark the ripe asci. As the asci mature they elongate above the general level of the hymenium. The ascus tips are phototropic and this ensures that when they explode the spores are thrown away from the dung. The ascospores have a mucilaginous epispore which aids attachment. In *A. immersus*, which has very large ascospores (about 70 μm × 30 μm), the epispores cause all the eight ascospores to adhere to form a single projectile about 250 μm long which is capable of being discharged for distances up to 30 cm horizontally. The spores attach themselves to herbage and when eaten by a herbivore germinate in the faeces. It is likely that digestion stimulates spore germination for most spores fail to germinate on nutrient media, but can be stimulated to do so by treatment with a hydroxide or by bile salts (Yu, 1954). The purple pigment in the spore wall develops late, and immature spores are colourless. The spore wall frequently bears longitudinal colourless striations.

A single ascospore culture of *A. scatigenus* (= *A. magnificus*) does not produce apothecia. Sex organs, coiled ascogonia, and antheridia are formed only when mycelia of different mating type are grown together. Ascogonia and antheridia develop on both strains, i.e. each strain is hermaphrodite, but it is self-incompatible: the antheridia of one strain do not fertilise the ascogonia borne on the same mycelium (Dodge, 1920; Gwynne-Vaughan & Williamson, 1932). The ascospores of this fungus are of two types, 'A' and 'a', and fertilisation can only occur between an 'A' ascogonium and an 'a' antheridium and reciprocally. There is thus a gene for mating type represented in two allelic states 'A' and 'a', and incompatibility is controlled by this gene irrespective of the presence of both types of sex organ on each strain. There is no morphological difference between the two compatible mating types and Whitehouse (1949*a*) has used the term 'two-allelomorph physiological heterothallism' to describe this situation.

A similar situation occurs in *A. furfuraceus* but here each strain first produces chains of arthrospores or oidia (see Fig. 162C). The oidia can germinate to form a fresh mycelium, and they also play a part in sexual reproduction. It has been shown that mites and flies may transport oidia of one strain to the mycelium of the alternate strain, and following this apothecia develop (Dowding, 1931). The process of fertilisation has been studied in more detail by Bistis (1956, 1957) and by Bistis & Raper (1963). If an 'A' oidium is transferred to an 'a' mycelium, the oidium fails to germinate and within 10 hr an ascogonial primordium appears on the 'a' mycelium (Fig. 162D). The ascogonium consists of a broad coiled base and a narrow apical trichogyne which shows chemotropic growth towards the oidium and eventually fuses with it. There is evidence that this sequence of events is under hormonal control, and it has been suggested that a fresh 'A' oidium is not immediately capable of inducing development of ascogonial primordia, but must itself

Figure 163. *Aleuria aurantia.* A mature apothecium and two developing apothecia.

be first sexually activated by a secretion from the 'a' mycelium. Following activation the oidium can induce ascogonial development. By substitution experiments it has been shown that an 'A' ascogonium can be induced to fuse with an 'A' oidium, i.e. an oidium of the same mating type, but apothecia fail to develop from such fusions. In compatible crosses fertile apothecia develop within about ten days of fertilisation, each ascus producing 4 'A' and 4 'a' spores. The development of apothecia of *A. furfuraceus* has been studied by Gamundi & Ranalli (1963), Gremmen (1955), and Corner (1929). The ascogonium becomes surrounded by sheath hyphae which develop from the ascogonial stalk, and the paraphyses and excipular tissues develop from the sheath hyphae. The asgonium gives rise to numerous ascogenous hyphae. Ascocarp development in *A. scatigenus* (Gwynne-Vaughan & Williamson, 1932; Wood, 1953) shows a broad similarity to that of *Pyronema*, and again, the earlier claims for double nuclear fusion have been refuted. Van Brummelen (1967) has distinguished two kinds of ascocarp development in *Ascobolus* (*gymnohymenial* and *cleistohymenial*). In gymnohymenial forms the hymenium is exposed from the first until the maturation of the asci. In cleistohymenial forms the hymenium is enclosed, at least during its early development.

Ascobolus furfuraceus and *A. immersus* are examples of cleistohymenial development.

A. immersus has proved a useful tool in interpreting the fine structure of the gene. Although the wild-type strains have purple ascospores, several series of

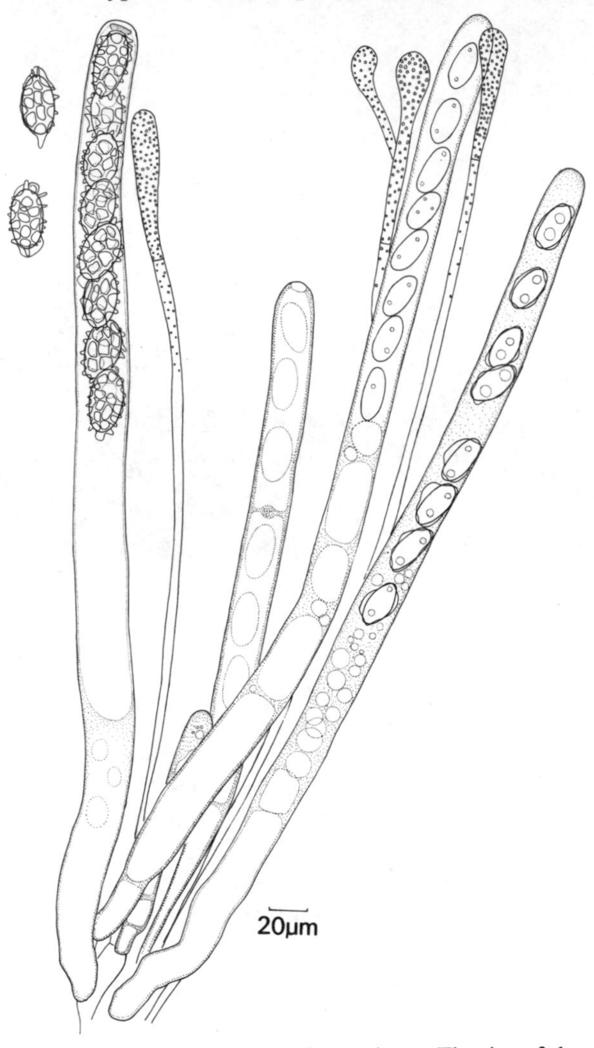

Figure 164. *Aleuria aurantia.* Asci, ascospores and paraphyses. The tips of the paraphyses are filled with orange granules.

mutants with pale spores have been found. When crosses are made between certain pairs of such mutants wild-type recombinants can result from two types of event: (a) crossing over, giving reciprocal recombinants and (b) conversion, yielding non-reciprocal recombinants corresponding to only one of the four products of meiosis.

Figure 165. A. *Morchella esculenta*. B. *Morchella elata*.

A conversion is detected by the presence of coloured and colourless spores in the ratio 6 : 2, whilst crossing over results in a 4 : 4 ratio. In certain series of mutants conversions are distinctly more frequent than crossing over, and this has led to the postulation of a genetic unit termed a *polaron*, a linear structure on which mutant sites are located and within which only non-reciprocal changes (conversions) can occur. Conversions have been interpreted as implying a double replication of one part of a chromatid whilst the corresponding part of the other is not replicated (Lissouba *et al.*, 1962; Fincham & Day, 1965).

OTHER MEMBERS OF THE PEZIZALES

There are numerous other common representatives of this group. *Coprobia granulata* is abundant on cow dung, forming orange apothecia about 3 mm in diameter. It is heterothallic, but sex organs are lacking; plasmogamy is brought about by fusion of mycelia of opposite mating type (Gwynne-Vaughan & Williamson, 1930). Species of *Peziza* form saucer-shaped apothecia up to 5 cm or more in diameter, growing on the ground, on rotting wood or dung. The ascus tip of *Peziza* stains blue with iodine and this character distinguishes it from *Aleuria*. *Aleuria aurantia*, sometimes called the orange-peel *Peziza*, forms bright orange saucer-shaped apothecia in the autumn up to 10 cm in diameter (Fig. 163). The ascospores have a coarsely reticulated

Figure 166. *Helvella crispa.*

surface (Fig. 164). The orange colour is caused by orange granules in the club-shaped tips of the paraphyses. *Morchella esculenta* (Fig. 165A) is edible and forms stalked apothecia about 10–15 cm high. It is found in woods or grassland on limestone in April and May. The hymenium lines the shallow depressions on the upper part of the fruit body, whilst the ridges between lack asci. The ascus tips of *Morchella* are curved so that the spores are projected outwards, and do not impinge on the opposite side of the depressions in which the asci are formed. *Helvella crispa* forms white stalked apothecia about 10 cm high in autumn. The hymenium is borne on a saddle-shaped head borne on a sterile stipe (Fig. 166).

HYPOGAEOUS DISCOMYCETES

TUBERALES

These are truffles, ascomycetes which form subterranean fruit-bodies, in which the hymenium is not open to the exterior. The asci do not discharge their spores violently. Many of the fruit-bodies have a strong smell and flavour, and are excavated and eaten by animals such as squirrels and rabbits, and it is therefore possible that dispersal is brought about in this way. The truffles of

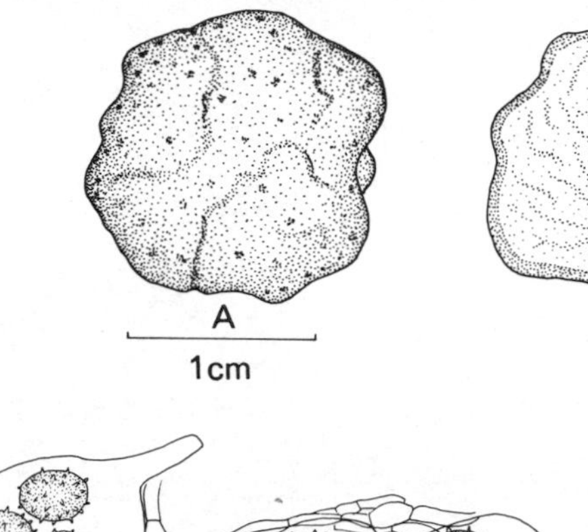

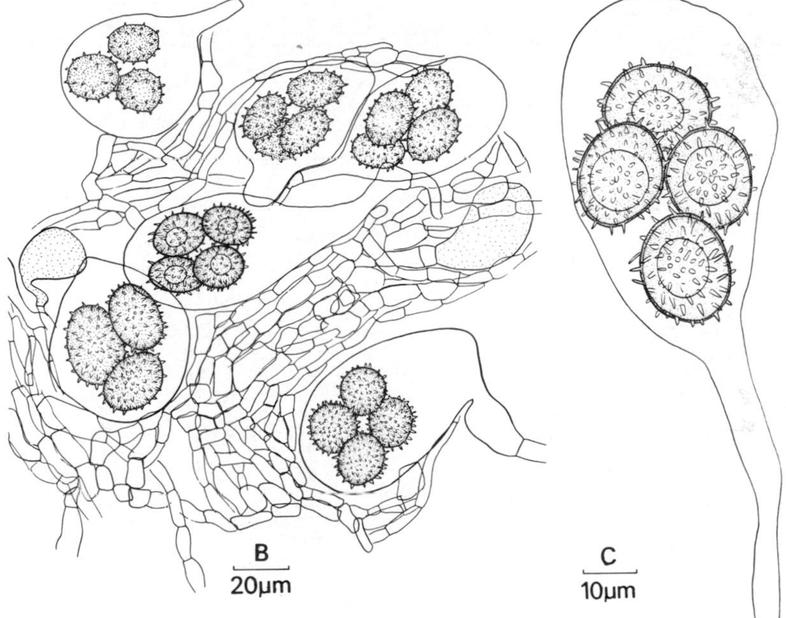

Figure 167. *Tuber rufum*. A. Fruit-body in surface view and in section showing the veins. B. Portion of hymenium. C. Ascus.

commerce are *Tuber melanosporum* (the Périgord truffle) and *T. magnatum* (the white truffle of Piedmont), and a number of other species.

The Périgord truffle is associated with the roots of *Quercus* spp. in France, and truffles are cultivated there by growing the appropriate species of oak (Malençon, 1938; Singer, 1961). Crops of truffles develop after about seven years, and are sometimes collected with the aid of trained pigs or dogs who

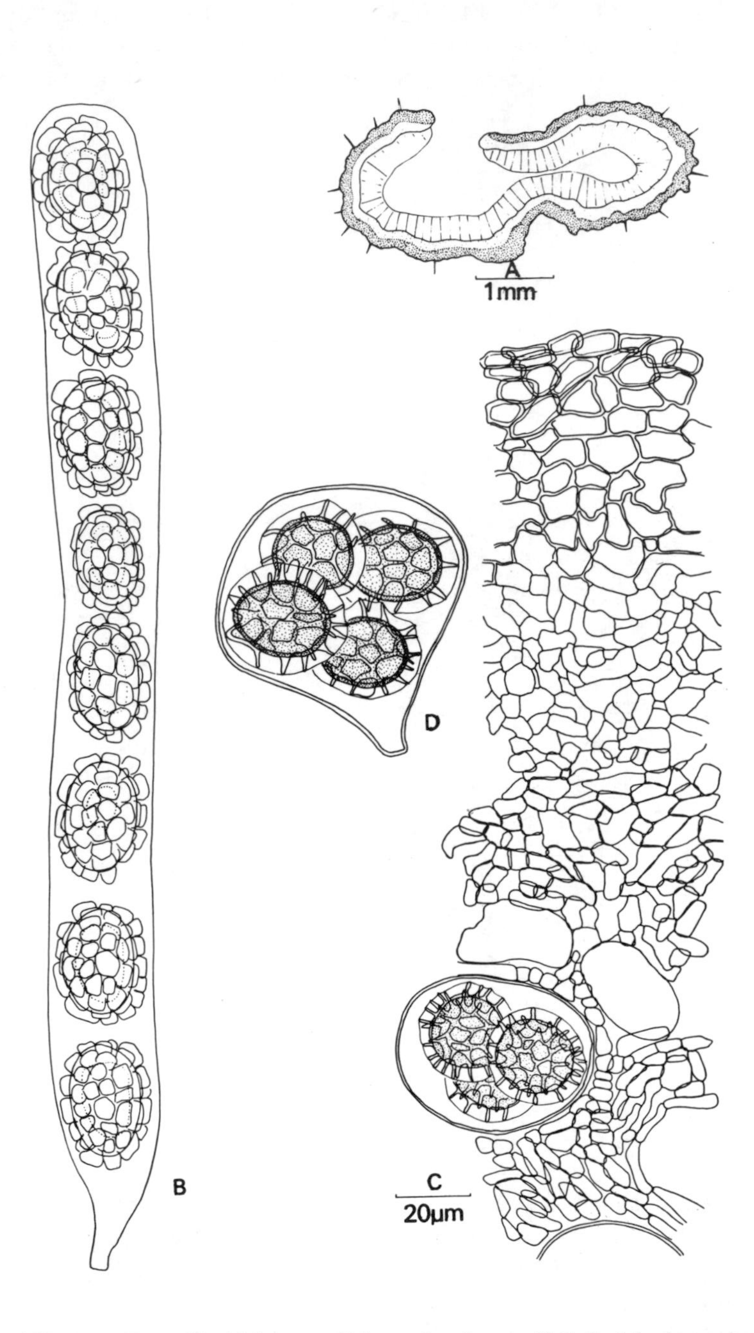

Figure 168. A, B. *Genea hispidula*. C, D. *Tuber puberulum*. A. V.S. fruit-body. B. Ascus. C. V.S. fruit-body. D. Ascus.

Figure 169. *Elaphomyces granulatus*. Two ascocarps, one cut open to show contents.

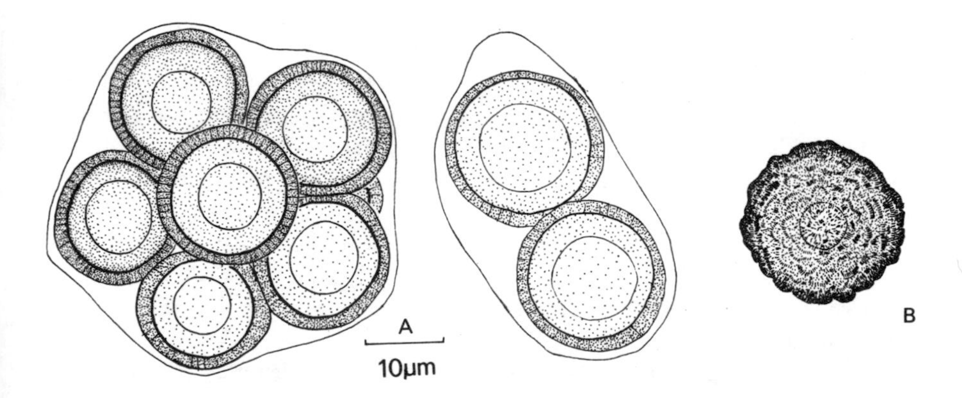

Figure 170. *Elaphomyces granulatus*. A. Asci with two and seven ascospores. B. Mature ascospore.

can detect the fruit-bodies by their smell. Skilled truffle collectors can also detect the position of truffles, guided by truffle flies, which also seek out the fruit-bodies (Ramsbottom, 1953).

There are several genera of Tuberales in Britain (Hawker, 1954). Whilst some authors include the genus *Elaphomyces* here, others place it in the Eurotiales.

Tuber

Fruit-bodies of *Tuber* spp. can be collected by raking the soil under trees. Scrapings of rabbits and squirrels are often a useful indication of likely sites. Common British species are *T. rufum*, *T. puberulum* and *T. excavatum*. The fruit-bodies are globose and up to about 3 cm in diameter. In section they consist of an outer peridium, often of thick-walled cells, and a gleba, or central fertile part, traversed by darker 'veins' which represent the hymenium. In some species the veins communicate with the exterior by one or more pores. For example in *T. excavatum* there is a basal cavity from which extensions protrude into the gleba. The asci are globose and often contain fewer than eight spores, often only two to four. The ascospores are thick-walled and ornamented by spines (e.g. *T. rufum*, Fig. 167) or by reticulate foldings of the outer wall of the spore (e.g. *T. puberulum*, Fig. 168D). Although there are reports of the germination of the ascospores Hawker (1954) writes that 'germination of the ascospores and the subsequent development of a mycelium has never been observed with certainty'.

The development of fruit-bodies of *T. excavatum* have been studied by Bucholtz (1897) and Hawker (1954). The young fruit-body is a disc-like mass of hyphae which becomes corrugated on the lower side, and by arching of the upper surface becomes globose with a large basal opening leading to a hollow chamber. The convoluted lower surface forms a system of complex branched channels, called the 'venae externae'. A palisade-like layer of paraphyses develops over the surface of the venae externae and also forms within the flesh of a fruit-body lining a system of internal cavities, the 'venae internae'. Beneath the palisade binucleate cells appear and give rise to ascogenous hyphae bearing the globose asci. The paraphyses grow out to form a loose weft of hyphae filling the venae externae. The cytology of ascus development has been followed in *T. brumale* and *T. aestivum* (Greis, 1938, 1939). The binucleate condition is established by fusion of two uninucleate cells, and from the fusion cell ascogenous hyphae develop containing numerous pairs of nuclei. Crozier formation occurs as in many other Ascomycetes.

It has been suggested (Fischer, 1938) that the Tuberales may have evolved from Ascomycetes with exposed hymenia as found in the existing Pezizales. Included in the Tuberales are genera such as *Genea* (see Fig. 168A,B) in which the fruit-body resembles an apothecium with a pore opening on the upper side. The asci of *Genea* are cylindrical and eight-spored, and in this respect resemble more closely the asci of epigaeous Discomycetes.

Elaphomyces

Elaphomyces, the hart's truffle, is probably the most common British hypogaeous fungus, and fruit bodies of *E. granulatus* (Figs. 169, 170) and *E. muricatus* can be collected throughout the year beneath the litter layer under various trees, but especially beech. *Elaphomyces muricatus* is often parasitised by *Cordyceps ophioglossoides* forming yellow mycelium around the subterranean fruit-bodies, and a club-shaped perithecial stroma above ground. The *Elaphomyces* fruit-bodies vary in size from about 1–4 cm. When cut open an outer rind (or cortex) can be distinguished from a central mass containing the globose asci, traversed by lighter sterile 'veins'. The asci in *E. granulatus* usually contain six spores and in *E. muricatus* two to four. The spores are dark brown and thick-walled when mature and the conditions necessary for their germination are not known.

5: LOCULOASCOMYCETES

The characteristic feature of this group is that the ascus is bitunicate; it has two separable walls. The outer wall does not stretch readily, but ruptures laterally or at its apex to allow the stretching of a thinner inner layer. The fruit-body within which the asci develop is regarded as an ascostroma. This has been defined as an aggregation of vegetative hyphae not resulting from a sexual stimulus (Wehmeyer, 1926), but Holm (1959) has questioned the accuracy of this definition since examples are now known where the ascocarps do develop following a sexual stimulus (Shoemaker, 1955). Within the developing ascocarp one or more locules are formed by the down-growth of pseudoparaphyses (see below) and the development of asci. One or more ostioles develop by the breakdown of a pre-formed mass of tissue. Where a single locule develops a structure resembling a perithecium results, and although this term is commonly used these structures should strictly be termed pseudothecia. The name given to the group was coined by Luttrell (1955) and corresponds to the Ascoloculares of Nannfeldt (1932). Several large orders are included (Luttrell, 1965*a*) but we shall consider only one, the Pleosporales (sometimes termed the Pseudosphaeriales).

PLEOSPORALES

This is a large group of ascomycetes including some economically important genera of plant pathogens such as *Cochliobolus* and *Pyrenophora* parasitic on grasses and cereals; *Ophiobolus*, *Pleospora* and *Leptosphaeria* common saprophytes or weak parasites of herbaceous plants; and *Sporormia* a saprophyte of dung.

The development of pseudothecia in these forms conforms to the *Pleospora*

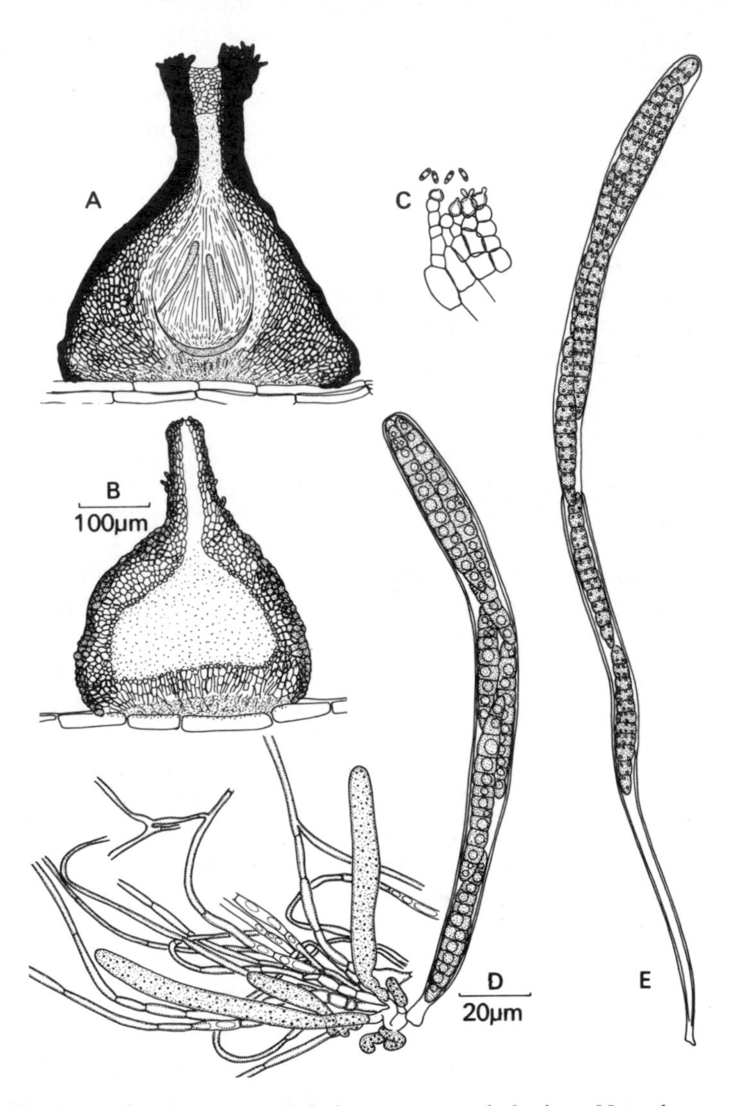

Figure 171. *Leptosphaeria acuta*. A. L.S. immature pseudothecium. Note the young asci (stippled) elongating between a pre-formed mass of pseudoparaphyses, and the thin-walled cells at this stage blocking the ostiole, later dissolved. The centrum subsequently enlarges, dissolving the pseudoparenchyma surrounding it. B. L.S. pycnidium. C. High-power drawing of cells lining the pycridium showing the origin of the conidia. D. Cluster of developing asci from a young perithecium. Note the branching of the pseudoparaphyses. E. Stretched bitunicate ascus showing rupture of the outer wall at its apex.

type of Luttrell (1951). Ascogonia arise within a stroma, and in the region of the ascogonia a group of vertically arranged septate hyphae appears, each one arising as an outgrowth from a stroma cell. These hyphae are capable of elongating by intercalary growth and are termed *pseudoparaphyses* (Luttrell, 1965*b*). Pseudoparaphyses arise near the upper end of the cavity and grow downwards. Their tips soon intertwine and push between the other cells of the stroma so that free ends are seldom found. They may thus be distinguished from *true paraphyses* formed in other fungi from hyphae attached at the base of the cavity, extending upwards and free at their upper

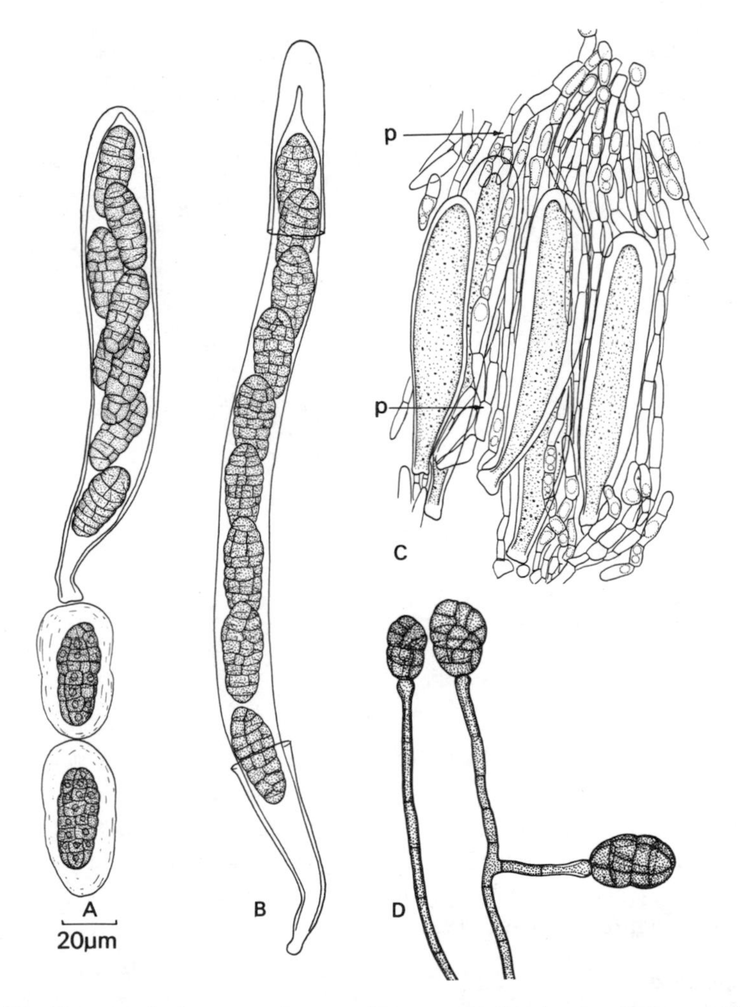

Figure 172. *Pleospora herbarum*. A. Ascus and ascospores showing mucilaginous epispore. B. Stretched bitunicate ascus showing rupture of outer wall. C. Developing asci and pseudoparaphyses. The arrows (p) indicate points of branching of ascending and descending pseudoparaphyses. D. Conidia of *Stemphylium botryosum* type.

ends. They may also be distinguished from *apical paraphyses* which are attached above, arising from a clearly defined meristem near the apex of a perithecium, forming a well-defined palisade of hyphae free at their lower ends (see the *Nectria* type of development, p. 212). In the *Pleospora* type of development asci arise amongst the pseudoparaphyses at the base of the cavity and grow upwards between them. The ostiole develops lysigenously,

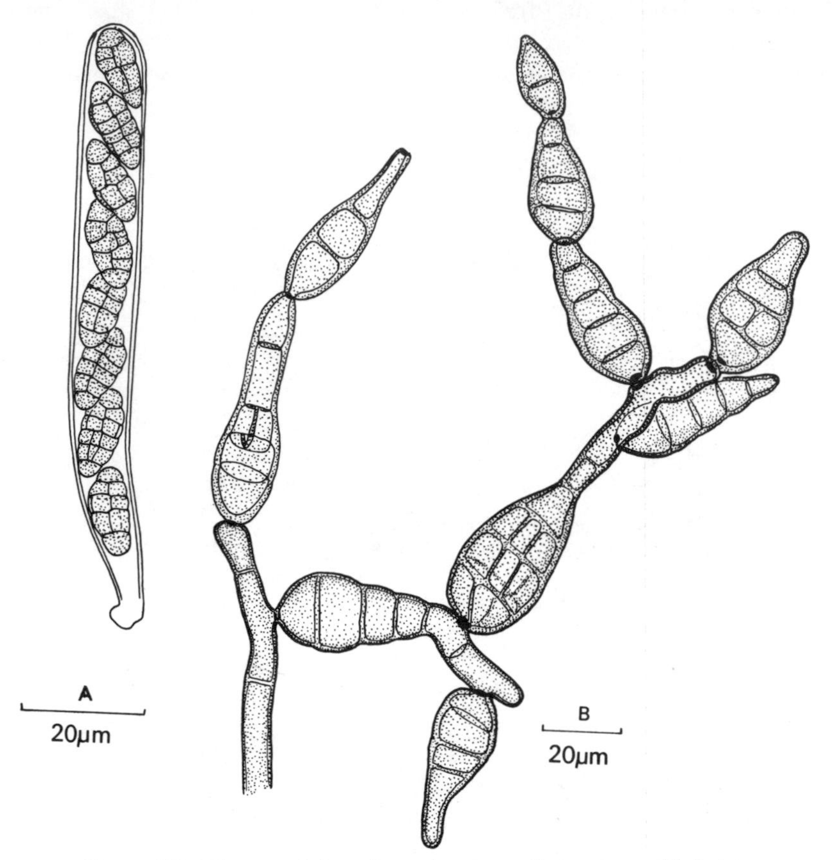

Figure 173. *Pleospora infectoria*. A. Ascus. B. *Alternaria* conidial state.

i.e. by breakdown or separation of pre-existing cells. Development of this general type has been described in *Pleospora herbarum* (Wehmeyer, 1955), *Leptosphaeria* (Dodge, 1937), *Sporormia* (Arnold, 1928; Morisset, 1963) and other fungi (see Luttrell, 1951).

Leptosphaeria

Leptosphaeria species grow on moribund leaves and stems of herbaceous plants. There are probably some 200 species, many growing on a wide range of hosts, but some confined to one host plant. Although most are saprophytic or only weakly parasitic, some are troublesome parasites, e.g. *L. avenaria* the cause of speckled blotch of oats and *L. coniothyrium* the cause of cane blight of raspberry. *Leptosphaeria acuta* grows in abundance at the base of overwintered stems of nettles (*Urtica dioica*). The black shining pseudothecia are somewhat conical and flattened at the base. The bitunicate asci elongate within a pre-formed group of branching pseudoparaphyses and close examination of the direction of growth and branching suggests that the pseudoparaphyses may be both ascending and descending. The ostiole of the perithecium is formed lysigenously by breakdown of a pre-existing mass of thin-walled cells (Fig. 171A).

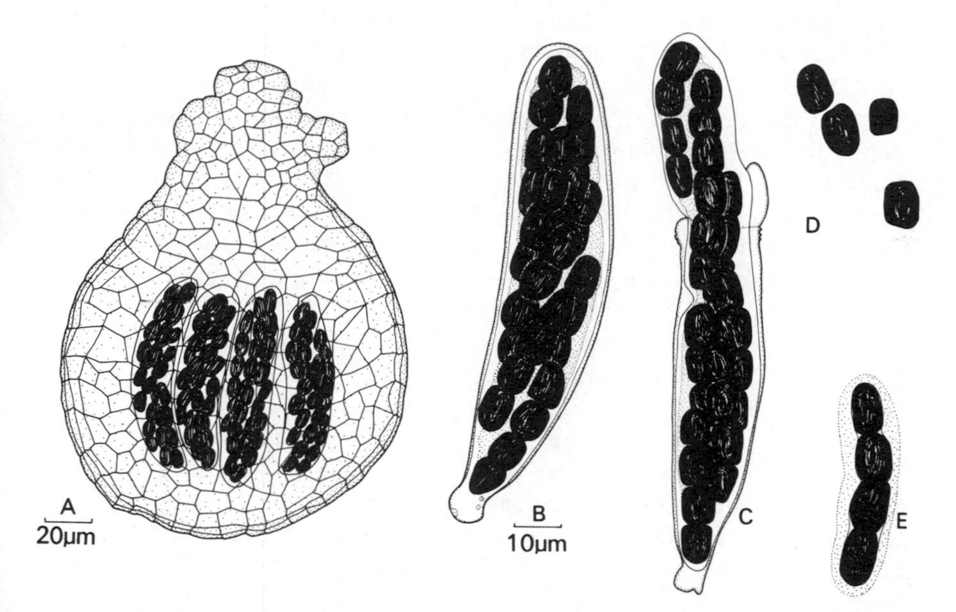

Figure 174. *Sporormia intermedia.* A. Pseudothecium with asci visible through the transparent wall. B. Ripe unextended ascus showing the double wall. C. Elongating ascus showing rupture of the outer wall and extension of the inner. D. Ascospore separated into its four component cells. E. Intact ascospore.

The bitunicate structure of mature asci is difficult to discern because as the ascus expands the inner wall protrudes through a thin area in the outer wall at the ascus tip (Fig. 171E) and then the inner wall extends. Thus the tip of the ascus in expanded asci is single-walled. The ascospores are discharged successively in the space of about 5 sec., and the ascus shows a slight decrease in length as each spore escapes (Hodgetts, 1917).

Associated with the thick-walled conical pseudothecia on the nettle stems are thinner-walled, slightly smaller globose pycnidia with cylindrical necks (Fig. 171B). The cavity of the pycnidia is lined by small spherical cells which give rise to numerous minute rod-shaped pycnospores which are capable of germination (Fig. 161c). Such pycnidia have been named *Phoma acuta*, and culture studies show that this stage is the conidial state of *L. acuta* (Müller & Tomasevic, 1957). Various different kinds of conidial apparatus have been found in other species of *Leptosphaeria* (Lucus & Webster, 1967; Lacoste, 1965).

Pleospora

Wehmeyer (1961) recognises over 100 species, but this is probably an underestimate. Most species form fruit-bodies on moribund herbaceous stems, apparently as saprophytes, but it is likely that some are weak parasites. Of these *P. bjoerlingii* (= *P. betae*) is the cause of black-leg of sugar beet. *Pleospora herbarum* attacks a wider range of cultivated hosts causing such diseases as net-blotch of broad bean and leaf spot of clover, lucerne and other hosts. This fungus is also abundant on overwintered herbaceous stems of numerous maritime plants. The large black somewhat flattened pseudothecia contain broad sac-like bitunicate asci, with eight yellowish-brown slipper-shaped ascospores with transverse and longitudinal septa (Fig. 172A). The conidial state, *Stemphylium botryosum*, is often associated with the perithecia. The conidia develop singly from the tips of conidiophores swollen at their apices, as blown-out ends. A narrow neck of cytoplasm connects the developing spore to its conidiophore through a pore, and Hughes (1953) has termed conidia of this type porospores (Fig. 172D). According to Meredith (1965) the conidia are violently jolted from the tip of the conidiophore.

Pleospora infectoria which forms pseudothecia on grass and cereal culms in spring has smaller ascospores. In culture this fungus forms branching chains of beaked spores, and new spores are formed at the tip of the chain (Fig. 173). Conidia of this type belong to the form genus *Alternaria*, and are also porospores. Spores of *Alternaria* are abundant in the air in late summer and autumn and may be a cause of inhalant allergy (Hyde & Williams, 1946).

Sporormia

Pseudothecia of *Sporormia* are common on dung of herbivores, and the fungus is also occasionally isolated from soil. *S. intermedia* is one of the most common species, and has thin transparent perithecial walls through which the asci can be seen (Fig. 174). The ascus has a well-defined double wall. The outer wall does not stretch readily and as the ascus becomes turgid the outer wall ruptures and the thin elastic inner wall stretches considerably. The tip of the ascus projects through the ostiole and the ascus then explodes. The ascospores of *S. intermedia* are four-celled and surrounded by a mucilaginous envelope. The spore may break up into individual cells, each capable of germination. Spore discharge is nocturnal (Walkey & Harvey, 1966*b*).

4

BASIDIOMYCOTINA (Basidiomycetes)

Many of the familiar larger fleshy fungi are members of this group, which includes the toadstools, bracket fungi, fairy clubs, puff-balls, stinkhorns, earth-stars, birds'-nest fungi and jelly fungi. Most of these are saprophytes, causing decay of litter, wood or dung, and some are serious agents of wood decay such as *Serpula lacrymans* (*Merulius lacrymans*) the dry-rot fungus (Cartwright & Findlay, 1958). Some of the toadstools which are associated with trees form mycorrhiza, a symbiotic association (Harley, 1969), but some are severe parasites, e.g. *Armillaria mellea*, the honey agaric, which destroys a wide range of woody and herbaceous plants. Whilst the fleshy fungi enjoy a notorious reputation for being poisonous, the majority of British toadstools are harmless, and several species besides the field mushroom are good to eat (Ramsbottom, 1953). Two important groups of plant pathogens, the rusts (Uredinales) and smuts (Ustilaginales), are also classified in the Basidiomycetes. In nature these organisms are confined to living host plants.

The characteristic spore-bearing structure is the basidium. In contrast with the endogenous spores of the ascus, basidia bear spores exogenously, usually on projections termed sterigmata. The number of spores per basidium is typically four, but two-spored basidia are quite common. In *Phallus impudicus*, the stinkhorn, there may be as many as nine spores per basidium. Basidia vary considerably in structure, and the form of the basidium is an important criterion in classification. In the toadstools and their allies the basidium is a single cylindrical cell, undivided by septa, typically bearing four basidiospores at its apex (Fig. 175). Such basidia are termed *holobasidia*. In the Uredinales and Ustilaginales the basidium develops from a thick-walled cell (teliospore or chlamydospore) and is usually divided into four cells by three transverse septa. Transversely segmented basidia are also found in the Auriculariaceae, but here the basidia do not arise from resting cells. In the Tremellaceae the basidia are longitudinally divided into four cells, whilst in the Dacrymycetaceae the basidium is unsegmented but forked into two long arms, to form the so-called tuning fork type of basidium. Segmented basidia are sometimes termed *phragmo-basidia* (or *heterobasidia*). Some of these different kinds of basidia are illustrated in Fig. 176.

DEVELOPMENT OF BASIDIA

The development of a basidium can be illustrated by reference to a gill-bearing fungus such as *Oudemansiella radicata* (= *Collybia radicata*) which grows on old stumps of deciduous trees (Figs. 175, 192A). The basidium arises as a terminal cell of a hypha making up the gill tissue. The basidia are packed together to form a fertile layer or hymenium. A basidium is at first densely packed with cytoplasm, but soon several small vacuoles appear. Later, a single large vacuole develops at the base of the basidium, and by the enlargement of this vacuole cytoplasm is pushed towards the end of the basidium. A clear cap is visible at the tip, and it is here that the sterigmata develop. Corner (1948) has postulated that there must be four elastic areas of the wall

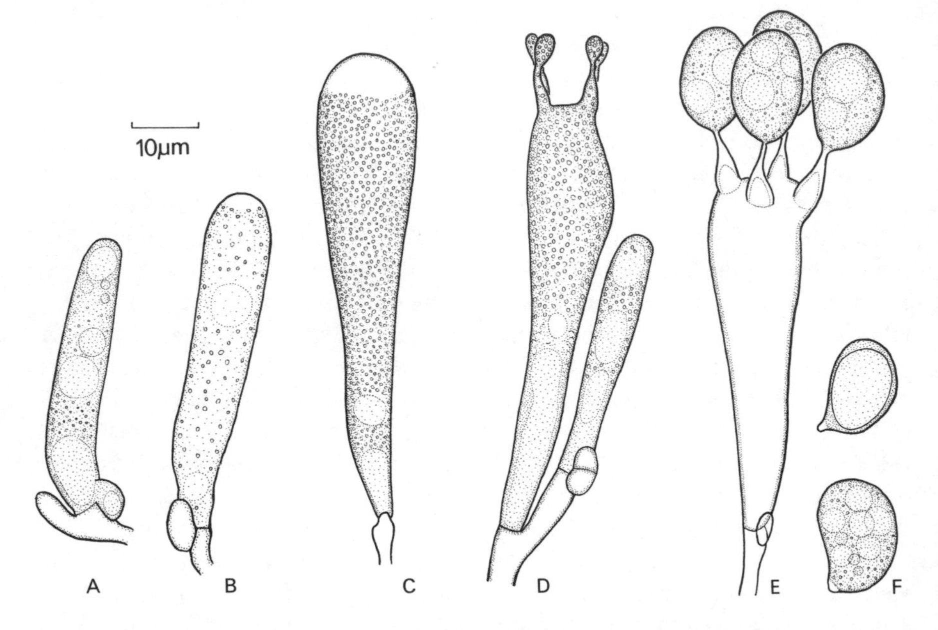

Figure 175. *Oudemansiella radicata*. Stages in development of basidia. A. Young basidium with numerous vacuoles. Note the clamp connection at the base and the formation of a further basidial initial. B. Later stage showing the development of a clear apical cap. C. Localisation of vacuoles towards the base of the basidium. D. Development of sterigmata and spore initials. Note the enlargement of the basal vacuole. E. Fully developed basidium. The spores are full of cytoplasm, whilst the body of the basidium contains only a thin lining of cytoplasm, surrounding an enlarged vacuole. F. Discharged basidiospores.

from which the sterigmata extend, and he has suggested that the force for the development of the basidia and the spores comes from the enlargement of the basal vacuole which acts like a piston ramming the cytoplasm into the spores. Ripe basidia thus contain very little cytoplasm, but contain merely a large vacuole (Fig. 175E).

Young basidia are binucleate, and nuclear fusion occurs here. The resulting fusion nucleus undergoes meiosis immediately so that four haploid daughter nuclei result, and one is distributed to each basidiospore. In some basidia meiosis is followed by a mitotic division, so that some basidiospores are binucleate. The plane of spindle-formation during meiosis may be transverse to the long axis of the basidium, and such fungi are said to be *chiastobasidial*. The contrasting condition in which the spindles are orientated parallel to the long axis of the basidium is termed *stichobasidial*. The division planes may have some taxonomic significance. For a fuller discussion of the cytology of basidia see Raper, (1966); Olive, (1953, 1965).

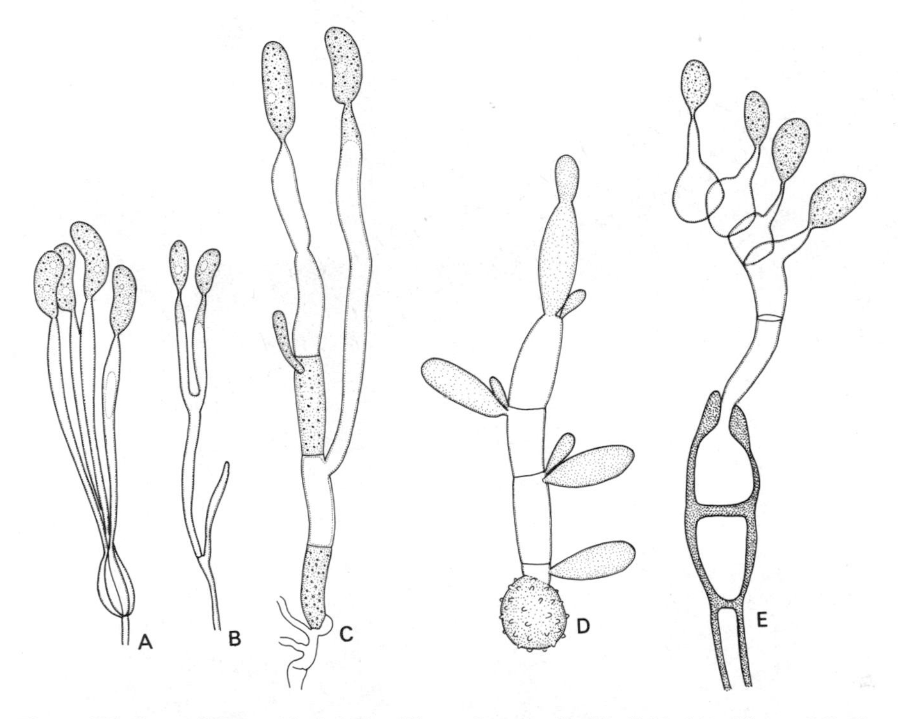

Figure 176. Some different kinds of basidia. A. Longitudinally divided basidium of *Exidia glandulosa* (Tremellaceae). B. Tuning-fork type basidium of *Calocera viscosa* (Dacrymycetaceae). C. Transversely divided basidium of *Auricularia auricula* (Auriculariaceae). D. Germinating chlamydospore of *Ustilago avenae* (Ustilaginales). E. Germinating teliospore of *Puccinia graminis* (Uredinales).

The range in structure of basidia has raised difficult problems of nomenclature. The terminology proposed by Martin (1957) has been followed:

Basidium. The organ of the Basidiomycetes which bears the basidiospores, either directly or on extensions from the primary basidial cell. Nuclear fusion and meiosis usually occur in the earlier stages but neither such occurrence nor the places in which they occur is an essential part of the definition.

Probasidium. The primary basidial cell. It may pass without great change into the nearly mature basidium or it may be represented by a thick-walled basidial cyst which gives rise to a thin-walled extension.

Metabasidium. A thin-walled extension from a basidial cyst. This is the 'promycelium' of the rusts, but is represented in a number of genera of the Auriculariaceae.

Sterigma. A conical or cornute, usually subaerial structure which bears the basidiospore and from which the basidiospore is commonly violently discharged.

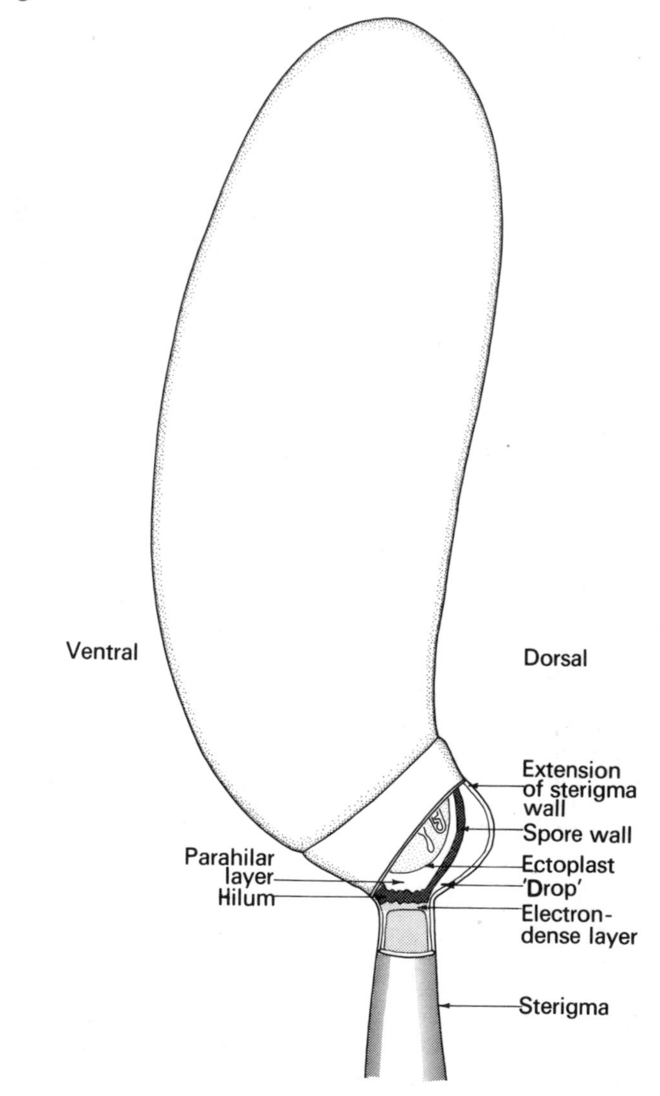

Figure 177. *Schizophyllum commune*. Reconstruction of the structure of the sterigma and basidiospore based on electron micrographs of thin sections (after Wells, 1965).

Epibasidium. A tubular or swollen extension from the body of the basidium to the hymenial surface, where it produces a sterigma.

For a fuller discussion see Martin (1957) and Talbot (1954).

THE MECHANISM OF BASIDIOSPORE DISCHARGE

Observations on ripe basidia show that the individual spores are in turn projected violently from the tip of the sterigma. The term ballistospore has been applied to violently projected spores (Derx, 1948). Whilst most basidiospores are ballistospores, some, e.g. those of the Gasteromycetes, are not. Shortly before discharge occurs a swelling appears on the adaxial side where the spore and the sterigma join. Although observed earlier by other workers, Buller (1909, 1922) drew especial attention to it, and it is often referred to as 'Buller's drop'. Buller assumed that the drop consisted largely of water. Prince (1943) showed that the drop could form under water, whilst Corner (1948) showed that the drop could persist in material preserved in alcohol–formalin. He also found that a drop may appear, increase in size, and then diminish and disappear before discharge. These observations suggest that the drop is surrounded by a membrane, and may consist of a ballooning outwards of the sterigma wall. Electron micrographs of thin longitudinal sections of sterigmata with attached spores (Wells, 1965) support this view (Fig. 177). Opinions vary on the content of the drop. Corner (1948), Müller (1954), and Wells (1965) conclude that the membrane contains liquid, whilst Olive (1964) believes that it contains gas. Evidence that the drop contains liquid is shown by the fact that a drop is often carried with the spore (Buller, 1922) and that spores immediately after discharge may be surrounded by liquid (Müller, 1954). Spore discharge may occasionally take place without the appearance of the drop, and in some species, e.g. in the rust-fungus *Cronartium ribicola*, the drop has not been observed (Bega & Scott, 1966). It is therefore possible that the drop is not an essential part of the spore discharge mechanism.

A number of possible mechanisms have been suggested for basidiospore discharge. In considering them, we cannot dismiss the possibility that several distinct mechanisms may operate.

1. *Rounding-off of turgid cells*: in his studies of spore development in the rust fungus *Gymnosporangium nidus-avis*, Prince (1943) illustrated the presence of a flat septum across the neck of the sterigma. Following spore release, the hilum of the spore and the end of the sterigma were convex (Fig. 178). It was therefore concluded that the basidiospore bounced off the sterigma by rounding-off of secondary walls laid down on either side of the original septum. Wells (1965) has illustrated the development of a flat septum formed by inward growth of wall material in electron micrographs of developing basidia of *Schizophyllum commune* (see Fig. 177). After discharge the apex of the

sterigma is rounded, although the hilum of the spore remains flat, and Wells concludes that turgor pressure within the basidium contributes to the force of basidiospore discharge. The rounded papilla-like hilum of the spore after discharge in the rust-fungus *Cronartium ribicola* has led Bega & Scott (1966) to suggest that a rounding-off mechanism may be involved in this fungus. This mechanism is known to occur in other fungi, e.g. the sporangia of some

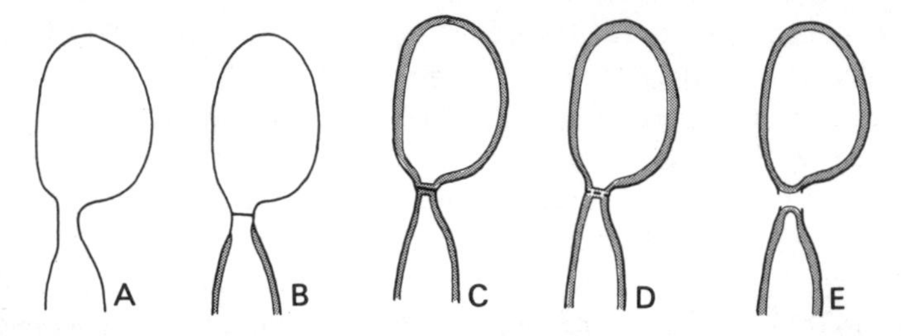

Figure 178. Basidiospore development and discharge in *Gymnosporangium nidus-avis* (after Prince, 1943). A. Basidiospore which has reached full size. B. Formation of primary cross-wall across the neck of the sterigma. C. Laying down of secondary walls on either side of the primary wall. D. Dissolution of primary wall. E. Spore bounces off sterigma due to rounding off of both turgid secondary walls.

Entomophthorales, the aeciospores of rust fungi and the conidia of some Fungi Imperfecti. Buller (1909), who first proposed it for basidiospores, termed it *jerking discharge*, to contrast it with *squirting discharge* as found for example in Ascomycetes.

2. *Jet propulsion*: Befeld (1877) claimed that basidiospores were propelled by tiny jets of liquid from the sterigmata, and Corner (1948) has supported the idea. Objections to the hypothesis are:

1. There is no appreciable reduction in basidial volume as spore discharge proceeds (Buller, 1922).
2. Since spores are discharged successively, for the continued maintenance of basidial turgor it is also necessary to postulate that the sterigma tip be sealed quickly after spore projection. Electron micrographs of sections through sterigmata after spore release show no evidence of a pore (Wells, 1965; Bega & Scott, 1966).

3. *Surface tension*: Buller (1922) raised the possibility that the surface tension of the drop might contain sufficient energy for spore projection. Ingold (1939) calculated the surface energy of the drop, assuming it to be water, and found it to be more than adequate to provide energy for spore release. It was not shown, however, how this energy, which is *potential energy*, could be transformed into the necessary *kinetic energy* to perform the work of spore projection, and the idea has received no support.

4. *Explosive discharge*: Olive (1964) studied spore release in the mirror-yeast *Sporobolomyces* (probably a Basidiomycete) and inferred that the bubble which appears at the junction of the sterigma and spore contains gas, possibly CO_2. This conflicts with Müller's (1954) observation on *S. salmonicolor* that the bubble contains liquid. Olive believes that the explosion of the gas-containing bubble causes the spore to be jolted from the sterigma. Support for the gas-explosion hypothesis comes from experiments by Ingold & Dann (1968) on the effect of external gas pressure on spore discharge in *Schizophyllum*. The number of spores discharged against a pressure of 50 atmospheres was negligible as compared with the numbers discharged at atmospheric pressure. In contrast spore discharge in *Entomophthora coronata*, which occurs by a rounding-off mechanism, is little affected by external pressure.

5. *Electrostatic repulsion*: If electrostatically charged plates are placed under basidiomycete fruit-bodies it is found that the majority of spores collect on one or other of the plates, indicating that the spores themselves carry an electrostatic charge (Buller, 1909; Gregory, 1957). It has been suggested that electrostatic repulsion may be partly responsible for spore projection. Savile (1965) believes that basidiospores are released by the gas-explosion proposed by Olive, and that electrostatic forces are also involved in directing accurately the path of the spore. Two objections can be raised against this hypothesis:

(a) The magnitude of the charge is probably inadequate to displace the spore laterally for a sufficient distance. Swinbank *et al.* (1964) determined the charge on the basidiospores of *Serpula lacrymans*, the dry-rot fungus, and concluded that it is unlikely that charges play any important part in the escape of spores from fruit-bodies.

(b) The charge on the spore is probably acquired in the process of separation from the sterigma. Before separation the spore and sterigma would have the same potential, and if the spore acquired its charge on separation, the sterigma would simultaneously acquire a charge of the same magnitude, but of opposite sign. Thus since spore and sterigma have charges of opposite sign they would tend to attract, rather than repel, each other.

There is still no general agreement on the mechanism of basidiospore discharge (see Ingold, 1965, 1966). At present the balance of the evidence is probably in favour of a rounding-off mechanism, preceded by the bursting of the sterigma wall. The tearing of the wall is necessary before the turgor mechanism can operate. In the face of observations that discharged spores are surrounded by a drop of liquid, more convincing evidence for the existence of a gas bubble will be needed before the gas-explosion hypothesis is generally accepted. Studies of the fine structure of the sterigmata of further Basidiomycetes may clarify the issue.

In many Basidiomycetes the hymenia are almost vertical, and the basidia lie horizontally. The spores on release are shot forward horizontally for a short distance, usually less than 1 mm, and then turn abruptly through a right angle and fall at a characteristic slow terminal velocity. The peculiar right-angled trajectory (Fig. 179) is a consequence of the large surface-to-volume ratio of basidiospores, which results in a disproportionately high wind resistance. This trajectory has been termed a sporabola by Buller (1909). The form of the trajectory has been discussed more fully by Ingold (1965).

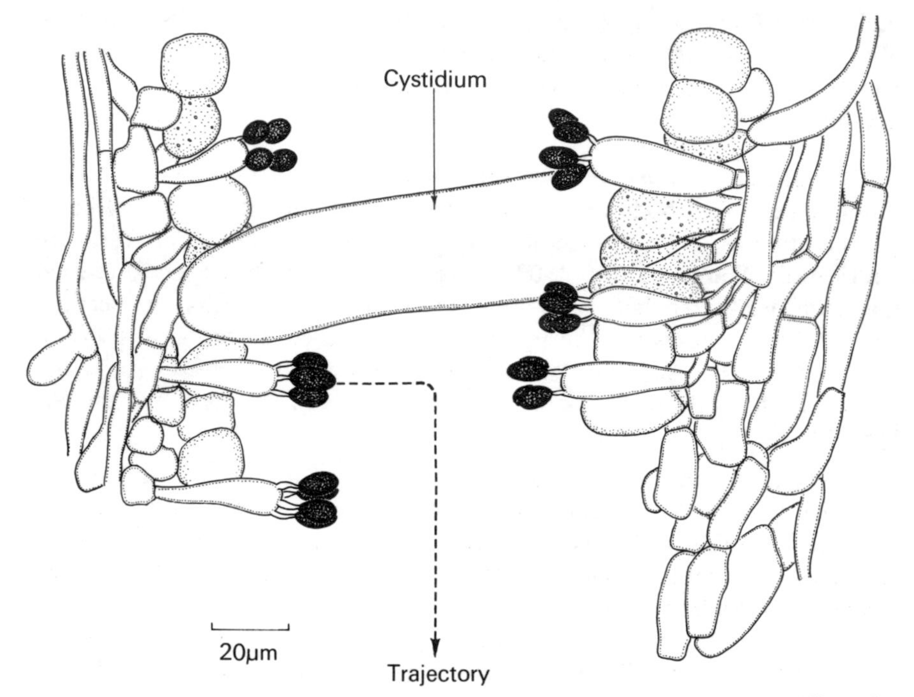

Figure 179. *Coprinus atramentarius*. Vertical section of gill showing ripe basidia and a cystidium extending from one gill to another. The arrow indicates the possible trajectory of the basidiospores.

NUMBERS OF BASIDIOSPORES

The number of basidiospores produced by fruit-bodies can be extremely high. It has been calculated (Buller, 1909) that the detached cap of a mushroom *Agaricus campestris* produced $1{\cdot}8 \times 10^9$ spores in two days, at an average rate of 40 million per hour. Estimates for some other fungi are given in the table opposite.

The mycelium of all these fungi is perennial, and a single mycelium can bear numerous fruit-bodies, so it is clear that any single basidiospore has an infinitesimal chance of successfully establishing a fruiting mycelium.

Number of spores produced by certain Basidiomycetes
(after Buller, 1922)

	Total number of spores	Spore fall period	Number of spores discharged per day
Calvatia gigantea	7×10^{12}	—	—
Ganoderma applanatum	5.46×10^{12}	6 months	3×10^{10}
Polyporus squamosus	5×10^{10}	14 days	$3{\cdot}5 \times 10^{9}$
Agaricus campestris	$1{\cdot}6 \times 10^{10}$	6 days	$2{\cdot}6 \times 10^{9}$
Coprinus comatus	$5{\cdot}2 \times 10^{9}$	2 days	$2{\cdot}6 \times 10^{9}$

THE STRUCTURE OF THE MYCELIUM

On germination many basidiospores produce a mycelium (the primary mycelium) which may at first be multinucleate, but later septa are laid down cutting the mycelium into uninucleate segments. The septa on the primary mycelium are simple cross-walls (Fig. 180A). The nuclei in a given haploid mycelium are derived from the original single nucleus which enters the basidiospore so that all are usually identical, or *homokaryotic*. Since each segment of the mycelium is uninucleate it can also be described as *monokaryotic* (Jinks & Simchen, 1966). In the majority of Basidiomycetes homokaryotic mycelia do not fruit, and before fruiting can occur fusion is necessary between two homokaryons of different mating type (see below). When two compatible homokaryotic mycelia come into contact breakdown of the walls separating them occurs to achieve cytoplasmic continuity. Nuclear migration follows. Lange (1966) has measured rates of nuclear movement as high as 300 μm/min between compatible monokaryons of *Coriolus versicolor*. Although plasmogamy occurs at this point, nuclear fusion or karyogamy is delayed, and a mycelium develops with binucleate segments, each segment containing two genetically distinct (i.e. compatible) nuclei. Such mycelia are *dikaryotic* (strictly *heterokaryotic dikaryons*). The two nuclei in each cell of a dikaryon usually divide simultaneously, and nuclear division is described as *conjugate*. In many cases, but by no means all, dikaryotic mycelia bear at each septum a characteristic lateral bulge termed a clamp connection, or clamp (Figs. 180, 181). Whilst it is reasonable to infer that mycelia which bear clamps are usually dikaryotic, the converse is not true: there are numerous fungi in which the dikaryotic mycelium does not bear clamps. This is often true of the hyphae making up a fruit-body (Furtado, 1966). The clamp connection is a device which ensures that when a dikaryotic mycelium segments, each segment contains two genetically distinct nuclei. In the absence of clamps or some other mechanism for re-arrangement of nuclei there would be a tendency for dikaryotic mycelia to segment at their apices into homokaryotic segments (see Fig. 181). Some authors regard the clamp connection of Basidiomycetes as homologous to the crozier at the tip of ascogenous hyphae in Ascomycetes, and have argued that the Basidiomycetes evolved from Ascomycete ancestors (Olive, 1965; Gaümann, 1964; Teixeira, 1962).

The fine structure of Basidiomycete septa is much more complex than that of an Ascomycete. Close to the limits of resolution of the light microscope it

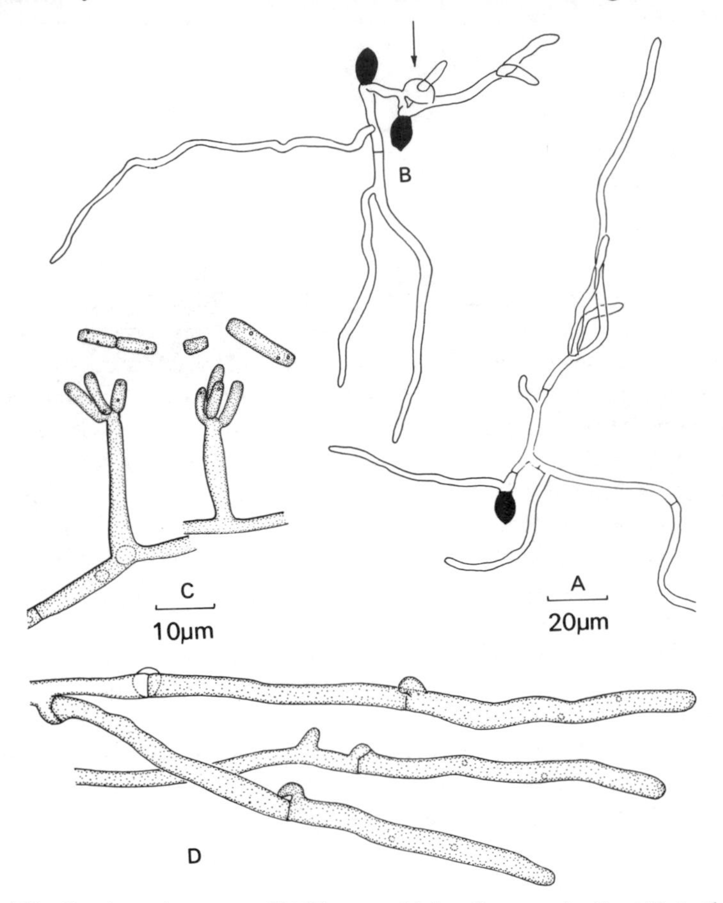

Figure 180. *Coprinus cinereus*. A. Basidiospore 24 hr after germination. Note the simple septa. B. Two basidiospores showing fusion of germ tubes (arrowed). C. Oidia formed on the monocaryotic mycelium. D. Dikaryotic mycelium showing stages in the development of clamp connections. A and B to same scale; C and D to same scale.

is possible to discern that in both monokaryotic and in dikaryotic mycelia each septum is pierced by a narrow septal pore, 0·1–0·2 μm wide, surrounded by a barrel-shaped thickening, the septal swelling (see Fig. 182, after Bracker & Butler, 1963). This kind of septum has been found in numerous Basidiomycetes (Auriculariaceae, Tremellaceae, Aphyllophorales, Agaricales) but not, apparently in the Ustilaginales or Uredinales. It is sometimes known as the *dolipore septum* (Moore & McAlear, 1962; Moore, 1965). The septal pore is overarched with a perforated cap which is an extension of the endoplasmic reticulum. This cap is sometimes termed the parenthosome. Despite these apparent barriers there is good cytoplasmic continuity between adjacent cells, and organelles such as mitochondria have been observed within the septal pore of *Rhizoctonia* (Bracker & Butler, 1964). However in *Polyporus rugulosus* the perforations of the pore cap are too small to allow the passage

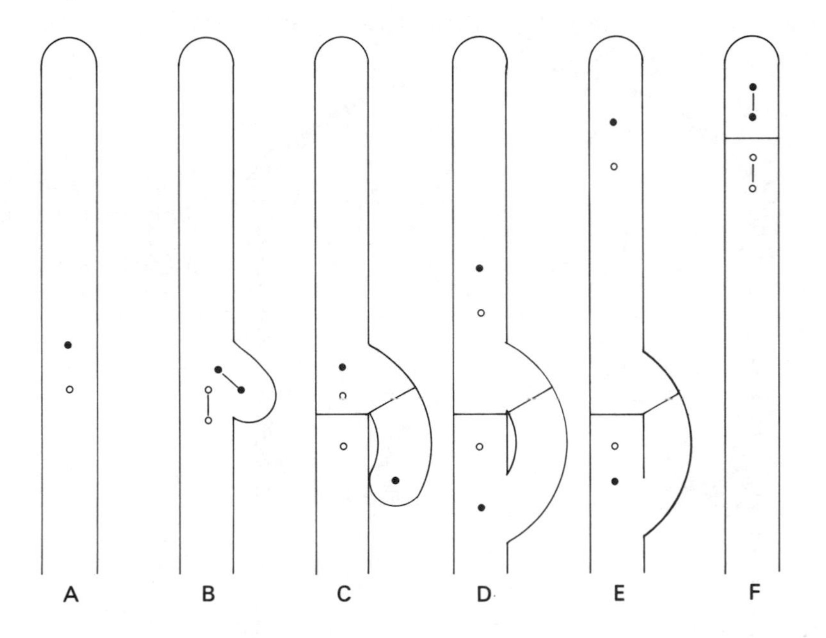

Figure 181. Diagrammatic representation of clamp formation. A. A dikaryotic hyphal tip. B. Simultaneous nuclear division and formation of a backwardly directed lateral branch into which one of the daughter nuclei passes. C. Formation of two cross-walls cutting off a terminal cell which contains two compatible nuclei; and the lateral branch with a single nucleus. D. Fusion of lateral branch with subterminal cell which is now dikaryotic. E. Later stage. F. Hypothetical nuclear division and segmentation in a dikaryotic hypha without clamp formation. Note that the hyphal tip would become homokaryotic.

of mitochondria (Wilsenach & Kessel, 1965*b*). Migration of nuclei from one cell to another is accompanied by a breakdown of the complex dolipore apparatus (Giesy & Day, 1965).

Cell walls of most Basidiomycetes that have been examined are composed of microfibrils of chitin (Aronson, 1965; Scurfield, 1967).

MATING SYSTEMS IN BASIDIOMYCETES

About 10% of Basidiomycetes which have been tested are homothallic (Raper, 1966). Three types of homothallic behaviour may be distinguished:

1. *Primary homothallism*: In *Coprinus sterquilinus* a single basidiospore germinates to form a mycelium which soon becomes organised into binucleate segments bearing clamp connections at the septa. There is no genetical distinction between the two nuclei in each cell, and this mycelium is capable of forming fruit-bodies.

2. *Secondary homothallism*: In *Coprinus ephemerus* f. *bisporus* the basidia bear only two spores, but the spores are heterokaryotic. After meiosis two nuclei enter each spore and a mitotic division may follow. On germination, a

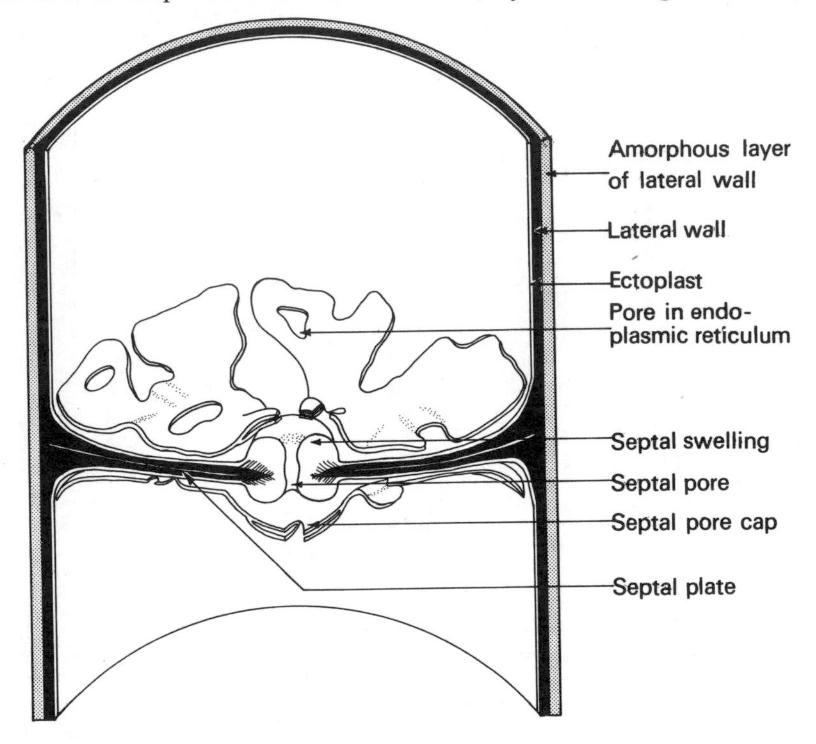

Figure 182. Fine structure of a basidiomycete septum (*Rhizoctonia solani*, after Bracker & Butler, 1963).

single spore germinates to form a dikaryotic mycelium capable of fruiting. Occasional spores give rise to non-clamped mycelia which individually do not fruit, but if the non-clamped mycelia are paired in certain combinations fruiting occurs, showing that the fungus is basically heterothallic. This situation is closely parallel to that found in certain four-spored Ascomycetes, such as *Neurospora tetrasperma*, and occurs in a number of other two-spored Basidiomycetes (Raper, 1966).

3. *Unclassified homothallism*: the four-spored wild mushroom *Agaricus*

campestris and the two-spored cultivated mushroom *Agaricus bisporus* are homothallic in the sense that a mycelium derived from a single spore is capable of fruiting. In both species there is nuclear fusion in the basidium, followed by two nuclear divisions, presumably meiotic. In *A. bisporus* two nuclei enter each spore, but in *A. campestris* only one. However, paired nuclei, conjugate nuclear divisions and clamp connections have not been observed. In *A. bisporus* the number of nuclei in the cells of the fruit-body may be 1–36 (mean 6·4) (Evans, 1959). Another species whose life-cycle is difficult to interpret is *Armillaria mellea*. Here also dikaryotic clamped hyphae have not been observed, and although single-spore mycelia may fruit, it is possible that neither meiosis nor nuclear fusion occurs—i.e. the life cycle may be entirely asexual (see Raper, 1966).

Amongst the remaining 90% of Basidiomycetes reported to be heterothallic we can distinguish two conditions.

1. *Bipolar*: in species such as *Coprinus comatus* (the shaggy ink-cap) and *Piptoporus betulinus* (the birch polypore) when mycelia obtained from single spores from any one fruit-body are mated together dikaryons are formed in half the crosses. This can be explained on the basis of a single gene with two alleles. Segregation of the two alleles at meiosis ensures that a single spore carries only one allele. Dikaryons are only formed between monokaryons carrying different alleles at the locus for incompatibility. In fact it is known that in a population of fruit-bodies collected over a wide area there may be numerous alleles at the incompatibility locus (see below). About 25% of Basidiomycetes examined have been shown to be bipolar. The Uredinales and most Ustilaginales have mating systems of this type.

2. *Tetrapolar*: In species such as *Coprinus cinereus* (often referred to as *C. lagopus*) and *Schizophyllum commune* when monosporous mycelia derived from a single fruit-body are intercrossed fertile dikaryons result in only one-quarter of the matings. The genetical basis proposed for this situation is that incompatibility is controlled by two genes, with two alleles at each locus. Thus we can denote the two genes as *A* and *B* and their two alleles as A_1, A_2 and B_1, B_2 respectively. Consider the cross of a monokaryon bearing A_1B_1 with another bearing A_2B_2. This would result in a fertile dikaryon which we can write as $(A_1B_1+A_2B_2)$. Such a dikaryon would form spores following meiosis and the spores would be of four kinds: A_1B_1, A_2B_2 (parentals) and A_2B_1 and A_1B_2 (recombinants). In most cases studied the proportions of the four kinds of spore are equal, showing that the *A* and *B* loci are not linked, but borne on different chromosomes. The result of crossing the four different types of monokaryotic mycelia is shown in the Table on p. 292.

It can be seen that fertile dikaryons are only formed when the alleles present at each locus in the opposing monokaryons differ—i.e. in crosses of the type $A_1B_1 \times A_2B_2$ or $A_2B_1 \times A_1B_2$. Where there is a common allele at one or both loci the cross is unsuccessful. Thus the success of inbreeding within the spores of any one fruit-body is only 25% in tetrapolar as compared with 50% in bipolar forms.

In both *Schizophyllum* and *C. cinereus* it was soon discovered that although a single spore from one fruit-body was compatible with only one-quarter of of its fellow spores, if crosses were made instead by crossing spores from different fruit-bodies, successful matings would often occur in every cross, i.e. a spore from one fruit-body could mate successfully with 100% of the spores from a different fruit-body. The explanation of this phenomenon is

TABLE

Crossing experiment with all four kinds of monokaryotic mycelium in a tetrapolar Basidiomycete. The + sign denotes the formation of a fertile dikaryon

	A_1B_1	A_2B_2	A_2B_1	A_1B_2
A_1B_1	−	+	−	−
A_2B_2	+	−	−	−
A_2B_1	−	−	−	+
A_1B_2	−	−	+	−

that instead of the single pair of alleles at each locus necessarily present in any one dikaryotic mycelium, in a population representing the species as a whole a large number of alleles is present. Suppose that a second fruit-body had the composition ($A_3B_3+A_4B_4$), then all the four kinds of spore it produced, A_3B_3, A_3B_4, A_4B_3 and A_4B_4 would be compatible with all the spores of the original fruit-body, on the assumption that the essential requirement for fertility is that in any cross both alleles should differ at both loci. On this basis although inbreeding would be only 25% successful, outbreeding would be much more successful, approaching 100%. This 100% successful outbreeding would imply the existence of an infinite number of alleles. Analysis of the number of A and B alleles in a world-wide sampling of 57 collections of *Schizophyllum commune* yielded 96 A factors and 56 B factors (Raper, Krongelb & Baxter, 1958). Extrapolation from these results led to the conclusion that the total numbers of incompatibility factors in the world population of *S. commune* are 339 (217 to 562) A factors, and 64 (53 to 79) B factors (Raper, 1966). For *C. cinereus* 164 A factors and 79 B factors have been estimated (Day, 1963), and similar numbers have been estimated for other tetrapolar Hymenomycetes. Much lower numbers of alleles have been estimated for the Gasteromycetes (e.g. *Cyathus striatus*, 4 A and 5 B; *Crucibulum vulgare*, 3 A and about 16 B), and it has been suggested that the small number of incompatibility factors may be related to the specialised method of dispersal of the basidiospores within peridiola (Whitehouse, 1949*b*).

The large number of alleles in *Schizophyllum* raises questions about the structure and function of the incompatibility genes. It has been shown that when certain strains are crossed pairs of 'new' alleles may appear in the progeny, and when the 'new' strains are inter-crossed, parental alleles may re-appear. From this evidence it was concluded that the 'A' locus in *Schizophyllum* consists of two sub-loci, α and β, with a number of alleles at each sub-locus. On the basis of this model we can suppose that A_1 has the structure of $\alpha_1-\beta_1$, and A_2 has the structure $\alpha_2-\beta_2$. Recombination within the A 'locus' would give $\alpha_1-\beta_2$ and $\alpha_2-\beta_1$, which behave effectively as 'new' mating-type factors. In order to explain the large number of A factors it is necessary to postulate nine different alleles at the A_α locus and 26 different alleles at the A_β locus (Raper, 1966). It has not proved possible to interpret the B locus in terms of two sub-loci, and it is likely that the B locus may be more complex, possibly containing at least three sub-loci. Separate functions have been ascribed to the A and B loci. Working with *Cyathus stercoreus* Fulton (1950) ascribed to the A locus the formation of clamp connections, since only when this locus is heterozygous are clamp-connections formed. The B locus controls nuclear migration, which only takes place if the mated mycelia have different B alleles. The same distinctive functions of the A and B loci have been claimed for other tetrapolar Basidiomycetes, including *S. commune* and *C. cinereus* (Raper, 1966; Fincham & Day, 1965), so that it is no longer necessary to allocate the symbols in an arbitrary way.

As yet there is no clear understanding of the reasons for the high degree of specificity of action of the large number of alleles at the two loci, but for a discussion of some hypotheses see Raper, 1966; Dick, 1965; Parag, 1965.

The Buller Phenomenon: Buller (1931) discovered that if a homokaryon of *C. cinereus* was opposed to a dikaryon, it was possible for the homokaryon to be converted to the dikaryotic state. The same phenomenon has been reported in other tetrapolar and in bipolar forms. Conversion or dikaryotisation is brought about by nuclear migration from the dikaryon into the monokaryon. Buller estimated that migration can occur at 0·5–1·0 mm/hr, which is several times greater than the growth rate of mycelium. For *S. commune* rates of 1·5–5·4 mm/hr have been estimated (Snider & Raper, 1958). A number of different kinds of combination are possible (Papazian, 1950*a*; Raper, 1966).

I. Legitimate
   A. Compatible: homokaryon compatible with both components of the dikaryon:
      e.g. bipolar—$(A_1+A_2)\times A_3$
      tetrapolar—$(A_1B_1+A_2B_2)\times A_3B_3$
   B. Hemicompatible: homokaryon compatible with only *one* of the components of the dikaryon:

e.g. bipolar—$(A_1+A_2)\times A_2$
tetrapolar—$(A_1B_1+A_2B_2)\times A_1B_1$.

II. Illegitimate (noncompatible): homokaryon compatible with neither component of the dikaryon:
tetrapolar—$(A_1B_1+A_2B_2)\times A_1B_2$ or A_2B_1.

A number of surprising features were discovered in such pairings.

In compatible pairings using *Schizophyllum* it was found that the selection of a compatible nucleus from the dikaryon was not a matter of chance. Consider the fully compatible 'di–mon' mating. $(A_1B_1+A_2B_2)\times A_3B_3$. If a conversion of the A_3B_3 homokaryon by one of the compatible nuclei of the dikaryon were entirely random, dikaryons of the type $(A_1B_1+A_3B_3)$ and $(A_2B_2+A_3B_3)$ would be equally frequent. In fact there is evidence of preferential selection of one mating type, but the reasons for the selection are obscure (Ellingboe & Raper, 1962; Raper 1966). A second unexpected feature is the discovery that dikaryotisation can occur in incompatible pairings. One reason for this phenomenon is that somatic recombination between the nuclei of the original dikaryon can occur to give rise to a nucleus compatible with the monokaryon (Raper, 1966).

EVOLUTIONARY CONSIDERATION OF MATING SYSTEMS IN BASIDIOMYCETES

Consideration has been given to the evolutionary relationships of the various kinds of mating system. Within a single genus such as *Coprinus* there are homothallic, bipolar and tetrapolar representatives. A close degree of similarity has been found in the details of incompatibility control in unrelated tetrapolar forms such as the Gasteromycetes, Polyporales and Agaricales. The genetical complexity of the tetrapolar condition makes it highly improbable that such a condition could have evolved independently on several occasions. For these and other reasons Raper (1966) has argued that the primitive condition is the tetrapolar one and that the bipolar and homothallic states are secondary.

CLASSIFICATION

Ainsworth (1966) has proposed the following classification of the Basidiomycetes, presented here in the form of a key.

A. Basidia arising from a thick-walled cell (teliospore or chlamydospore) Hemibasidiomycetes
  I. Basidia bearing two to four basidiospores **Uredinales** (rusts)
  II. Basidia bearing numerous basidiospores **Ustilaginales** (smuts)

B. Basidia not arising from a thick-walled cell

I. Hymenium freely exposed at maturity; basidiospores function as ballistospores **Hymenomycetes**

(a) Basidia not segmented or forked

1. Fruit-bodies fleshy **Agaricales** (agarics and boleti)
2. Fruit-body usually leathery, corky or woody **Aphyllophorales** (polypores, etc.)

(b) Basidia segmented or forked **Tulasnellales** (jelly fungi)

II. Hymenium enclosed; basidiospores not violently projected **Gasteromycetes**

(a) Fruit-bodies not opening at maturity, mostly subterranean **Hymenogastrales**

(b) Fruit-bodies open at maturity, usually visible above ground

(i) Basidiospores contained within a specialised peridiole or glebal mass which is dispersed as a unit **Nidulariales** (birds'-nest fungi, etc.)

(ii) Basidiospores not enclosed in a specialised peridiole

(a) Basidiospores presented in a sticky mass dispersed by insects **Phallales** (stinkhorns, etc.)

(b) Basidiospores dispersed dry:

1. Fruit-body with a central sterile columella and capillitial threads **Lycoperdales** (puff-balls, earth-stars, etc.)
2. Fruit-body lacking a columella and capillitial threads **Sclerodermatales** (earth-balls, etc.)

It is not possible in an abbreviated key to present all the distinctions between the various groups. This classification is by no means a natural one and its origin can be traced to Elias Fries, a nineteenth-century Swedish mycologist who was largely dependent on macroscopic criteria. For example, it is well known that a group such as the Gasteromycetes is not made up of closely related forms. It has been suggested that some Gasteromycetes are closely related to the agarics *Russula* and *Lactarius*, whilst the Gasteromycete *Rhizopogon* is closely related to *Boletus* (see Heim, 1948 and Corner, 1954 for discussion). Much research will be necessary to delimit the truly natural groupings amongst the Basidiomycetes, and for the present a Friesian system is being adopted as a framework. Corner (1966) has summed up the position admirably:

It is now agreed that the Friesian classification is artificial and, for the world flora, unworkable. It was based on gross features which microscopic study has shown to be the result of parallel evolution as well as of common heritage. It mixed artificial grades and natural series. In its place, more fundamental methods are being developed by microscopic anatomy and microchemistry. Probably, however, because of the labour which these methods involve, not one quarter of the world flora has been studied and confirmed in the necessary detail. Therefore, there is no alternative comprehensive classification.

1 HYMENOMYCETES

This is the largest group of the basidiomycetes, and includes many of the well-known toadstools, bracket fungi, fairy clubs, jelly fungi and the like. The basidia are often arranged in a palisade-like fashion to form a hymenium which is fully exposed at maturity, in contrast with the Gasteromycetes where it is enclosed. Three large classes have been distinguished: Agaricales, Aphyllophorales and Tulasnellales. The first two classes have holobasidia whilst the Tulasnellales have heterobasidia (the group is sometimes termed the Heterobasidiae). An important distinction between the Agaricales and Aphyllophorales is that the fruit bodies of the Agaricales are fleshy, being usually composed of thin-walled hyphae which inflate. Such construction is termed *monomitic*. In contrast the fruit bodies of many Aphyllophorales are often more complex, being composed of thin-walled *generative hyphae*, which may be accompanied by either thick-walled unbranched *skeletal hyphae*, or by thick-walled much-branched *binding hyphae* or both. Their construction may thus be monomitic, *dimitic* or *trimitic*.

In many Agaricales the developing hymenophore is surrounded by one or more veils, but these are not present in the Aphyllophorales. A further distinction is that in the Agaricales the nuclear spindles in the basidia are transverse (chiastobasidial) whilst in the Aphyllophorales forms with longitudinally orientated spindles (stichobasidial) occur.

AGARICALES

Singer (1962) has classified the group into some 16 families, and his concept has been followed broadly except that *Schizophyllum* has been considered along with the Aphyllophorales. I have also not followed Singer in placing the Polyporaceae in the Agaricales. Traditionally all the gill-bearing Hymenomycetes were placed in a single family, the Agaricaceae, but modern taxonomic treatments have resulted in subdivision into more homogeneous families. Agarics are mostly saprophytic, and play an important role in the decay of woodland and grassland litter, wood, dung and composts. A few are parasitic, e.g. *Armillaria mellea* (= *Armillariella mellea*), a serious pathogen of woody hosts. Many agarics and boleti form mycorrhiza with forest trees (Harley, 1969). Most of the fleshy forms are edible, and some are specially cultivated for food, notably *Agaricus bisporus* the cultivated white

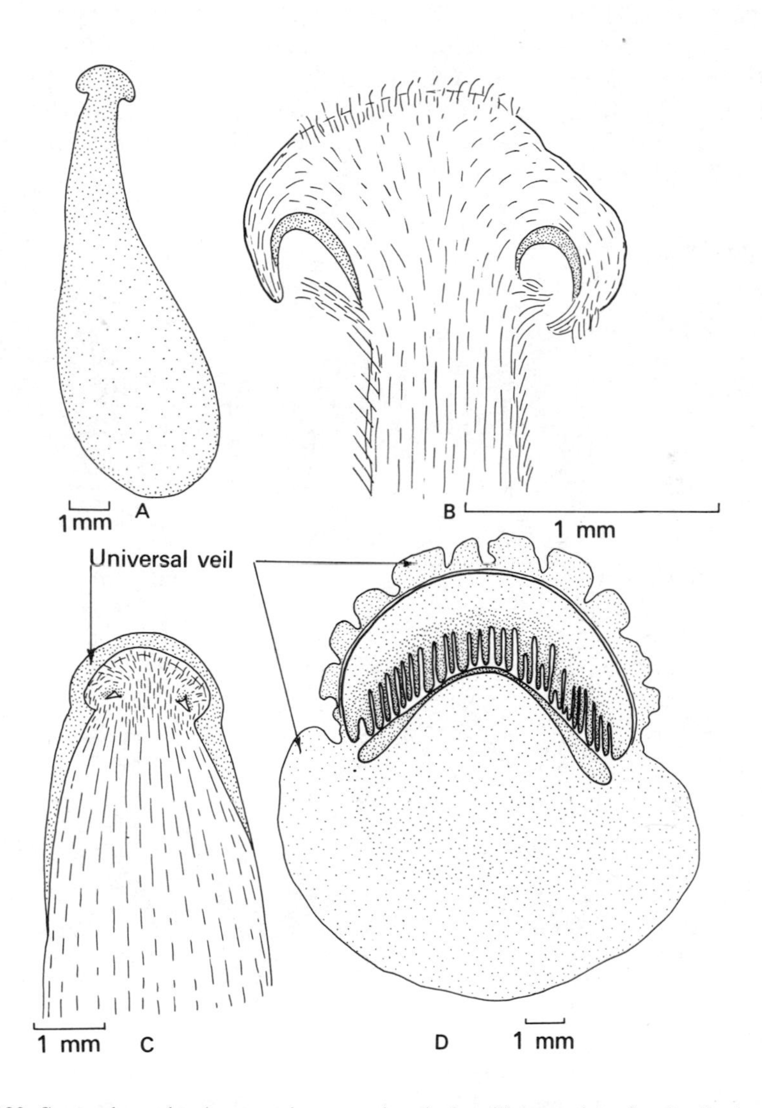

Figure 183. Sporophore development in some Agaricales, illustrated by longitudinal sections (after Reijnders, 1963). A. *Clitocybe clavipes*. Gymnocarpic development. B. *Lentinus tigrinus*. Secondary angiocarpy resulting from extension of hyphae from pileus margin and stipe to enclose the previously differentiated hymenophore. C. *Stropharia semiglobata*. Primary angiocarpy. Note the universal veil enclosing the upper part of the primordium. In mature fruit-bodies this becomes gelatinous. The hymenophore is also enclosed by a partial veil. D. *Amanita rubescens*. Tangential section. Note the break-up of the universal veil to form scales on the surface of the pileus. The gill chamber is enclosed by a partial veil.

mushroom in Europe and North America, *Volvariella volvacea* the Padi straw mushroom in the tropics, and *Lentinus edodes* the Shiitake of East Asia (Singer, 1961; Atkins, 1966). Some are poisonous, especially *Amanita phalloides* and *A. verna* (Ramsbottom, 1953), whilst others have hallucinogenic properties, notably *Psilocybe mexicana* and related species (Heim & Wasson, 1958; Singer, 1962; Heim, 1963).

Fruit-bodies of agarics usually arise from a dikaryotic mycelium which may be short-lived or perennial. In some cases the individual hyphae may be aggregated into mycelial strands or rhizomorphs. The general form of an

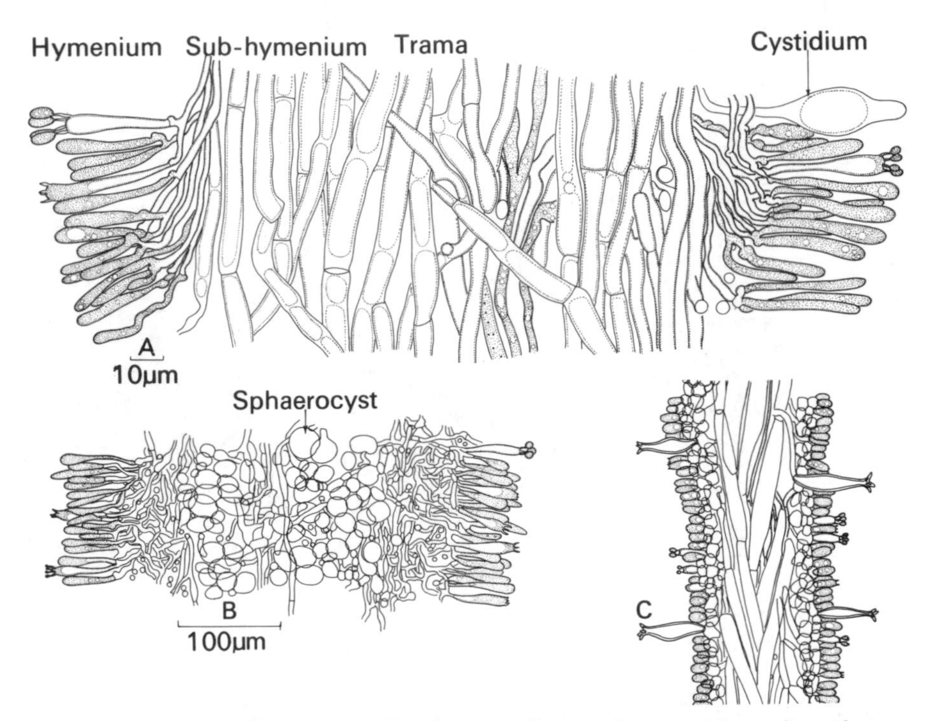

Figure 184. L.S. gills of various agarics of the aequihymenial type. A. *Flammulina velutipes*. B. *Russula cyanoxantha*. Note the globose sphaerocysts. C. *Pluteus cervinus*. Note the characteristic V arrangement of the cells of the trama (inverted trama). Note also the hooked cystidia. B and C to same scale.

agaric fruit body is umbrella-shaped, with a central stalk or stipe, supporting a cap or pileus with numerous radially arranged gills of lamellae on the lower side of the cap. In some species, especially those growing on wood, the stipe may be excentric (lateral) or absent. In the related Boletaceae the lower side of the cap bears a series of tubes opening to the exterior by individual pores. The hymenium covers the face of the gills or lines the tubes. The young fruit-body may be enclosed in a mass of tissue called the *universal veil* which is broken as the pileus expands leaving a basal cup-like *volva*, and sometimes scales on the pileus as in *Amanita* spp. In some agarics the hymenium may be protected by a *partial veil* stretching from the edge of the cap to the stem. Where the partial veil is thin and diaphanous as in *Cortinarius* spp. it is termed the *cortina*, but in some genera, e.g. *Agaricus*, *Armillaria* and *Amanita*, the partial veil is composed of a firmer tissue which remains as a distinct ring or *annulus* attached to the stem (see Fig. 183).

The arrangement of the veils reflects the development of the fruit-body. Reijnders (1963) has given a comprehensive survey of development. Several kinds of development occur:

1. *Gymnocarpic*: the hymenophore is naked from the time of its first appearance and is never enclosed by tissue. The pileus develops at the tip of the stipe and the hymenophore differentiates on the lower side. Gymnocarpic development is found in several unrelated genera, e.g. *Cantharellus cibarius*, *Xerocomus subtomentosus* (*Boletus subtomentosus*), *Russula emetica*, *Lactarius rufus and Clitocybe clavipes* (Fig. 183A).

2. *Angiocarpic*: here the hymenophore, or the part of the primordium from which the hymenophore will be formed, is enclosed by tissue during part of its development. Two kinds of angiocarpic development have been distinguished:

(a) *Primary angiocarpy*: the pileus margin, the hymenophore, and sometimes the pileus and stipe differentiate beneath the surface of the primary tissue of the primordium (protenchyma). *Stropharia semiglobata* and *Amanita rubescens* are primarily angiocarpic (Fig. 183C,D).

(b) *Secondary angiocarpy*: the hyphae from an already differentiated surface grows out towards the exterior to enclose the primordium or part of it. The hyphae may extend from the margin of the pileus towards the stipe, or from the stipe to the pileus, or both.

In *Lentinus tigrinus* hyphae from both the pileus margin and the stipe extend to enclose the developing gills (Fig. 183B).

Reijnders (1963) has classified the different kinds of development further and has provided further terms to describe them. He has argued that gymnocarpic development is more primitive than angiocarpic.

FRUIT-BODY ANATOMY

The tissue of the sporophore in the Agaricales is pseudoparenchymatous, consisting of an aggregation of hyphae. The hyphae are usually thin-walled and dikaryotic (generative hyphae), and may or may not bear clamp connections. The fruit-body expands due to inflation of the cells. Although no differentiation into skeletal or binding hyphae occurs, specialised tissues or cells may arise. In a study of the fine structure of the sporophore of *Agaricus campestris*, Manocha (1965) has shown that the stipe contains two kinds of cells: wide inflated cells and narrower thread-like cells. A similar differentiation is found in *Coprinus cinereus*. According to Borriss (1934) when portions of stipe tissue are planted on to suitable media it is only the thinner hyphae which give rise to vegetative growth. A further example of differentiation is seen in *Lactarius* which has a system of laticiferous hyphae containing latex which exudes if the flesh is broken (Lentz, 1954). Characteristic swollen spherical cells (sphaerocysts) give the flesh of genera such as *Lactarius* and *Russula* a spongy texture (Fig. 184B). The surface layers of the cap often show considerable modification.

Two main types of gill structure have been distinguished:

1. *Aequi-hymeniiferous* (aequi-hymenial): most agaric genera have gills of this type. In longitudinal section the gills are wedge-shaped. The term aequi-hymeniiferous refers to the fact that the hymenium develops in an equal manner all over the surface of the gill—i.e. basidial development is not

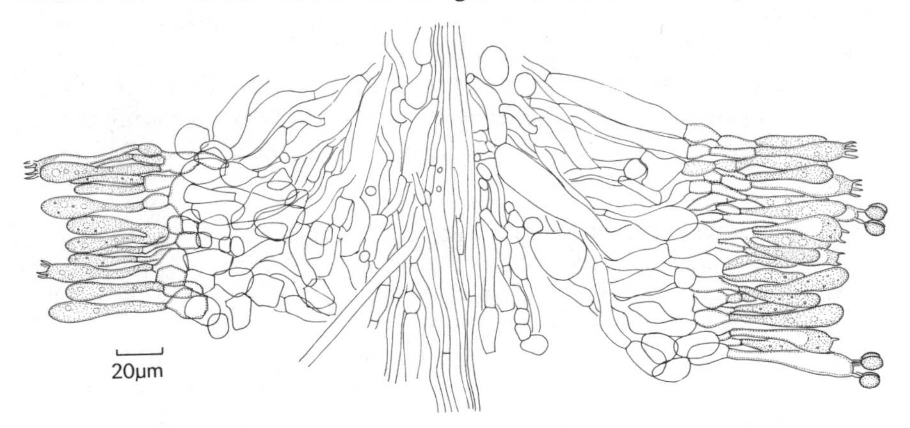

Figure 185. V.S. gill of *Amanita rubescens*, showing the bilateral hymenophoral trama.

localised at any one point on the gill. The wedge-shaped section may be an adaptation to minimise wastage of spores should the fruit-body be tilted from the vertical. Buller (1909) calculated that for the field mushroom *Agaricus campestris*, a displacement of 2°30′ from the vertical would still allow all the spores to escape. Slight adjustments in the orientation of the stipe and of the gills themselves may further help to minimise wastage (Fig. 186).

2. *Inaequi-hymeniiferous* (inaequi-hymenial): this type of gill is characteristic of the genus *Coprinus* (popularly known as ink-caps) where the gills are not wedge-shaped in section, but parallel-sided. The term inaequi-hymeniiferous refers to the fact that the hymenium develops in an *unequal* manner, with basidia ripening in *zones*. In *Coprinus* a wave of gill maturation begins at the base and passes slowly upwards. After the basidia at the lower edge of the gill have discharged their spores the gill tissue undergoes digestion (deliquescence) into an inky black liquid which drips away from the cap. The

geotropic *gill* curvature characteristic of aequi-hymenial types is absent, but the stipe curves to bring the gills into an approximately vertical position.

The structure of an aequi-hymeniiferous gill as seen in longitudinal section is shown in Fig. 184. There is a central group of longitudinally running threads termed the *trama* (*hymenophoral trama*). In some cases the trama may consist of more than one layer (Singer, 1962; Douwes & von Arx, 1965). Then, the layer immediately outside the central layer of the trama is termed the hymenopodium, and the trama is said to be bilateral (see *Amanita rubescens*, Fig. 185). In forms with a simple trama the ends of the tramal hyphae turn outwards to form a distinct layer of shorter cells, the *subhymenium*, which lies immediately beneath the *hymenium*, consisting of a palisade-like layer of ripe basidia, developing basidia (basidioles) and sometimes other

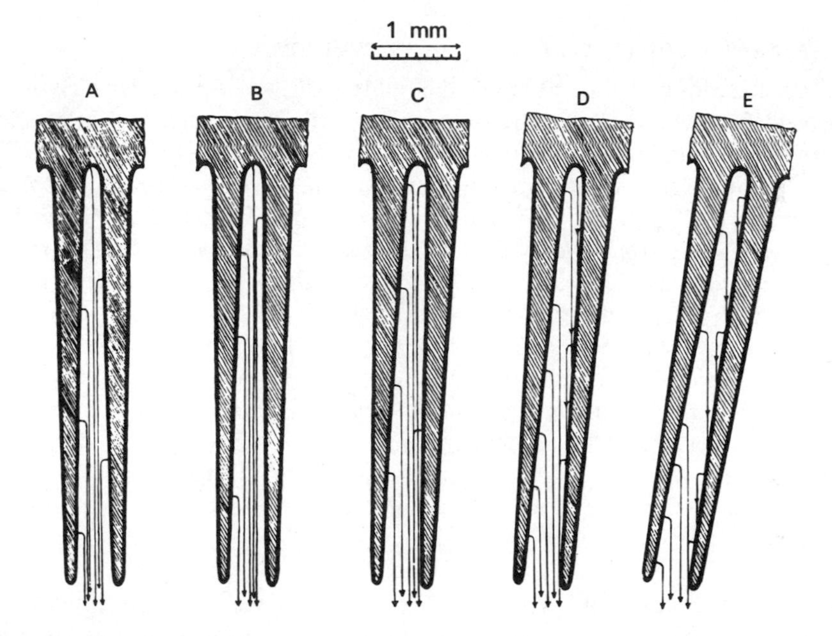

Figure 186. The effect of tilting an agaric fruit-body with wedge-shaped gills. Note that appreciable tilting of the fruit-body can take place before the trajectories of most spores are interrupted (after Buller, 1909).

structures such as cystidioles and cystidia. All these represent the terminal cells of hyphae making up the fruit-body. Possibly they are homologous with basidia. Fusion of paired nuclei within cystidia has been reported (Lentz, 1954). Cystidioles are thin-walled, sterile elements of the hymenium, about the same diameter as the basidia, and usually protruding only slightly from the hymenial surface. Cystidia are more varied. They are often enlarged conical or cylindrical cells which may arise in the hymenium along with the basidia (hymenial cystidia) or sometimes deeper, for example in the trama (tramal cystidia). In many species of *Coprinus* (inaequi-hymenial) the cystidia may stretch across the space between the gills (Fig. 179) and probably serve to space the gills apart. In *Pluteus* the tramal cystidia bear hook-like tips (Fig. 184c) whose function is not understood. The suggestion that they might deter animals such as slugs from eating the gill tissue is not supported by feeding experiments (Buller, 1924). Various terms have been used to describe cystidia. Those on the gill face are termed pleurocystidia: those on the gill margin cheilocystidia. But cystidia are not confined to the hymenium. Similar structures have been found on the surface of the pileus (pileocystidia) and the stipe (caulocystidia). We are still ignorant of the function of these structures, although some are concerned with secretion. For a fuller discussion see Romagnesi (1944); Lentz (1954) and Singer (1962).

THE PHYSIOLOGY OF FRUIT-BODY DEVELOPMENT

Much remains to be learnt about the control of morphogenesis in Hymenomycete fruit-bodies. Some of the problems will be discussed in relation to three agarics and one polypore which have been intensively investigated. The agarics are *Agaricus bisporus* (the cultivated white mushroom), *Coprinus cinereus* (= *C. lagopus*) a common coprophilous fungus, and *Flammulina velutipes* (= *Collybia velutipes*) a lignicolous toadstool which fruits throughout the winter. The polypore is *Polyporellus brumalis* (= *Polyporus brumalis*) also lignicolous, fruiting in winter and spring, and somewhat unusual in having a centrally stalked fruit-body (Fig. 187). With the exception of *A. bisporus* which fruits well only on compost, the other species fruit readily, even on synthetic liquid media, and this enables the chemical composition of the medium to be varied. By changing the level of CO_2 and humidity of the atmosphere and the light intensity it is possible to control the differentiation of the fruit-body. The effects of nutrition will, however, not be considered here.

1. *Light*: Light intensity, duration and wavelength are important components of any treatment. Light may have an effect on induction of sporophore initials, a tropic effect, or an effect on fruit-body form, e.g. ratio of stipe length to pileus diameter. In *C. cinereus* fruiting may occur both in light and in darkness, but takes place earlier in the light (after 10 days in continuous light and 15 days in continuous darkness according to Madelin, 1956). Sensitivity of fruiting to the effect of light occurs after about seven days' growth, and exposure to white light as brief as 1 sec. at 250 foot-candles or 5 sec. at 0·1 foot-candles was effective. Further increase in duration had no further effect on the numbers of sporophores produced. This behaviour contrasts with that of *F. velutipes* where light also stimulates sporophore production. Here the yield of sporophores increases with increasing duration of light (Plunkett, 1953). Curiously, the influence of light on sporophore

development in *C. cinereus* can be replaced by mechanical agitation (Stiegel, 1952).

It has long been known that low light intensity, or absence of light, may result in sporophores of curious shape, often with elongated stipes and poor pileus development. Fruit-bodies of this kind, have been found on timber in

Figure 187. Fruit-bodies of some Hymenomycetes, not to same scale. A. *Flammulina velutipes*. B. *Coprinus cinereus*. C. *Polyporellus brumalis*. D. *Agaricus bisporus*.

mine workings and caves. Light is necessary for pileus formation in *F. velutipes* (Plunkett, 1953, 1956).

Light may also stimulate directional growth of the stipes of many lignicolous and coprophilous hymenomycetes, e.g. *C. cinereus*, *P. brumalis*, whilst in terrestrial agarics such as *Agaricus campestris* sporophores at all stages of development appear to be non-phototropic (Buller, 1909). For fungi developing in crevices in dung or wood the phototropic stimulus for stipe growth followed by a light-operated stimulus to cap development or expansion would probably be of survival value.

2. *The aeration complex*

(a) *CO_2 concentration*: the CO_2 content of the atmosphere may have a profound influence on sporophore development. In *A. bisporus* the concentration of CO_2 in the substratum of commercial compost beds rarely falls below 0·3%, about 10 times the normal atmospheric concentration, and may rise to 20% or higher during the growth of the mycelium. Concentrations above 1·5% stimulated stipe elongation, but prevented cap expansion. Normal sporophore development occurs at about 0·2% CO_2 (Tschierpe, 1959; Tschierpe & Sinden, 1964). It has been suggested that CO_2 levels within the compost would stimulate stipe elongation which would carry the cap up into regions where the CO_2 concentration was sufficiently low to allow cap expansion and spore release. Expansion of the cap and release of spores within the compost would clearly be disadvantageous. In *F. velutipes* increasing CO_2 concentration above normal concentrations up to 5% causes a reduction in pileus diameter, but also results in shorter stipes. The presence of light and low CO_2 levels are necessary for normal sporophore development (Plunkett, 1956).

(b) *Humidity and evaporation*: in controlled conditions the rate of evaporation from developing sporophores may affect sporophore shape. In *P. brumalis* cap diameter is reduced as transpirational water loss is decreased. Increased transpiration results in more rapid translocation rates into the fruit-body as indicated by the movement of dyes (Plunkett, 1956, 1958). In comparable experiments with *F. velutipes* quite low rates of transpiration inhibited sporophore expansion.

3. *Gravity*: The hymenium of most agarics and polypores must be arranged more or less vertically in order to ensure maximum efficiency of spore release (Buller, 1909). Many fruit-bodies can adapt to changes in position either by growth of new tubes in the case of polypore fruit-bodies on wood, or by adjustments in the curvature of the stipe. In some cases, e.g. *P. brumalis*, the fruit-body may respond to a phototropic stimulus during the early part of its development, and apparently acquire a capacity for geotropic adjustment relatively late. It is of interest to enquire how the changing response of the sporophore to these two different stimuli is achieved. It is likely that both

stimuli are operative throughout development. Perception and response to both stimuli occur in the growing apical part of the stipe. Once the pileus initial has formed it shades the stipe so that the response to gravity becomes relatively more important during the later stages of fruit body development (Plunkett, 1961). By arranging to illuminate cultures from below Plunkett has induced pileus formation to occur upside down, i.e. with the pores facing upwards. In such fruit bodies a normal hymenium developed.

4. *Temperature*: There is usually a range of temperature between which sporophore development occurs. Increase of temperature within the lower part of this range hastens development. For *C. cinereus* the optimum temperature for mycelial growth is about 37°C, but fruit bodies are not formed above about 30°C. In *F. velutipes* the optimum temperature for growth is 25°C, but sporulation does not occur at this temperature. Continuous incubation in the light at 20°C induces fruit-body formation. A short period of incubation (12 hr at 15°C) can stimulate fruit-body initiation and following such treatment fruit-body development may occur at 25°C (Kinugawa & Furukawa, 1965).

FRUIT-BODY EXPANSION

The rapid expansion of mushroom fruit-bodies is a well-known phenomenon. The force of expansion can be considerable, *Coprinus atramentarius* is capable of cracking asphalt paving, and Buller (1931) showed that *C. sterquilinus* can lift a weight over 200 grams, many times its own weight. Studies of the growth of various agarics, e.g. *Agaricus bisporus* (Bonner *et al.*, 1956), *C. cinereus* (Borriss, 1934) have shown that the final rapid stage of expansion is due almost entirely, if not entirely, to cell extension, i.e. of a number of cells originally laid down in the young primordium. The most rapid zone of extension is in the zone of the stipe immediately beneath the cap (Figs. 188, 189). Glycogen is present in the cells of the young stipe of *C. cinereus* and tends to disappear as it matures (Borriss, 1934). The osmotic concentration of the cell sap of the stipe cells actually shows a decrease as they mature. There is an increase in elasticity of the stipe cells, but this seems hardly adequate to explain the 50-fold increase in growth rate measured during the final stages of stipe extension. Possibly there is an increase of the permeability of the stipe cells to water. There is evidence that extension of the stipe is under hormonal control, and that the maturing gills produce a hormone which induces cell extension. In *A. bisporus* fruit-bodies in which all but a narrow strip of pileus tissue, and all the attached gill, is cut away, the attachment of agar blocks containing an alcoholic extract of gill tissue causes positive curvature of the stipe (Fig. 190). The unilateral removal of part of the pileus also causes curvature, and this effect has also been shown in a number of other agarics (e.g. *Flammulina velutipes*). In others, e.g. *Coprinus sterquilinus*, there

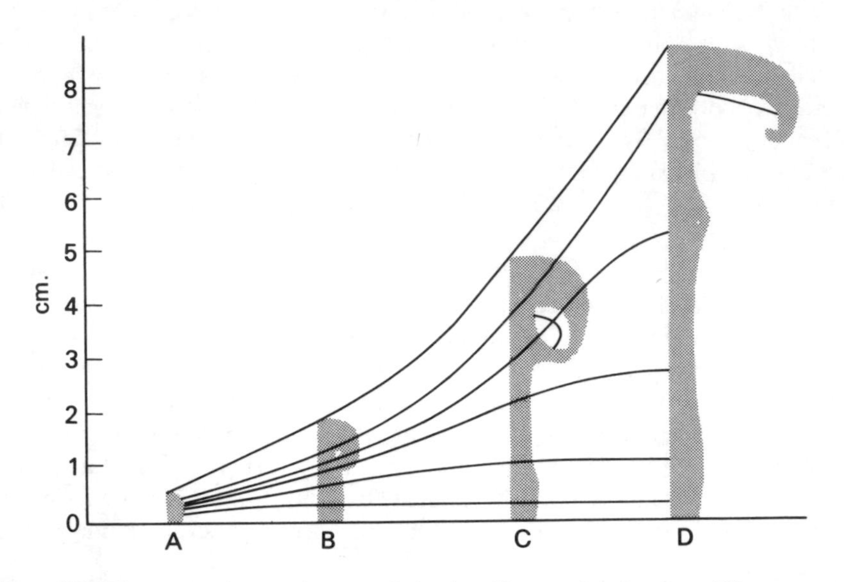

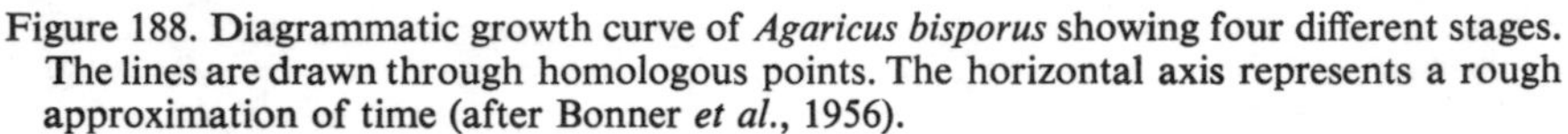
Figure 188. Diagrammatic growth curve of *Agaricus bisporus* showing four different stages. The lines are drawn through homologous points. The horizontal axis represents a rough approximation of time (after Bonner *et al.*, 1956).

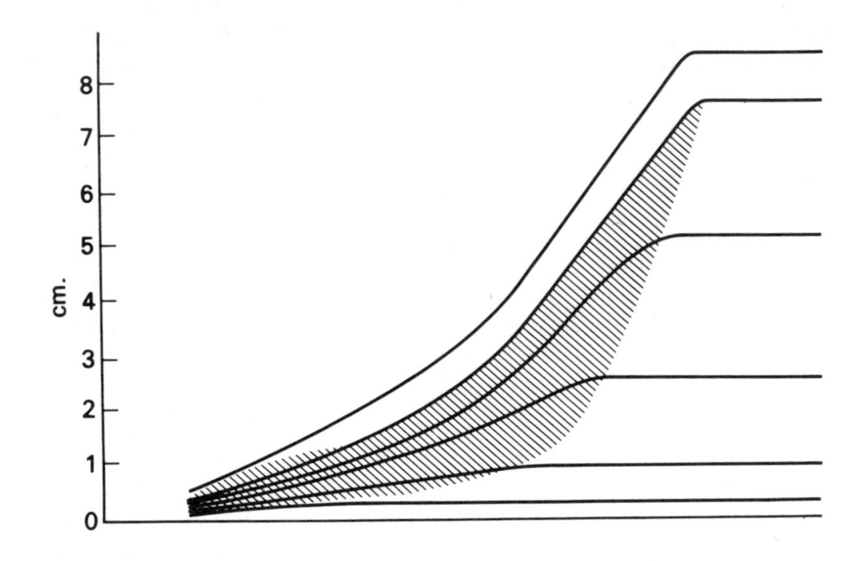

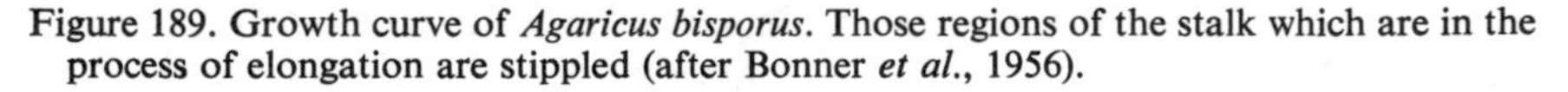
Figure 189. Growth curve of *Agaricus bisporus*. Those regions of the stalk which are in the process of elongation are stippled (after Bonner *et al.*, 1956).

is no evidence of hormonal mediation of stipe curvature (Jeffreys & Greulach, 1956). The nature of the hormone remains to be determined. It is unlikely to be one of the naturally occurring higher plant hormones since the addition of these compounds does not yield comparable results (Hagimoto & Konishi, 1959, 1960; Hagimoto, 1963; Gruen, 1959, 1963).

FRUIT-BODY FORM AND FUNCTION

It is a common observation that the larger agaric sporophores tend to have stipes which seem disproportionately wide when compared with those of smaller fruit-bodies. The mass of the cap of an agaric is proportional to its

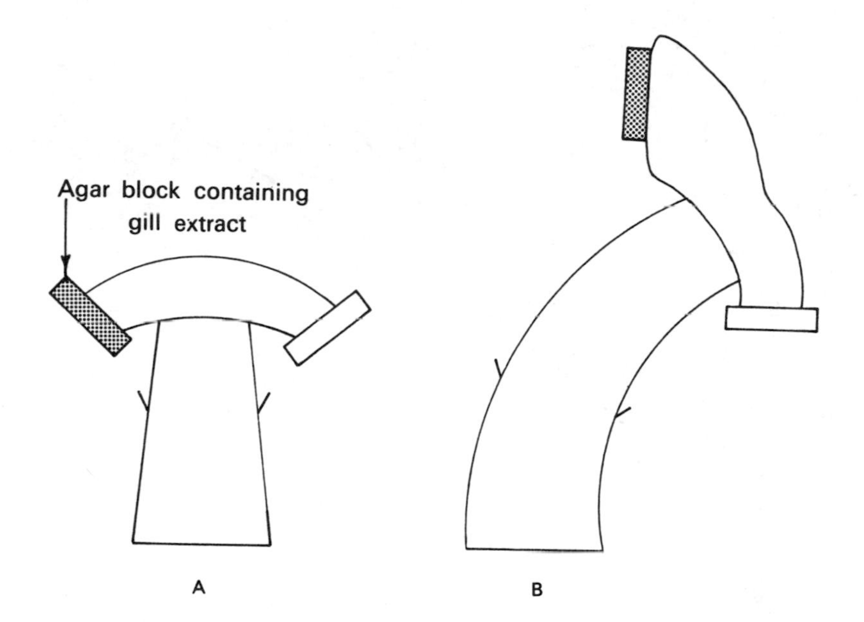

Figure 190. Evidence for hormonal control of stipe and pileus expansion in *Agaricus bisporus*. A fruit-body is prepared (A) by removal of all the pileus tissue except for a segment across the diameter. All the gill tissue is also removed. An agar block containing an alcoholic extract of gill tissue is attached to one side of the prepared pileus, and a similar agar block lacking the extract is placed opposite. Expansion of the stipe and pileus tissue on the side to which the test substance is applied is shown in B (after Hagimoto & Konishi, 1960).

volume, i.e. is proportional to the *cube* of its radius, and this mass must be supported by the cross-sectional area of the stipe which is in proportion to the *square* of its radius. This is in accord with the principle of similitude (D'Arcy Thompson, 1961). If this principle holds good for agaric sporophores a plot of pileus diameter (y) against stipe diameter (x) should not be linear but should have the form $y^3 = ax^2$. Ingold (1946*b*; 1965) and Bond (1952) have shown that the relationship holds good (Fig. 191).

SOME COMMON AGARICS AND BOLETI

A detailed survey of the Agaricales would be out of place here, and those interested are referred to more specialised texts such as Singer (1962), Kühner & Romagnesi (1953), Lange (1935–40), Lange & Hora (1963) and Moser (1967). A check list giving modern nomenclature of British agarics is available (Dennis, Orton & Hora, 1960). A list of some important families is given below together with the names of some common genera. Criteria used in classification are the presence or absence of annulus and volva, spore colour,

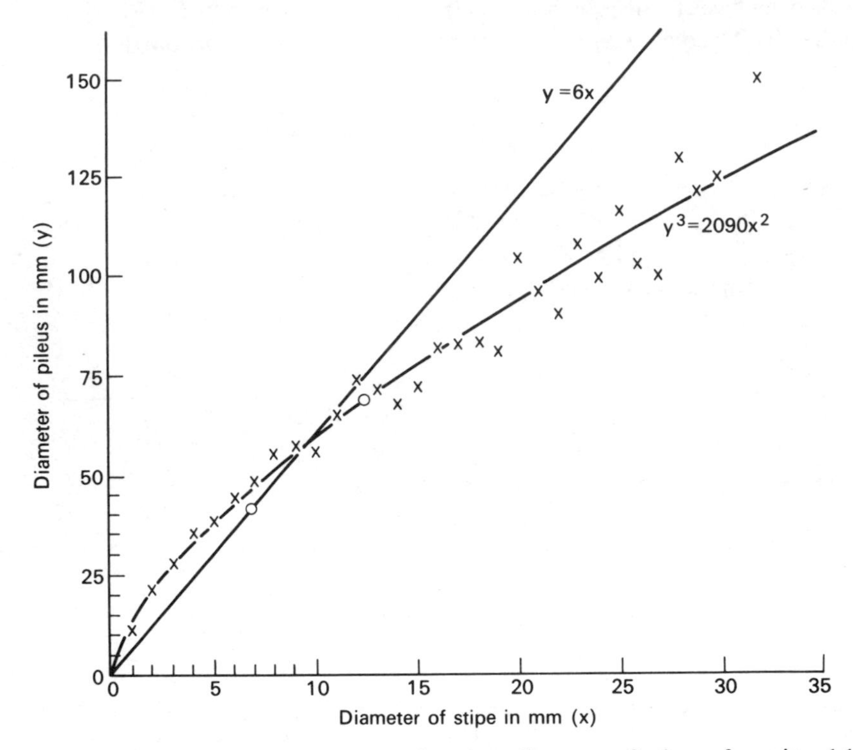

Figure 191. Diameters of pileus averaged against diameters of stipes of agarics; 1,166 determinations. The straight line is a plot connecting the origin to the arithmetic mean value, giving a line whose equation is $y = ax$. The curve is of the form $y^3 = ax^2$, which is in accord with the Principle of Similitude (after Bond, 1952).

method of attachment of gills to stem, the structure of the trama, presence or absence of sphaerocysts, cystidia and other specialised structures, and the chemical reactions of the flesh and spores.

1. Hygrophoraceae

Species of *Hygrophorus* are often brightly coloured and viscid, with a waxy texture. They are especially common in pastures.

2. Tricholomataceae

Common genera: *Laccaria, Clitocybe, Tricholoma, Armillaria* (*Armillariella*), *Collybia, Oudemansiella, Marasmius and Mycena. Armillaria mellea*, the honey agaric, has already been mentioned as a parasite of woody hosts. It is a root parasite and can spread by root-to-root contact or by the extension of bootlace-like rhizomorphs from infected stumps. Similar, but flattened, rhizomorphs form beneath the bark showing where the cambium has been destroyed.

Rhizomorph formation occurs readily in culture (Snider, 1959). Details of the life cycle are still obscure. Curiously, this destructive parasite is a mycorrhizal associate of the colourless orchid *Gastrodia elata*, within the tuber of which the fungal hyphae undergo digestion.

Oudemansiella radicata (= *Collybia radicata*) grows on stumps and may form a long tapering root (pseudorhiza) which extends downwards through the soil to buried wood (Buller, 1934). The basidia and spores are large and convenient for the study of development (Fig. 175).

Marasmius oreades is often called the fairy ring agaric because this and other species often fruit in circles. Even when fruit bodies are not apparent dark green circles of grass may indicate the presence of the fungus. The circular form results from radial outgrowth from a central point, with the mycelium towards the centre dying off. Measurements of the annual rate of radial advance and of the diameter of fairy rings show that some are probably several centuries old (Shantz & Piemeisel, 1917; Bayliss-Elliott, 1926; Ramsbottom, 1953; Parker-Rhodes, 1955). Analysis of the mating-types found in isolations from sporophores all collected from a common ring show that they are identical, which tends to confirm the origin of the ring from a single point (Burnett & Evans, 1966). The appearance of the ring varies throughout the year, but at certain times there may be two dark green rings separated by a brown ring where the grass is dying. The dead zone might be due to harmful effects of the mycelium on the grass roots, or to the lower water content of the soil in this region. It is of interest that HCN, a substance commonly found in basidiomycete sporophores, has been obtained from mycelia of *M. oreades* in pure culture and from affected turf (Lebeau & Hawn, 1963; Ward & Thorn, 1965; Filer, 1966). Studies of the soil fungi associated with the mycelial zone of *M. oreades*, which may be confined to the top few inches of soil, show that fewer species of fungi can be isolated from this zone than from the soil beyond it (Warcup, 1951). The fruit-bodies of *M. oreades* are edible and can be dried.

Mycena species have delicate brittle sporophores which are common in woodland litter growing amongst grass or on decaying leaves, twigs or stumps (*M. galericulata*). A characteristic feature of the genus is that the gill edge bears cystidia (cheilocystidia). In some species, e.g. *M. galopus*, the broken stipe exudes latex.

3. Amanitaceae

Amanita

Several deadly poisonous toadstools belong to this genus, notably *A. phalloides* (the death cap), *A. verna* and *A. virosa*. The poisonous principle has been termed amanitatoxin, which is a complex of three substances, α amanitine, β amanitine and phalloidine. Amanitine is a polypeptide with the empirical formula $C_{33}H_{45(47)}O_{12}N_7S$, and phalloidine is a hexapeptide with the formula $C_{30}H_{39}O_9N_7S$ (see Ramsbottom, 1953; Heim, 1963). Some other species, whilst definitely poisonous, are less deadly. They include *A. muscaria*, the fly-agaric, which has hallucinatory properties. The active principle muscarine has the formula $C_8H_{20}O_2N$ (Singer, 1962; Heim, 1963). Some other species such as *A. rubescens*, *A. vaginata* and *A. fulva* are good to eat. The last two species were earlier classified in a separate genus (*Amanitopsis*) because of the absence of an annulus, but the distinction is no longer maintained. Many species of *Amanita* are mycorrhizal. A characteristic feature is the bilateral hymenophoral trama (see Figs. 185, 192).

Agaricaceae

Agaricus

The best-known representatives are *A. campestris* the field mushroom and *A. bisporus* the cultivated mushroom (Fig. 187D). The cultivated species was derived from naturally growing mycelium which spawn-collectors were able to distinguish from other kinds (Ramsbottom, 1953; Singer, 1961; Bohus, 1962). The pink colouration of the young gills is due to cytoplasmic pigment in the spores. Later the gills turn a purplish-brown due to the deposition of dark pigments in the spore wall. Whilst most of the larger specimens of *Agaricus* are edible, *A. xanthodermus* is slightly poisonous. The generic name *Psalliota* has earlier been used for these species.

Lepiota

L. procera and *L. rhacodes* are two handsome scaly toadstools (parasol mushrooms) which are edible. A feature of both is that the ring is free to move up and down the stipe (Fig. 192D).

Coprinaceae

Coprinus

Some aspects of the biology of the ink-caps have already been studied. The characteristic feature of the genus is the parallel-sided (inaequi-hymenial) gill which in most species undergoes deliquescence from the base upwards,

Figure 192. Fruit-bodies of some common agarics. A. *Amanita excelsa*. Note the cracking of the volva on the cap of the youngest specimen, and the annulus on the stem. B. *Amanita fulva* (the tawny grisette). Note the cup-like volva at the base of the fruit-body. Here there is no annulus. C. *Armillaria mellea* (the honey agaric). Cluster of fruit-bodies near the base of a living birch tree. Note the annulus. D. *Lepiota procera* (the parasol mushroom). In mature specimens the ring on the stem is movable.

immediately following spore discharge. In *C. curtus* and some other species the cap expands by widening of downwardly directed grooves which divide the gills into Y-shaped forms. Autodigestion of the basal part of the vertical limb also occurs (Buller, 1931). In *C. plicatilis* in which the cap expands in this way there is no deliquescence (Buller, 1909). Many species, e.g. *C. cinereus* (Fig. 187B) are coprophilous, but there are forms which grow on wood, e.g. *C. micaceus* above ground, or from buried wood, e.g. *C. atramentarius*. *Coprinus comatus* is a large terrestrial species (the shaggy ink-cap of lawyer's wig) which is edible. In this species large cystidia line the inner free edge of the gill. Buller (1924) has shown that in many species of *Coprinus* the basidia are dimorphic, either long or short. The short forms are not merely immature basidia, but it appears that basidiospores ripen at two separate levels on the hymenium. This arrangement makes it possible to crowd a large number of basidia in a given area without interference.

Panaeolus

Most species grow on dung or on rich soil. The fruit-bodies resemble those of

Coprinus but the gills are aequi-hymenial and do not deliquesce. The surface of the gill appears mottled because the basidia in a given area tend to ripen together (Buller, 1922). Common species on dung are *P. campanulatus*, *P. sphinctrinus* and *P. semi-ovatus* (sometimes placed in a separate genus *Anellaria*).

Strophariaceae

Stropharia

The best-known species is *S. semi-globata* which is a very common coprophilous agaric with a yellow hemispherical viscid cap. *Stropharia aeruginosa* (the verdigris-cap) is an attractive bluish-green agaric with white scales, found amongst grass in pastures and woods.

Hypholoma

H. fasciculare, the sulphur tuft forms clumps of yellow fruit-bodies at the base of dead deciduous tree stumps. It is distinguishable from *H. sublateritium*, which grows in the same situations, because the latter has a larger brick-red cap. The generic name *Naematoloma* is sometimes used.

Psilocybe

Mention of *Psilocybe* has already been made in connection with the hallucinogenic properties of certain species. *Psilocybe semilanceata* is widespread in lawns and pastures, but is innocuous.

Pholiota

Pholiota squarrosa grows parasitically, and forms clumps of large yellowish-brown scaly fruit-bodies at the base of trees such as ash and elm. It is somewhat reminiscent of *Armillaria mellea* from which it is distinguished by its rusty brown spores.

Cortinariaceae

Cortinarius

Over 200 species are known in Britain alone. Almost all species are terrestrial and many are mycorrhizal associates of trees. Although determination of species is difficult, the genus is easy to recognise by the clay-coloured gills and remnants of the cortina on the stem.

Boletaceae

Boletus

This is a large genus of fleshy toadstools with a series of tubes on the lower side of the cap. Most species are mycorrhizal associates. *Boletus elegans*

Figure 193. Fruit-bodies of some common agarics and boleti. A. *Oudemansiella radicata*. This fungus grows attached to tree stumps and has a long tapering rooting base to the stem. B. *Paxillus involutus*. Note the decurrent gills and inrolled cap margin. C. *Boletus elegans*. Note the ring on the stem and the pores on the underside of the cap. This fungus is a mycorrhizal associate of *Larix*. D. *Boletus chrysenteron*. Here there is no ring on the stem.

(Fig. 193) is associated exclusively with larch, *B. bovinus* with pines, *B. scaber* and *B. testaceoscaber* with birches. Other common species which are not so closely asscoiated with any particular host are *B. chrysenteron* and *B. subtomentosus*. *Boletus parasiticus* is exceptional in growing parasitically and fruiting on *Scleroderma*. Some are gymnocarpic in development, but in others, e.g. *B. elegans*, there is a clearly defined partial veil. Both edible and poisonous species are known. One of the best edible species is *B. edulis* whilst *B. satanas* is poisonous. The flesh of several boleti turns blue when exposed to the air or when bruised and this is due to the oxidation of an anthraquinone pigment to a blue colour (boletoquinone) by the enzyme laccase. The species are sometimes disposed into several genera such as *Suillus*, *Ixocomus*, *Xerocomus* and *Leccinum* (Singer, 1962).

Paxillus

Some authorities classify the gill-bearing genus *Paxillus* in the Boletaceae, whilst others place it in a separate family, the Paxillaceae. Apart from the close structural similarity it is interesting that both *Paxillus* and *Boletus* are parasitised by a mould, *Apiocrea chrysosperma*, which does not attack other agarics. *Paxillus involutus* is a common associate of birch, but according to Singer (1962) there is no evidence of a true mycorrhizal relationship with forest trees.

Russulaceae (*Russula and Lactarius*)

These two genera are very closely related to each other. This is shown by the close similarity of their spores, the presence of sphaerocysts, and of laticiferous hyphae in the flesh (Fig. 184). When the flesh of a *Lactarius* sporophore is broken latex exudes which may be colourless or brightly-coloured. In some cases the latex changes colour, e.g. in *L. deliciosus*, the edible saffron milk-cap where the latex is at first bright orange, but turns green. Latex does not exude from *Russula* sporophores. Numerous species of both genera are abundant in woodland where many of them are mycorrhizal associates of trees. *Russula ochroleuca* is common in mixed deciduous woodland, whilst *R. fellea*, distinguished from the former by its bitter taste and its honey-coloured gills, is confined to beech woods. *Lactarius turpis* is a mycorrhizal associate of birch, whilst *L. rufus*, distinguished from other reddish-brown species by its peppery taste, is associated with conifers.

APHYLLOPHORALES

Donk's (1964) concept of this group has been adopted. He has defined it as

> an artificial order of holobasidious Hymenomycetes, opposed to the Agaricales, forming distinct fruit bodies. Fruit body developing centrifugally with one-sided hymenophores, or clavarioid with amphigenous hymenium, not developing within a universal veil, the hymenium not covered by a partial veil and exposed during the maturation of the basidiospores. Hymenophore smooth (hymenium may be folded), toothed or tubulate, exceptionally, and then mostly imperfectly lamellate.

In place of the traditional Friesian system of classification based on gross macroscopic features of the fruit-body and details of hymenial arrangement, modern systems of classification lay stress on microscopic features, especially the details of anatomy, colour and staining reactions of hyphae and spores. Critical examination of Basidiomycete fruit-bodies in this way demonstrates close relationships between fungi hitherto classified in quite separate groups under the Friesian system. The emergence of more natural groupings of species and genera is reflected in the arrangement of families and creation of new genera. Donk has recognised over 20 families within the Aphyllophorales, but the arrangement is still tentative and as a result of further research it may well prove desirable to create further families. We shall study examples from only a few of these families, viz.: Auriscalpiaceae, Clavariaceae, Ganodermataceae, Hydnaceae, Polyporaceae, Schizophyllaceae and Stereaceae.

Polyporaceae

In older systems of classification, Hymenomycetes, in which the hymenium lined tubes opening to the exterior by means of pores, were all classified here. It was later recognised that there is only a limited number of ways in which a

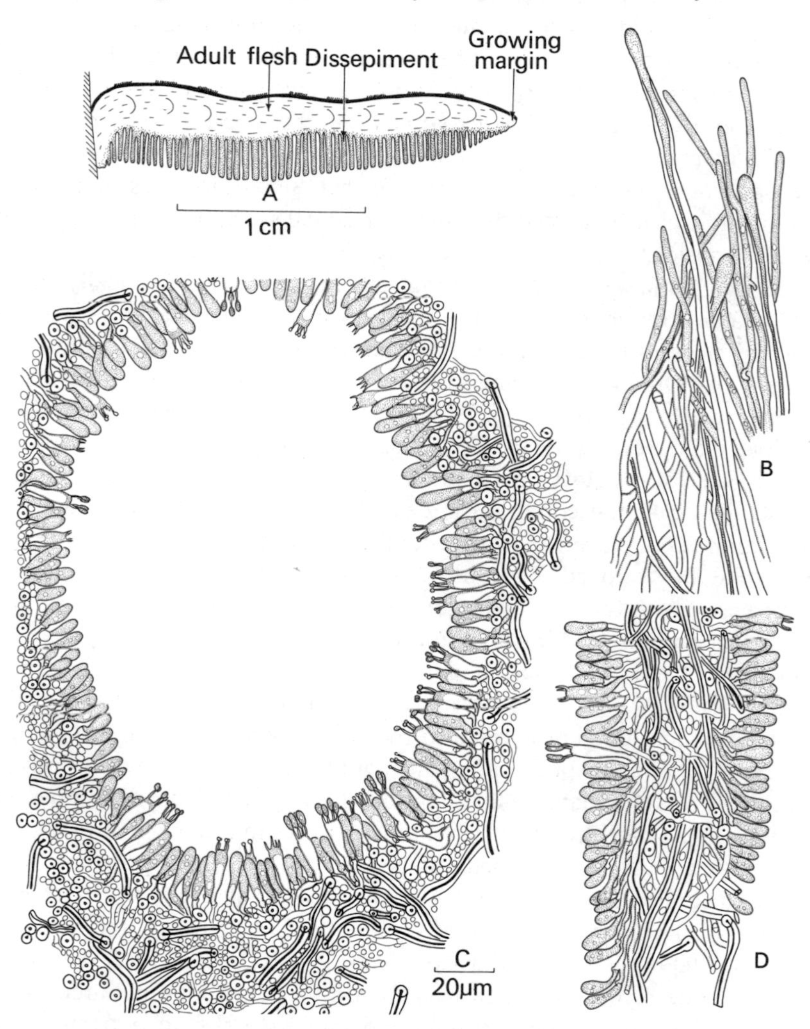

Figure 194. *Coriolus versicolor*. A. Vertical section through sporophore. B. Group of hyphae teased from the growing margin. Only generative and skeletal hyphae are found here. In the adult flesh binding hyphae are also present (see Fig. 195). C. Transverse section across a pore showing the hymenium. D. Longitudinal section of part of a dissepiment. The tissue contains only generative and skeletal hyphae. B, C, D to same scale.

hymenium can be arranged so that basidiospores can be discharged successfully, and it became clear that the tubular arrangement of the hymenium had probably evolved independently several times. Thus *Boletus* is now classified in the Agaricales. A striking demonstration of convergent evolution was the discovery that *Aporpium caryae* which for over 120 years had been treated as a polypore (*Polyporus caryae*) had cruciate–septate basidia characteristic of the Tremellaceae.

Microscopic analysis and studies of fruit-body development have led to splitting of the old heterogeneous Polyporaceae into a number of smaller, more homogeneous, families. For example, *Fistulina hepatica*, the beef-steak fungus parasitic chiefly on *Quercus* and *Castanea*, is now placed in a separate family, the Fistulinaceae. *Ganoderma* has been removed to a separate family, the Ganodermataceae. The residual Polyporaceae is undoubtedly still heterogeneous and we may expect further separations as relationships become more clearly defined. It is in this residual sense that the Polyporaceae are here considered.

The Polyporaceae or bracket fungi are important economically because they include a number of serious pathogens of coniferous trees, e.g. *Heterobasidion annosum* (= *Fomes annosus*) the cause of heart-rot or butt-rot of conifers and *Piptoporus betulinus* (= *Polyporus betulinus*) the cause of heart rot of birch. Many are also responsible for decay of timber (Cartwright & Findlay, 1958; Fergus, 1960). Two types of decay have been recognised. In fungi capable of cellulose decay, the brown lignin content of the wood is left undestroyed, and the condition is described as a brown rot. Where lignin destruction takes place the rotted wood has a white appearance so that the term white rot is applied. The ability to degrade lignin by a fungus appears to be correlated with the production of an extracellular oxidase in pure culture, and this can be detected by the oxidation of gallic or tannic acid to a brown colour or by the rapid oxidation of an alcoholic solution of guiacum to a blue colour (Nobles, 1958*a*,*b*, 1965). Analysis of sound and decayed wood shows that the cellulose content is reduced in both brown and white rots, and the lignin content remains practically constant in the brown rots but is decreased in the white rots (Findlay, 1940). Isolation of fungi from decaying timber often results in monokaryotic growth in culture, suggesting that the monokaryotic mycelium may have a prolonged independent existence. All polypores which have been studied are heterothallic, about 44% bipolar and 56% tetrapolar. Multiple alleles have been demonstrated in both bipolar and tetrapolar forms (Whitehouse, 1949*b*; Burnett, 1956; Nobles, 1958*a*,*b*; Raper, 1966). The dikaryotic mycelium which may or may not bear clamp connections is often perennial in large tree trunks and may give rise to a fresh crop of sporophores annually. In some cases, e.g. species of *Fomes*, the fruit-body itself may be perennial. Typically the fruit-bodies develop as fan-shaped brackets lacking stipes, but there are forms with lateral, i.e. excentric stipes, e.g. *Polyporus squamosus* and a few with centrally stalked fruit-bodies, e.g.

Polyporellus brumalis (= *Polyporus brumalis*) and *Coltricia perennis*. The last is unusual in being terrestrial, not lignicolous. When fruit-bodies of polypores and other wood-rotting fungi develop on the underside of logs they may be appressed to the surface of the wood and are then described as resupinate.

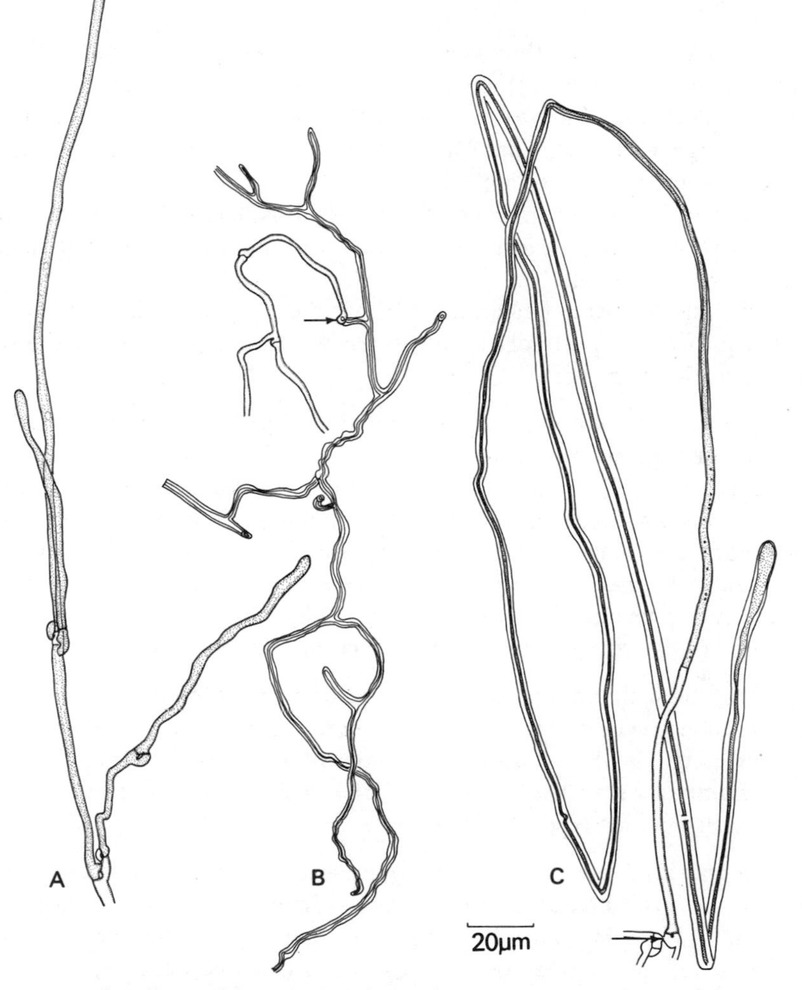

Figure 195. Hyphae dissected from the fruit-body of a trimitic polypore (*Coriolus versicolor*). A. Generative hyphae, characterised by thin walls, dense cytoplasmic contents and clamp connections. B. Binding hypha, branched and thick-walled. The arrow shows the origin from a generative hypha. C. A skeletal hypha, unbranched and thick-walled. The arrow shows the origin from a generative hypha.

In order to understand the structure of a polypore fruit body it is necessary to tease out a thick section of the flesh with fine needles under a dissecting microscope. This technique of hyphal analysis was developed by Corner

(1932, 1953) and has also been described by Teixeira (1962). It may be necessary to study hyphae from various parts of the fruit-body; the growing margin, the adult flesh, the context immediately above the tubes and the dissepiments, i.e. the tissues separating the tubes (Fig. 194). Corner has described three distinct kinds of hyphae, although not all may be present in any species:

1. **Generative hyphae:** thin-walled near the growing margin, often thicker-walled behind, with or without clamps, usually with distinct cytoplasmic contents. This kind of hypha is universally present in all polypore fruit-bodies at some stage of development. The generative hyphae give rise to basidia and also to two other kinds of hyphae.
2. **Skeletal hyphae:** unbranched thick-walled hyphae with a narrow lumen which arise as lateral branches of the generative hyphae. The skeletal hyphae form a rigid framework.
3. **Binding hyphae:** much-branched, narrow, thick-walled hyphae of limited growth. These hyphae tend to weave themselves between the other hyphae of the flesh.

The three kinds of hyphae are illustrated in Fig. 195. Where all three kinds are present together the fruit body is said to be trimitic (Greek μιτοζ, a thread of the warp). *Coriolus versicolor* (*Polyporus*, *Polystictus*, *Trametes versicolor*) is a good example of a trimitic polypore (Fig. 196A). The growing margin and the tissue of the dissepiments are dimitic, consisting of generative and skeletal hyphae only. Binding hyphae are only found in the adult flesh some distance behind the growing margin (Fig. 194).

There are two types of dimitic construction in polypores:

(*a*) Dimitic with binding hyphae (i.e. generative and binding hyphae). This kind of construction is found in *Laetiporus sulphureus* (= *Polyporus sulphureus*) (Fig. 196B). The dissepiments are, however, monomitic.

(*b*) Dimitic with skeletal hyphae: the fruit-bodies of *Heterobasidiom annosum* are of this type (Fig. 196C).

Where generative hyphae only are present, the construction is described as monomitic. Fruit-bodies of *Bjerkandera adusta* (= *Polyporus adustus*) are monomitic. Here the walls of the generative hyphae may thicken with age.

The distinction between the different kinds of construction is best appreciated by attempting to tear the fruit bodies of these fungi apart. *Coriolus versicolor* tears with difficulty in contrast to the cheese-like consistency of *B. adusta*. Various modifications to the different hyphal systems may occur with age. For example in *L. sulphureus* the generative hyphae may become inflated. In *Polyporus squamosus* the binding hyphae arise relatively late following inflation of the generative hyphae, converting the sappy flesh of the fully grown fruit body to a drier firmer texture. In *Piptoporus betulinus* also, binding hyphae arise very late, entirely replacing the generative hyphae. The

Figure 196. Fruit-bodies of some Aphyllophorales. A. *Coriolus versicolor* as seen from above. B. *Laetiporuss ulphureus*. C. *Heterobasidium annosum* at the base of living pine tree. D. *Piptoporus betulinus* on dead birch trunk.

dissepiments show a different construction, being dimitic with skeletal hyphae.

The Polyporaceae is a large group, probably containing over 1,000 species. Over 40 genera occur in Britain, and Pegler (1966, 1967) has given keys. Notes on interesting features of some common polypores are given below. *Coriolus versicolor* (Fig. 196A) is a common saprophyte on hardwood stumps and logs, causing a white rot. Both the mycelium and the fruit-bodies are resistant to desiccation. The annual trimitic fruit bodies have a zoned velvety upper surface which readily absorbs rain. No conidial state has been reported. This species is tetrapolar with multiple alleles. Extensive studies of its structure have been made by both light- and electron-microscopy (Girbardt, 1958, 1961).

Heterobasidium annosum (Fig. 196C) is the cause of heart-rot or butt-rot of conifers, and occasionally of deciduous trees such as birch. The disease is especially common on alkaline soils and is the most important cause of heart-rot in coniferous woods in Britain (Peace, 1962). The perennial fruit-bodies are formed close to the ground at the base of stumps and are readily identified by their orange-brown colour with a white margin. They are dimitic with skeletals. Spores are produced throughout the year. The conidial state (Fig. 197) has been called *Oedocephalum lineatum*. The focus of the

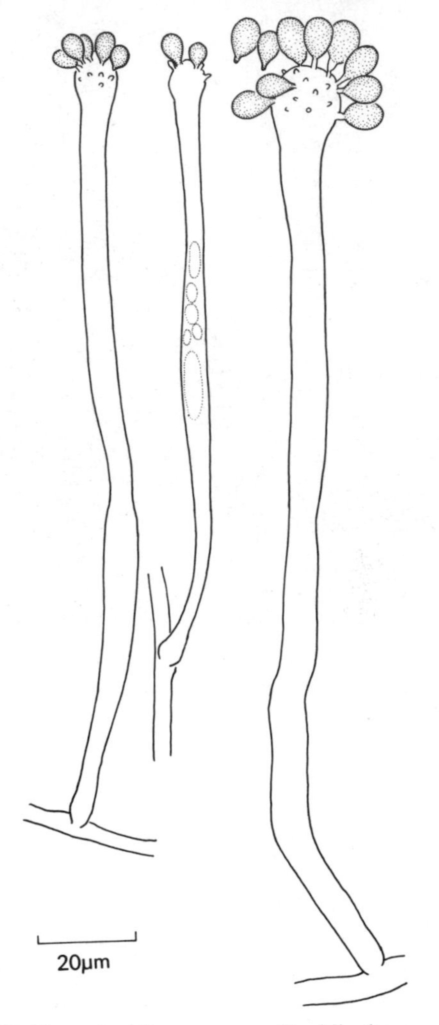

Figure 197. *Heterobasidiom annosum*. Conidiophores and conidia.

disease in plantations is often from stumps which have become infected from spores. Colonisation of the roots of the stump is followed by spread to adjacent healthy roots by root to root contact: the fungus does not grow freely in the soil. Modern methods of control are aimed at preventing stump

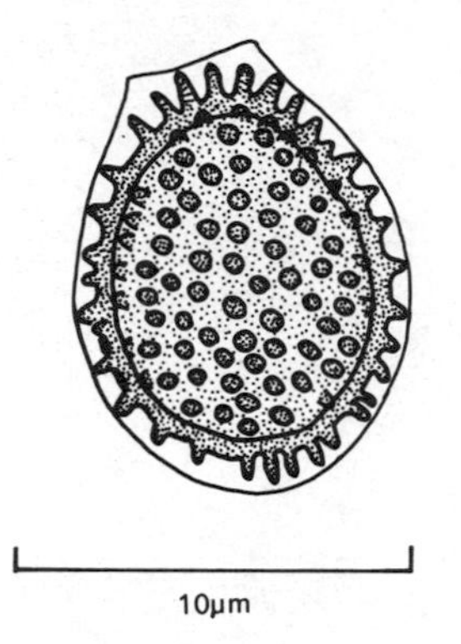

Figure 198. *Ganoderma applanatum*. Basidiospore. The truncate portion is the apex of the spore.

infection by treatment with creosote, sodium nitrite or other disinfectants. An interesting alternative treatment is to inoculate stumps with spores of saprophytic basidiomycetes such as *Peniophora gigantea* which competes with the parasite in pine, rotting the stumps and preventing successful colonisation (Anon., 1967*a*; Rishbeth, 1952, 1961).

Piptoporus betulinus is a common sight on dead and dying birch trees (Fig. 196D). It is probably a wound parasite, basidiospores entering where branches have broken off. Infected trees show a brown rot of the heart wood at first with cubical cracking, later powdery. Although the brown rot indicates cellulose decay, there is also evidence of oxidase activity which may also suggest lignin breakdown (MacDonald, 1937). The fungus is bipolar with about 30 alleles at the mating-type locus. Studies of the distribution of alleles in successive annual crops of sporophores on the same birch trunk show that several dikaryotic mycelia may exist side by side in the same trunk, and that each mycelium does not always produce a fruit-body in any one year (Burnett & Partington, 1957; Burnett, 1965).

Polyporus squamosus is a wound parasite of deciduous trees such as elm and sycamore. The mycelium may persist on dead stumps and logs, and form successive annual crops of sporophores during the early summer. The fruit-bodies are distinctly fleshy, and Singer (1962) includes this fungus in the Agaricales.

Ganodermataceae

This family is usually included in the Polyporaceae. The distinguishing feature is that the spore is double-walled with a dark-coloured inner layer bearing an ornamentation which pierces the hyaline outer one, so that the

Figure 199. *Ganoderma applanatum*. A. Two sporophores attached to a beech tree. B. Detached sporophore split vertically to show two layers of hymenial tubes.

Figure 200. *Auriscalpium vulgare*. Two fruit-bodies growing from a pine cone.

spore appears to have a spiny surface (Fig. 198) (Donk, 1964; Heim, 1962; Furtado, 1962). The hyphal structure of the fruit-body is trimitic. A characteristic feature is that the skeletal hyphae are of two types: (a) *arboriform*, showing an unbranched basal part with a branched tapering end, and (b) *aciculiform*, unbranched and usually with a sharp tip (Furtado, 1965; Teixiera & Furtado, 1965; Hansen, 1958). The best-known example is *Ganoderma applanatum*, a wound parasite which causes active heart rot of beech and other trees (Fig. 199). Both cellulose and lignin are attacked. The sporophores are perennial, brown, woody fan-shaped brackets often 50 cm

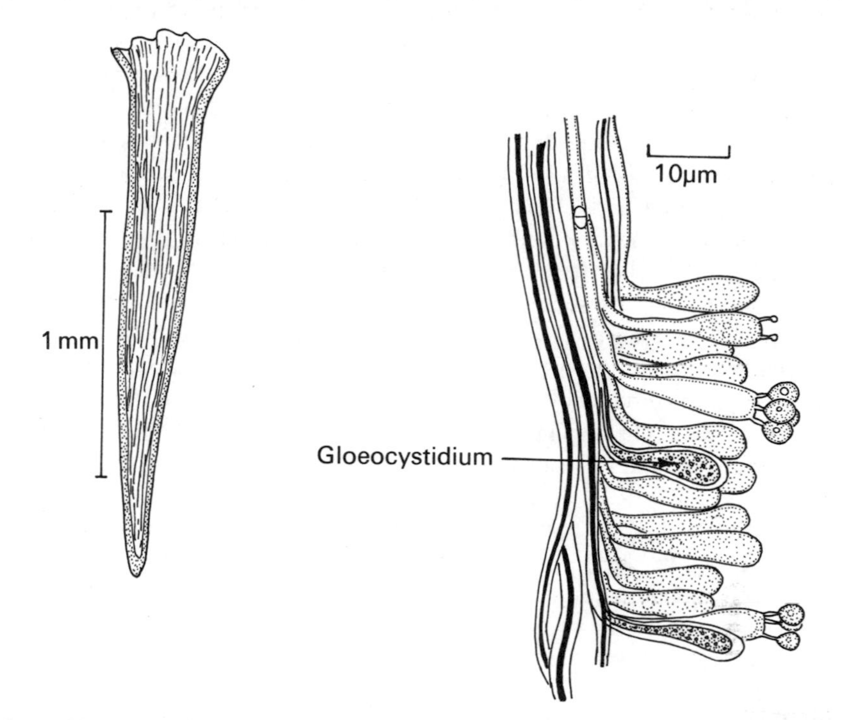

Figure 201. *Auriscalpium vulgare*. A. Spine from underside of cap, bearing hymenium. B. Portion of hymenium showing gloeocystidia.

and occasionally almost 1 m across (Herrick, 1953). A new layer of hymenia tubes is formed each year beneath the layer of the previous year. The tubes may be up to 2 cm in length and about 0·1 mm in diameter. Thus the tube may be 200 times as long as broad, and the fall of the spore down this tube raises problems. The hard rigid construction of the sporophore minimises lateral disturbance from the vertical alignment of the tubes. Gregory (1957) has shown that the majority of the spores carry a positive electrostatic charge, but whether this charge has any relevance to the positioning of the spores during their fall seems doubtful. It has been calculated that a large specimen may release as many as 20 million spores per minute during the five or six months from May to September (Buller, 1922). Spore discharge can continue even during periods of drought, doubtless associated with uptake of water from the tree host (Ingold, 1954, 1957). Spores plated on to media suitable for germination may take 6–12 months to develop germ tubes. Pairings between monoporous mycelia show that the fungus is tetrapolar, with multiple alleles (Aoshima, 1953).

Auriscalpiaceae

Hymenomycetes in which the hymenium covers teeth or spines were classified by Fries in the Hydnaceae. It has since become clear that this gross arrangement occurs in a number of unrelated forms. This is well shown by *Pseudohydnum gelatinosum* (Fig. 216) whose basidia are longitudinally divided (cruciate), and which therefore has affinities with the Tremellaceae. Some genera formerly placed in the Hydnaceae are now placed in families such as the Auriscalpiaceae and Hericiaceae (Donk, 1964) so that the Hydnaceae now contains only *Hydnum* and a few residual genera which may eventually be accommodated elsewhere. The Auriscalpiaceae includes not only forms with toothed hymenium such as *Auriscalpium*, but also *Lentinellus* which has gills (Maas-Geesteranus, 1963*a*). *Auriscalpium vulgare* (sometimes called *Hydnum auriscalpium*) grows on buried pine cones forming stalked one-sided brown hairy fruit-bodies at the surface of the ground during autumn and winter (Fig. 200). The hyphal construction is dimitic with skeletals (Ragab, 1953). The hymenium is formed on vertical finger-like downgrowths from the underside of the pileus. Interspersed amongst the basidia are irregularly enlarged thin-walled hyphal tips with highly refractile contents termed gloeocystidia (Fig. 201) (Lentz, 1954). The sporophores are capable of proliferation and show rapid readjustment to the vertical position if displaced laterally (Harvey, 1958).

Hydnaceae

Fruit-bodies of *Hydnum repandum* (Fig. 202A) and *H. rufescens* grow in woodland. They are more or less mushroom-shaped, with a stalked cap which may be central or lateral. The construction is monomitic, with generative hyphae which become inflated, giving the fruit-body a fleshy texture (Maas Geesteranus, 1963*b*). The hymenium covers the tapering spines which develop from the lower side of the cap. In contrast with *Auriscalpium*, gloeocystidia are absent.

Schizophyllaceae

Traditionally *Schizophyllum* has been classified in the Agaricales but details of fruit-body development (Essig, 1922; Wessels, 1965) and the development

Figure 202. A. *Hydnum repandum*. Portion of lower side of pileus showing the spines which bear the hymenium. B. *Schizophyllum commune*. Lower side of fruit-body showing the split 'gills'.

of the 'gills' are 'so peculiar that homology with the agaric gill is out of the question' (Donk, 1964). *Schizophyllum commune* grows saprophytically or parasitically on woody substrata forming fan-shaped, laterally attached fruit-bodies with a furry upper surface. The name *Schizophyllum* refers to the longitudinally split gills which may be a xeromorphic adaptation (Fig. 202B). In dry weather the gills curve inwards so that the hymenial surface is protected by a series of adjoining folds. The curvature is believed to be due to the shrinkage of the hymenial layers on drying. Since the remaining tissue of the gill is composed of thick-walled clamped hyphae which do not shrink so readily, inward curvature follows. In the region of the split the cells are thinner-walled (Fig. 203). Fruit-bodies which have been kept dry for two years rapidly take up water through the hairy upper surface and within two to three hours the gills straighten out as the hymenium expands. Spore discharge commences after three to four hours (Buller, 1909). Material which Buller subjected to freeze-drying in 1910 and 1912 has been re-wetted, and after 52 years some of the sporophores have revived and produced spores (Ainsworth, 1965).

Schizophyllum has been studied intensively by workers interested in its mating system (Raper, 1966), genetics (Raper & Miles, 1958), in nuclear migration (Snider, 1965), morphogenesis (Wessels, 1965) and taxonomy (Linder, 1933; Cooke, 1961). Dikaryotic mycelia fruit readily in culture (Papazian, 1950*b*).

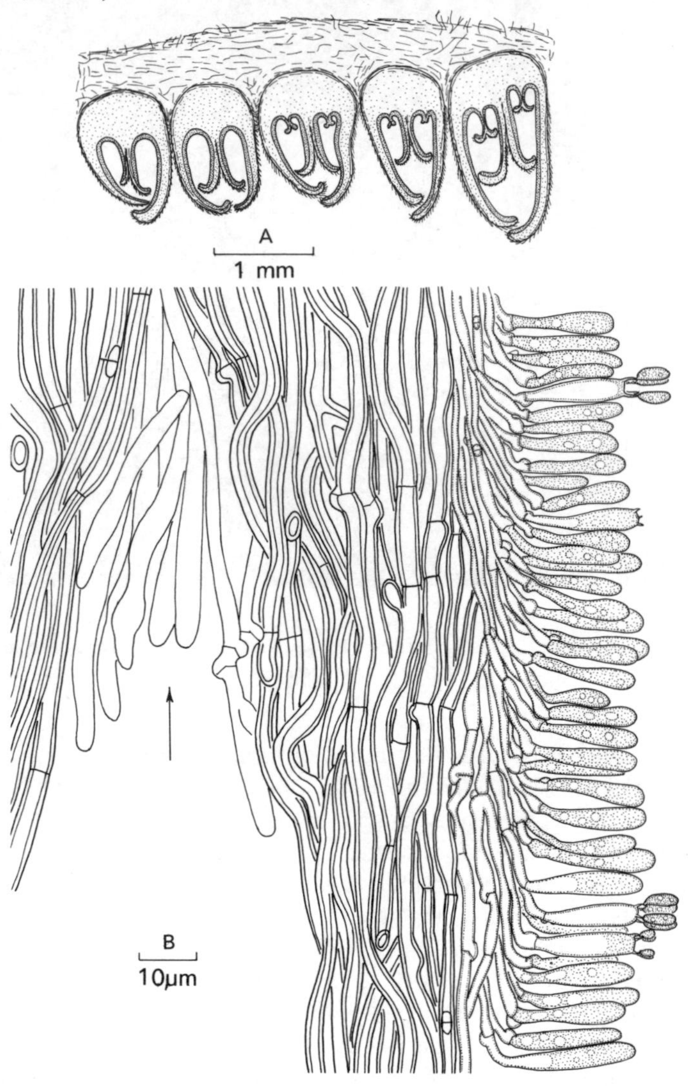

Figure 203. *Schizophyllum commune.* A. V.S. portion of sporophore in the dried state showing the divided inrolled 'gills'. B. High-power drawing of part of a 'gill' in the region of the split (arrowed). Note the thin-walled hyphae in this region, contrasting with the thicker-walled hyphae making up the rest of the flesh.

The fruit-body develops as an inverted cup with a hymenium developing over the entire lower surface, and by more rapid growth on one side the fruit-body may become fan-shaped. The split gills arise by marginal proliferation, and their number is increased by downgrowths from the flesh of the fruit-body. The cytology of basidial development shows no unusual features. Mature basidiospores are binucleate, following mitosis of the nucleus which enters each spore (Ehrlich & McDonough, 1949).

Coniophoraceae

Serpula lacrymans causes dry rot, and is one of the most serious agents of timber decay in buildings (Cartwright & Findlay, 1958). Both hardwoods and softwoods are attacked. Only wood with a moisture content above about 20–25% of the oven dry weight is liable to attack by the fungus. Properly dried and seasoned timber has a moisture content of 15–18% and in a properly ventilated house this soon falls to 12–14% or lower (Findlay & Savory, 1960). If woodwork becomes wet through contact with damp masonry, through faulty construction or poor ventilation then infection from air-borne basidiospores is likely to follow. The mycelium within the wood develops chiefly at the expense of the cellulose, and water produced by the breakdown of cellulose (sometimes termed the water of metabolism) may render the timber on which it is growing moister, which in turn stimulates further growth. The epithet '*lacrymans*' (weeping) refers to the beads of moisture sometimes found on decaying timber. Well-rotted timber is shrunken with transverse cracks and has a dry crumbly texture. Sheets of mycelium may extend over the timber and adjacent brickwork, and the fungus is also capable of spreading over long distances (several metres) by means of strands or rhizomorphs up to 5 mm in diameter. The rhizomorphs can penetrate mortar and spread from room to room of houses. The internal hyphae are modified for rapid conduction, enabling water to be transported, and enabling colonisation of relatively dry timber. Even if infected timber is removed rhizomorphs can initiate fresh infections.

Fruit-bodies develop as flat fleshy surface growths. On the lower side a brown honeycomb-like arrangement of shallow pores supports the hymenium (Fig. 204). The construction is monomitic. Immense numbers of rusty-red basidiospores can be produced which may form deposits visible to the naked eye.

In the control of dry rot it is important to strip out all infected timber and to sterilise adjoining brickwork. New timber can be treated with disinfectants. The most important measure, however, is to ensure, by proper construction, that the moisture level of the timber remains below the point at which infection is likely.

Serpula lacrymans is essentially a fungus of buildings and is rarely found away from human habitation. It has, however, been collected on spruce logs in the Himalayas, at an altitude of 8,000–10,000 ft (Bagchee, 1954). It is widely distributed throughout northern Europe, but is confined to the cool temperate zone. Its optimum temperature (23°C) is rather low, and its maximum tem-

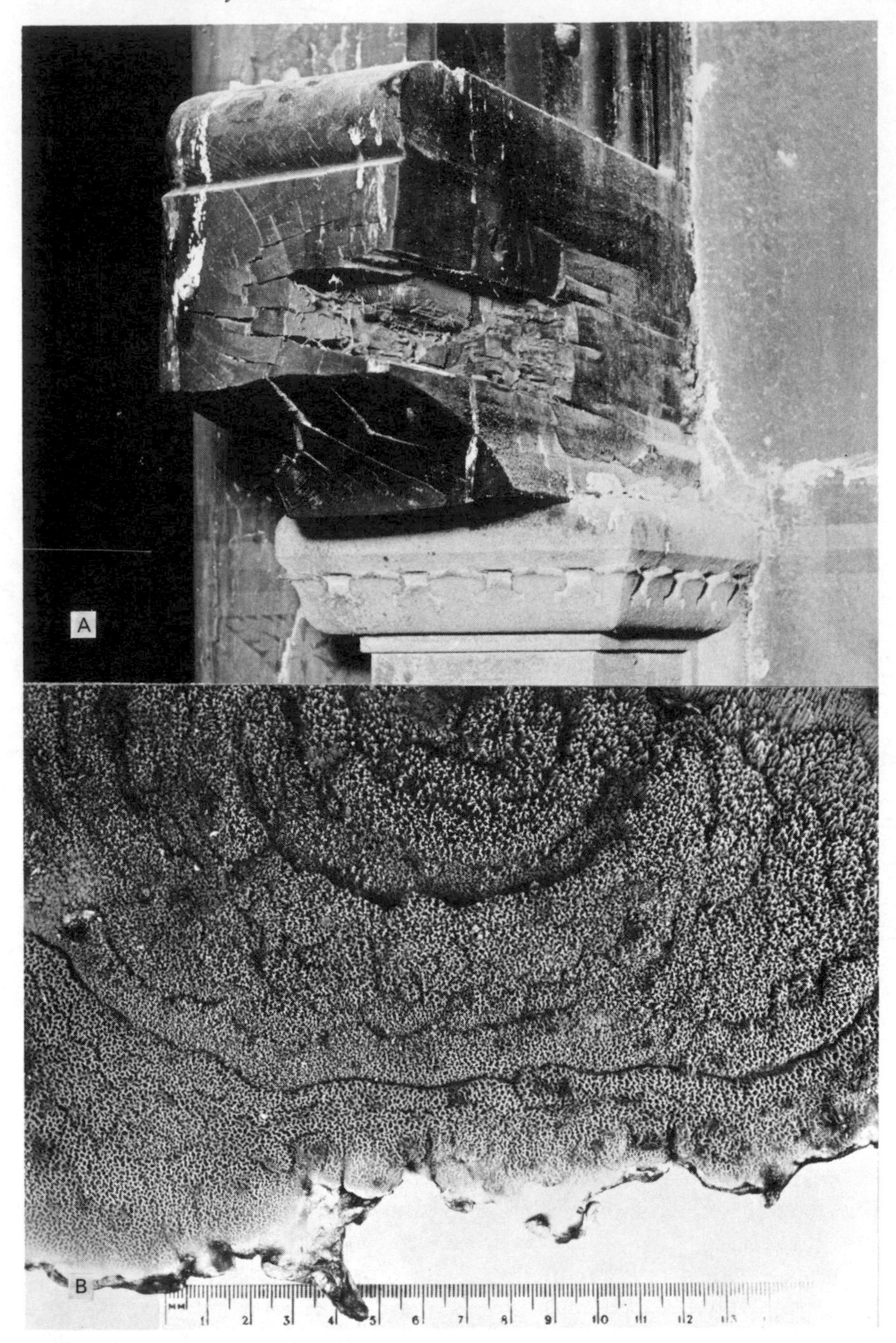

Figure 204. *Serpula lacrymans*, the dry-rot fungus. A. Beam supporting a roof arch of a church showing the typical cracking transverse to the grain of the wood. The wood also shows shrinkage. B. Fruit-body seen from the lower side showing the shallow pores.

perature is about 26°C. These temperature characteristics may explain its distribution and why it is not found on exposed timber (Cartwright & Findlay, 1958).

Clavariaceae

Hymenomycetes with branched or unbranched cylindrical or clavate fructifications were previously aggregated into this family, but again, microscopical analysis shows that such an arrangement groups together unrelated forms, and it is clear that the Clavarioid type of fructification has evolved independently in several unrelated basidiomycete groups. Corner (1950) has monographed the Clavarioid fungi and has attempted to sort them into natural (i.e. related) groups. Microscopic structure shows that relatives of certain Clavarioid forms lie with Hydnoid, Polyporoid and Agaricoid genera.

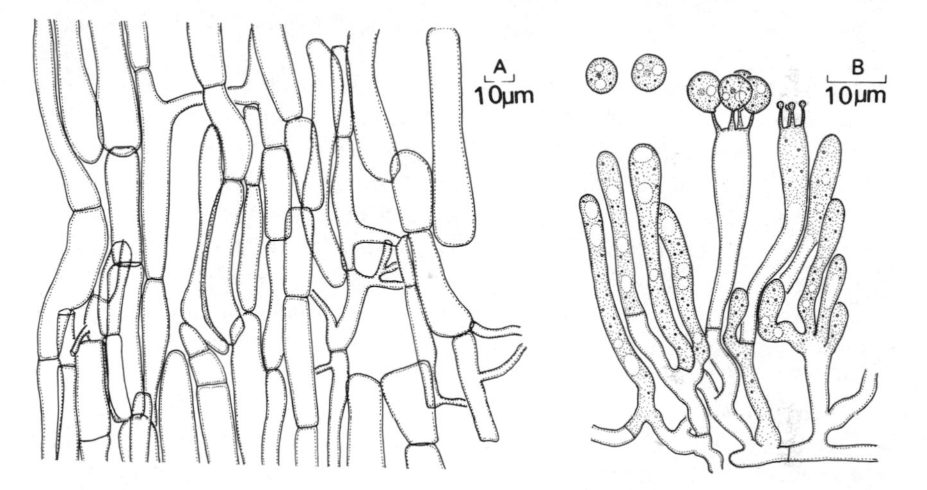

Figure 205. *Clavaria vermicularis*. A. Hyphae from the flesh. Note the absence of clamp connections. B. Part of the hymenium. There are no cystidia. Drawings prepared from var. *sphaerospora*.

Clavaria

This is a large genus of pasture and woodland fungi with cylindrical or club-shaped branched or unbranched fructification. The flesh is made up of thin-walled hyphae which lack clamp connections, and may become inflated and develop secondary septa (Fig. 205). The hymenium which covers the whole surface of the fruit-body usually consists of four-spored basidia with or without basal clamps, bearing colourless spores. A typical species is *C. vermicularis* (Fig. 206A) which grows in grassland forming tufts of whitish simple fruit-bodies. *Clavaria argillacea* forms yellow club-shaped fructifications on moors and heaths and peat bogs. Here the basidia bear a wide loop-like

Figure 206. A. *Clavaria vermicularis* fruit-bodies. B. *Clavulina cristata* fruit-bodies.

clamp at the base. There are numerous common representatives of the Clavariaceae (Lange & Hora, 1963). *Clavariadelphus pistillaris* which forms exceptionally large club-shaped fruit-bodies (7–30 × 2–6 cm) grows in deciduous woods. The construction is monomitic, with clamps at the septa. As the fruit-body matures the hymenium becomes thicker by the development of further layers of basidia. *Clavulinopsis corniculata* with branched yellow fruit-bodies, and *Clavulinopsis fusiformis* with simple yellow fruit-bodies are frequently found in acid grassland. The flesh is composed of clamped hyphae without secondary septa. The hymenium becomes thickened with age, and the basidia are typically four-spored.

Clavulina cristata is a very variable fungus with highly branched fructifications (Fig. 206B). In microscopical details its construction is very similar to *Clavulina rugosa* which is also variable but generally forms whitish, less richly branched fruit-bodies. A characteristic feature of the genus is that the basidia are two-spored, narrowly cylindrical, and often septate after spore discharge. The hymenium thickens with age. The construction is monomitic, with clamped inflated hyphae (Fig. 207). Donk (1964) places *Clavulina* in a separate family, the Clavulinaceae.

Some of the more richly branched fairy clubs are placed in the genus *Ramaria*, distinguished by the tougher flesh and yellow- to brown-coloured basidiospores, which are often rough. *Ramaria stricta* is unusual in growing on rotten wood. The hyphae of the flesh have thick walls. The mycelial hyphae are dimitic with skeletals.

Stereaceae

In this family the fruit-body is flattened, appressed or resupinate, sessile or stalked, with the hymenium smooth (unfolded) and on one side of the fructification. The construction is usually dimitic with skeletals, but monomitic and trimitic forms are known. There are many genera. Most species are lignicolous and saprophytic, but some are important parasites.

Stereum

Species of *Stereum* (Figs. 208, 209) are common on decaying stumps and branches. *Stereum hirsutum* forms clusters of yellowish fan-shaped leathery brackets (Fig. 208A), on various woody hosts and is important as a cause of decay of sapwood of oak logs after felling. *Stereum gausapatum* is another

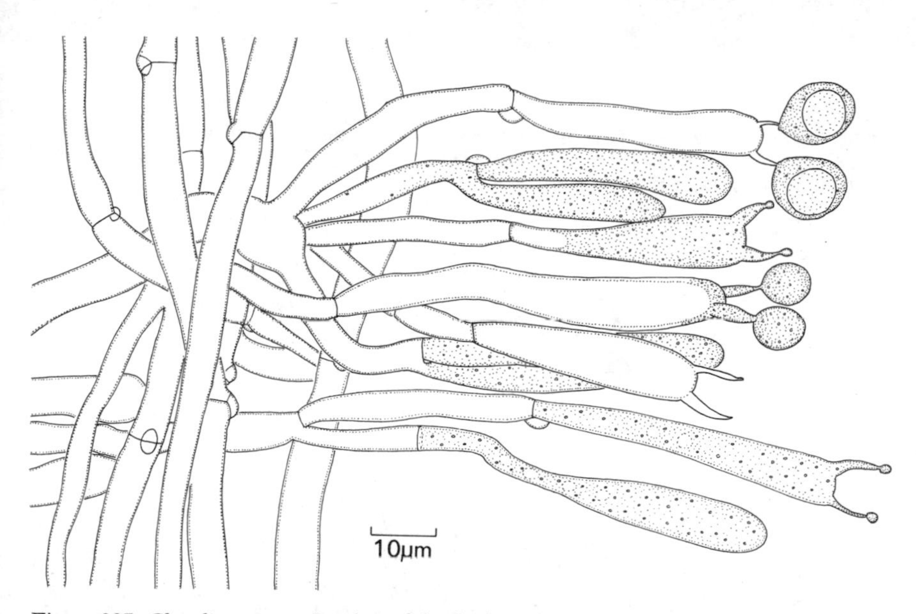

Figure 207. *Clavulina rugosa*. Portion of the flesh and hymenium. Note the clamped hyphae and the narrow two-spored basidia.

common fungus on oak, which can grow parasitically on living trees and cause 'pipe rot' of the heartwood. When bruised the fruit bodies 'bleed', i.e. exude a red latex. This phenomenon is also found in *S. rugosum*, on broad-leaved trees and *S. sanguinolentum*, on conifers. In all these cases there are specialised laticiferous hyphae in the flesh which may extend through to the hymenium (see Fig. 209 of *S. rugosum*). These hyphae are interpreted as modified skeletal hyphae.

Stereum purpureum is an important parasite of Rosaceae including fruit trees such as plum and cherry causing 'silver leaf' disease. The silver sheen on the leaves is caused by the separation of the epidermis from the palisade mesophyll due to toxic secretions from the mycelium of the fungus, not in the leaves themselves, but in the branches beneath. Infection is through wounds.

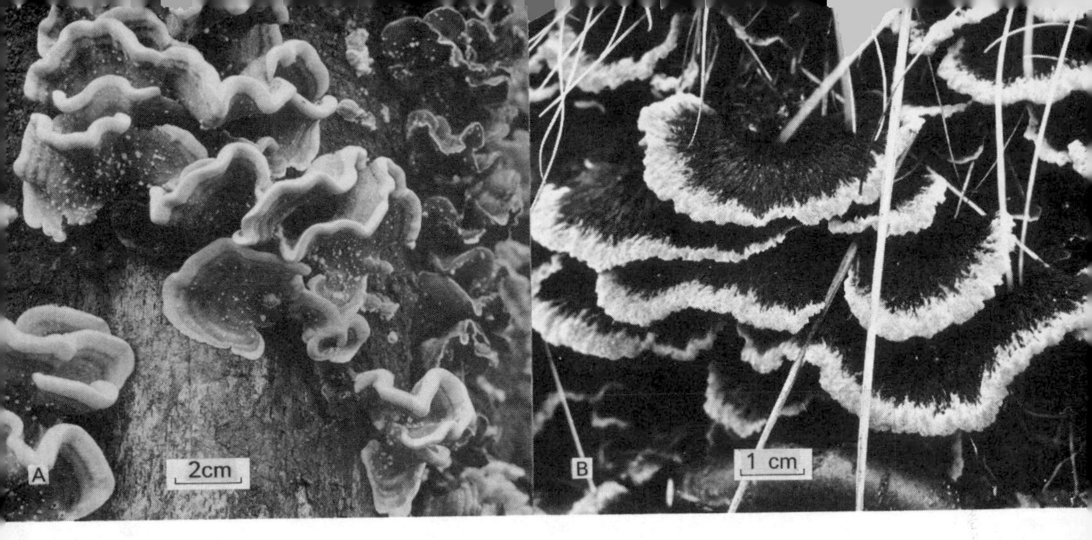

Figure 208. A. *Stereum hirsutum*. Sporophores on beech stump. The hymenial surface is on the lower face of the fruit-body. B. *Thelephora terrestris*. Fruit-bodies seen from above. The hymenium is on the lower side.

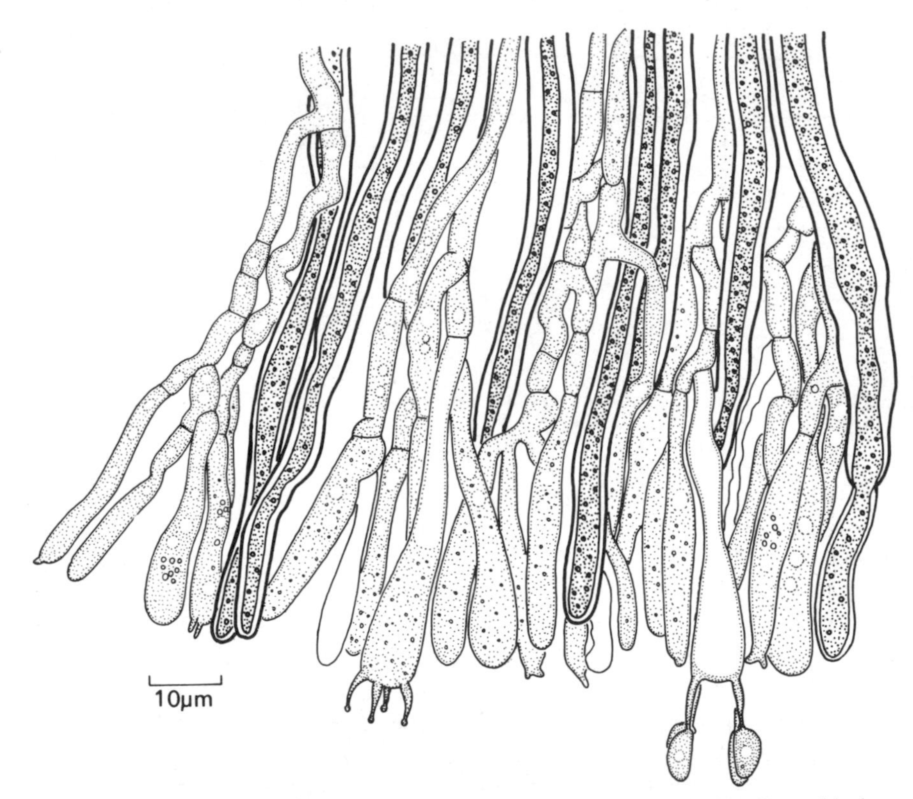

Figure 209. *Stereum rugosum*. Section of hymenium. The thick-walled hyphae with dense contents are laticiferous hyphae which exude a red latex when damaged causing the fruit body to 'bleed'.

Control measures include the protection of pruning cuts, and the removal of all wood bearing sporophores from orchards (Anon., 1967*b*).

Some authorities classify *Stereum* in the Thelephoraceae (see below).

Thelephoraceae

Here the fruit-body is generally monomitic in contrast with the prevailing dimitic construction of the Stereaceae, and cystidia are usually absent. Most of the genera are terrestrial but some are lignicolous.

Thelephora

Thelephora terrestris grows on light acid soil, sometimes running over twigs and other debris. The fan-shaped fructification superficially resembles a *Stereum*, but is made up of clamped generative hyphae only. The basidiospores are brown and warty (Figs. 208B, 210).

TULASNELLALES

In contrast with the Agaricales and Aphyllophorales which have undivided basidia (holobasidia), the Tulasnellales have basidia which are divided by longitudinal or transverse septa (phragmobasidia) or are forked. The name

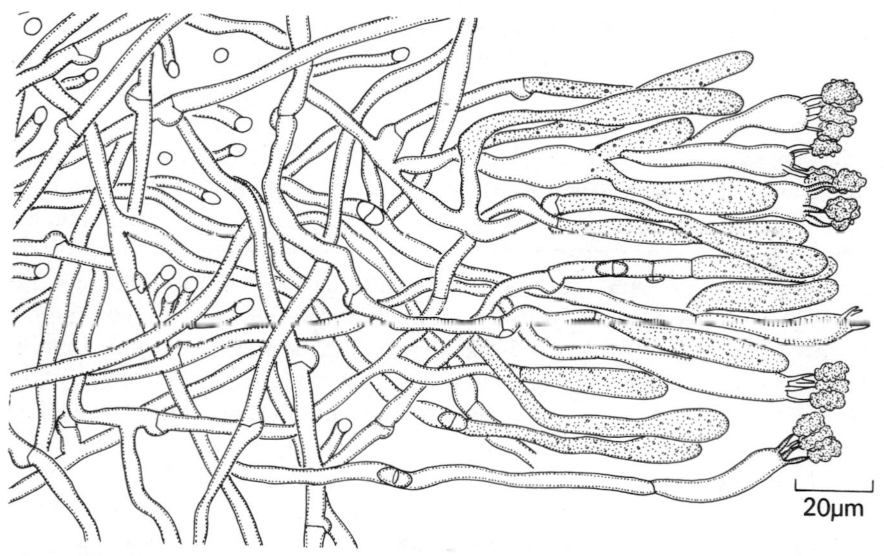

Figure 210. *Thelephora terrestris*. Section of fruit-body and hymenium. Note the monomitic construction.

Heterobasidiae has sometimes been used for them. They mostly have gelatinous fructifications and are sometimes called jelly fungi. They are rarely of economic significance, being generally saprophytic on woody substrata. Martin (1952) has given an account of the group. We shall consider representatives from three families: Auriculariaceae, Tremellaceae and Dacrymycetaceae. The taxonomic position of the Sporobolomycetaceae is uncertain, but arguments can be made for regarding the representatives of this family as reduced Hymenomycetes, possibly related to the Tulasnellales. A common feature of the Tulasnellales is that the basidiospores commonly germinate by *repetition*, i.e. by producing a second spore which is projected violently (ballistospore), or by budding. Both these types of germination are common in *Sporobolomyces*.

Auriculariaceae

In this family the basidium is divided by transverse septa. Apart from the genus *Auricularia*, described below, other genera of interest are *Septobasidium* and *Helicobasidium*. Species of *Septobasidium* grow symbiotically and parasitically on scale insects. The septate part of the basidium arises from a swollen thick-walled cell sometimes called the probasidium (Couch, 1938). *Helicobasidium purpureum* is the cause of violet root-rot of various cultivated plants, and may also cause tuber-rot of potato. The basidium is curved or coiled.

Auricularia

The Jew's ear fungus *A. auricula* (sometimes known as *Hirneola auricula-judae*) forms rubbery, ear-shaped fruits on elder branches (Fig. 211), and is a weak parasite. A wide range of other hosts has been reported. Duncan & MacDonald (1967) divide *A. auricula* into three 'units'; a European deciduous unit, a North American deciduous unit, and a North American coniferous unit. Interfertility studies between these populations show that the European populations are intersterile with the American deciduous unit, and virtually intersterile with the American coniferous unit. Crosses between the two American units show that they are only partially interfertile. A section through the flesh shows a hairy upper surface, a central gelatinous layer containing narrow clamped hyphae and a broad hymenium on the lower side. Details of the anatomy of the fruit body are useful in classification (Lowry, 1951, 1952). The fruit body can dry to a hard brittle mass, but on wetting it quickly absorbs moisture and discharges spores within a few hours. The basidia are cylindrical and become divided by three transverse septa into four cells (Fig. 212). Each cell of the basidium develops a long cylindrical *epibasidium* which extends to the surface of the hymenium and terminates in a conical sterigma bearing a basidiospore, which is projected (ballistospore). On germination a second ballistospore may be formed: alternatively sickle-shaped conidia or a germ tube may develop. The fungus is heterothallic, and there are indications of multiple alleles. Barnett (1937) claimed that it was bipolar, but Banerjee (1956) states that it is tetrapolar. Another common species is *A. mesenterica* which forms thicker hairy fan-shaped fruit bodies on old stumps and logs of

Figure 211. *Auricularia auricula.* A. Fruit-body on branch of elder, as seen from above. B. Fruit-body seen from the lower, hymenial surface.

elm and other trees. It causes active wood decay, and may occasionally be parasitic.

Dacrymycetaceae

Here the basidium consists of a cylindrical base which forks into two tapering arms, the so-called 'tuning fork' type of basidium. The form of the fruit-body varies from a flat crust to cushion-shape, cup-like or clavarioid. Most members of the group are saprophytic on wood. Kennedy (1958*a*) has given keys to nine genera.

Dacrymyces

The orange gelatinous cushions about 1–5 mm in diameter so common on damp rotting wood are the fructifications of *D. deliquescens* (Fig. 213). Close inspection with a hand lens reveals that the cushions are of two kinds: soft, bright orange, hemispherical cushions, and firmer, pale yellow, flatter cushions. The bright orange cushions are conidial pustules which consist of hyphae whose tips are branched and fragment into numerous multicellular conidia or *arthrospores* (Fig. 214C). The cells are packed with oil globules containing carotene. Such conidia are readily dispersed by rain-splash and are obviously similar in function to the splash-dispersed conidia of *Nectria cinnabarina.*

The basidial cushions are attached centrally to the woody substratum. The surface layers are composed of clusters of forked basidia (Fig. 214B) which bear pairs of spores. At the time of discharge the basidiospores are unicellular, but before germination they usually develop three transverse septa. Germination may be by means of germ tubes from any of the cells, or small conidia may develop on short conidiophores formed on germination (Fig. 214A). Similar conidia may also arise on older hyphae. For a key to other species of *Dacrymyces* see Kennedy (1958*b*).

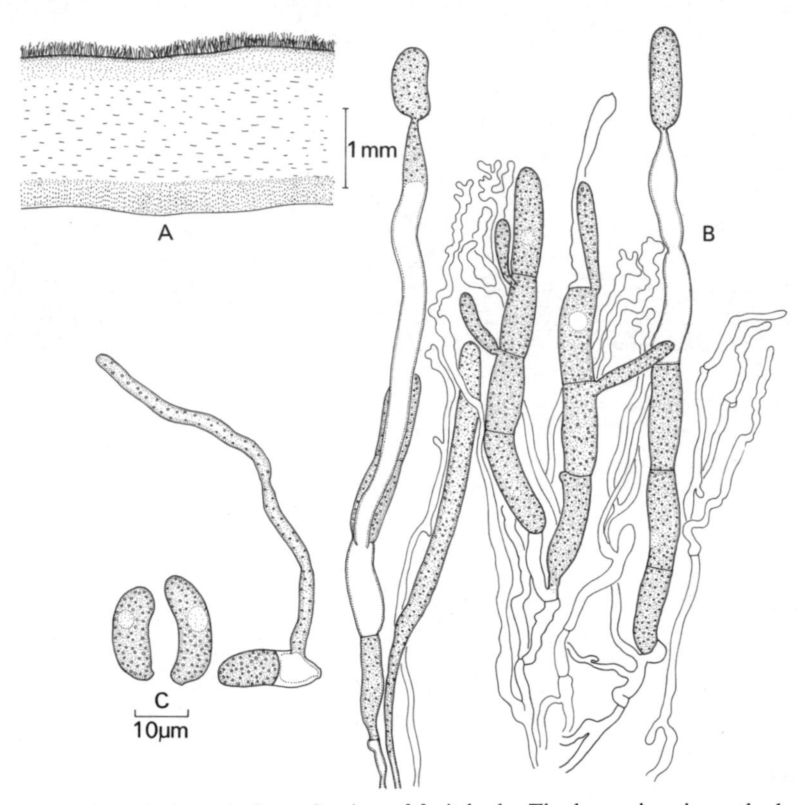

Figure 212. *Auricularia auricula*. A. Section of fruit-body. The hymenium is on the lower side. B. Squash preparation showing basidia. Note the transverse segmentation and the long epibasidia. The basidia are associated with branched hyphae. C. Basidiospores, one germinating. Note the formation of a septum.

Calocera

At first sight the cylindrical orange outgrowths of *C. viscosa* from coniferous logs, or the smaller *C. cornea* from hardwood logs could be mistaken for species of *Clavaria* (Figs. 214B,C). The gelatinous consistency and the characteristic forked basidia (Fig. 215) place them in the Dacrymycetaceae. At maturity the basidiospores of both species are one-septate. They germinate by the formation of germ tubes or globose conidia (McNabb, 1965).

Tremellaceae

The longitudinally divided basidium (Figs. 217, 218) distinguishes this group from the other Tulasnellales. Again the fructifications are usually gelatinous,

Figure 213. A. *Dacrymces deliquescens*, basidial cushions. B. *Calocera viscosa* growing from a buried conifer stump. C. *Calocera cornea*.

and often consist of flat folds with the hymenium on one or both faces. In *Pseudohydnum gelatinosum* (sometimes called *Tremellodon gelatinosum*) the fruit-body has a short excentric stalk and a pileus on the lower side of which conical teeth project, resembling those of a *Hydnum* (Fig. 216). Polyporoid fructifications with Tremellaceous basidia are found in *Aporpium caryae*. Common representatives are *Exidia* and *Tremella*.

Exidia

Exidia glandulosa, sometimes called Witches' butter, forms black rubbery fructifications on decaying branches of various woody hosts, especially lime (Figs. 216B; 217). The hymenium is borne on the lower side of the fructification and in this species is studded with small black warty outgrowths. Wells (1964*a,b*) has described the fine structure of *E. nucleata*.

Tremella

Here the fructification consists of flattened contorted folds, with the hymenium on both faces. *Tremella frondosa* grows on oak and beech stumps forming

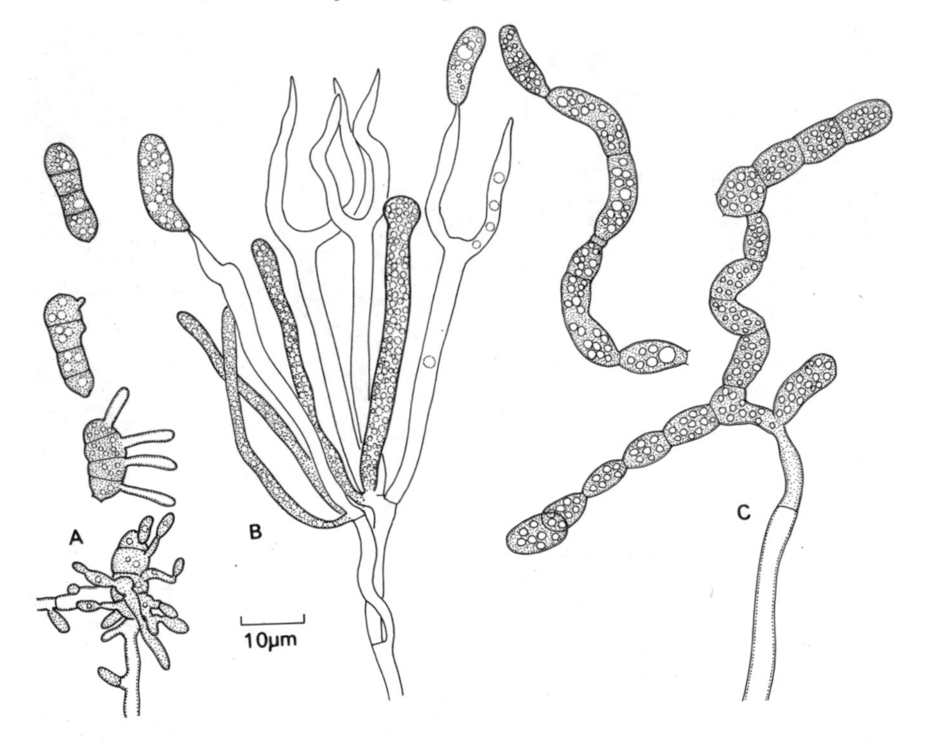

Figure 214. *Dacrymyces deliquescens*. A. Basidiospores showing germination. The basidiospore at the bottom is producing blastospores. B. Basidia. Note that the attached basidiospores are unicellular. They become three-septate on germination. C. Arthrospores from a conidial pustule.

flesh-coloured to pale brown fruit-bodies. The basidiospores may germinate by repetition, by yeast-like budding (Fig. 218C,D) or by germ tube. *Tremella mesenterica* forms yellow to orange lobed fructification on various woody hosts such as willow and beech. Heterothallism in *T. mesenterica* is tetrapolar, and there is evidence of a hormone-like substance which stimulates directional growth of compatible germ tubes (Bandoni, 1963, 1965).

Sporobolomycetaceae

A group of yeast-like fungi which discharge ballistospores is classified here. *Sporobolomyces* has pink colonies and asymmetrical spores, whilst *Bullera* forms whitish colonies and has symmetrical spores.

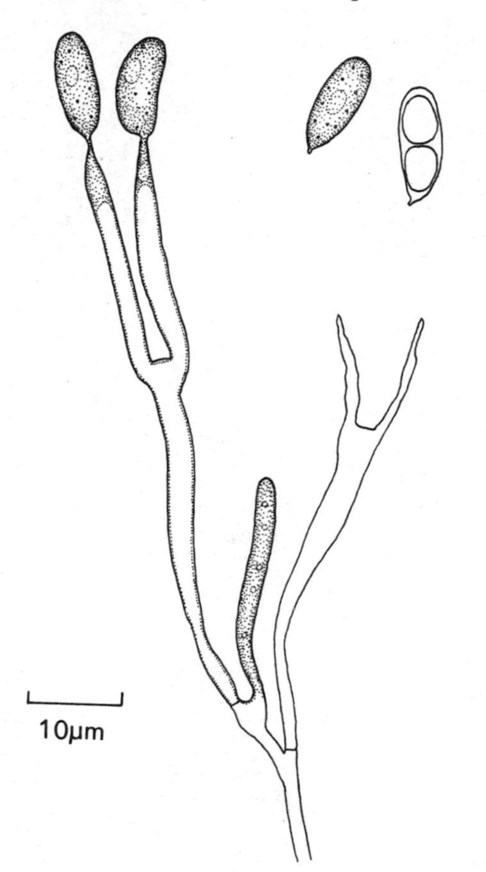

Figure 215. *Calocera viscosa*. Basidia and basidiospores.

Sporobolomyces roseus is abundant on moribund vegetation, and can be isolated readily by placing senescent plant material (e.g. a grass leaf) over nutrient agar. Ballistospores projected from the surface of the leaf colonise the agar, and yeast-like colonies develop by budding. Within a few days, further ballistospores are formed. Spores of Sporobolomycetaceae are frequent in the air, especially on warm nights in the summer, and concentrations

Figure 216. A. *Tremella frondosa*: fruit-body on oak stump. The hymenium is on both faces. B. *Exidia glandulosa*: fruit-body on lime. The hymenial surface bears black warts and is on one face of the fruit-body. C. *Pseudohydnum gelatinosum*: fruit-bodies seen from below and above. The hymenium is borne on the spines.

of hyaline ballistospores probably belonging to *Sporobolomyces* and *Tilletiopsis* may reach a concentration of 10^5–$10^6/m^3$ (Last, 1955). The budding phase of *S. roseus* is uninucleate and nuclear division occurs at about the time of bud formation (Buller, 1933). When spores are formed a conical sterigma

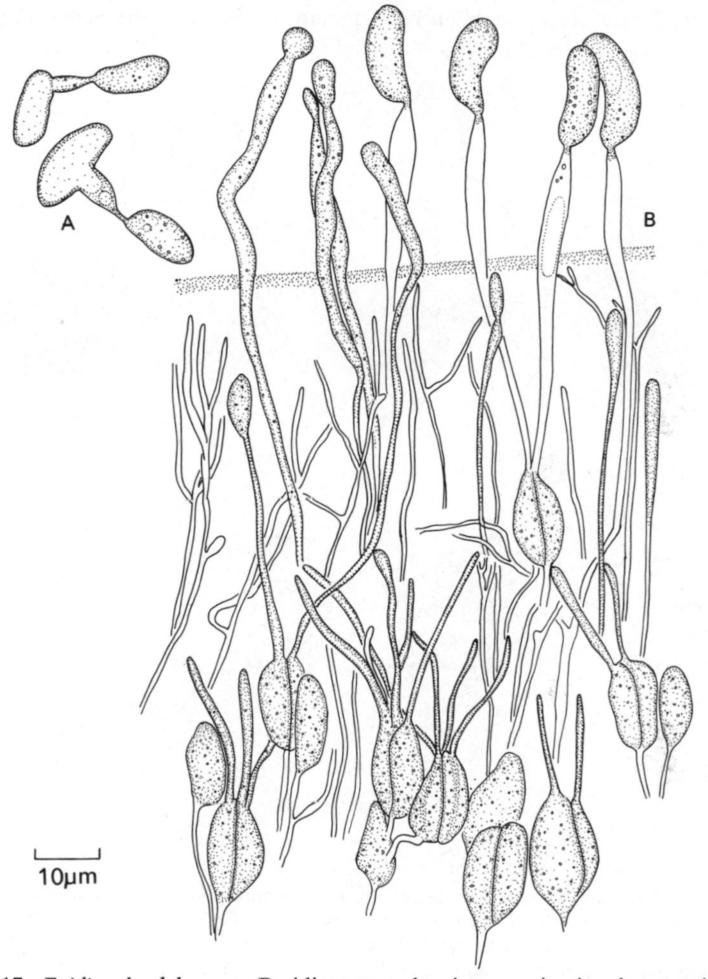

Figure 217. *Exidia glandulosa*. A. Basidiospores showing germination by repetition. B. Section of hymenium. Note the longitudinally divided basidia with long epibasidia penetrating to the surface.

develops vertically and bears a sausage-shaped spore into which a daughter nucleus passes (Fig. 219). The spore is thrown for a distance of about 0·1 mm apparently by the same mechanism as in many Basidiomycetes (Müller, 1955). Electron micrographs of detached spores show a cylindrical projection with a torn rim (Ingold, 1966). A single sterigma may form a second or even a third spore, and occasionally two sterigmata may arise from one cell. The spore deposit on a Petri-dish lid reflects the pattern of the colony from which it was thrown, and for this reason the Sporobolomycetaceae are called mirror yeasts. Reports of nuclear fusion and reduction division in *Sporobolomyces* (see Laffin & Cutter, 1959) require confirmation. In *S. roseus* and *S. salmonicolor* a true mycelium has been reported in addition to the yeast-like phase.

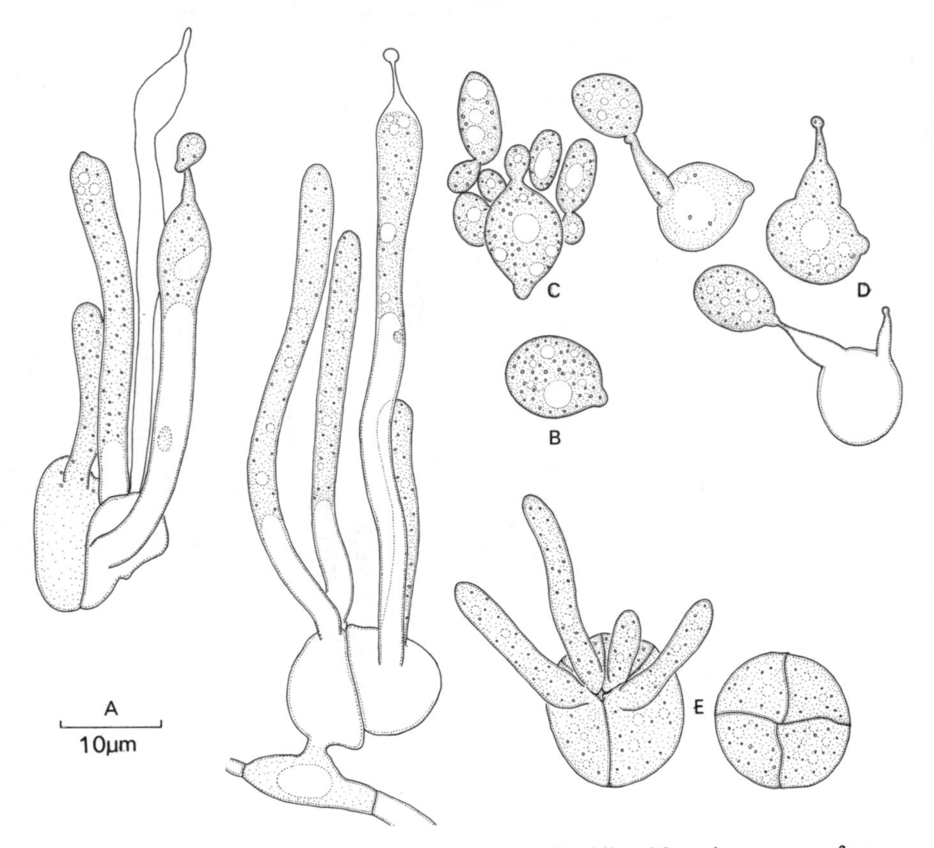

Figure 218. *Tremella frondosa*. A. Basidia showing epibasidia with various stages of spore development. B. Freshly discharged basidiospore. C. Basidiospore germinating on malt extract agar to form yeast-like cells which also undergo budding. D. Basidiospores germinating in water by repetition, i.e. by producing ballistospores. E. Immature basidia seen from above, showing the division of the basidium into four cells by longitudinal septa.

In *Sporidiobolus*, a possible relative of *Sporobolomyces*, there is a clamped mycelium with predominantly binucleate segments. Ballistospores are formed which may bud or may germinate to form a mycelium. Chlamydospores are also formed, containing paired nuclei which fuse (Nyland, 1949; Laffin & Cutter, 1959). In *Itersonilia* and *Tilletiopsis* mycelium is produced and reproduction is by ballistospores, but in neither genus is there a yeast-like phase.

Clamp connections have been found in *Itersonilia* (Derx, 1930, 1948; Olive, 1952).

Many members of the Sporobolomycetaceae appear to be associated with lesions caused by other plant parasites. *Itersonilia perplexans* was first described from rusted hollyhock leaves whilst *Sporidiobolus johnsoni* was isolated from rusted leaves of raspberry. Increased numbers of *Sporobolomyces* colonies are also associated with damage of foliage from nematodes and mites (Last & Deighton, 1965).

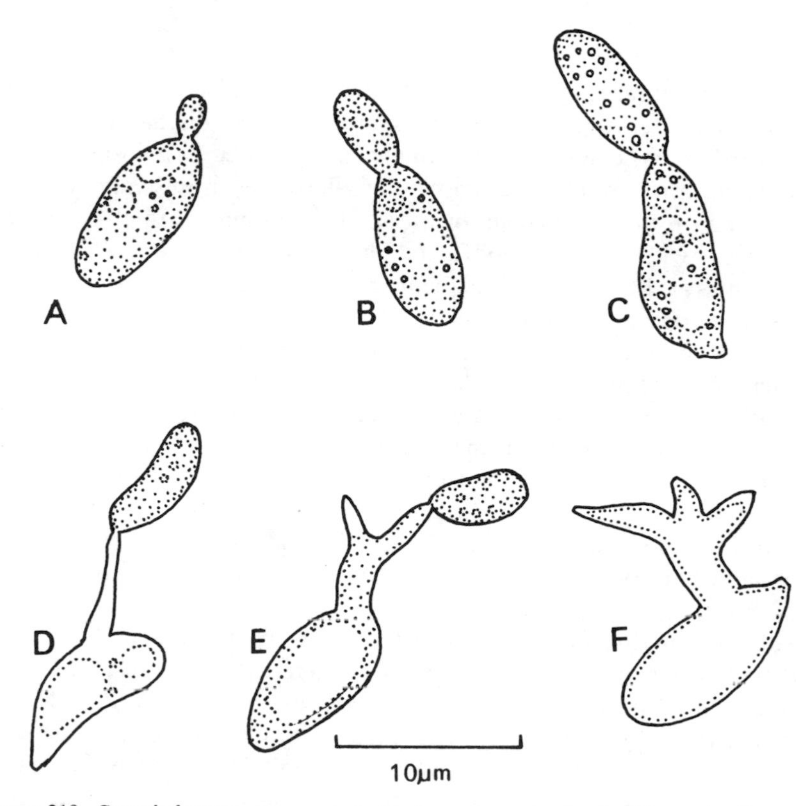

Figure 219. *Sporobolomyces roseus*. A–C. Various stages in the budding of cells. D. Cell bearing a sterigma and a ballistospore. E. Cell showing two sterigmata. F. Cell with three sterigmata.

The relationships of the Sporobolomycetaceae are not generally agreed. Because the spore discharge resembles that of Basidiomycetes, Buller and other workers regard *Sporobolomyces* as a reduced Basidiomycete, but this view is not universally accepted (see Buller, 1933; Lodder & Kreger-van Rij, 1952). It is now known that ballistospores are not confined to Basidiomycetes. They have also been reported in the Protostelida, a group of Mycetozoa (Olive & Stoianovitch, 1966; Olive, 1967). Studies of the fine structure of

S. roseus (Prusso & Wells, 1967) do not clarify the relationships. It has been suggested that the family is related to the Dacrymycetaceae (Bulat, 1953) Ustilaginaceae (Nyland, 1950) or the Tremellaceae (Olive, 1952; Martin, 1952).

2: GASTEROMYCETES

This is an unnatural assemblage of basidiomycetes which share the common negative character that their basidiospores are not discharged violently from their basidia. Instead of the asymmetrically poised basidiospores found in the Hymenomycetes, Gasteromycete basidiospores are usually symmetrically poised on their sterigmata or are sessile. Commonly the basidia open into cavities within a fruit body, and the basidiospores are released into these cavities as the tissue between them breaks down or dries out. Sometimes, as in *Lycoperdon*, the fruit-body opens by a pore through which the spores escape, but in forms with subterranean fruit-bodies there is no special opening, and it is possible that the spores are dispersed by rodents or other soil animals. In *Phallus* and its allies the spores are exposed in a sticky mass attractive to insects, whilst in the Nidulariaceae the spores are enclosed in separate glebal masses or peridioles which are dispersed as units.

All members of the group are saprophytic, and grow on such substrata as soil, rotting wood and other vegetation, and on dung. *Rhizopogon*, which forms subterranean fruit-bodies, and *Scleroderma* can form mycorrhiza with forest trees.

Evidence from the microscopic structure of the fruit-bodies and spores has led to the conclusion that certain genera of Gasteromycetes are related to agaric genera. It is believed that the Gasteromycete forms may have evolved from Hymenomycete ancestors, possibly as an adaptation to xerophytic conditions. Ingold (1965) regards the Gasteromycetes as a biological group which, having lost the explosive spore discharge mechanism of their Hymenomycete ancestors, have attempted a remarkable series of experiments in spore liberation. It should be stressed that these ideas are essentially speculative. For a discussion of the problems of Gasteromycete phylogeny see Heim (1948) and Singer & Smith (1960).

Ainsworth (1966) has proposed that the group should be subdivided into five orders; Hymenogastrales, Lycoperdales, Nidulariales, Phallales and Sclerodermatales. Keys separating the orders, families and genera have been provided by Zeller (1949) who recognises nine orders. Fischer (1933) has provided a general descriptive account. The Hymenogastrales will not be considered further. They are made up almost entirely of subterranean forms sometimes referred to as false truffles (Fig. 222B). An account of the British species is provided by Hawker (1954).

LYCOPERDALES

Common examples of this group are the puff-balls *Lycoperdon*, *Calvatia* and *Bovista*, and the earth-star *Geastrum*. Their fruit bodies may begin development beneath the surface of the soil, but mature above ground.

Lycoperdon (Figs. 220A,B; 221)

The fruit-bodies are pear-shaped or top-shaped. Most species grow on the ground, but *L. pyriforme* occurs on old stumps, rotting wood and sawdust heaps. Fruit-bodies commonly arise on mycelial cords or rhizomorphs. The individual cells or the mycelium usually contain paired nuclei, but clamp connections are absent (Dowding & Bulmer, 1964). A longitudinal section of a young fruit-body of *L. pyriforme* (Fig. 221A,B) shows that it is surrounded by a two-layered rind or peridium, but as the fruit-body expands the outer pseudoparenchymatous exoperidium may slough off or crack into numerous scales or warts whilst the tougher endoperidium made up of both thick-walled and thin-walled hyphae remains unbroken, apart from a pore at the apex of the fruit-body. The tissue within the peridium is termed the *gleba*. It is differentiated into a non-sporing region or *sub-gleba* at the base which extends as a columella into the sporing region, or fertile part of the gleba in the upper part of the fructification. The glebal tissue is sponge-like, containing numerous small cavities, and in the upper fertile part the cavities are lined by the hymenium. The tissue separating the hymenial chambers is made up of thick- and thin-walled hyphae. The thin-walled hyphae break down as the fruit-body ripens, but the thick-walled hyphae persist to form the *capillitium* threads between which the spores are contained. The basidia lining the cavities of the gleba are rounded and bear one to four basidiospores symmetrically arranged on sterigmata of varying length (Fig. 221C). Young basidia are binucleate, and nuclear fusion and meiosis occur in the usual way (Ritchie, 1948; Dowding & Bulmer, 1964). The spores are not violently projected from the sterigmata. As the glebal tissue breaks down and dries the spores are left as a dusty mass inside the fruit-body. The thin upper layer of the endoperidium is elastic and acts as a bellow. When rain-drops impinge on this layer small clouds of spores are puffed out (Gregory, 1949). Little is known of the mating behaviour of *Lycoperdon*.

Calvatia (Fig. 220C)

C. gigantea forms fruit-bodies about the size of a rugby football. There is no definite pore; the endoperidium breaks away to leave a brown spore mass. Buller (1909) has estimated that the output of a specimen measuring 40×28

Figure 220. A. *Lycoperdon pyriforme*. Fruit-bodies growing from a stump buried beneath leaves. B. *Lycoperdon perlatum*. Fruit-bodies amongst grass. C. *Calvatia gigantea*. Fruit-bodies growing with *Urtica dioica*. The coin is 2·5 cm in diameter. D. *Scleroderma aurantium*.

× 20 cm was 7×10^{12} spores. Even larger specimens, 120–150 × 60 cm, have been recorded (Kreisel, 1961).

When attempts are made to germinate the spores of this and other puff-balls in the laboratory the percentage germination is extremely low; often fewer than 1 spore in 1,000 germinates. Germination may take several weeks, and is stimulated by the growth of yeasts (Bulmer, 1964; Wilson & Beneke, 1966). Interest in this fungus has been aroused by claims that an extract of it, calvacin, can inhibit the growth of certain types of tumour (Beneke, 1963). Because meiosis in the entire fruit-body is almost simultaneous it has been possible to follow enzymic changes in tissue extracts during and after meiosis (Bulmer & Li, 1966).

Geastrum

Species of *Geastrum*, earth-star (Fig. 222A), produce fruit-bodies that begin development beneath soil or litter, or at the soil surface. *G. triplex* grows amongst the leaves of beech, sycamore and pine. The young fruit-body is onion-shaped. The exoperidium is more complex than that of *Lycoperdon*, consisting of a brown outer layer made of narrow hyphae mostly running

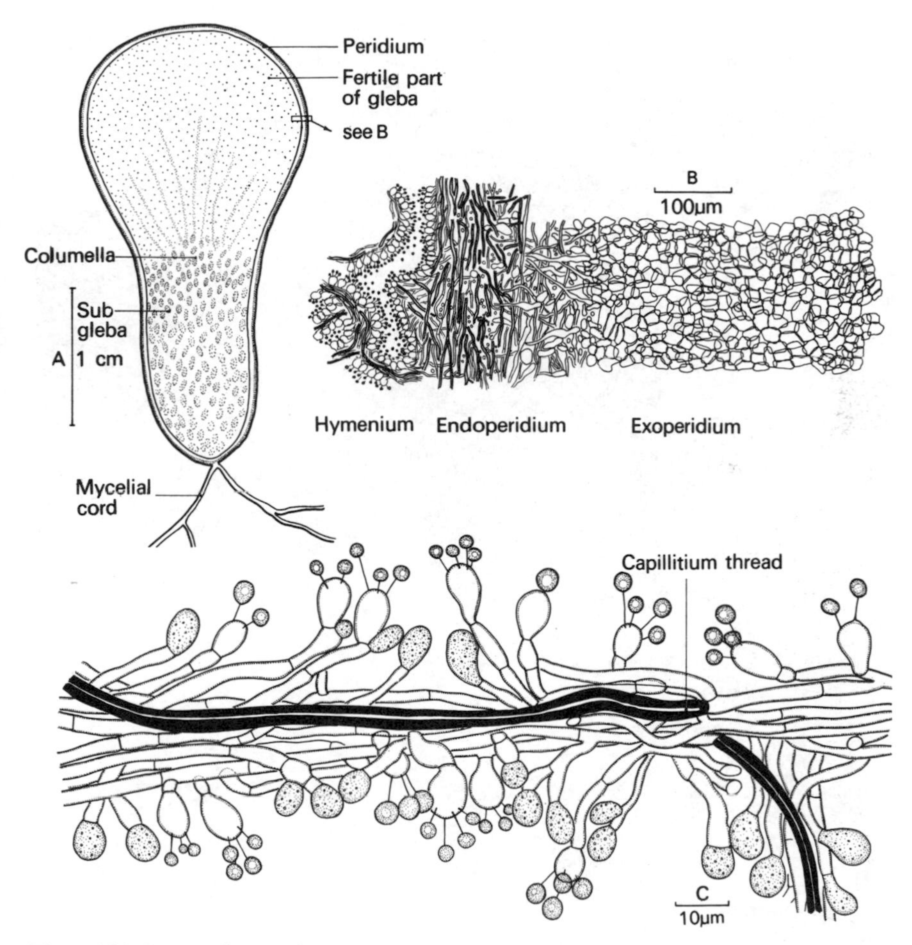

Figure 221. *Lycoperdon pyriforme*. A. L.S. fruit-body. B. Portion of peridium and gleba. Note the pseudoparenchymatous exoperidium and the fibrous endoperidium. C. Portion of gleba showing basidia, thin-walled hyphae and capillitium threads.

longitudinally and a paler pseudoparenchymatous inner layer. As the fruit-body ripens the whole of the exoperidium splits open from the tip in a stellate fashion and, due to swelling of the pseudoparenchyma cells of the exoperidium, the triangular flaps curve outwards and make contact with the soil, lifting the inner part of the fruit-body into the air (Fricke & Handke, 1962). The endoperidium is thin and papery and opens by an apical pore. Spores are puffed out when rain-drops strike it (Ingold 1965). The gleba contains a columella (sometimes termed a pseudocolumella) and capillitium much as in *Lycoperdon*. Basidial development can only be observed in young unexpanded fruit-bodies. The basidia are pear-shaped, with four to six (or possibly eight) spores borne on a knob-like extension of the pointed end (Palmer, 1955).

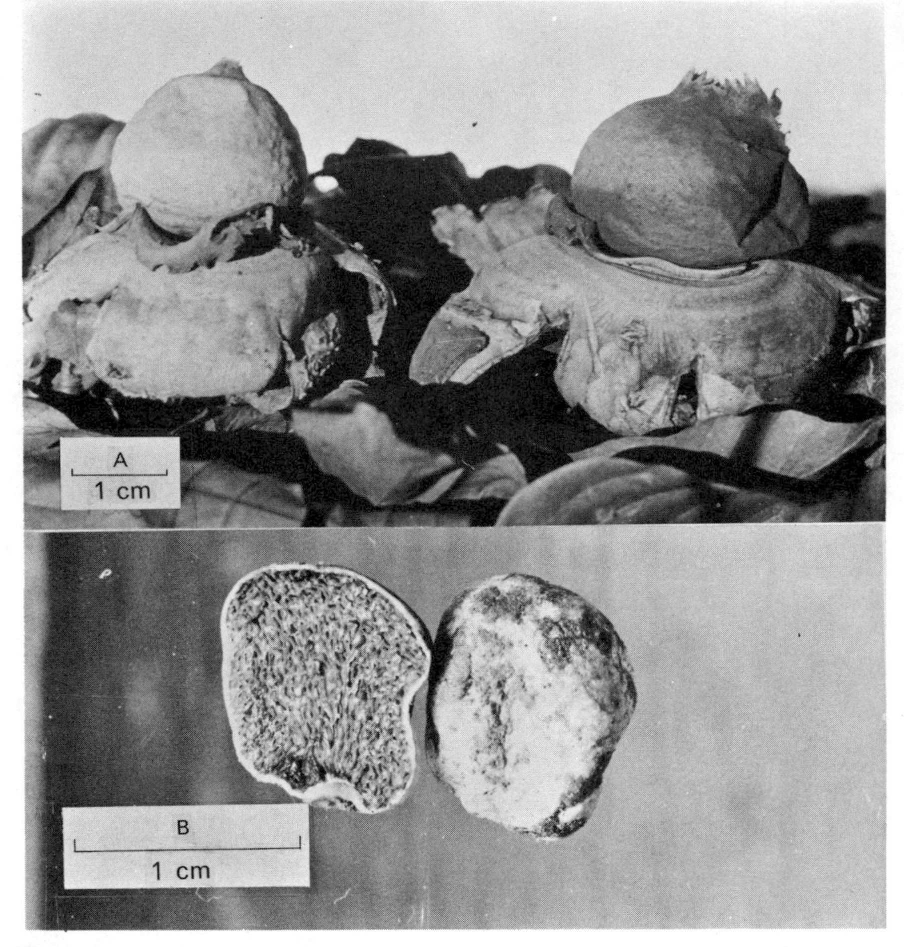

Figure 222. A. *Geastrum triplex*. Note the recurved exoperidium. B. *Hymenogaster tener*. The fruit-bodies are subterranean. One has been cut open and shows the hymenial chambers.

NIDULARIALES

Here the fruit-bodies are globose or funnel-shaped and the gleba is differentiated into one or more spherical to lens-shaped *peridioles* or glebal masses which contain the basidiospores. Common examples are *Cyathus*, *Crucibulum* and *Sphaerobolus*.

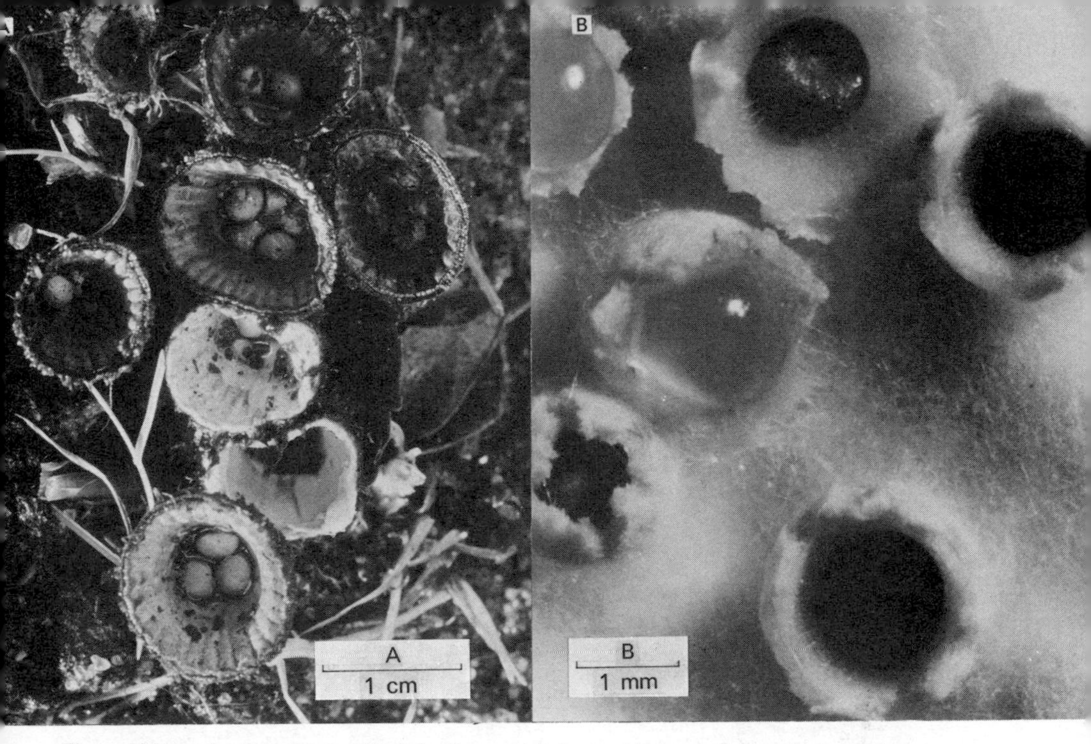

Figure 223. A. *Cyathus striatus*. Fruit-bodies containing peridiola. B. *Sphaerobolus stellatus*. Fruit-bodies at various stages of development. Four still contain glebal masses, and two (left) have discharged.

Cyathus (Figs. 223A; 224)

The funnel-shaped fruit-bodies of *C. olla* can be found in autumn growing amongst cereal stubble. *C. striatus*, recognised by the furrowed inner wall of its cup, grows on old stumps and twigs, whilst *C. stercoreus* grows on old dung patches. This last species can be made to fruit readily on chaff cased in soil (Warcup & Talbot, 1962), and since the peridioles retain viability for many years it provides a convenient example to study. Light is essential for fruit-body formation. The first sign of development is the appearance of brown mycelial strands at the soil surface, on which knots of hyphae differentiate. In young fruit-bodies the mouth of the funnel is closed over by a thin papery epiphragm (Fig. 224A), which ruptures as the fruit-body expands. Within the funnel the peridioles develop. They are lens-shaped, slate-blue in colour and attached to the peridium by a complex *funiculus*. In earlier stages of development in this and other species of *Cyathus* the peridioles are separated by thin-walled hyphae which disappear at maturity (Walker, 1920). The peridioles are surrounded by a *tunica* made up of loosely interwoven hyphae and a dark thick-walled *cortex*, in turn lined by a mass of very thick-walled hyaline cells (Fig. 224G). The inner part of the peridiolum is made up of thin-walled hyphae between which basidia develop. The basidia form four to eight basidiospores. The basidia disappear soon after the formation of spores, but the spores continue to enlarge and become thick-walled (Fig.

224G). Most of the thin-walled hyphae also break down, possibly providing nutrients for the enlarging spores. Martin (1927) calls the thin-walled cells 'nurse hyphae'. Nuclear fusion and meiosis occur in the basidia (Lu & Brodie, 1964). It is reported that *C. stercoreus* is probably tetraploid (Lu, 1964).

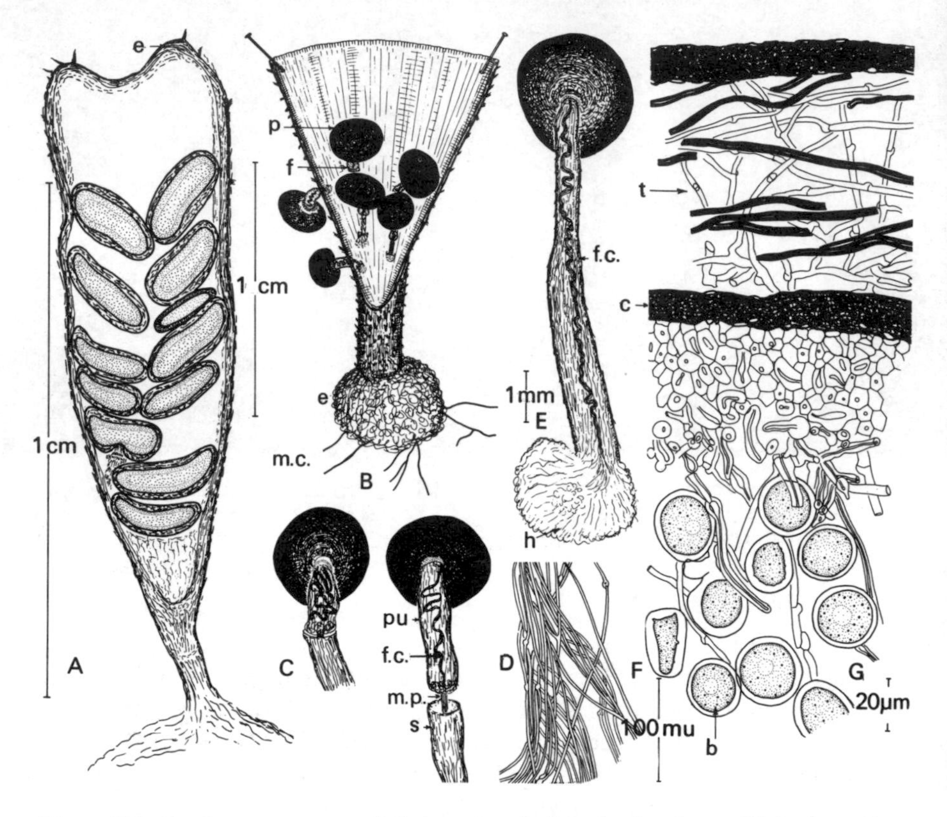

Figure 224. *Cyathus stercoreus*. A. L.S. immature fruit-body showing peridioles in section. B. Fruit-body cut open and pinned back to show the attachment of the peridioles. C–E. Details of structure of funiculus. C. Condition of funiculus before stretching. D. Stretched funiculus. Note the funicular cord coiled within the purse. E. Funicular cord extended after rupture of the purse. The base of the funicular cord is frayed out to form the hapteron. F. Portion of funicular cord. Note the spirally coiled hyphae. The thickenings are modified clamp connections. G. Detail of peridiole wall and contents. (b, basidiospore; c, cortex; e, epiphragm; em, emplacement; f, funiculus; f.c., funicular cord; h, hapteron; m.c., mycelial cords; m.p., middle-piece; s, sheath; p, peridiole; pu, purse; t, tunica.)

The fruit bodies of *Cyathus* and *Crucibulum* have been aptly termed splash-cups. The peridioles are splashed out by the action of rain drops to distances of over 1 m. The key to understanding the mechanism of discharge lies in the structure of the funiculus. In *Cyathus* (Fig. 224C,D,E), the funiculus is made up of the following structures (Brodie, 1951, 1956, 1963).

(a) **Sheath:** a tubular network of hyphae attached to the inner wall of the peridium.

(b) **Middle piece:** the innermost hyphae of the sheath unite to form a short cord termed the middle piece.

(c) **Purse:** the middle piece flares out at its top where its hyphae are attached to a cylindrical sac, the purse, which is firmly attached to the peridiole at a small depression.

(d) **Funicular cord:** folded up within the purse is a long strand of spirally coiled hyphae (Fig. 224F).

(e) **Hapteron:** the free end of the funicular cord is composed of a tangled mass of adhesive hyphae.

Rain-drops, which may be as much as 4 mm in diameter and have a terminal velocity of about 6 m/sec., fall into the cup. The force creates a strong upward thrust which tears open the purse. The spirally wound funicular cord swells explosively, stretching to a length of about 2–3 mm in *C. stercoreus*, whilst in *C. striatus* it may be as long as 4–12 cm. The sticky hapteron at the base of the flunicular cord helps to attach the peridiole to surrounding vegetation and the momentum of the peridiole may cause the funicular cord to wrap around objects. The peridioles of *C. stercoreus* are presumably eaten by herbivorous animals and it is known that the basidiospores on release from the peridiole are stimulated to germinate by incubation at about body temperature, but whether animals play a significant role in the dispersal of peridioles of other birds'-nest fungi is uncertain. The funiculus of *Crucibulum* is different from that of *Cyathus*, with a longer middle piece, a very short purse and a funicular cord which is composed of relatively few hyphae only slightly coiled.

Both *Cyathus* and *Crucibulum* show tetrapolar heterothallism with relatively few alleles (generally not more than 15) at each locus and this is the most usual condition found within the Nidulariales (Burnett & Boulter, 1963).

Sphaerobolus (Figs. 223B; 225)

S. stellatus forms globose orange fruit bodies about 2 mm in diameter attached to rotten wood, rotting herbaceous stems, sacking and old dung of herbivores such as the cow and the rabbit. Ripe fruit-bodies open to form a star-like arrangement of two cups fitting inside each other attached only by the triangular tips of their teeth (Fig. 223B). Within the inner cup is a single brown peridiole or glebal mass about 1 mm in diameter. By sudden eversion of the inner cup the glebal mass is projected for a considerable distance. Buller (1933) has given a detailed account of glebal discharge. He showed that the glebal mass could be projected vertically for more than 2 m, and horizontally for over 4 m (see also Ingold, 1965). The fungus can readily be cultivated if a glebal mass is placed in a plate of oat agar and incubated for about three weeks in the light. A section through an almost mature, but unopened, fruit-body is shown in Fig. 225A. The glebal mass is surrounded by a peridium in which six distinct layers can be distinguished. Three of these layers form the structure of the outer cup. The three layers making up the inner cup consist of an outer layer of tangentially arranged interwoven hyphae, a layer of radially elongated cells forming a kind of palisade, and a thin layer

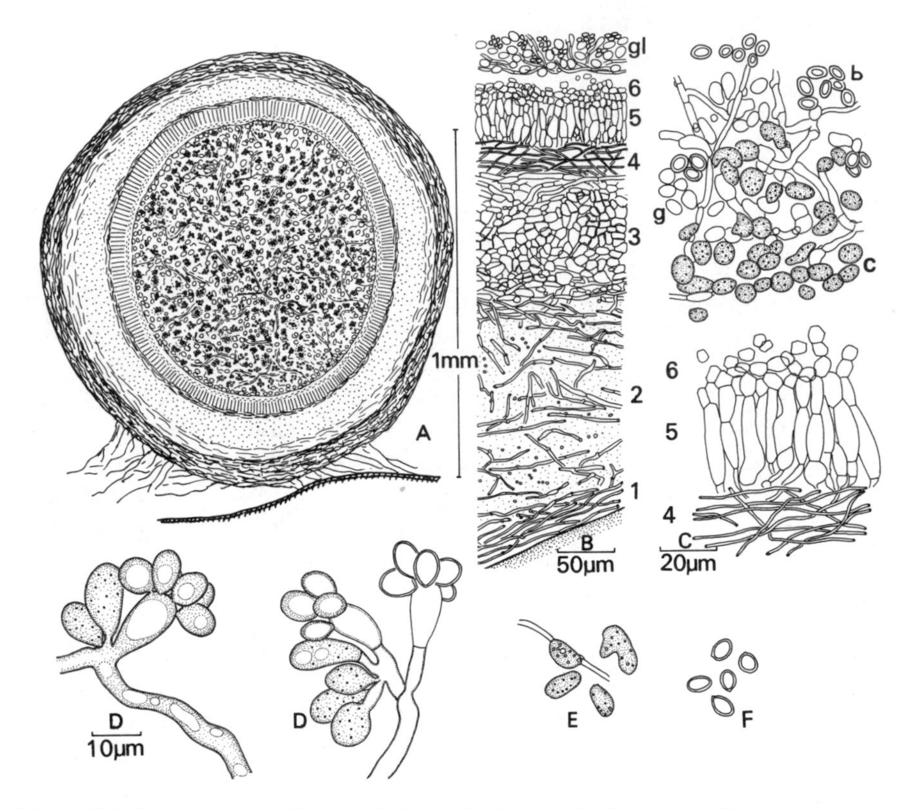

Figure 225. *Sphaerobolus stellatus*. A. V.S. nearly ripe fruit-body showing the central glebal mass surrounded by a six-layered peridium. B. Details of peridial layers: 1, outermost layer composed of interwoven hyphae; 2, layer in which the hyphae are separated by extensive mucilage; 3, pseudoparenchymatous layer; 4, fibrous layer; 5, palisade layer; 6, layer of lubricating cells; gl, outer layers of glebal mass. C. Enlarged portion of layers 4, 5 and 6 and portion of glebal mass: c, cystidia; g, gemmae; b, basidiospores. D. Clusters of basidia from unripe fruit bodies. There are usually 4–6 basidiospores. E. Gemmae. F. Basidiospores. C, E and F to same scale.

of pseudoparenchyma whose cells undergo deliquescence before glebal discharge to form a liquid which bathes the gleba and lies in the bottom of the inner cup. Before the fruit-body opens the cells of the palisade layer are rich in glycogen, but this disappears during ripening and is converted to glucose (Engel & Schneider, 1963), which causes the osmotic concentration of the cells to rise, so that they absorb water and become more turgid. The swelling of the palisade layer is restrained by the tangentially arranged hyphae, and this sets up strains within the tissues of the inner cup which are only released by its turning inside out. Light is necessary for development, and the opening of the fruit-body is phototropic, ensuring that the glebal mass is projected towards the light (Alasoadura, 1963). Peridiole discharge follows a diurnal rhythm, discharge occurring during the light. In continuous light rhythmic discharge ceases, but in continuous darkness a culture which has previously been exposed to alternate periods of 12 hours of light and darkness continues to discharge peridioles rhythmically at times corresponding to the previous light periods, indicating an endogenous rhythm (Engel & Friederichsen, 1964).

The spherical glebal mass is surrounded by a dark-brown sticky coat derived from the breakdown of the cells of the innermost peridial layer. Immediately within the brown outer coat are layers of rounded cells sometimes termed cystidia (Fig. 225C). Apparently these cells are incapable of germination and their function is not known. The rest of the glebal mass consists of thick-walled oval basidiospores and thinner walled gemmae. Four to eight basidiospores develop on the basidia about two days before discharge (Fig. 225D), but as the glebal mass ripens the basidia disappear. Gemmae arise either terminally or in an intercalary position on hyphae within the glebal mass. Fat-containing cells are also present. The sticky glebal mass adheres readily to objects on which it is impacted and after drying it is very difficult to dislodge, even by a jet of water. Glebal masses are viable for several years. Projectiles adhere to herbage and may be eaten by animals. This may explain the presence of fruit-bodies on dung.

On germination the glebal masses give rise to clamped hyphae which usually arise directly from the gemmae and not from the basidiospores. The mating system of *Sphaerobolus* is not quite clear. Clamps have been reported on mycelia derived from cells which had the appearance of spores (Walker, 1927) and occasionally monosporous mycelia will fruit (Fries, 1948). However, most basidiospores germinate to give mycelia with simple septa. Pairings of monosporous mycelia have indicated that the fungus is usually heterothallic, but the results have been interpreted as indicating a bipolar condition (Lorenz, 1933) and a tetrapolar condition (Fries, 1948). It would be of interest to see if multiple alleles exist.

SCLERODERMATALES

Scleroderma (Figs. 220D; 226)

Two common species of earth-ball are *S. aurantium* and *S. verrucosum*. Earlier, before the distinction between the two species was understood, the name *S. vulgare* was used to include both. Earth-balls are found in autumn in acid woodland and heaths growing with such trees as *Pinus*, *Betula*,

Quercus and *Fagus* with which they may form mycorrhiza (Fries, 1942, Harley, 1969). The peridium is composed of a single fairly thick layer. Although the glebal mass may be traversed by a system of sterile veins, there is no columella and no capillitium. The basidiospores are sessile (Fig. 226). When the fruit-body is ripe it cracks open irregularly and the dry spores escape. There is no well-developed bellows mechanism as in *Lycoperdon* or *Geastrum*.

PHALLALES

Common examples of this order are *Phallus impudicus*, the stinkhorn, and *Mutinus caninus* the dog's stinkhorn.

Phallus (Figs. 227B; 228)

In late summer and autumn stinkhorns can be detected readily by their smell. The fruit-bodies arise from 'eggs' about 5 cm in diameter which in turn

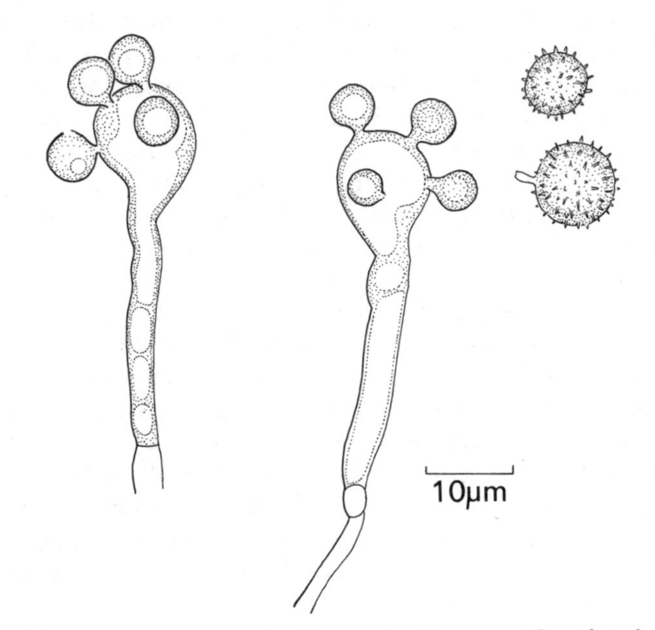

Figure 226. *Scleroderma verrucosum*. Basidia and basidiospores. Note that the spores are almost sessile.

develop on extensive white mycelial cords or rhizomorphs which can usually be traced underground to a buried stump (Grainger, 1962). A longitudinal section of an 'egg' (Fig. 228A) shows a thin papery outer and inner peridium and a wider mass of jelly making up the middle peridium. The central part of the fruit-body is differentiated into a cylindrical hollow stipe and a folded honeycomb-like receptacle which bears the fertile part of the gleba. Within the young gleba are cavities lined by basidia, bearing up to nine spores (Fig. 228B), but as the glebal mass ripens the basidia disintegrate. Fruit-bodies expand very rapidly: within a few hours the stipe may elongate from about

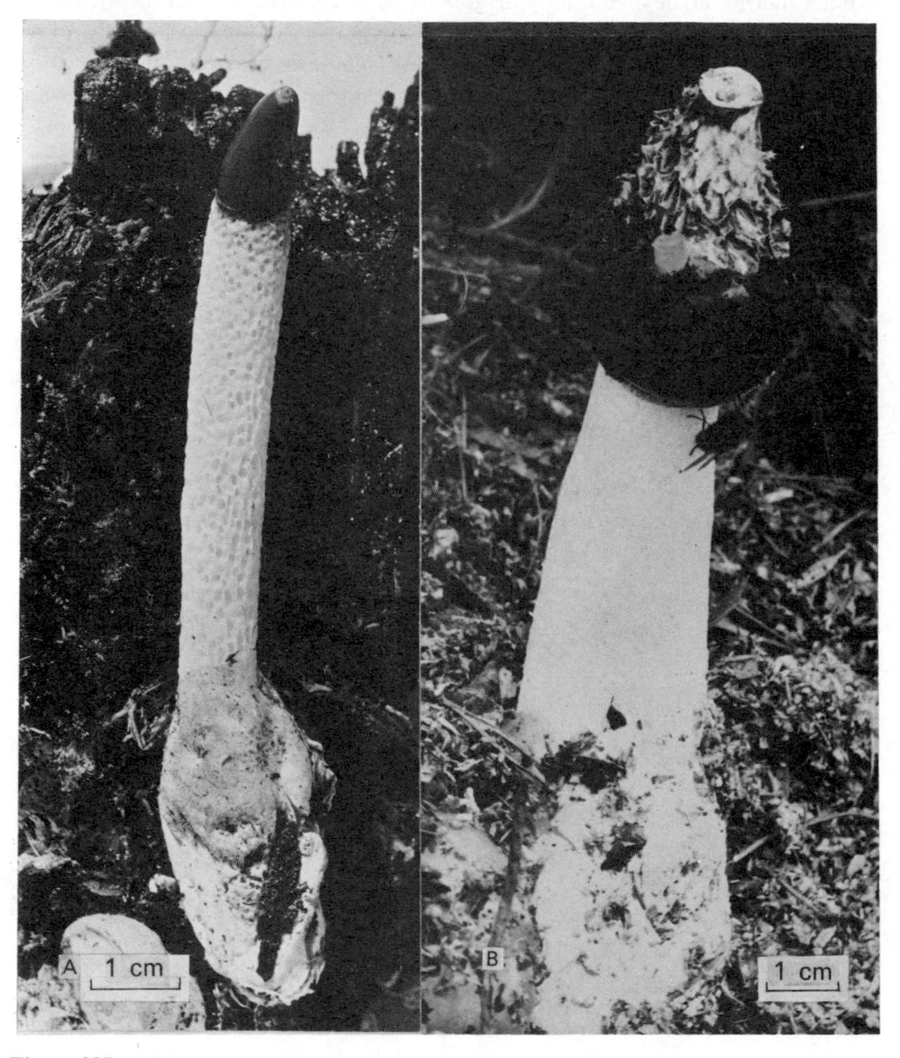

Figure 227. A. *Mutinus caninus*. Expanded fruit body attached to rotting wood. B. *Phallus impudicus*. Ripe fruit body with blue-bottle feeding on the gluten.

5 cm to a length of about 15 cm. This sudden expansion is probably at the expense of water stored within the jelly in the middle peridium. The mean weight of expanded stipes is more than twice that of unexpanded ones (Ingold, 1959). Expansion of the stipe is accompanied by breakdown of glycogen and its conversion to sugar (Buller, 1933). A similar conversion has been reported

in the related fungus *Dictyophora indusiata*. In the unexpanded stipe the cells are folded, but expand to almost 12 times their original volume as the stipe elongates (Kinugawa, 1965). About the same time as the stipe of *Phallus* is elongating the fertile glebal mass begins to secrete a strong-smelling substance together with sugar, both of which are attractive to flies, especially blue-bottles and other insects which normally feed on carrion and on dung (Schremmer, 1963). Within a few hours the green spore mass is removed.

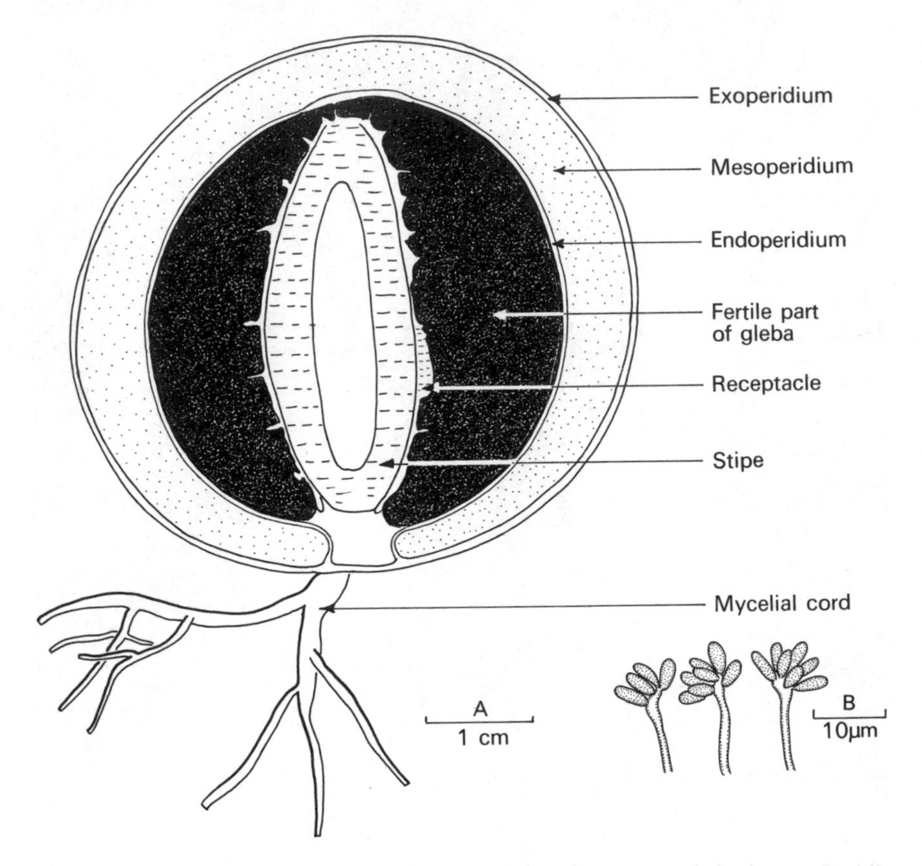

Figure 228. *Phallus impudicus*. A. L.S. 'egg' showing the unexpanded stipe. B. Basidia.

The spores are defaecated, apparently unharmed, on to surrounding vegetation and the soil, often within a short time of ingestion, but the details of their germination are unknown. How the mycelium succeeds in colonising tree stumps is not understood.

The general form of *Mutinus* is similar to *Phallus*, but the fruit-bodies are smaller (Fig. 227A). The upper part of the stipe is orange in colour, and the smell is less obvious. The receptacle bearing the glebal mass is not reticulate.

3: HEMIBASIDIOMYCETES

Two important groups of plant pathogens are included here: the Ustilaginales (or smut fungi) and the Uredinales (or rust fungi). It is doubtful if they are closely related. In contrast with most other Basidiomycetes the basidia do not arise on well-developed sporophores or even from a diffuse mycelium. Instead they arise on germination of a thick-walled spore termed a chlamydospore (Ustilaginales), or a teliospore (Uredinales). The basidium (sometimes termed the promycelium) is often, but not invariably, transversely septate. In the Ustilaginales the number of basidiospores which can develop from a basidium is indefinite. In the Uredinales there are normally four basidiospores per basidium. Whilst the Ustilaginales have a capacity for saprophytic growth in culture, the Uredinales have been cultivated only with difficulty in tissue-culture, or recently in synthetic media.

The Ustilaginales are confined to Angiosperms, but the Uredinales parasitise ferns, Gymnosperms and Angiosperms, and in many cases complete their life cycles on two unrelated hosts. The Uredinales have a long fossil history, and were probably parasitic on ferns in Carboniferous times.

USTILAGINALES

The Ustilaginales or smut fungi are parasitic on Angiosperms, where they often cause diseases of economic significance. Certain families of flowering plants seem particularly favourable hosts for smut fungi, notably the Cyperaceae and Gramineae (Fisher & Holton, 1957), and it is as parasites of cereals that these fungi are especially important. Ainsworth & Sampson (1950) have described 74 British species. The term smut refers to the mass of dark powdery chlamydospores (or brand spores) formed in sori on the leaves, stems or in the flowers or fruits of the host plant (see Fig. 229). Young chlamydospores are dikaryotic and arise on a dikaryotic mycelium which is often systemic. Usually there are no specialised haustoria, but short branches of the mycelium may enter host cells. The two nuclei in the chlamydospore fuse, and at maturity the spore contains a single diploid nucleus. The chlamydospores are commonly dispersed by wind and germinate by a variety of methods. Those of *Ustilago avenae*, the cause of loose smut of oats and false-oat grass (*Arrhenatherum elatius*) produce a germ tube or promycelium into which the diploid nucleus passes and undergoes meiosis. Thereafter the promycelium develops three transverse septa to cut off four cells each containing a single haploid nucleus. Within each cell the nucleus divides, and a daughter nucleus passes into a spore budded off from the cell. Nuclear division may be repeated and each cell may form numerous spores. Detached spores may also form further spores by budding (Fig. 230A). In this yeast-like phase many smuts are capable of prolonged saprophytic growth, but parasitic growth by the haploid phase does not usually occur (Hüttig, 1931; Holton, 1936). The four-celled promycelium is regarded as equivalent to a basidium, and the spores (often referred to as sporidia) as basidiospores. *Ustilago avenae* like most other smuts is heterothallic, and is bipolar, i.e. the basidiospores are of two mating

Figure 229. Symptoms of smut diseases. A. *Ustilago violacea* anther smut of Caryophyllaceae on *Silene alba*. B. *Ustilago hypodytes*, stem smut of grasses on *Elymus arenarius*. C. *Ustilago nuda*, loose smut of barley. D. *Ustilago avenae*, loose smut of oats on *Arrhenatherum elatius*. E. *Tilletia caries*, bunt or stinking smut of wheat. F. *Urocystis anemones*, anemone smut on *Ranunculus repens*.

types + and − (Whitehouse, 1951). Basidiospores of opposite mating type fuse by means of a short conjugation tube (Fig. 230C) to initiate a dikaryotic mycelium, which is capable of infecting a new host. Haploid basidiospores or mycelia are incapable of infection. In some promycelia the dikaryon may be initiated by fusion of adjacent cells of the promycelium (Fig. 230B).

In *Ustilago maydis*, maize smut, sporidial fusion and the formation of a dikaryon usually occur as in *U. avenae*, but it has also been shown that mycelia derived from single sporidia can bring about infection. Such infections give rise to chlamydospores which form sporidia of two mating types

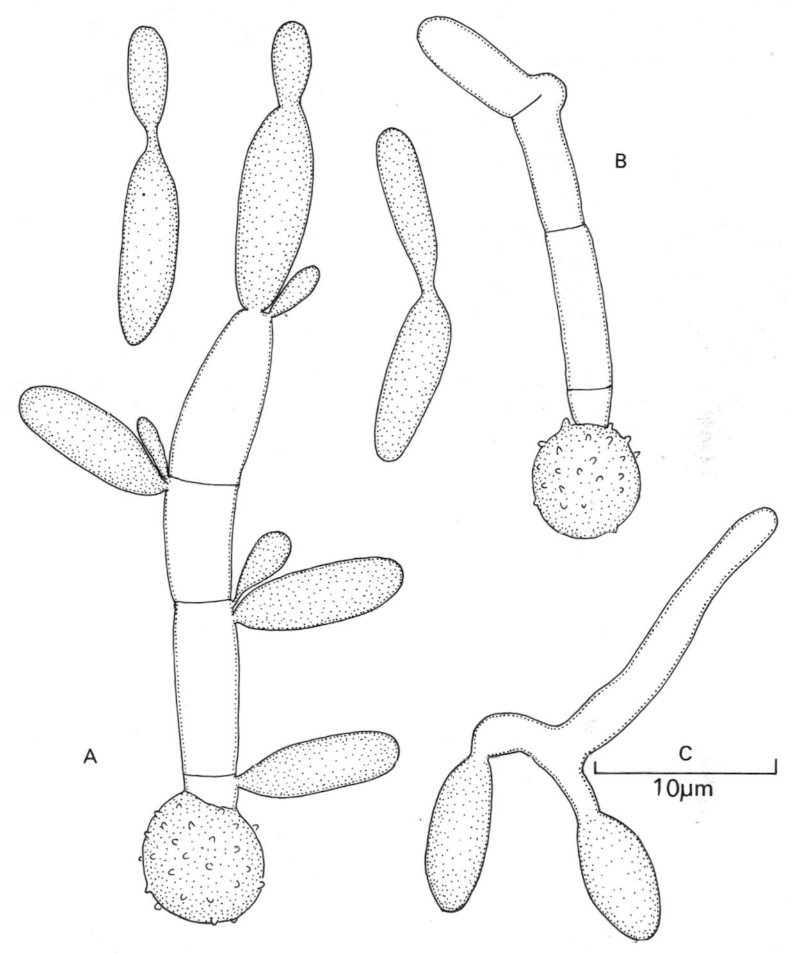

Figure 230. *Ustilago avenae*. A. Germinating chlamydospore showing the four-celled promycelium (basidium) each cell of which is producing sporidia (basidiospores). Budding of detached basidiospores is also shown. B. Germinating chlamydospore showing fusion of the two terminal cells to initiate a dicaryon. C. Fusion of germ tubes from two basidiospores to initiate a dicaryon.

and it is therefore possible that the sporidia from which these so-called solopathogenic lines developed were not homokaryotic: possibly they were dikaryotic or diploid (Christensen, 1932; Whitehouse, 1951; Fischer & Holton, 1957). A more complicated type of heterothallism occurs in this species. Incompatibility may be controlled by two loci, and there is also evidence suggesting the existence of multiple alleles at one of the loci. It is

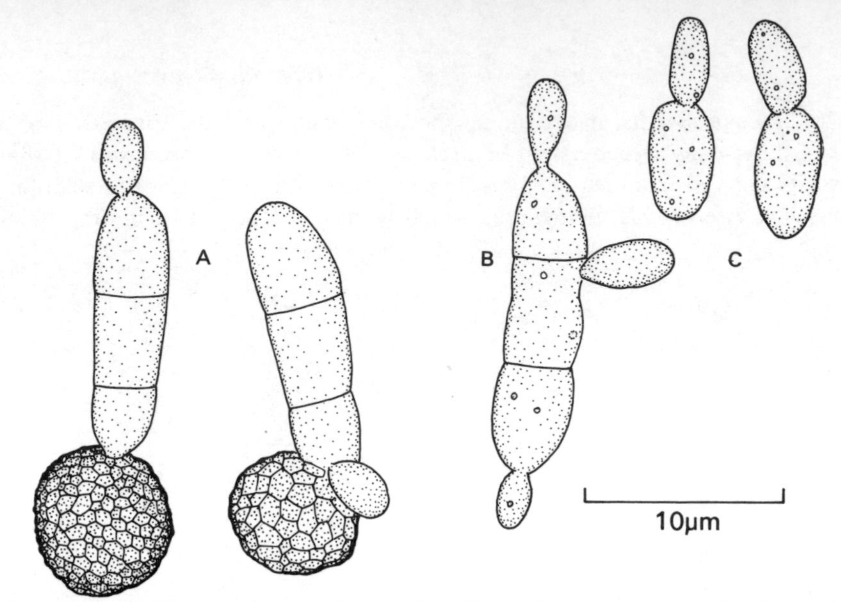

Figure 231. *Ustilago violacea*. A. Germinating chlamydospores showing the three-celled promycelium. B. Detached promycelium producing basidiospores. C. Detached basidiospores producing buds.

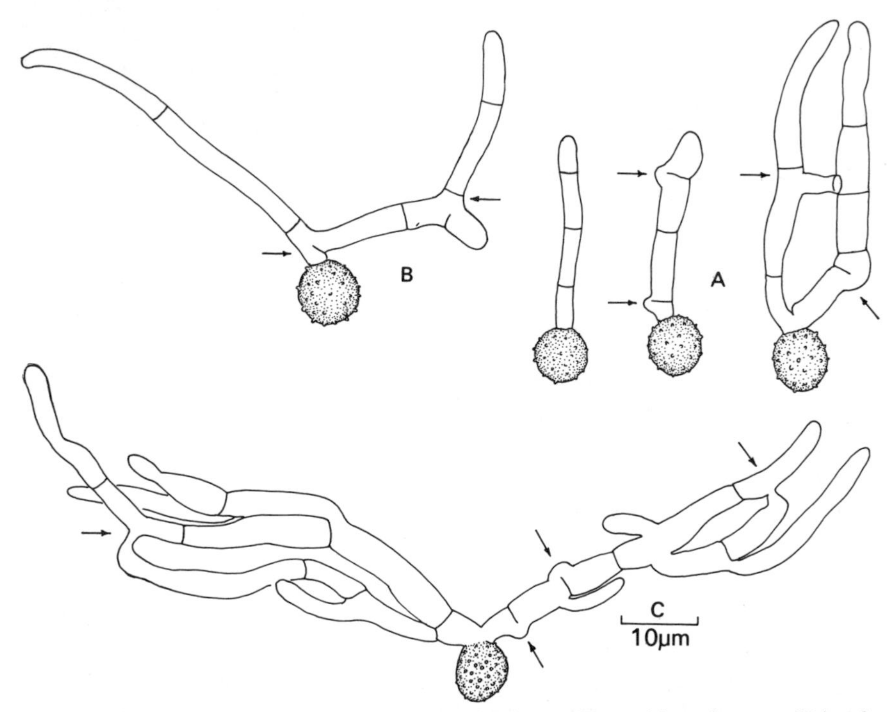

Figure 232. *Ustilago nuda*. Chlamydospore germination. A. Three chlamydospores 20 hr after germination showing various stages of development of the promycelium. The arrows indicate points where cell fusion has occurred. B. A later stage showing the extension of mycelium from the fusion cells. C. Two-day-old germinating chlamydospore showing repeated cell fusions.

not at present possible to determine whether the second locus is concerned with sexual compatibility or with pathogenicity of the dikaryon (Holliday, 1961*a*; Halisky, 1965; Christensen, 1963). A further feature of interest in this fungus is the demonstration of parasexual recombination (Holliday, 1961*b*).

The chlamydospores of *U. violacea* the cause of anther smut of Caryophyllaceae (Fig. 229A) germinate differently. The promycelium is here often

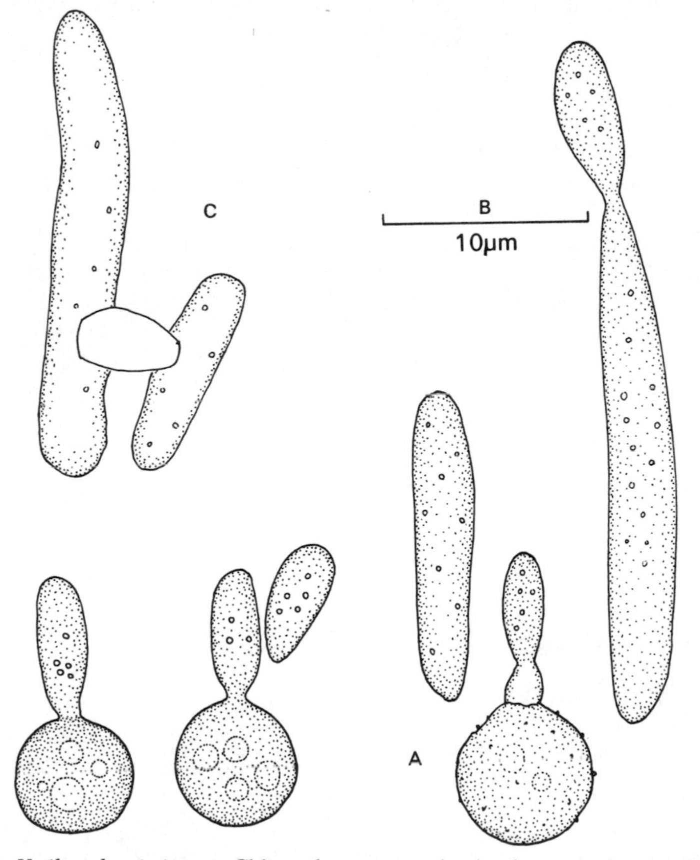

Figure 233. *Ustilago longissima*. A. Chlamydospore germination by successive development of sporidia. Note the absence of an extended promycelium. The first-formed sporidia are short. B. Sporidium showing budding. C. Sporidia conjugating. A dikaryotic mycelium arises following conjugation.

only three-celled, and is readily detached from the chlamydospore and may continue to develop basidiospores after separation (Fig. 231). A succession of promycelia may be formed from one chlamydospore (Wang, 1932). The bipolar nature of incompatibility in smuts was first demonstrated in this fungus (Kniep, 1919).

In *Ustilago nuda* the cause of loose smut of wheat and barley (Fig. 229C) there are no basidiospores. A promycelium develops on germination and becomes septate, but the dikaryophase is established by fusion of germ tubes derived from the individual uninucleate cells of the promycelium (Fig. 232) (Paravicini, 1917; Hüttig, 1931). Possibly this kind of development is connected with the biology of infection (see below).

Ustilago longissima the cause of leaf stripe of *Glyceria* spp. has chlamydospores which, on germination, do not produce an obvious promycelium, but merely a short tube from which sporidia are budded off successively (Paravicini, 1917; Wang, 1934) (Fig. 233).

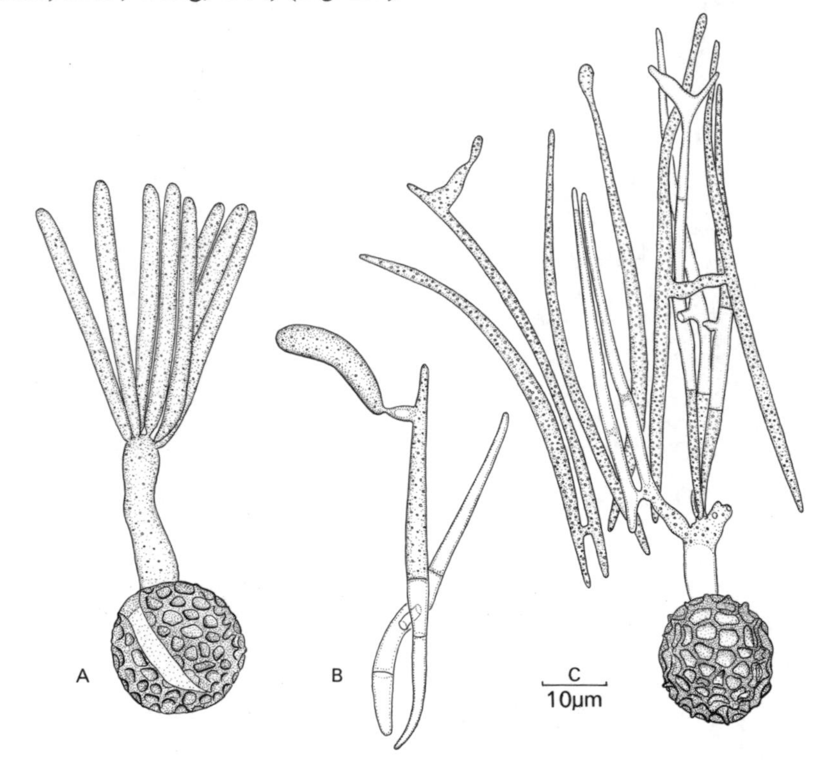

Figure 234. *Tilletia caries*. A. Germinating chlamydospore showing a non-septate promycelium and a crown of primary sporidia. B. Two detached primary sporidia showing conjugation. A secondary sporidium has developed from one of the primary sporidia. C. Primary sporidia attached to the promycelium showing conjugation.

Spore germination in *Tilletia caries*, the cause of bunt or stinking smut of wheat (Fig. 229E), follows yet another pattern. The young spores are again binucleate, and the two nuclei fuse to form the single diploid nucleus of the mature spore. The fusion nucleus divides meiotically and one or more mitotic divisions follow so that 8–16 nuclei are formed. The promycelium is often, but not invariably, non-septate, and from its tip arise narrow curved uninucleate primary sporidia, corresponding in number with the number of nuclei in the young promycelium (Fig. 234). The primary sporidia conjugate in pairs by means of short conjugation tubes, often whilst still attached to the tip of the promycelium (Fig. 234C). Detached primary sporidia may also conjugate. During conjugation a nucleus from one primary sporidium passes into the other sporidium which therefore becomes binucleate. Each H-shaped pair of primary sporidia develops a single lateral sterigma on which a curved binucleate spore develops (Fig. 234B). This spore is projected violently from

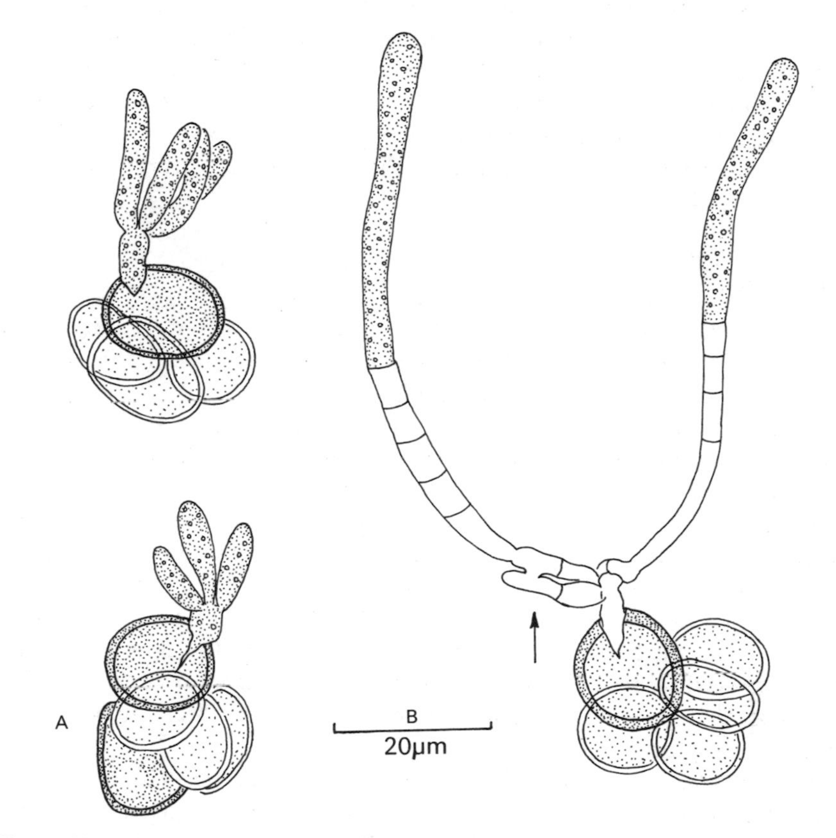

Figure 235. *Urocystis anemones*. A. Spore balls showing germination. A promycelium develops from a fertile cell, bearing a crown of three of four sporidia. 24 hr. B. Later stage of development. The sporidia have fused (arrow) and a septate mycelium develops from the fusion cell. 48 hr.

the sterigma, and a water-drop has been found associated with the spore shortly before discharge. The violently projected spore is sometimes referred to as the secondary sporidium or secondary conidium. Buller & Vanterpool (1933) have argued that because of its characteristic method of discharge this spore is to be interpreted as a basidiospore, and they use the term primary sterigma for the primary sporidia, and secondary sterigma for the structure subtending the secondary sporidium. It is these spores which bring about infection of the host.

Whilst the chlamydospores of *Ustilago* and *Tilletia* are unicellular those of *Urocystis* consist of one or more central fertile cells surrounded by a group of thin-walled sterile cells (Fig. 235). In *Urocystis agropyri*, the cause of a leaf-stripe smut of wheat and *Agropyron*, the fertile cell develops a non-septate promycelium into which the fusion nucleus passes, and then undergoes meiosis. A whorl of four sporidia develops at the tip of the promycelium, and these sporidia conjugate in pairs whilst still attached, as in *Tilletia*. After conjugation the two nuclei migrate into an infection hypha which develops from the conjugation tube (Thirumalachar & Dickson, 1949). In *U. anemones* which causes smut of *Anemone* and *Ranunculus* leaves (Fig. 229F) three or four sporidia develop at the tip of a non-septate promycelium. Fusion of the sporidia in pairs precedes the development of branches from their tips (Fig. 235B).

INFECTION

Infection of the host is by means of the dikaryotic mycelium, initiated by sporidial fusion or by fusion of branches of the promycelium. The actual route of infection varies as selected examples will show.

Loose smuts of cereals. The smutted heads of wheat or barley arise on plants infected systemically with *Ustilago nuda*. The embryos of the seeds contain mycelium of the fungus and infection occurs at flowering time. It was earlier thought that the point of entry of the mycelium was the stigma of healthy flowers, but it has been shown that the normal entry point is the young tissue at the base of the ovary (Batts, 1955). The brand-spores of *U. nuda* are short-lived, rarely surviving more than a few days under normal conditions. *Ustilago avenae*, the cause of loose smut of oats, can infect at flowering time, but in this case if spores are dusted on to healthy seed, infection of the seedling may occur as it germinates. Spores of this fungus have been shown to be viable after 13 years.

Covered smuts of cereals. The chlamydospores of *Tilletia caries* are surrounded by the pericarp of the wheat grain and are not released until the grain is threshed, when they are dusted over the surface of the grain. The brand spores are viable for up to 15 years and when grain bearing chlamydospores is sowed, the spores germinate along with the grain, and the coleoptiles of the seedlings are infected following penetration of the germ tube of the secondary sporidium. Infection is systemic, the mycelium growing in the tissues of the shoot. Although infected plants may grow less vigorously than uninfected ones they show no obvious outward sign of infection until the ears are almost ripe. The glumes of infected heads are pushed wider apart than healthy glumes by the rounded bunted grains which have a bluish appearance (Fig. 229E). Crushed bunt balls have a fishy smell caused by the presence of trimethylamine. For this reason the disease is sometimes termed 'stinking smut'.

Control of loose and covered smuts present very different problems. Whilst the surface of grain is merely *contaminated* with the spores of covered smuts, in the case of loose smuts the ripe grain is already *infected* by a mycelium within the embryo. Control of covered smuts by means of fungicidal dusts is therefore simple, and it is standard practice for seed grain to be treated by seed merchants in this way. In most countries with well-developed agriculture, bunt of wheat is now a rare disease. The incidence of bunt balls in seed samples sent to the Official Seed Testing Station at Cambridge fell from 12–33% in the period 1921–5 to 0·2–0·3% in 1955–7 (Marshall, 1960*b*). Control of loose smuts at first depended on the discovery by Jensen that the mycelium of *Ustilago nuda* could be killed by hot water treatment which the grain itself could survive. Barley or wheat grain was soaked in cold water for 5 hr, followed by a dip in hot water; 10 min. at 54°C for wheat and 15 min. at 52°C for barley (Fischer & Holton, 1957). The method, although efficacious, is obviously risky, and has the further disadvantage that the grain must be dried or sowed immediately afterwards. Recently, progress has been made in chemical treatment of loose smut mycelium in barley (Wagner, 1963; Edgington & Reinbergs, 1966).

Since infection of next season's grain occurs at flowering one obvious method of control is to inspect crops grown for seed at flowering time and assess the incidence of smutted heads. Only crops which contain fewer than one smutted ear in 10,000 ears are approved for use as seed stocks (Doling, 1966). It is also possible to detect the presence of loose smut mycelium within the embryos by microscopic examination following embryo extraction, and in this way the health of seed samples can be checked (Popp, 1958; Morton, 1961). A further method of control is to select for cleistogamous flowers. If the flowers do not open, the possibility of embryo infection is reduced since chlamydospores cannot penetrate the cereal flower (Macer, 1959). As with rust fungi the control of smuts by breeding resistant host varieties is complicated by the existence of several physiological races of the pathogens (Johnson, 1960; Halisky, 1965).

Anther smut of Caryophyllaceae. One of the most curious smut diseases is caused by *Ustilago violacea* which forms purple brand-spores in the anthers of certain Caryophyllaceae. When female plants of the dioecious red or white campion (*Silene dioica* and *S. alba*) are infected, the flowers are stimulated to produce anthers, which are absent from healthy female plants. The anthers of the infected females become filled with chlamydospores and the ovaries may be poorly developed. Male plants of these hosts are also infected. This phenomenon of induced hermaphroditism has aroused much interest (for references see Fischer & Holton, 1957). It has been suggested that sex-expression depends upon local concentrations of auxin or auxin-inhibitors (Heslop-Harrison, 1957; Baker, 1947), but no evidence has yet been presented to show that hormone or hormone-oxidase levels of flower tissues are significantly affected by infection (Garay & Sagi, 1960). Attempts to extract sex hormones from culture filtrates of the fungus have likewise proved negative (Erlenmeyer & Geiger-Huber, 1935). Although the chlamydospores replace the pollen in infected plants there is no evidence that the disease is seed-borne. When chlamydospores are transferred to healthy flowers, mycelium extends beyond the flower into other host tissues and may become systemic. Later-formed flowers are often smutted. Infection of young seedlings, underground shoots and axillary buds can also result in a systemic perennial infection (for references see Ainsworth and Sampson, 1950; Baker, 1947).

UREDINALES

The popular name for the Uredinales is the 'rust fungi', which relates to the reddish-brown colour of some of the spores. Rusts are ecologically obligate parasites of Angiosperms, Gymnosperms and Pteridophytes, and until recently only limited success has resulted from attempts to grow them in pure culture or in tissue culture (for references see Staples & Wynn, 1965; Brian, 1967). It was therefore surprising when it was shown that *Puccinia graminis* could be grown on Czapek-Dox, yeast extract agar (Williams *et al.*, 1966).

There are numerous rusts of wild plants (Wilson & Henderson, 1966; Gaümann, 1959; Cummins, 1959). Some species attack cultivated crops, causing diseases of considerable economic importance, especially of cereals. Many rust fungi have complex life cycles involving five distinct types of spore, although some show modifications in which one or more of these stages may be absent. In naming the various kinds of spore pustule in rusts the nomenclature proposed by Laundon (1967) has been followed, and commonly used alternatives have been indicated in brackets. Another distinctive feature is that many rusts complete their life cycle on two different host plants which are usually quite unrelated to each other. For example *Puccinia graminis* attacks grasses and cereals, but also grows on *Berberis* spp. Rusts with two different host plants are *heteroecious*. Others complete the entire life cycle on a single host, e.g. *P. menthae* mint rust, and are described as *autoecious*.

Puccinia

Puccinia graminis is the cause of black stem-rust of cereals and grasses. In Britain it is not a severe parasite, possibly because the summer temperatures are not high. Occasional epidemics of the disease on cereals in the south and west of Britain probably originate from spores which have blown across the sea from Spain and Portugal, and there is no strong evidence that the fungus overwinters on cereals in Britain (Ogilvie & Thorpe, 1966). The most common cereal host is wheat; occasionally oats, barley and rye. Grass hosts include *Agropyron*, *Agrostis*, and *Dactylis*. A symptom of infection on wheat leaves (Fig. 236B) is the appearance of brick-red pustules, uredia (uredosori), between the veins. These pustules contain stalked one-celled spores or urediospores which burst through the host epidermis (Fig. 237A,B). The urediospores are dikaryotic and arise from a dikaryotic mycelium which is intercellular, forming spherical intracellular haustoria. Electron migrographs of haustoria (Ehrlich & Ehrlich, 1963) show that they are not surrounded by the host plasma membrane, nor do they lie freely in the host cytoplasm. They are surrounded by a distinct encapsulation, possibly consisting of metabolic products either of fungal or of host origin. Small vesicles are also found in the host cytoplasm surrounding haustoria. The haustoria are often closely appressed to the host cell nucleus, which often breaks down (Manocha &

Figure 236. *Puccinia graminis*. A. Wheat straw showing telia as black raised pustules. B. Wheat leaf showing uredia which appear as reddish-brown powdery masses.

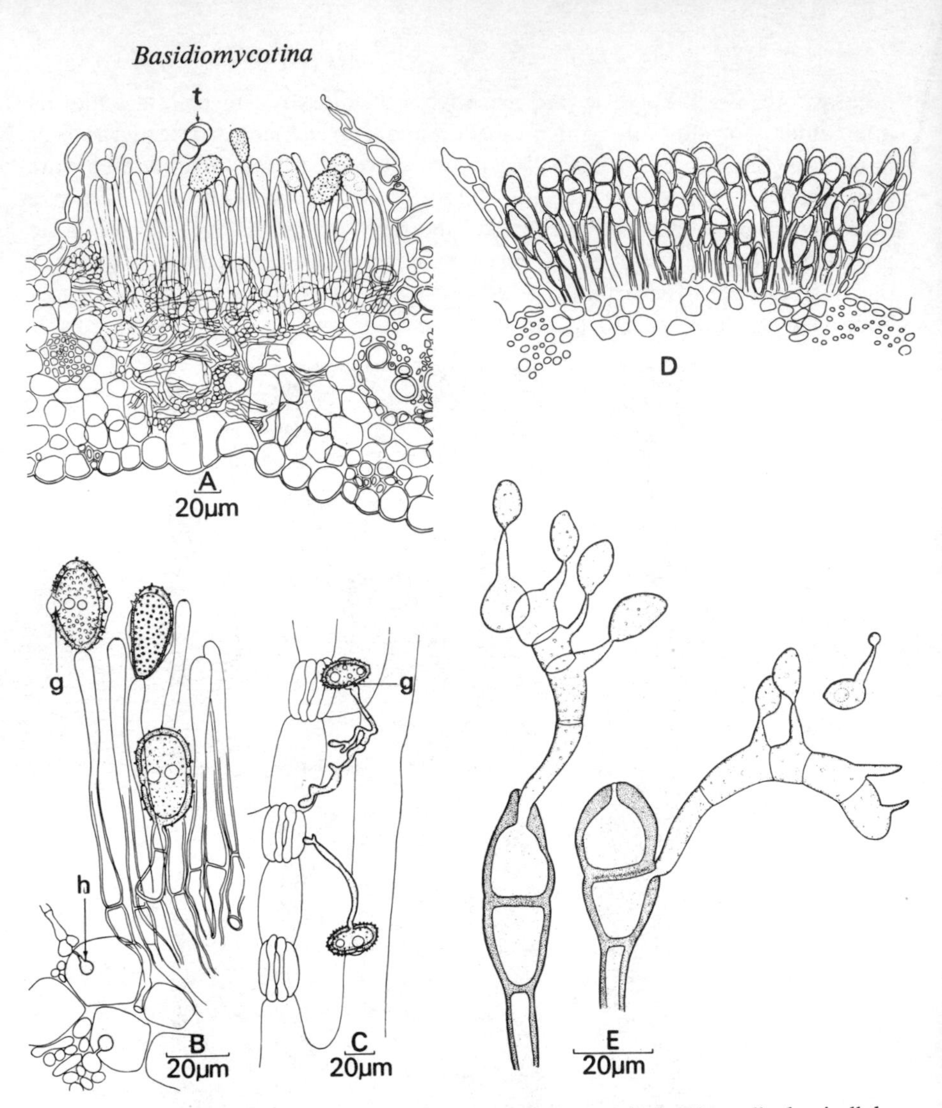

Figure 237. *Puccinia graminis*. A. T.S. wheat leaf through a uredium. The stalked unicellular urediospores are protruding through the ruptured host epidermis. A bicellular teliospore (*t*) is also present. B. Higher-power detail of urediospores. Note the germ pores (*g*) and the haustoria (*h*) in the host cells. C. Germination of two urediospores on wheat leaf. Note the directional growth of the germ tubes towards a stoma. D. T.S. leaf sheath through a telium. The stalked teliospores are projecting through the ruptured epidermis. Drawing to same scale as A. E. Germination of teliospores to form basidia bearing sterigmata and basidiospores. One basidiospore is giving rise to a secondary spore.

Shaw, 1966). In contrast with many Basidiomycetes the septa in the intercellular mycelium contain simple pores. The urediospores have a spiny wall. Near the middle of the spore the wall has four thinner areas or germ pores. The urediospores are detached by wind and blown to fresh wheat leaves upon which they germinate by extruding a germ tube from one of the germ pores

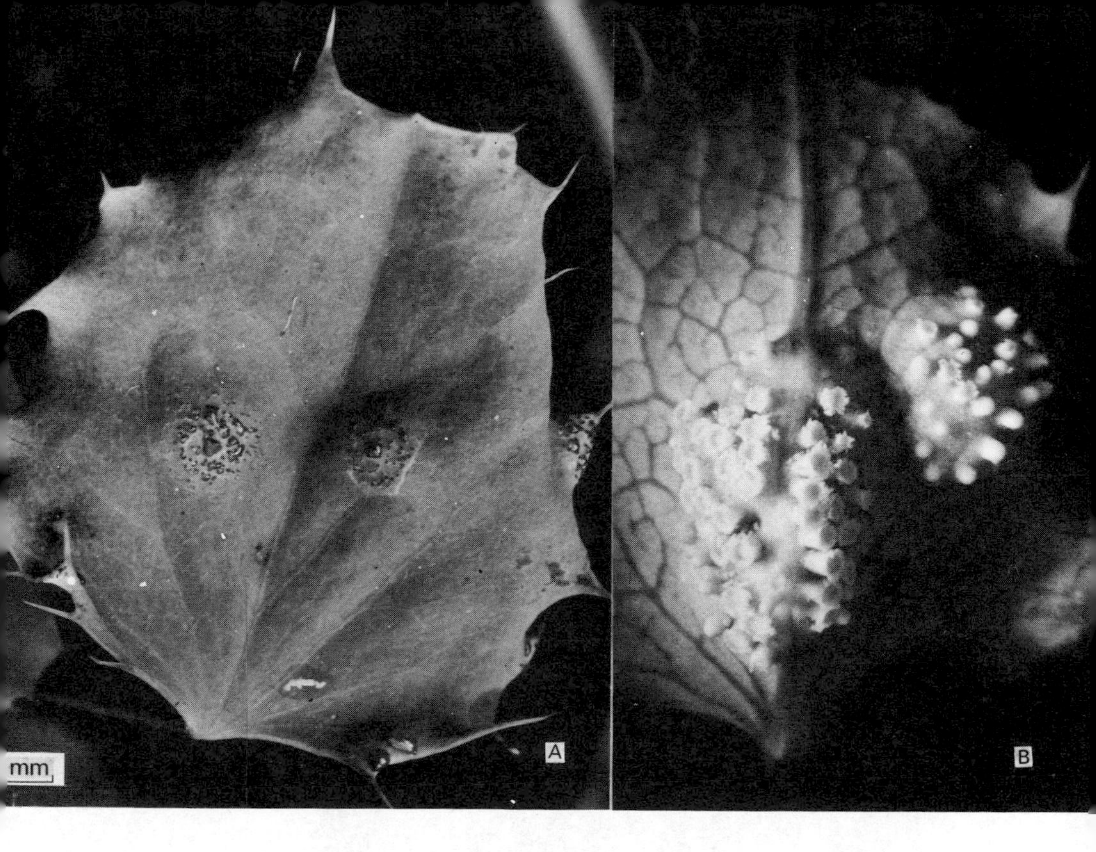

Figure 238. *Puccinia graminis*. A. Pycnial pustules on the upper surface of a leaf of *Berberis vulgaris*. Note the drops of nectar. B. Aecia on the lower side of a *Berberis* leaf. The outer frilly layer is the white peridium, within which is a mass of orange-coloured aeciospores.

(Fig. 237C). The germ tube usually penetrates the leaf through a stoma, and beneath the stoma the tip of the germ tube expands to form a sub-stomatal vesicle from which branches develop to give rise to intercellular mycelium and haustoria. Within about 7–21 days of infection a new crop of urediospores is formed and thus can cause a rapid build-up of infection within a crop. A single uredium may contain from 50,000 to 400,000 spores and there may be four to five generations of urediospores in the growing period of a wheat crop. Later in the season a second kind of spore may be visible along with the urediospores. These spores are thicker-walled and two-celled (Fig. 237A,D). They are called teliospores (teleutospores). Eventually pustules are formed containing teliospores only. These pustules are termed telia (teleutosori) and appear as black raised streaks along leaf-sheaths and stems of infected plants (Fig. 236A). The individual cells of the teliospores each contain a pair of nuclei, and during the later stages of development these two nuclei fuse to form a diploid nucleus. The teliospores represent the overwintering stage and only develop further after a period of maturation corresponding to winter dormancy. They survive the winter on stubble and the following spring (April or May) they germinate. Each cell of the teliospore emits a curved four-celled basidium, and each cell of the basidium bears a single basidiospore on a sterigma (Fig. 237E). Prior to the development of the basidium the diploid nuclei of the teliospore divide meiotically to give four haploid nuclei, one of which enters each basidiospore. The basidiospores are projected from the basidium, but are incapable of infecting wheat. Instead, they infect young leaves of the alternate host, barberry, *Berberis vulgaris*. Infection of a *Berberis* leaf by a single basidiospore results in the formation of a haploid mycelium which appears as a yellowish circular pustule (Fig. 238A). On the upper surface of the leaf, penetrating the epidermis, several yellowish flask-shaped structures termed pycnia (pycnidia, spermogonia) develop. The mouth of the flask is lined by a bunch of unbranched tapering, pointed, orange-coloured hairs, the periphyses, and

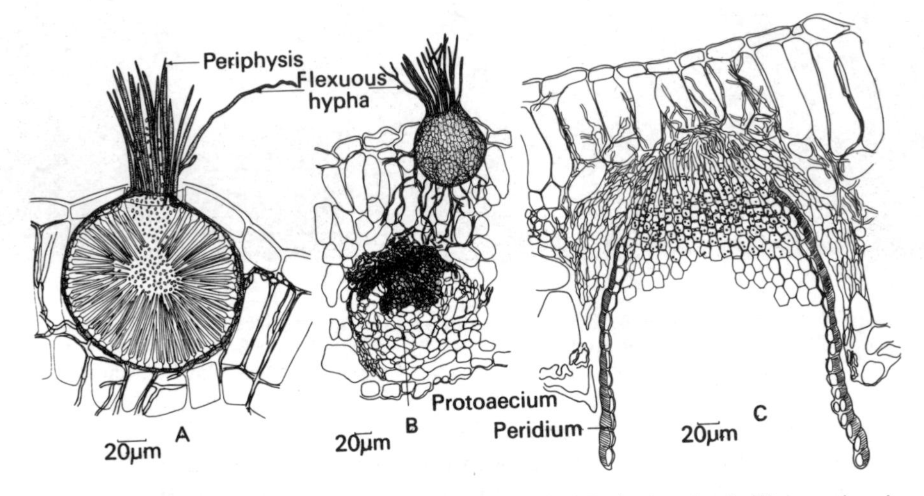

Figure 239. ***Puccinia graminis***. A. T.S. pycnium on leaf of ***Berberis vulgaris***. The pycnium is penetrating the upper epidermis. The wall of the pycnium is lined by tapering cells which give rise to pycnospores. B. T.S. leaf of *Berberis* showing a pycnium and a proto-aecium which has become dikaryotised. This is shown by the presence of binucleate cells. C. T.S. leaf of *Berberis* showing an aecium in section. The aecium has burst through the lower epidermis of the host leaf. Note the chains of cells consisting of alternate large and small cells. The large cells are the aeciospores.

among the periphyses are thinner-walled branched hyphae, the flexuous hyphae. Lining the body of the pycnium are tapering cells which give rise to minute uninucleate pycnospores (spermatia) exuded through the mouth of the pycnium and held in a sticky sweet-smelling drop of liquid by the periphyses (Figs. 238A; 239). Within the mesophyll of the barberry leaf the haploid mycelium gives rise to several spherical structures or proto-aecia (proto-aecidia). The proto-aecia are mostly made up of large-celled pseudoparenchyma, but along the upper wall is a crescent-shaped mass of smaller, denser cells.

Single haploid pustules are incapable of further development, unless cross-fertilisation occurs. The sweet-smelling nectar containing pycnospores attracts insects which feed on the nectar. Insects visiting several pustules transfer pycnospores from one to another. The haploid pustules are of two mating types (+ and −), and if a + pycnospore is brought close to a − flexuous hypha, a short germ tube is formed which anastomoses with the flexuous hypha (Craigie, 1927*a*,*b*; Buller, 1950). Nuclear transfer from the pycnospore to the flexuous hyphae then occurs, and migration and multiplication of the introduced nucleus follows (Craigie & Green, 1962). This

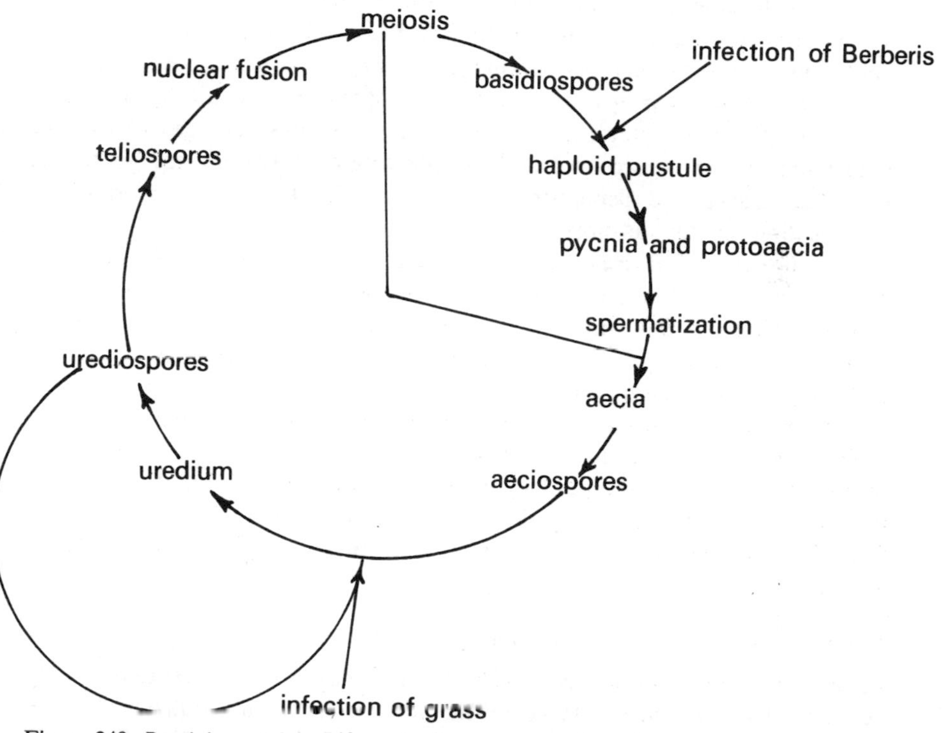

Figure 240. *Puccinia graminis*. Life cycle diagram. The life cycle can be divided into two parts, a haploid stage on *Berberis*, and a dikaryotic stage initiated by the formation of aecia on *Berberis* and continued on the grass host. The haploid part of the life cycle is included in the smaller sector, and the dikaryotic part is included in the larger sector of the circle.

results in the dikaryotisation of the original haploid pustule, and about three days after transfer of pycnospores binucleate cells become clearly visible in the deep staining tissue which forms the roof of the proto-aecium (Fig. 239B). The binucleate cells now give rise to chains of cells which are also binucleate, but composed of alternately long and short cells. The longer cells enlarge and become aeciospores but the short cells become crushed and flattened as the spore chain develops (Fig. 239C). During the development of the spore chain the large pseudoparenchymatous cells of the proto-aecium are also crushed and pushed aside. Surrounding the chains of spores is a specially differentiated layer of cells homologous with the spore chain, whose outer walls are thick and fibrous. This layer of cells forms a clearly defined border or peridium surrounding the spores. Eventually the peridium and spore chain burst through the lower epidermis. The spores are visible as orange-coloured cells enclosed by the white cup-like peridium, (Fig. 238B). The cup-like sori are termed aecia and several of them are usually clustered together beneath a pustule, so that this stage is popularly known as the cluster-cup stage. In a section through the centre of a group of young aecia it is usually possible to find the pycnia penetrating the upper epidermis and the aecia penetrating the lower. The aeciospores (aecidiospores) are violently projected from the end of the spore chain by rounding-off of the flattened interface between adjacent spores (Dodge, 1924). The dikaryotic aeciospores are incapable of infecting barberry, but can germinate on wheat leaves and penetrate them, giving rise to infections on which urediospores shortly develop.

The life cycle of *P. graminis* thus occurs on two hosts. On *Berberis* the haploid infection, following spermatisation, gives rise to dikaryotic aeciospores, which can only infect Gramineae. On the grass or cereal host the dikaryotic mycelium gives rise to two kinds of spore: urediospores, which can be thought of as dikaryotic conidia; and teliospores, which are not dispersed but are the organs within which nuclear fusion and meiosis occur (Fig. 240).

The effects of infection on cereal hosts are increased transpiration and respiration, and decreased photosynthesis, resulting in a marked deterioration in quality and quantity of grain. Control of the disease by application of fungicides, though possible, is not economic. Instead two main methods are adopted:

1. ERADICATION OF THE ALTERNATE HOST

Before the connection between *Berberis* and wheat rust was established (by de Bary, 1861–5) the incidence of the disease in wheat growing close to barberry had been noted and legislation had been introduced (in Rouen in 1660, and in Massachusetts in 1755 according to Large, 1958) for the eradication of barberry. This method is only efficacious in areas where the urediospores cannot survive the winter, but in areas with a mild winter urediospores may survive and bring about infection the following spring.

Another difficulty which renders barberry eradication ineffective is the long-distance transport of urediospores already mentioned. In the U.S.A. clouds of urediospores have been tracked from the Mississippi delta into the northern states and Canada, travelling distances of up to 2,000 miles (Craigie, 1945; Stakman & Christensen, 1946; Stakman & Harrar, 1957). Urediospores can retain their viability for several months given suitable conditions of temperature and humidity.

2. BREEDING FOR RESISTANCE

Different grasses and cereals react in different ways to a given strain of *Puccinia graminis*, and on the basis of host susceptibility it is possible to classify *Puccinia gramminis* into six varieties (Eriksson & Henning, 1894):

1. *P. g. tritici* on wheat.
2. *P. g. avenae* on oats and certain grasses.
3. *P. g. secalis* on rye and certain grasses.
4. *P. g. agrostidis* on *Agrostis*.
5. *P. g. poae* on *Poa*.
6. *P. g. airae* on *Deschampsia* (*Aira*) *caespitosa*.

Slight but significant differences in dimensions of urediospores and teliospores from these different hosts occur. For example the urediospores of var. *tritici* are about 30×18 μm whilst those of var. *poae* are about 21×14 μm (Batts, 1951). Within a single variety of the rust fungus further specialisation is found. If the spores of a strain of *P. graminis tritici* are inoculated on to a range of wheat varieties these hosts will differ in their response. Some may prove to be resistant, others highly susceptible, whilst yet others may be intermediate in their reaction. Spores from a second race may give an entirely different pattern of response. Using the reactions of differential wheat varieties it has proved possible to classify strains of *P. graminis tritici* into over 300 physiological races. The existence of this large range of rust races complicates the task of the plant breeder, but fortunately not all these races are prevalent in an area at any one time so that the task of breeding resistant varieties is practicable and profitable (Johnson, 1953, 1961). The frequency of different rust races varies over the years reflecting changes in the wheat varieties planted (see Fig. 241). A new race may suddenly build up in frequency as shown by the appearance of Race 15 (actually a sub-race or biotype referred to as Race 15B) in 1950 in Canada. The appearance of new races presents problems to plant breeders, and it is clear that the task of breeding resistant varieties is a continuing one. The origins of the large number of rust races are:

(a) *Recombination during sexual reproduction.* New races of rust are often found adjacent to barberry bushes; and following inoculation with a single race a number of variants may be found among the aeciospore progeny.

(b) *A mechanical mixing of nuclei of different races* may occur if urediospores from the different strains germinate sufficiently close together on a susceptible host for anastomosis of germ tubes or hyphae to occur. Nuclei of different origin may then come together in a common cytoplasm and may be associated in new urediospores. It has been found that races with distinctive pathogenicity may arise in this way (for references see Ellingboe, 1961).

(c) *Parasexual recombination.* When Race 11 and Race 121 were inoculated simultaneously on to a susceptible wheat variety at least 15 different races were identified in the urediospore progeny, which is more than would be expected from simple nuclear re-assortment (Bridgmon, 1959). One possible explanation is that parasexual recombination has occurred, but the possibili-

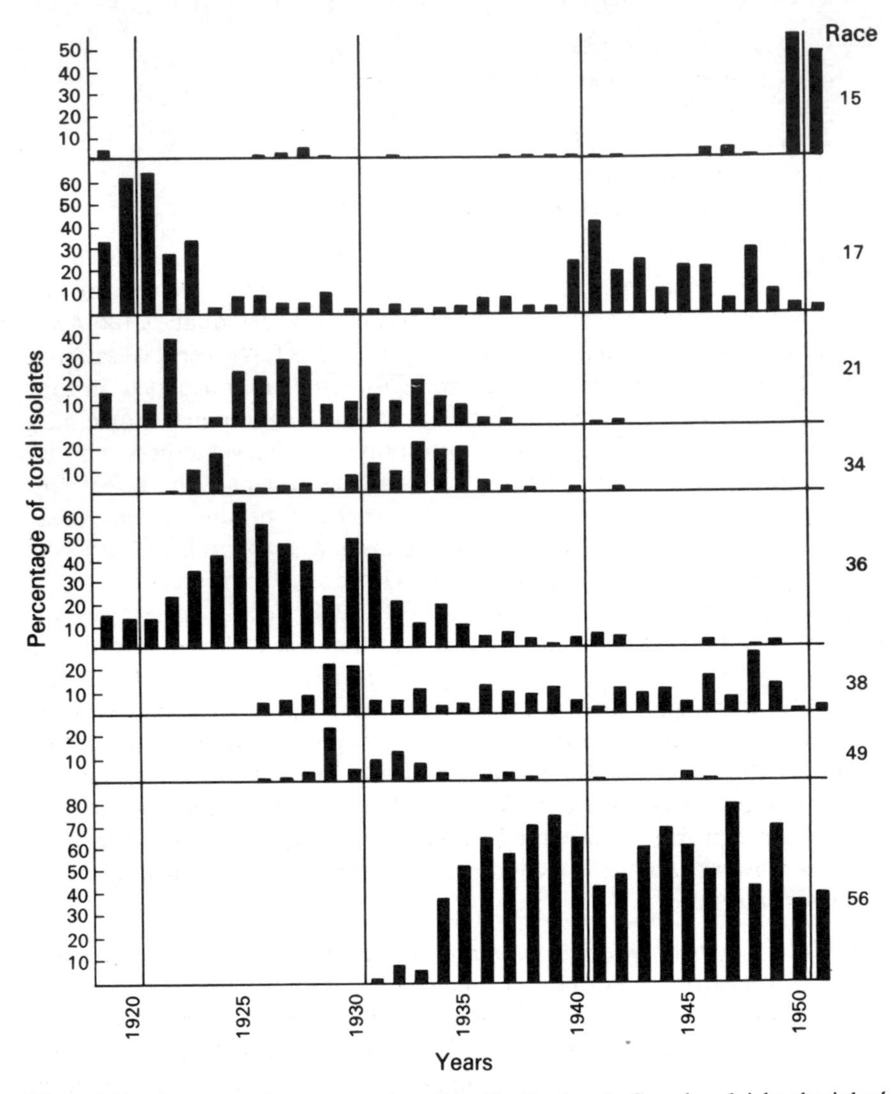

Figure 241. Diagrammatic representation of the distribution, in Canada, of eight physiological races of wheat-stem rust during the period 1919–51 (after Johnson, 1953).

ties of gene suppression in the parental strains or of cytoplasmic effects cannot be excluded (Watson, 1957).

(d) *Mutation.* Several examples are known where mutation has occurred giving rise to progeny with increased virulence (Stakman *et al.*, 1930).

When the urediospores of *P. graminis* germinate on an unsuitable host a variety of reactions occur. In some cases the host cells die immediately after infection, and the only externally visible signs of infection are small flecks, which are islands of dead tissue. Because the fungus can only grow on living cells this reaction limits the extension of the infection. The phenomenon is termed *hypersensitivity*. It is not confined to rusts but occurs with other obligate pathogens.

As already stated *P. graminis* is not normally a troublesome pathogen in Britain. The most harmful rust on wheat in Britain is *P. striiformis* (= *P. glumarum*) commonly known as yellow rust because of its bright yellow uredia. No aecial host has yet been discovered. The disease can appear very early in the season and it is likely that the fungus can overwinter as mycelium on winter cereals infected in the previous season. *Puccinia recondita* f. sp. *tritici*, brown rust or leaf rust, is also common on wheat, but is relatively harmless in Britain. In America it is of considerable significance (Chester, 1946). The aecia which grow on *Thalictum* and *Isopyrum* in Europe and Asia have not been found in Britain. Other heteroecious rusts with life cycles resembling *Puccinia graminis* are *P. caricina* with uredia and telia on *Carex* and pycnia and aecia on *Urtica*; and *P. poarum* with uredia and telia on *Poa* and pycnia and aecia on *Tussilago*. This species is unusual in having two aecial generations in one year (Wilson & Henderson, 1966). *Puccinia menthae* also has all five spore stages, but is autoecious (i.e. has no alternate host) and is systemic.

Species such as *P. graminis* which have a complete life cycle are said to be *macrocyclic*, and the prefix 'eu' is sometimes attached to the generic name as a descriptive epithet, i.e. *P. graminis* is an *eu-Puccinia*. Many variants from this type of life cycle are known. One of the best-known is shown by the thistle rust *P. punctiformis* (syn. *P. suaveolens*, *P. obtegens*). This is a systemic autoecious rust attacking *Cirsium arvense*. In spring infected plants are clearly distinguishable by their yellowish appearance and appressed leaves, and by the strong sweet smell associated with numerous pycnia which develop all over the infected shoots. The infections develop from a mycelium perennating in the rootstock. Although the overwintering mycelium is dikaryotic, the mycelium from which the pycnia develop is haploid, and segregation of the two nuclei of the dikaryon presumably occurs as the new shoots are infected. Transfer of pycnospores to compatible flexuous hyphae results in dikaryotisation, but the resulting structures which develop do not resemble normal aecia, but resemble uredia, and form chocolate-brown masses on the leaves. These sori are best regarded as aecia, and have been termed uredinoid aecia: the term primary uredosorus is sometimes used, but it is best avoided (Wilson & Henderson, 1966). The spores from the aecia can infect healthy thistles and normal uredia develop on these infections. Later, teliospores develop (Buller, 1950; Menzies, 1953). Rusts with uredinoid aecia are brachycyclic (i.e. *P. punctiformis* is a *brachy-Puccinia*). Infected shoots of thistles have a higher endogenous gibberellin content than healthy plants (Bailiss & Wilson, 1967).

Another modification of the life cycle involves the absence of urediospores. This kind of life-cycle is found in heteroecious forms in *Gymnosporangium*, and amongst autoecious forms in *Xenodocus carbonarius*. Rusts showing this kind of life cycle are said to be *demicyclic*.

Some rusts have only telia, with or without pycnia, and are said to be *microcyclic*. A common example is *Puccinia malvacearum*, hollyhock rust, which attacks Malvaceae, especially *Althaea* and *Malva*. In this rust only teliospores and basidiospores are known. The teliospores germinate readily, and a single basidiospore can form an infection on which telia arise, i.e. this species is homothallic. The basidiospores give rise to a haploid mycelium with uninucleate segments, and the dikaryotic condition arises shortly before the development of teliospores, either by cell fusions or by cell division not accompanied by the formation of septa (Ashworth, 1931; Buller, 1950). Buller (1950) has argued that all rusts lacking pycnia should prove to be homothallic.

There are other variations in the life cycles of rusts. It is generally believed that the heteroecious macrocyclic forms represent the primitive condition, and that the autoecious condition arose later in evolution. It is also believed that forms with shorter life cycles arose during the course of evolution from macrocyclic ancestors. For a discussion of the evidence for these beliefs see Buller, (1950) and Jackson, (1931).

OTHER COMMON RUST GENERA

Wilson & Henderson (1966) have classified British rusts into three families.

1. Pucciniaceae

In this family the teliospores are stalked. In *Puccinia* the teliospores are two-celled, whilst in *Uromyces* they are one-celled (Fig. 242A). The separation of these two genera is, however, probably unjustified, and there is evidence that they are very closely related. Common species of *Uromyces* are *U. ficariae*, a microcyclic species forming brown telia on *Ranunculus ficaria*, *U. muscari* on *Endymion non-scriptus*, and *U. dactylidis* with teliospores and urediospores on *Dactylis*, *Festuca* and *Poa*, and aecia on *Ranunculus* spp. *Triphragmium* has three-celled stalked teliospores (Fig. 242B). The commonest species is *T. ulmariae*, a brachy-form with large bright orange uredinoid aecia on *Filipendula ulmaria*.

In *Phragmidium* the teliospore has several cells (Fig. 242C). *Phragmidium violaceum* is easily recognisable on leaves of *Rubus fruticosus* agg. by the violet leaf-spots it causes. All species of *Phragmidium* are autoecious and confined to Rosaceae.

Xenodocus has teliospores consisting of long chains of up to about 20 dark cells (Figs. 242D). No uredia occur. The only British species is *X. carbonarius* which forms bright orange-red aecia on *Sanguisorba officinalis*. *Gymnosporangium* forms teleutosori on *Juniperus*. On this host the mycelium is perennial and is associated with swelling of the branches. In spring the swollen

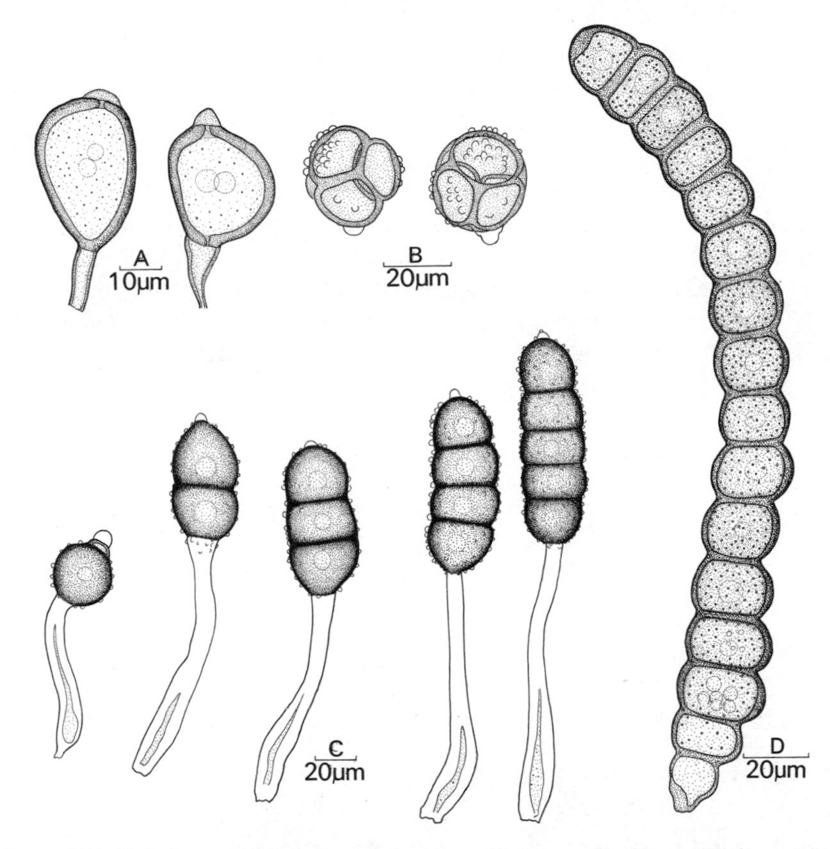

Figure 242. Teliospores of various rust fungi. A. *Uromyces ficariae*. B. *Triphragmium ulmariae*. C. *Phagmidium violaceum*. D. *Xenodocus carbonarius*.

shoots produce *Clavaria*-like projections which contain numerous thin-walled two-celled teleutospores.

In *G. clavariiforme* the basidiospores infect *Crataegus* which then bears cylindrical aecia (roestelia).

2. Coleosporiaceae

Here the teliospores are sessile and form a layer of cells often united laterally. The spores have thin side walls. Before germination the whole spore becomes divided into four cells arranged in a row, each of which produces a long sterigma bearing a single basidiospore. Although it forms teliospores on a wide range of hosts such as *Tussilago*, *Senecio* and *Melampyrum* the various forms are now regarded as belonging to a single species, *C. tussilaginis*. The pycnia and aecia grow on needles of *Pinus* spp.

3. Melampsoraceae

Here the unicellular teleutospores are also sessile and often form a subepidermal crust. Germination is by an external basidium of the usual type. The aecia lack peridia so that they are diffuse instead of cup-shaped. Such diffuse aecia are termed caeomata (sing. caeoma). Many members of the family grow on ferns and conifers. *Melampsora lini* is an autoecious rust common on *Linum catharticum*. A variety *liniperda* is parasitic on cultivated flax, *Linum usitatissimum*. Some other species of *Melampsora* are parasitic on *Salix*, whilst *M. populnea* forms urediospores and teliospores on *Populus*, and pycnia and aecia on *Mercurialis perennis*. *Melampsoridium betulinum* is abundant on leaves of *Betula* where it forms urediospores and teliospores. The alternate host is *Larix*.

References

Abbayes, H. des (1951). Traité de Lichénologie. *Encyclopédie biologique*, **41**, 217 pp. Paris: Lechevalier.

Abe, M. & Yamatodani, S. (1964). Preparation of alkaloids by saprophytic culture or ergot fungi. *Progress in Industrial Microbiology*, **5**, 203–29.

Abrams, E. (1950). Microbiological deterioration of cellulose during the first 72 hours of attack. *Text Res. J.* **20**, 71.

Adams, A. M. (1949). A convenient method of obtaining ascospores from bakers' yeast. *Can. J. Res.* C, **27**, 179–89.

Agar, D. D. & Douglas, H. C. (1955). Studies of budding and cell wall structure in yeast. Electron microscopy of thin sections. *J. Bact.* **70**, 427–34.

Agar, D. D. & Douglas, H. C. (1957). Studies on the cytological structure of yeast: Electron microscopy of thin sections. *J. Bact.* **73**, 365–75.

Agarwal, P. N., Verma, G. M., Verma, R. K. & Sahgal, D. D. (1963*a*). Decomposition of cellulose by the fungus *Chaetomium globosum*. Part I. Studies on enzyme activity. *Indian J. exp. Biol.* **1**, 46–50.

Agarwal, P. N., Verma, G. M., Verma, R. K. & Rastogi, V. K. (1963*b*). Decomposition of cellulose by the fungus *Chaetomium globosum*. Part III. Factors affecting elaboration of cellulolytic enzymes. *Indian J. exp. Biol.* **1**, 229–30.

Ahmad, M. (1965). Incompatibility in yeasts. In *Incompatibility in Fungi*, 13–23. Editors, Esser, K. & Raper, J. R. Berlin: Springer.

Ahmadjian, V. (1962). Investigations on lichen synthesis. *Am. J. Bot.* **49**, 277–83.

Ahmadjian, V. (1965) Lichens. *A. Rev. Microbiol.* **19**, 1–20.

Ahmadjian, V. (1967*a*). *The Lichen Symbiosis*. 152 pp. Waltham, Mass: Blaisdell.

Ahmadjian, V. (1967*b*). A guide to the algae occurring as lichen symbionts: Isolation, culture, cultural physiology and identification. *Phycologia*, **6**, 127–60.

Ainsworth, G. C. (1952). *Medical Mycology. An Introduction to its Problems.* 105 pp. London: Pitman.

Ainsworth, G. C. (1965). Longevity of *Schizophyllum commune*. II. *Nature, Lond.* **195**, 1120–1.

Ainsworth, G. C. (1966). A general purpose classification of fungi. *Bibliography of Systematic Mycology* (1966), 1–4. Kew, Surrey: Commonwealth Mycological Institute.

Ainsworth, G. C. & Austwick, P. K. C. (1959). *Fungal Diseases of Animals.* Commonwealth Agricultural Bureau.

Ainsworth, G. C. & Sampson, K. (1950). *The British Smut Fungi* (*Ustilaginales*). 137 pp. Kew, Surrey: Commonwealth Mycological Institute.

Alasoadura, S. O (1963) Fruiting in *Sphaerobolus* with special reference to light. *Ann. Bot. Lond.* N.S. **27**, 123–45.

Alexopoulos, C. J. (1962). *Introductory Mycology*. 2nd edition. 613 pp. New York and London: Wiley.

Alexopoulos, C. J. (1963). The Myxomycetes. II. *Bot. Rev.* **29**, 1–78.

Allen, P. J. (1942). Changes in the metabolism of wheat leaves induced by infection with powdery mildew. *Am. J. Bot.* **29**, 425–35.

Ames, L. M. (1934). Hermaphroditism involving self-sterility and cross-fertility in the Ascomycete *Pleurage anserina*. *Mycologia*, **26**, 392–414.

Ames, L. M. (1961). A monograph of the Chaetomiaceae. *U.S. Army Res. Dev. Ser.* No. 2, 125 pp.

Andrus, C. F. & Harter, L. L. (1937). Organization of the unwalled ascus in two species of *Ceratostomella*. *J. agric. Res.* **54**, 19–46.

Anon. (1967*a*). *Fomes annosus*. A fungus causing butt rot, root rot and death of conifers. Forestry Commission leaflet No. 5. 11 pp. H.M.S.O.

Anon. (1967*b*). *Silver Leaf Disease of Fruit Trees*. Advisory leaflet 246, Ministry Agric. Fisheries & Food. 7 pp. H.M.S.O.

Aoshima, K. (1953). Sexuality of *Elfingia applanata* (*Fomes applanatus*). *Mycol. J. Nagao Inst.* **3**, 5–11.

Apinis, A. E. (1964). Revision of British Gymnoascaceae. *Mycol. Pap. C.M.I.* **96**, 1–56.

Aragno, M. (1967). Formation et évolution de l'asque chez *Rhytisma acerinum* (Pers.) Fr. *Ber. Schweiz. bot. Ges.* **77**, 173–86.

Arnold, C. A. (1928). The development of the perithecium and spermagonium of *Sporormia leporina*. *Am. J. Bot.* **15**, 241–5.

Aronson, J. M. (1965). The cell wall. In *The Fungi: An Advanced Treatise*, **1**, 49–76. Editors, Ainsworth, G. C. & Sussman, A. S. New York and London: Academic Press.

Aronson, J. M., Cooper, B. A. & Fuller, M. S. (1967). Glucans of oomycete cell walls. *Science, N.Y.* **155**, 332–5.

Aronson, J. M. & Preston, R. (1960). An electron microscopic and X-ray analysis of the walls of selected lower Phycomycetes. *Proc. R. Soc.* B, **152**, 346–52.

Arx, J. A. von & Müller, E. (1954). Die Gattungen der amerosporen Pyrenomyceten. *Beitr. KryptogFlora Schweiz.* Bd. **11**, (1), 1–434.

Ashworth, D. (1931). *Puccinia malvacearum* in monosporidial culture. *Trans. Br. mycol. Soc.* **16**, 177–202.

Atkins, F. C. (1968). *Mushroom Growing Today*. 188 pp. London: Faber & Faber.

Bachmann, B. J. & Strickland, W. N. (1965). *Neurospora Bibliography and Index*. 225 pp. Yale University Press.

Backus, M. P. (1939). The mechanics of conidial fertilisation in *Neurospora sitophila*. *Bull. Torrey bot. Club*, **66**, 63–76.

Bagchee, K. (1954). *Merulius lacrymans*. (Wulf.) Fr. in India. *Sydowia*, **8**, 80–5.

Bailiss, K. W. & Wilson, I. M. (1967). Growth hormones and the creeping thistle rust. *Ann. Bot. Lond.* N.S. **31**, 195–211.

Baker, H. G. (1947). Infection of species of *Melandrium* by *Ustilago violacea* and the transmission of the resultant disease. *Ann. Bot. Lond.* N.S. **11**, 333–48.

Bakshi, B. K. (1951). Development of perithecia and reproductive structures in two species of *Ceratocystis*. *Ann. Bot. Lond.* N.S. **15**, 53–61.

Banbury, G. H. (1955). Physiological studies in the Mucorales. II. The zygotropism of zygophores of *Mucor mucedo* Brefeld. *J. exp. Bot.* **6**, 235–44.

Banbury, G. H. (1959). Phototropism of lower plants. In *Encyclopedia of Plant Physiology*. 17 (1) 530–78. Editor, Ruhland, W. Berlin: Springer.

Bandoni, R. J. (1963). Conjugation in *Tremella mesenterica*. *Can. J. Bot.* **41**, 467–74.

Bandoni, R. J. (1965). Secondary control of conjugation in *Tremella mesenterica*. *Can. J. Bot.* **43**, 627–30.

Banerjee, S. (1956). Heterothallism in *Auricularia auricula-judae* (Linn). Schroet. *Sci. Cult.* **21**, 549–50.

Barger, G. (1931). *Ergot and Ergotism*. London and Edinburgh: Gurney and Jackson.

Barksdale, A. W. (1960). Interthallic sexual reactions in *Achlya*, a genus of the aquatic fungi. *Am. J. Bot.* **47**, 14–23.

Barksdale, A. W. (1962). Effect of nutritional deficiency on growth and sexual reproduction of *Achlya ambisexualis. Am. J. Bot.* **49,** 633–8.
Barksdale, A. W. (1963*a*) The uptake of exogenous hormone A by certain strains of *Achlya. Mycologia,* **55,** 164–71.
Barksdale, A. W. (1963*b*). The role of hormone A during sexual conjugation in *Achlya ambisexualis. Mycologia,* **55,** 627–32.
Barksdale, A. W. (1965). *Achlya ambisexualis* and a new cross-conjugating species of *Achlya. Mycologia,* **57,** 493–501.
Barksdale, A. W. (1966). Segregation of sex in the progeny of a selfed heterozygote of *Achlya bisexualis. Mycologia,* **58,** 802–4.
Barnett, H. L. (1937). Studies in the sexuality of the Heterobasidiae. *Mycologia,* **29,** 626–49.
Barton, A. A. (1950). Some aspects of cell division in *Saccharomyces cerevisiae. J. gen. Microbiol.* **4,** 84–6.
Bartholomew, J. W. & Mittwer, T. (1953). Demonstration of yeast bud scars with the electron microscope. *J. Bact.* **65,** 272–5.
Bartnicki-Garcia, S. & Nickerson, W. J. (1962*a*). Induction of yeast like development in *Mucor* by carbon-dioxide. *J. Bact.* **84,** 829–40.
Bartnicki-Garcia, S. & Nickerson, W. J. (1962*b*). Nutrition, growth and morphogenesis of *Mucor rouxii. J. Bact.* **84,** 841–58.
Bartnicki-Garcia, S. & Nickerson, W. J. (1962*c*). Isolation, composition and structure of cell walls of filamentous and yeast-like forms of *Mucor rouxii. Biochim. Biophys. Acta,* **58,** 102–19.
Basu, S. N. (1951). Significance of calcium in the fruiting of *Chaetomium* species, particularly *Chaetomium globosum. J. gen. Microbiol.* **5,** 231–8.
Batra, L. R. (1963). Contributions to our knowledge of Ambrosia fungi. II. *Endomycopsis fasciculata* nom. nov. (Ascomycetes). *Am. J. Bot.* **50,** 481–7.
Batts, C. C. V. (1951). Physiologic specialization of *Puccinia graminis* Pers. in South-East Scotland. *Trans. Br. mycol. Soc.* **34,** 533–8.
Batts, C. C. V. (1955). Observations on the infection of wheat by loose smut (*Ustilago tritici* (Pers.) Rostr.). *Trans. Br. mycol. Soc.* **38,** 456–75.
Bayliss-Elliott, J. S. (1926). Concerning 'fairy rings' in pastures. *Ann. appl. Biol.* **13,** 277–88.
Beadle, G. W. (1959). Genes and chemical reactions in *Neurospora. Science, N.Y.* **129,** 1715–19.
Beadle, G. W. & Coonradt, V. L. (1944). Heterokaryosis in *Neurospora crassa. Genetics,* **29,** 291–308.
Beaumont, A. (1947). Dependence on the weather of the date of potato blight epidemics. *Trans. Br. mycol. Soc.* **31,** 45–53.
Bega, R. V. & Scott, H. A. (1966) Ultrastructure of the sterigma and sporidium of *Cronartium ribicola. Can. J. Bot.* **44,** 1726–7.
Beneke, E. S. (1963). Calvatia, calvacin and cancer. *Mycologia,* **55,** 257–70.
Benjamin, C. R. (1955). Ascocarps of *Aspergillus* and *Penicillium. Mycologia,* **47,** 669–87.
Benjamin, C. R. & Hesseltine, C. W. (1959). Studies on the genus *Phycomyces. Mycologia,* **51,** 751–71.
Benjamin, R. K. (1956). A new genus of the Gymnoascaceae with a review of the other genera. *Aliso,* **3,** 301–28.
Benjamin, R. K. (1959). The merosporangiferous Mucorales. *Aliso,* **4,** 321–433.
Benjamin, R. K. (1962). A new *Basidiobolus* that forms microspores. *Aliso,* **5,** 223–33.
Benjamin, R. K. (1966). The merosporangium. *Mycologia,* **58,** 1–42.

Berkson, B. M. (1966). Cytomorphological studies of the ascogenous hyphae in four species of *Chaetomium*. *Mycologia*, **58**, 125–30.

Berlin, J. D. & Bowen, C. C. (1964). The host-parasite interface of *Albugo candida* on *Raphanus sativus*. *Am. J. Bot.* **51**, 445–52.

Berry, C. R. & Barnett, H. C. (1957). Mode of parasitism and host range of *Piptocephalis virginiana*. *Mycologia*, **49**, 374–86.

Bessey, E. A. (1950). *Morphology and Taxonomy of Fungi*. 791 pp. Philadelphia and Toronto: Blakiston.

Bistis, G. N. (1956). Sexuality in *Ascobolus stercorarius*. I. Morphology of the ascogonium; plasmogamy; evidence for a sexual hormonal mechanism. *Am. J. Bot.* **43**, 389–94.

Bistis, G. N. (1957). Sexuality in *Ascobolus stercorarius*. II. Preliminary experiments on various aspects of the sexual process. *Am. J. Bot.* **44**, 436–43.

Bistis, G. N. & Raper, J. R. (1963). Heterothallism and sexuality in *Ascobolus stercorarius*. *Am. J. Bot.* **50**, 880–91.

Black, W. (1952). A genetical basis for the classification of strains of *Phytophthora infestans*. *Proc. R. Soc. Edinb.* B, **65**, 36–51.

Black, W. (1960) Races of *Phytophthora infestans* and resistance problems in potatoes. *Scott. Pl. Breeding Sta. Rep.* 1960, 29–38.

Blackwell, E. M. (1934*a*). The life history of *Phytophthora cactorum* (Leb. & Cohn) Schroet. *Trans. Br. mycol. Soc.* **26**, 71–89.

Blackwell, E. M. (1934*b*). Presidential Address. On germinating the oospores of *Phytophthora cactorum*. *Trans. Br. mycol. Soc.* **26**, 93–103.

Blakeslee, A. F. (1906). Zygospore germinations in the Mucorineae. *Annls. mycol.* **4**, 1–28.

Blumer, S. (1967). *Echte Mehltaupilze* (*Erysiphaceae*). *Ein Bestimmungsbuch für die in Europa vorkommenden Arten*. 436 pp. Jena: Fischer.

Bohus, M. G. (1962). Der Formenkreis des *Agaricus* (*Psalliota*) *bisporus* (Lange). Treschow und die Benützung der wildwachsenden Formen (Sorten) beim Veredelungsverfahren. *Schweiz. Z. Pilzk.* **40**, 1–7.

Bond, T. E. T. (1952). A further note on size and form in agarics. *Trans. Br. mycol. Soc.* **35**, 190–4.

Bonner, J. T. (1967). *The Cellular Slime Molds*. 2nd edition. 205 pp. Princeton University Press.

Bonner, J. T., Kane, K. K. & Levey, R. H. (1956). Studies on the mechanics of growth in the common mushroom, *Agaricus campestris*. *Mycologia*, **48**, 13–19.

Booth, C. (1959). Studies of Pyrenomycetes: IV. *Nectria* (Part I). *Mycol. Pap. C.M.I.* **73**, 1–115.

Booth, C. (1960). Studies of Pyrenomycetes: V. Nomenclature of some *Fusaria* in relation to their Nectrioid perithecial states. *Mycol. Pap. C.M.I.* **74**, 1–16.

Booth, C. (1966). The genus *Cylindrocarpon*. *Mycol. Pap. C.M.I.* **104**, 1–56.

Borriss, H. (1934). Beiträge zur Wachstums- und Entwicklungsphysiologie der Fruchtkörper von *Coprinus lagopus*. *Planta*, **22**, 28–69.

Bracker, C. E. (1966). Ultrastructural aspects of sporangiospore formation in *Gilbertella persicaria*. In *The Fungus Spore*, 39–59. Editor, Madelin, M. F. Colston Papers No. 18. London: Butterworths.

Bracker, C. E. (1968). Ultrastructure of the haustorial apparatus of *Erysiphe graminis* and its relationship to the epidermal cell of Barley. *Phytopathology*, **58**, 12–30.

Bracker, C. E. & Butler, E. E. (1963). The ultrastructure and development of septa in hyphae of *Rhizoctonia solani*. *Mycologia*, **55**, 35–58.

Bracker, C. E. & Butler, E. E. (1964). Function of the septal pore apparatus in *Rhizocionia solani* during streaming. *J. Cell. Biol.* **21**, 152–7.

Bradshaw, A. D. (1959). Population differentiation in *Agrostis tenuis* Sibth. II. The incidence and significance of infection by *Epichlöe typhina*. *New Phytol.* **58,** 310–15.

Brefeld, O. (1877). Botanische Untersuchungen über Schimmelpilze. III. Heft. Basidiomyceten. I. Leipzig.

Brefeld, O. (1881). Botanischen untersuchungen über Schimmelpilze. Untersuchungen aus dem Gesammtgebiet der Mykologie. IV. 4. *Pilobolus*. Leipzig.

Brian, P. W. (1960). Presidential address. Griseofulvin. *Trans. Br. Mycol. Soc.* **43,** 1–13.

Brian, P. W. (1967). Obligate parasitism in fungi. *Proc. R. Soc.* B, **168,** 101–8.

Bridgmon, G. H. (1959). Production of new races of *Puccinia graminis* var. *tritici* by vegetative fusion. *Phytopathology*, **49,** 386–8.

Brierley, W. B. (1918). The microconidia of *Botrytis cinerea*. *Kew Bull.* 129–46.

Brodie, H. J. (1951). The splash-cup dispersal mechanism in plants. *Can. J. Bot.* **29,** 224–34.

Brodie, H. J. (1956). The structure and function of the funiculus of the Nidulariaceae. *Svensk. bot. Tidskr.* **50,** 142–62.

Brodie, H. J. (1963). Twenty years of Nidulariology. *Mycologia*, **54,** 713–26.

Brodie, H. J. & Gregory, P. H. (1953). The action of wind in the dispersal of spores from cup-shaped plant structures. *Can. J. Bot.* **31,** 402–10.

Brown, A. H. S. & Smith, G. (1957). The genus *Paecilomyces* Bainer and its perfect stage *Byssochlamys* Westling. *Trans. Br. mycol. Soc.* **40,** 17–89.

Brummelen, J. van (1967). A world monograph of the genera *Ascobolus* and *Saccobolus* (Ascomycetes, Pezizales). *Persoonia*, Supplement Vol. I, 260 pp.

Bucholtz, F. (1897). Zur Entwicklungsgeschichte der Tuberaceen. *Ber. dt. bot. Ges.* **15,** 211–26.

Bucholtz, F. (1912). Beiträge zur Kenntniss der Gattung *Endogone* Link. *Beih. bot. Zbl.* **39,** 147–224.

Bulat, T. J. (1953). Cultural studies of *Dacrymyces ellisii*. *Mycologia*, **45,** 40–5.

Buller, A. H. R. (1909). *Researches on Fungi.* **1,** 274 pp. London: Longmans, Green & Co.

Buller, A. H. R. (1922). *Researches on Fungi.* **2,** 492 pp. London: Longmans, Green & Co.

Buller, A. H. R. (1924). *Researches on Fungi.* **3,** 611 pp. London: Longmans, Green & Co.

Buller, A. H. R. (1931). *Researches on Fungi.* **4,** 329 pp. London: Longmans, Green & Co.

Buller, A. H. R. (1933). *Researches on Fungi.* **5,** 416 pp. London: Longmans, Green & Co.

Buller, A. H. R. (1934). *Researches on Fungi.* **6,** 513 pp. London: Longmans, Green & Co.

Buller, A. H. R. (1950). *Researches on Fungi.* **7,** 458 pp. Toronto University Press.

Buller, A. H. R. & Vanterpool, T. C. (1933). The violent discharge of the basidiospores (secondary conidia) of *Tilletia tritici*. In Buller, A. H. R. *Researches on Fungi*, **5.** London: Longmans, Green & Co.

Bulmer, G. S. (1964). Spore germination of forty-two species of puffballs. *Mycologia*, **56,** 630–2.

Bulmer, G. S. & Li, Y.-T. (1966). Enzymic activities in *Calvatia cyathiformis* during and after meiosis. *Mycologia*, **58,** 555–61.

Burgeff, H. (1914). Untersuchungen über Variabilität, Sexualität und Erblichkeit bei *Phycomyces nitens*. I. *Flora*, **107,** 259–316.

Burgeff, H. (1920). Über den Parasitismus des *Chaetocladium* und die heterokaryotische Natur der von ihm auf Mucorineen erzeugten Gallen. *Z. Bot.* **12,** 1–35.

Burgeff, H. (1924). Untersuchungen über Sexualität und Parasitismus bei Mucorineen I. *Bot. Abh.* **4,** 1–135.
Burgeff, H. (1925). Über die Arten und Artkreuzung in der Gattung *Phycomyces. Flora,* **18,** 40–6.
Burnett, J. H. (1956). The mating systems of fungi. I. *New Phytol.* **55,** 50–90.
Burnett, J. H. (1965). The natural history of recombination systems. In *Incompatibility in Fungi,* 98–113. Editors, Esser, K. & Raper, J. R. Berlin: Springer.
Burnett, J. H. & Boulter, M. E. (1963). The mating systems of fungi. II. Mating systems of the Gasteromycetes *Mycocalia denudata* and *M. duriaeana. New Phytol.* **62,** 217–35.
Burnett, J. H. & Evans, E. J. (1966). Genetical homogeneity and the stability of mating-type factors of 'fairy rings' of *Marasmius oreades. Nature, Lond.* **210,** 1368–9.
Burnett, J. H. & Partington, D. (1957). Spatial distribution of fungal mating-type factors. *Proc. R. phys. Soc. Edinb.* **26,** 61–8.
Buston, H. W. & Khan, A. H. (1956). The influence of certain micro-organisms on the formation of perithecia by *Chaetomium globosum. J. gen. Microbiol.* **14,** 655–60.
Buston, H. W., Jabbar, A. & Etheridge, D. E. (1953). The influence of hexose phosphates, calcium and jute extract on the formation of perithecia by *Chaetomium globosum. J. gen. Microbiol.* **8,** 302–6.
Buxton, E. W. (1960). Heterokaryosis, saltation and adaptation. In *Plant Pathology: An Advanced Treatise,* **2,** 359–405. Editors, Horsfall, J. G. & Dimond, A. E. New York and London: Academic Press.
Cain, R. F. (1962). Studies of coprophilous ascomycetes. VIII. New species of *Podospora. Can. J. Bot.* **40,** 447–90.
Campbell, L. P. & Tyner, L. E. (1959). Comparison of degree and duration of susceptibility of barley to ergot and true loose smut. *Phytopathology,* **49,** 348–9.
Campbell, W. A. & Hendrix, F. F. (1967). A new heterothallic *Pythium* from Southern United States. *Mycologia,* **59,** 274–8.
Campbell, W. P. (1957). Studies on ergot infection in gramineous hosts. *Can. J. Bot.* **35,** 315–20.
Campbell, W. P. (1958). Infection of barley by *Claviceps purpurea. Can. J. Bot.* **36,** 615–19.
Campbell, W. P. (1959). Inhibition of starch formation by *Claviceps purpurea. Phytopathology,* **49,** 451–2.
Canter, H. M. & Willoughby, L. G. (1964). A parasitic *Blastocladiella* from Windermere plankton. *Jl R. micros. Soc.* **83,** 365–72.
Cantino, E. C. (1950). Nutrition and phylogeny in the water molds. *Q. Rev. Biol.* **25,** 269–77.
Cantino, E. C. (1955). Physiology and phylogeny in the water-molds: a re-evaluation. *Q. Rev. Biol.* **30,** 138–49.
Cantino, E. C. & Horenstein, E. A. (1954). Cytoplasmic exchange without gametic copulation in the water mold *Blastocladiella emersonii. Am. Nat.* **88,** 143–54.
Cantino, E. C. & Horenstein, E. A. (1956). Gamma and the cytoplasmic control of differentiation in *Blastocladiella. Mycologia,* **48,** 443–6.
Cantino, E. & Hyatt, M. T. (1953). Carotenoids and oxidative enzymes in the aquatic Phycomycetes *Blastocladiella* and *Rhizophlyctis. Am. J. Bot.* **40,** 688–94.
Cantino, E. C. & Lovett, J. S. (1964). Non-filamentous aquatic fungi: model systems for biochemical studies of morphological differentiation. *Adv. in Morphogenesis,* **3,** 33–93.
Cantino, E. C. Lovett, J. S., Leak, L. V. & Lythgoe, J. (1963). The single mitochondrion, fine structure, and germination of the spore of *Blastocladiella emersonii. J. gen. Microbiol.* **31,** 393–404.

Cantino, E. C., Truesdell, L. C. & Shaw, D. S. (1968). Life history of the motile spore of *Blastocladiella emersonii*: a study in cell differentiation. *J. Elisha Mitchell scient. Soc.* **84,** 125–46.

Cantino, E. C. & Turian, G. F. (1959). Physiology and development of lower fungi (Phycomycetes). *A. Rev. Microbiol.* **13,** 97–124.

Caporali, L. (1964). La biologie du *Taphrina deformans* (Berk.) Tul: relations entre l'hôte et le parasite. *Rev. gén. Bot.* **71,** 241–82.

Carlile, M. J. (1962). Evidence for a flavoprotein photoreceptor in *Phycomyces. J. gen. Microbiol.* **28,** 161–167.

Carlile, M. J. (1965). The photobiology of fungi. *A. Rev. Pl. Physiol.* **16,** 175–202.

Carlile, M. J. & Machlis, L. (1965*a*). The response of male gametes of *Allomyces* to the sexual hormone sirenin. *Am. J. Bot.* **52,** 478–83.

Carlile, M. J. & Machlis, C. (1965*b*). A comparative study of the chemotaxis of the motile phases of *Allomyces. Am. J. Bot.* **52,** 484–6.

Carr, A. J. H. & Olive, L. S. (1958). Genetics of *Sordaria fimicola.* II. Cytology. *Am. J. Bot.* **45,** 142–50.

Carroll, G. C. (1967). The ultrastructure of ascospore delimitation in *Saccobolus kerverni. J. Cell. Biol.* **33,** 218–24.

Cartwright, K. St. G. & Findlay, W. P. K. (1958). *Decay of Timber and its Prevention.* 2nd edition. 332 pp. London: H.M.S.O.

Castle, E. S. (1933*b*). The physical basis for the positive phototropism of *Phycomyces. J. gen. Physiol.* **17,** 49–62.

Castle, E. S. (1942). Spiral growth and reversal of spiraling in Phycomyces, and their bearing on primary wall structure. *Am. J. Bot.* **29,** 664–72.

Castle, E. S. (1953). Problems of oriented growth and structure in *Phycomyces. Q. Rev. Biol.* **28,** 364–72.

Castle, E. S. (1966). Light responses of *Phycomyces. Science, N.Y.* **154,** 1416–20.

Catcheside, D. G. (1951). Genetics of micro-organisms. 223 pp. London: Pitman.

Caten, C. E. & Jinks, J. L. (1966). Heterokaryosis: its significance in wild homothallic ascomycetes and fungi imperfecti. *Trans. Br. mycol. Soc.* **49,** 81–93.

Chadefaud, M. (1960). *Les végétaux non vasculaires (Cryptogamie)* Tome. I. In *Traité de Botanique Systématique*, by Chadefaud, M. & Emberger, L. Paris: Masson.

Chadefaud, M. & Avellanas, L. (1967). Remarques sur l'ontogenie et la structure des perithèces des *Chaetomium. Botaniste*, Sér. 50, 1–6, 59–87.

Chambers, T. C., Markus, K, & Willoughby, L. G. (1967). The fine structure of the mature zoosporangium of *Nowakowskiella profusa. J. gen. Microbiol*, **46,** 135–41.

Chapman, J. A. & Vujičic, R. (1965). The fine structure of sporangia of *Phytophthora erythroseptica* Pethybr. *J. gen. Microbiol.* **41,** 275–82.

Chaudhuri, H., Kuchhar, P. L., Lotus, S. S., Banerjee, M. L. & Khan, A. H. (1947). *A Handbook of Indian Water Moulds.* Lahore: Panjab University Botanical Publications.

Chester, K. S. (1946). *The Nature and Prevention of the Cereal Rusts as exemplified in the Leaf Rust of Wheat.* 269 pp. Waltham, Mass.: Chronica Botanica Company.

Chesters, C. G. C. & Greenhalgh, G. N. (1964). *Geniculosporium serpens* gen. et sp. nov., the imperfect state of *Hypoxylon serpens. Trans. Br. mycol. Soc.* **47,** 393–401.

Christensen, C. M. (1965). *The Molds and Man.* 3rd edition. 284 pp. University of Minnesota Press.

Christensen, J. J. (1932). Studies on the genetics of *Ustilago zeae. Phytopath. Z.* **4,** 129–88.

Christensen, J. J. (1963). Corn smut caused by *Ustilago maydis. Monogr. Amer., phytopath, Soc.* **2,** 41 pp.

Ciferri, R. & Tomaselli, R. (1955). The symbiotic fungi of lichens and their nomenclature. *Taxon*, **4**, 190–2.

Cochrane, V. W. (1958). *Physiology of Fungi*. 524 pp. New York: Wiley.

Coker, W. C. (1923). *The Saprolegniaceae, with Notes on other Water Molds*. University of North Carolina Press.

Coker, W. C. & Matthews, V. D. (1937). Saprolegniales. *North American Flora*, **2**, 15–76.

Colhoun, J. (1958). Club root disease of crucifers, caused by *Plasmodiophora brassicae* Woron. *Phytopath. Pap. C.M.I.* **3**, 108 pp.

Colhoun, J. (1966). The biflagellate zoospore of aquatic Phycomycetes with particular reference to *Phytophthora* spp. In *The Fungus Spore*, 85–92. Editor, M. F. Madelin, Colston Papers No. 18. London: Butterworths.

Colson, B. (1934). The cytology and morphology of *Neurospora tetrasperma*. *Ann. Bot. Lond.* **48**, 211–25.

Conti, S. F. & Naylor, H. B. (1959). Electron microscopy of ultrathin sections of *Schizosaccharomyces octosporus*. I. Cell division. *J. Bact.* **78**, 868–77.

Conti, S. F. & Naylor, H. B. (1960). Electron microscopy of ultrathin sections of *Schizosaccharomyces octosporus*. II. Morphological and cytological changes preceding ascospore formation. *J. Bact.* **79**, 331–40.

Cook, A. H. (1958). *The Chemistry and Biology of Yeasts*. 763 pp. New York.

Cook, W. R. I. & Schwartz, E. J. (1930). The life history, cytology and method of infection of *Plasmodiophora brassicae* Woron., the cause of finger-and-toe disease of cabbages and other crucifers. *Phil. Trans. R. Soc.* B, **218**, 283–314.

Cooke, R. C. & Mitchell, D. T. (1966). Sclerotium size and germination in *Claviceps purpurea*. *Trans. Br. mycol. Soc.* **49**, 95–100.

Cooke, W. B. (1961). The genus *Schizophyllum*. *Mycologia*, **53**, 575–99.

Corlett, M. (1966). Perithecium development in *Chaetomium trigonosporum*. *Can. J. Bot.* **44**, 155–62.

Corner, E. J. H. (1929). Studies in the morphology of Discomycetes. I–II. *Trans. Br. mycol. Soc.* **14**, 263–75; 275–91.

Corner, E. J. H. (1932). A *Fomes* with two systems of hyphae. *Trans. Br. mycol. Soc.* **17**, 51–81.

Corner, E. J. H. (1948). Studies in the basidium. I. The ampoule effect, with a note on nomenclature. *New Phytol.* **47**, 22–51.

Corner, E. J. H. (1950). *A Monograph of Clavaria and Allied Genera*. 740 pp. Oxford University Press.

Corner, E. J. H. (1953). The construction of polypores. I. Introduction: *Polyporus sulphureus*, *P. squamosus*, *P. betulinus* and *Polystictus microcyclus*. *Phytomorphology*, **3**, 152–67.

Corner, E. J. H. (1954). The classification of the higher fungi. *Proc. Linn. Soc. Lond.* **165**, 4–6.

Corner, E. J. H. (1966). *A Monograph of Cantharelloid Fungi*. 255 pp. Oxford University Press.

Couch, J. N. (1926). Heterothallism in *Dictyuchus*, a genus of the water moulds. *Ann. Bot. Lond.* **40**, 849–81.

Couch, J. N. (1938). The genus *Septobasidium*. 480 pp. University of North Carolina Press.

Couch, J. N. (1939). Heterothallism in the Chytridiales. *J. Elisha Mitchell scient. Soc.* **55**, 409–14.

Cox, A. E. & Large, E. C. (1960). *Potato Blight Epidemics throughout the World*. U.S. Dept. of Agriculture Handbook No. 174, 230 pp. Washington.

Crady, E. E. & Wolf, F. T. (1959). The production of indole acetic acid by *Taphrina deformans* and *Dibotryon morbosum*. *Physiol. Pl.* **12**, 526–33.

Craigie, J. H. (1927*a*). Experiments on sex in rust fungi. *Nature, Lond.* **120,** 116–17.

Craigie, J. H. (1927*b*). Discovery of the function of the pycnia of the rust fungi. *Nature, Lond.* **120,** 765–7.

Craigie, J. H. (1945). Epidemiology of stem rust in Western Canada. *Scient. Agric.* **25,** 285–401.

Craigie, J. H. & Green, G. J. (1962). Nuclear behaviour leading to conjugate association in haploid infections of *Puccinia graminis. Can. J. Bot.* **40,** 163–78.

Croxall, H. E., Collingwood, C. A., & Jenkins, J. E. E. (1951). Observations on brown rot (*Sclerotinia fructigena*) of apples in relation to injury by earwigs (*Forficula auricularia*). *Ann. appl. Biol.* **38,** 833–43.

Cummins, G. B. (1959). *Illustrated Genera of Rust Fungi.* 129 pp. Minneapolis: Burgess Publishing Company.

Curtis, K. M. (1921). The life history and cytology of *Synchytrium endobioticum* (Schilb.) Perc., the cause of wart disease in potato. *Phil Trans. R. Soc.* B, **210,** 409–78.

Cutter, V. M. (1942*a*). Nuclear behaviour in the Mucorales. I. The *Mucor* pattern. *Bull. Torrey bot. Club,* **69,** 480–508.

Cutter, V. M. (1942*b*). Nuclear behaviour in the Mucorales. II. The *Rhizopus, Phycomyces* and *Sporodinia* patterns. *Bull. Torrey bot. Club,* **69,** 592–616.

Daniels, J. (1961). *Chaetomium piluliferum* sp. nov., the perfect state of *Botryotrichum piluliferum. Trans. Br. mycol. Soc.* **44,** 79–86.

D'Arcy Thompson, W. (1961). *On Growth and Form.* Abridged edition. Editor, Bonner, J. T. 346 pp. Cambridge University Press.

Davey, C. B. & Papavizas, G. C. (1962). Growth and sexual reproduction of *Aphanomyces euteiches* as affected by the oxidation state of sulfur. *Am. J. Bot.* **49,** 400–4.

Davies, B. H. (1961). The carotenoids of *Rhizophlyctis rosea. Phytochemistry,* **1,** 25–9.

Davies, E. E. (1967). Zygospore formation in *Syzygites megalocarpus. Can. J. Bot.* **45,** 531–2.

Davis, R. H. (1966). Mechanisms of inheritance. 2. Heterokaryosis. In *The Fungi: An Advanced Treatise,* **2,** 567–88. Editors, Ainsworth, G. C. & Sussman, A. S. New York and London: Academic Press.

Day, P. R. (1963). The structure of the A mating-type factor in *Coprinus lagopus*: wild alleles. *Genet. Res. Camb.* **4,** 323–5.

Delamater, E. D., Yaverbaum, S. & Schwartz, L. (1953). The nuclear cytology of *Eremascus albus. Am. J. Bot.* **40,** 475–92.

Delay, C. (1966). Étude de l'infrastructure de l'asque d'*Ascobolus immersus* Pers., pendant la maturation des spores. *Ann. Sci. nat. Bot.* Sér. 12. **7,** 361–420.

Dengler, I. (1937). Entwicklungsgeschichtliche Untersuchungen an *Sordaria macrospora* Auersw., *S. uvicola* Viala et Mars., und *S. brefeldii* Zopf. *Jb. wiss. Bot.* **84,** 427–48.

Dennis, R. W. G. (1956). A revision of the British Helotiaceae in the herbarium of the Royal Botanic Gardens, Kew, with notes on related European species. *Mycol. Pap. C.M.I.* **62,** 216 pp.

Dennis, R. W. G. (1958*a*). *Xylaria* versus *Hypoxylon* and *Xylosphaera. Kew Bull.* **13,** 101–6.

Dennis, R. W. G. (1958*b*). Some Xylosphaeras from tropical Africa. *Revta Biol., Lisb.* **1,** 175–208.

Dennis, R. W. G. (1968). *British Ascomycetes.* 455 pp. Stuttgart: Cramer.

Dennis, R. W. G., Orton, P. D. & Hora, F. B. (1960). New check list of British Agarics and Boleti. Supplement to *Trans. Br. mycol. Soc.* **43,** 225 pp.

Derx, H. G. (1930). Étude sur les Sporobolomycètes. *Annls. mycol.* **28,** 1–23.

Derx, H. G. (1948). *Itersonilia*, nouveau genre de sporobolomycètes à mycélium bouclé. *Bull. bot. Gdns Buitenz.* III, **18,** 465–72.

Dick, M. W. (1962). The occurrence and distribution of Saprolegniaceae in certain soils of south-east England. II. Distribution within defined areas. *J. Ecol.* **50,** 119–27.

Dick, M. W. (1966). The Saprolegniaceae of the environs of Blelham Tarn: sampling techniques and the estimation of propagule numbers. *J. gen. Microbiol.* **42,** 257–82.

Dick, M. W. & Newby, H. V. (1961). The occurrence and distribution of Saprolegniaceae in certain soils of south-east England. I. Occurrence. *J. Ecol.* **49,** 403–19.

Dick, S. (1965). Physiological aspects of tetrapolar incompatibility. In *Incompatibility in Fungi*, 72–80. Editors, Esser, K. & Raper, J. R. Berlin: Springer.

Dodge, B. O. (1920). The life history of *Ascobolus magnificus*. *Mycologia*, **12,** 115–34.

Dodge, B. O. (1924). Aecidiospore discharge as related to the character of the spore wall. *J. agric Res.* **27,** 749–56.

Dodge, B. O. (1927). Nuclear phenomena associated with heterothallism and homothallism in the ascomycete *Neurospora*. *J. agric. Res.* **35,** 289–305.

Dodge, B. O. (1937). The perithecial cavity formation in a *Leptosphaeria* on *Opuntia*. *Mycologia*, **29,** 707–16.

Dodge, B. O. (1942). Heterocaryotic vigor in *Neurospora*. *Bull. Torrey bot. Club*, **69,** 75–91.

Doguet, G. (1957). Organogénie du *Creopus spinulosus* (Fuck.) Moravec. Organogénie comparée de quelques Hypocréales du même type. *Bull. Soc. mycol. Fr.* **73,** 144–64.

Doguet, G. (1960). Morphologie, organogénie et évolution nucléaire de l'*Epichloe typhina*. La place des Clavicipitaceae dans la classification. *Bull. Soc. mycol. Fr.* **76,** 171–203.

Doling, D. A. (1966). Loose smut in wheat and barley. *Agriculture*, **73,** 523–7.

Donk, M. A. (1964). A conspectus of the families of Aphyllophorales. *Persoonia*, **3,** 199–324.

Douwes, G. A. C. & von Arx, J. A. (1965). Das hymenophorale Trama bei den Agaricales. *Acta bot. néerl.* **14,** 197–217.

Dowding, E. S. (1931). The sexuality of *Ascobolus stercorarius* and the transportation of oidia by mites and flies. *Ann. Bot. Lond.* **45,** 621–37.

Dowding, E. S. (1958). Nuclear streaming in *Gelasinospora*. *Can. J. Microbiol*, **4,** 295–301.

Dowding, E. S. (1966). The chromosomes in *Neurospora* hyphae. *Can. J. Bot.* **44,** 1121–5.

Dowding, E. S. & Bakerspigel, A. (1954). The migrating nucleus. *Can. J. Microbiol.* **1,** 68–78.

Dowding, E. S. & Bulmer, G. S. (1964). Notes on the cytology and sexuality of puffballs. *Can. J. Microbiol.* **10,** 783–9.

Drayton, F. L. (1934). The sexual mechanism of *Sclerotinia gladioli*. *Mycologia*, **26,** 46–72.

Drayton, F. L. & Groves, J. W. (1952). *Stromatinia narcissi*, a new, sexually dimorphic discomycete. *Mycologia*, **44,** 119–40.

Drechsler, C. (1956). Supplementary developmental stages of *Basidiobolus ranarum* and *Basidiobolus haptosporus*. *Mycologia*, **48,** 655–76.

Drechsler, C. (1960). Two root rot fungi closely related to *Pythium ultimum*. *Sydowia*, **14,** 107–15.

Duncan, E. G. & MacDonald, J. A. (1967). Micro-evolution in *Auricula auricula*. *Mycologia*, **59,** 803–18.

Duncan, U. K. (1959). *A Guide to the Study of Lichens.* 164 pp. Arbroath: Buncle.
Duncan, U. K. (1963). *Lichen illustrations. Supplement to A Guide to the Study of Lichens.* 144 pp. Arbroath: Buncle.
Eddy, A. A. (1958*a*). The structure of the yeast cell wall. *Proc. R. Soc.* B, **149,** 425–40.
Eddy, A. A. (1958*b*). Aspects of the chemical composition of yeast. In *The Chemistry and Biology of Yeasts*, 157–249. Editor, Cook, A. H.
Edgington, L. V. & Reinbergs, E. (1966). Control of loose smut in barley with systemic fungicides. *Can. J. Pl. Sci.* **46,** 336.
Ehrlich, H. G. & McDonough, E. S. (1949). The nuclear history in the basidia and basidiospores of *Schizophyllum commune* Fries. *Am. J. Bot.* **36,** 360–3.
Ehrlich, H. G. & Ehrlich, M. A. (1963). Electron microscopy of the host-parasite relationships in stem rust of wheat. *Am. J. Bot.* **50,** 123–30.
Ehrlich, M. A. & Ehrlich, H. G. (1966). Ultrastructure of the hyphae and haustoria of *Phytophthora infestans* and hyphae of *P. parasitica. Can. J. Bot.* **44,** 1495–503.
El-Abyad, M. S. H. & Webster, J. (1968*a*). Studies on pyrophilous Discomycetes. I. Comparative physiological studies. *Trans. Br. mycol. Soc.* **51,** 353–67.
El-Abyad, M. S. H. & Webster, J. (1968*b*). Studies on pyrophilous Discomycetes. II. Competition. *Trans. Br. mycol. Soc.* **51,** 369–75.
Ellingboe, A. H. (1961). Somatic recombination in *Puccinia graminis* var. *tritici. Phytopathology*, **51,** 13–15.
Ellingboe, A. H. & Raper, J. R. (1962). The Buller Phenomenon in *Schizophyllum commune*: nuclear selection in fully compatible dikaryotic-homokaryotic matings. *Am. J. Bot.* **49,** 454–9.
Elliott, M. E. (1964). Self-fertility in *Botryotrinia porri. Can. J. Bot.* **42,** 1393–5.
Embree, R. W. (1959). *Radiomyces*, a new genus of the Mucorales. *Am. J. Bot.* **46,** 25–30.
Emerson, R. (1941). An experimental study of the life cycles and taxonomy of *Allomyces. Lloydia*, **4,** 77–144.
Emerson, R. (1954). The biology of water molds. In *Aspects of Synthesis and Order in Growth*, 171–208. Editor, Rudnick, D. Princeton University Press.
Emerson, R. (1958). Mycological organisation. *Mycologia*, **50,** 589–621.
Emerson, R. & Wilson, C. M. (1954). Interspecific hybrids and the cytogenetics and cytotaxonomy of Eu-Allomyces. *Mycologia*, **46,** 393–434.
Emmons, C. W. (1935). The ascocarps in species of *Penicillium. Mycologia*, **27,** 128–50.
Emmons, C. W. & Bridges, C. H. (1961). *Entomophthora coronata*, the etiologic agent of a Phycomycosis of horses. *Mycologia*, **53,** 307–12.
Ende, H. van der (1967). Sexual factor of the Mucorales. *Nature, Lond.* **215,** 211–12.
Engel, H. & Friederichsen, I. (1964). Der Abschluss der Sporangiolen von *Sphaerobolus stellatus* (Thode) Pers., in kontinuerlicher Dunkelheit. *Planta*, **61,** 361–70.
Engel, H. & Schneider, J. C. (1963). Die Umwandlung von Glykogen in Zucker in den Fruchtkörpern von *Sphaerobolus stellatus* (Thode) Pers., vor ihrem Abschluss. *Ber. dt. bot Ges.* **75,** 397–400.
Eriksson, J. & Henning, E. (1894). Die Hauptresultate einer neuen Untersuchung über die Getreideroste. *Z. PflKrankh.* **4,** 66–73.
Erlenmeyer, H. & Geiger-Huber, M. (1935). Notiz über die durch einen Brandpilz verursachte Geschlechtsummstimmung bei *Melandrium album. Helv. chim. Acta*, **18,** 921–3.
Esser, K. (1965). Heterogenic incompatibility. In *Incompatibility in Fungi*, 6–13. Editors, Esser, K. & Raper, J. R. Berlin: Springer.
Esser, K. & Kuehnen, R. (1967). *Genetics of Fungi.* Transl. by E. Steiner. 500 pp. Berlin: Springer.

Essig, F. M. (1922). The morphology, development and economic aspects of *Schizophyllum commune* Fries. *Univ. Calif. Publs. Bot.* **7,** No. 14, 447–98.

Evans, H. J. (1959). Nuclear behaviour in the cultivated mushroon. *Chromosoma* (Berl.), **10,** 115–35.

Farr, D. F. & Lichtwardt, R. W. (1967). Some cultural and ultrastructural aspects of *Smittium culisetae* (Trichomycetes) from Mosquito larvae. *Mycologia,* **59,** 172–82.

Fergus, C. L. (1960). *Illustrated Genera of Wood Decay Fungi.* 132 pp. Minneapolis: Burgess Publishing Company.

Filer, T. H. (1966). Effect on grass and cereal seedlings of hydrogen cyanide produced by mycelium and sporophores of *Marasmius oreades. Pl. Dis. Peptr.* **50,** 264–6.

Fincham, J. R. S. & Day, P. R. (1965). *Fungal Genetics.* 2nd edition. 326 pp. Oxford: Blackwell.

Findlay, W. P. K. (1940). Studies in the physiology of wood-destroying fungi. III. Progress of decay under natural and controlled conditions. *Ann. Bot. Lond.* N.S. **4,** 701–12.

Findlay, W. P. K. & Savory, J. G. (1960). *Dry Rot in Wood.* 6th edition. 36 pp. London: H.M.S.O.

Fischer, A. (1892). Die Pilze. IV. Abt. Phycomycetes. In *Kryptogamenflora von Deutschland, Oesterreich und der Schweiz.* Editor, Rabenhorst, L. Leipzig.

Fischer, E. (1933). Gastromyceteae. In *Die natürlichen Pflanzenfamilien.* 2nd edition, 7a: 1–122. Editors, Engler, A. & Prantl, K. Leipzig: Engelmann.

Fischer, E. (1938). Tuberineae. In *Die natürlichen Pflanzenfamilien.* 2nd edition, 5b, VIII: 1–42. Editors, Engler, A. & Prantl, K. Leipzig: Engelmann.

Fischer, F. G. & Werner, G. (1955). Eine Analyse des Chemotropismus einiger Pilze, insbesondere der Saprolegniaceen. *Hoppe-Seyl. Z.* **300,** 211–36.

Fischer, F. G. & Werner, G. (1958*a*). Die Chemotaxis der Schwärmsporen von Wasserpilzen (Saprolegniaceen). *Hoppe-Seyl. Z.* **310,** 65–91.

Fischer, F. G. & Werner, G. (1958*b*). Über die Wirkungen von Nicotinsäureamid auf die Schwärmsporen wasserbewohnende Pilze. *Hoppe-Seyl. Z.* **310,** 92–6.

Fischer, G. W. & Holton, C. S. (1957). *Biology and Control of the Smut Fungi.* 622 pp. New York: Ronald Press Company.

Fleming, A. (1944). The discovery of penicillin. *Br. med. Bull.* **2,** 4–5.

Foster, J. W. (1949). *Chemical Activities of Fungi.* 648 pp. New York and London: Academic Press.

Fothergill, P. G. & Child, J. H. (1964). Comparative studies of the mineral nutrition of three species of *Phytophthora. J. gen. Microbiol.* **36,** 67–78.

Fothergill, P. G. & Hide, D. (1962). Comparative nutritional studies of *Pythium* spp. *J. gen. Microbiol.* **29,** 325–34.

Fowell, R. R. (1952). Sodium acetate agar as a sporulation medium for yeast. *Nature, Lond.* **170,** 578.

Fox, R. A. (1953). Heterothallism in *Chaetomium. Nature, Lond.* **172,** 165–6.

Franke, G. (1957). Die Zytologie der Ascusentwicklung von *Podospora anserina. Z. indukt. Abstamm.-u. Vererblehre,* **88,** 159–60.

Fraymouth, J. (1956). Haustoria of the Peronosporales. *Trans. Brit. mycol. Soc.* **39,** 79–107.

Fricke, S. & Handke, H. H. (1962). Untersuchungen zur Offnungswerke der Geastracee-Fruchtkorper. *Z. Pilzk.* **27,** 113–22.

Fries, N. (1942). Einspormyzelien einiger Basidiomyceten als Mykorrhiza-bildner von Kiefer und Fichte. *Svensk. Bot. Tidskr.* **36,** 151–6.

Fries, N. (1948). Heterothallism in some Gasteromycetes and Hymenomycetes. *Svensk. bot. Tidskr.* **42,** 158–68.

Fuller, M. S. (1966). Structure of the uniflagellate zoospores of aquatic Phycomycetes. In *The Fungus Spore* 67–84. Editor, Madelin, M. F. Colston Papers No. 18. London: Butterworths.

Fulton, I. W. (1950). Unilateral nuclear migration and the interactions of haploid mycelia in the fungus *Cyathus stercoreus*. *Proc. natn Acad. Sci. U.S.A.* **36,** 306–12.

Furtado, J. S. (1962). Structure of the spore of the Ganodermoideae Donk. *Rickia, Archos Bot. Est. S. Paulo*, **1,** 227–41.

Furtado, J. S. (1965). Relation of microstructures to the taxonomy of the Ganodermoideae (Polyporaceae) with special reference to the cover of the pilear surface. *Mycologia*, **57,** 588–611.

Furtado, J. S. (1966). Significance of the clamp-connection in the Basidiomycetes. *Persoonia*, **4,** 125–44.

Galindo, J. & Gallegly, M. E. (1960). The nature of sexuality in *Phytophthora infestans*. *Phytopathology*, **50,** 123–8.

Galindo, J. A. & Zentmeyer, G. A. (1967). Genetical and cytological studies of *Phytophthora* strains pathogenic to pepper plants. *Phytopathology*, **57,** 1300–4.

Galloway, L. D. (1936). Report of the Imperial Mycologist. *Sci. Rep. agric. Res. Inst. Pusa*, 1934–5, 120–30.

Gams, W. & Williams, S. T. (1963). Heterothallism in *Mortierella parvispora* Linnemann. I. Morphology and development of zygospores and some factors influencing their formation. *Nova Hedwigia*, **5,** 347–57.

Gamundi, I. J. & Ranalli, M. E. (1963). Apothecial development in *Ascobolus stercorarius*. *Trans. Br. mycol. Soc.* **46,** 393–400.

Ganesan, A. T. (1959). The cytology of *Saccharomyces*. *C. r. Trav. Lab. Carlsberg*, **31,** 149–74.

Garay, A. & Sagi, F. (1960). Untersuchungen über die Geschlechtsumwandlung bei *Melandrium album* Mill., nach Infektion mit *Ustilago violacea* (Pers.) Fuckl., unter besonderer Berücksichtung der Auxinoxydase und Flavonoide. *Phytopath. Z.* **38,** 201–8.

Garnjobst, L. (1955). Further analysis of genetic control of heterokaryosis in *Neurospora crassa*. *Am. J. Bot.* **42,** 444–8.

Garnjobst, L. & Wilson, J. F. (1956). Heterokaryosis and protoplasmic incompatibility in *Neurospora crassa*. *Proc. natn. Acad. Sci. U.S.A.* **42,** 613–18.

Gauger, W. C. (1961). The germination of zygospores of *Rhizopus stolonifer*. *Am. J. Bot.* **48,** 427–9.

Gauger, W. L. (1965). The germination of zygospores of *Mucor hiemalis*. *Mycologia*, **57,** 634–41.

Gäumann, E. (1959). *Die Rostpilze Mitteleuropas*. *Beitr. Kryptogfl. Schweiz*. Bd. XII. 1407 pp. Bern: Büchler.

Gäumann, E. (1964). *Die Pilze. Grundzuge ihrer Entwicklungsgeschichte und Morphologie*. 541 pp. Basel and Stuttgart: Birkhauser.

Gay, J. L. & Greenwood, A. D. (1966). Structural aspects of zoospore production in *Saprolegnia ferax* with particular reference to the cell and vacuolar membranes. In *The Fungus Spore*, 95–108. Editor, Madelin, M. F. Colston Papers No. 18. London: Butterworths.

Gerdemann, J. W. & Nicolson, T. H. (1963). Spores of mycorrhizal *Endogone* species extracted from soil by wet sieving and decanting. *Trans. Br. mycol. Soc.* **46,** 235–44.

Giesy, R. M. & Day, P. R. (1965). The septal pores of *Coprinus lagopus* (Fr.) sensu Buller in relation to nuclear migration. *Am. J. Bot.* **52,** 287–93.

Gilles, A. (1947). Évolution nucléaire et développement du périthèce chez *Nectria flava*. *Cellule*, **51,** 371–400.

Girbardt, M. (1958). Über die Substruktur von *Polyporus versicolor* L. *Arch. Mikrobiol.* **28,** 255–69.

Girbardt, M. (1961). Licht- und elektronenoptische Untersuchungen an *Polystictus versicolor* (L). VII. Lebendbeobachtung und Zeitdauer der Teilung des vegetativen Kernes. *Exp. Cell Res.* **23,** 181–94.

Godfrey, R. M. (1957*a*). Studies of British species of *Endogone*. I. Morphology and taxonomy. *Trans. Br. mycol. Soc.* **40,** 117–35.

Godfrey, R. M. (1957*b*). Studies on British species of *Endogone*. II. Germination of spores. *Trans. Br. mycol. Soc.* **40,** 203–10.

Goldie-Smith, E. K. (1954). The position of *Woronina polycystis* in the Plasmodiophoraceae. *Am. J. Bot.* **41,** 441–8.

Goldstein, B. (1923). Resting spores of *Empusa muscae*. *Bull. Torrey. bot. Club,* **50,** 317–27.

Goldstein, S. (1960*a*). Physiology of aquatic fungi. I. Nutrition of two monocentric chytrids. *J. Bact.* **80,** 701–7.

Goldstein, S. (1960*b*). Factors affecting the growth and pigmentation of *Cladochytrium replicatum*. *Mycologia,* **52,** 490–8.

Goldstein, S. (1961). Studies of two polycentric chytrids in pure culture. *Am. J. Bot.* **48,** 294–8.

Gordon, C. C. (1966). A re-interpretation of the ontogeny of the ascocarp of species of the Erysiphaceae. *Am. J. Bot.* **53,** 652–62.

Graham, K. M. (1955). Distribution of physiological races of *Phytophthora infestans* (Mont.) de Bary in Canada. *Am. Potato J.* **32,** 277–82.

Grainger, J. (1962). Vegetative and fructifying growth in *Phallus impudicus*. *Trans. Br. mycol. Soc.* **45,** 147–55.

Gray, W. D. (1941). Studies on alcohol tolerance of yeasts. *J. Bact.* **42,** 561–74.

Gray, W. D. (1959). *The Relation of Fungi to Human Affairs.* 510 pp. New York: Holt, Rinehart and Winston.

Greenhalgh, G. N. & Evans, L. V. (1967). The structure of the ascus apex in *Hypoxylon fragiforme* with reference to ascospore release in this and related species. *Trans. Br. mycol. Soc.* **50,** 183–8.

Greer, D. L. & Friedman, L. (1966). Studies on the genus *Basidiobolus* with reclassification of the species pathogenic for man. *Sabouraudia,* **4,** 231–41.

Gregory, P. H. (1949). The operation of the puff-ball mechanism of *Lycoperdon perlatum* by raindrops shown by ultra-high-speed Schlieren cinematography. *Trans. Br. mycol. Soc.* **32,** 11–15.

Gregory, P. H. (1957). Electrostatic charges on spores of fungi in air. *Nature, Lond.* **180,** 330.

Gregory, P. H., Guthrie, E. J. & Bunce, M. E. (1952). Experiments on splash dispersal of fungus spores. *J. gen. Microbiol.* **20,** 328–54.

Greis, H. (1937). Entwicklungsgeschichte von *Sordaria fimicola* (Rob.). *Bot. Arch.* **38,** 113–51.

Greis, H. (1938). Die sexualvorgange bei *Tuber aestivum* und *T. brumale*. *Biol. Zbl.* **58,** 617–31.

Greis, H. (1939). Ascusentwicklung von *Tuber aestivum* und *T. brumale*. *Z. Bot.* **34,** 129–78.

Gremmen, J. (1955). Über Apothezienbildung bei *Ascobolus stercorarius* (Bull.) Schroet. *Schweiz. Z. Pilzk.* **33,** 42–5.

Griffiths, E. (1959). The cytology of *Gloeotinia temulenta* (blind seed disease of rye-grass). *Trans. Br. mycol. Soc.* **42,** 132–48.

Groves, J. W. & Drayton, F. L. (1939). The perfect stage of *Botrytis cinerea*. *Mycologia,* **31,** 485–9.

Groves, J. W. & Loveland, C. A. (1953). The connection between *Botryotinia fuckeliana* and *Botrytis cinerea*. *Mycologia*, **45**, 415–25.

Gruen, H. E. (1959). Auxins and fungi. *A. Rev. Pl. Physiol.* **10**, 405–40.

Gruen, H. E. (1963). Endogenous growth regulations in carpophores of *Agaricus bisporus*. *Pl. Physiol.* **37**, 652–66.

Gustafsson, M. (1965). On species of the genus *Entomophthora* Fres. in Sweden. II. Cultivation and physiology. *Lantbrukhögskolans Annlr.* **31**, 405–57.

Guttenberg, H. von & Schmoller, H. (1958). Kulturversuche mit *Peronospora brassicae* Gäum. *Arch. Mikrobiol.* **30**, 268–79.

Gwynne-Vaughan, H. C. I. & Broadhead, Q. E. (1936). Contributions to the study of *Ceratostomella fimbriata*. *Ann. Bot. Lond.* **50**, 747–58.

Gwynne-Vaughan, H. C. I. & Williamson, H. S. (1930). Contributions to the study of *Humaria granulata* Quél. *Ann. Bot. Lond.* **44**, 127–45.

Gwynne-Vaughan, H. C. I. & Williamson, H. S. (1932). The cytology and development of *Ascobolus magnificus*. *Ann. Bot. Lond.* **46**, 653–70.

Hagedorn, H. (1964). Die Feinstruktur der Hefezellen. I. Zellwand, Sporen, Cytoplasma, endoplasmatisches Retikulum. *Protoplasma*, **58**, 250–68. II. Kern, Mitochondrien, Reservestoffe. *Protoplasma*, **58**, 269–85.

Hagimoto, H. (1963). Studies on the growth of fruit body of fungi. IV. The growth of the fruit body of *Agaricus bisporus* and the economy of the mushroom growth hormone. *Bot. Mag. Tokyo*, **76**, 256–63.

Hagimoto, H. & Konishi, M. (1959). Studies on the growth of fruit body of fungi. I. Existence of a hormone active to the growth of fruit body in *Agaricus bisporus* (Lange.) Sing. *Bot. Mag. Tokyo*, **72**, 359–66.

Hagimoto, H. & Konishi, M. (1960). Studies on the growth of fruit body of fungi. II. Activity and stability of growth hormone in the fruit body of *Agaricus bisporus* (Lange) Sing. *Bot. Mag. Tokyo*, **73**, 283–7.

Hale, M. E. (1967). *The Biology of Lichens*. 176 pp. London: Arnold.

Halisky, P. M. (1965). Physiologic specialization and genetics of the smut fungi. III. *Bot. Rev.* **31**, 114–50.

Hall, I. M. & Halfhill, J. C. (1959). The germination of resting spores of *Entomophthora virulenta* Hall and Dunn. *J. econ. Entomol.* **52**, 30–5.

Hanlin, R. T. (1961). Studies in the genus *Nectria*. II. Morphology of *N. gliocladioides*. *Am. J. Bot.* **48**, 900–8.

Hanlin, R. T. (1965). Morphology of *Hypocrea schweinitzii*. *Am. J. Bot.* **52**, 570–9.

Hansen, L. (1958). On the anatomy of the Danish species of *Ganoderma*. *Bot. Tidsskr.* **54**, 333–52.

Harley, J. L. (1969). *The Biology of Mycorrhiza*. 2nd edition. pp. 334. London: Leonard Hill.

Harrold, C. E. (1950). Studies in the genus *Eremascus*. I. The re-discovery of *Eremascus albus* Eidam and some new observations concerning its life history and cytology. *Ann. Bot. Lond.* N.S. **14**, 127–48.

Harvey, R. (1958). Sporophore development and proliferation in *Hydnum auriscalpium* Fr. *Trans. Br. mycol. Soc.* **41**, 325–34.

Hashimoto, T., Conti, S. F. & Naylor, H. B. (1959). Studies on the fine structure of microorganisms. IV. Observations on budding *Saccharomyces cerevisiae* by light and electron microscopy. *J. Bact.* **77**, 344–54.

Hashimoto, T., Gerhardt, P., Conti, S. F. & Naylor, H. B. (1960). Studies on the fine structure of microorganisms. V. Morphogenesis of nuclear and membrane structures during ascospore formation in yeast. *J. biophys. biochem. Cytol.* **7**, 305–8.

Haskins, R. H. (1939). Cellulose as a substratum for saprophytic Chytrids. *Am. J. Bot.* **26**, 635–9.

Haskins, R. H. & Weston, W. H. (1950). Studies in the lower Chytridiales. I. Factors affecting pigmentation, growth and metabolism in a strain of *Karlingia* (*Rhizophlyctis*) *rosea*. *Am. J. Bot.* **37,** 739–50.

Hawker, L. E. (1954). British hypogeous fungi. *Phil. Trans. R. Soc.* B, **237,** 429–46.

Hawker, L. E. (1955). Hypogeous fungi. *Biol. Rev.* **30,** 127–58.

Hawker, L. E. (1957). *The Physiology of Reproduction in Fungi.* 128 pp. Cambridge University Press.

Hawker, L. E. & Abbott, P. McV. (1963*a*). The fine structure of vegetative hyphae of *Rhizopus*. *J. gen. Microbiol.* **30,** 401–8.

Hawker, L. E. & Abbott, P. McV. (1963*b*). An electron microscope study of maturation and germination of sporangiospore of two species of *Rhizopus*. *J. gen. Microbiol.* **32,** 295–8.

Heim, P. (1955). Le noyau dans le cycle évolutif de *Plasmodiophora brassicae* Woron. *Rev. mycol. Paris*, **20,** 131–57.

Heim, P. (1956*a*). Remarques sur le cycle évolutif du *Synchytrium endobioticum*. *C. r. Acad. Sci. Paris*, **42,** 2759–61.

Heim, P. (1956*b*). Remarques sur le développement, les divisions nucléaires et sur le cycle évolutif du *Synchytrium endobioticum* (Schilb.) Perc. *Rev. mycol. Paris*, **21,** 93–100.

Heim, P. (1960). Évolution du *Spongospora* parasite des racines du Cresson. *Rev. mycol. Paris*, **25,** 3–12.

Heim, R. (1948). Phylogeny and natural classification of macro-fungi. *Trans. Br. mycol. Soc.* **30,** 161–78.

Heim, R. (1962). L'organisation architecturale des spores de Ganodermes. *Rev. mycol. Paris*, **27,** 199–212.

Heim, R. (1963). *Les champignons toxiques et hallucinogènes.* Paris: Boubée.

Heim, R. & Wasson, R. G. (1958). Les champignons hallucinogenes du Mexique. *Archs. Mus. natn. Hist. nat., Paris*, 7 sér **5,** IV–VIII, 322 pp.

Herrick, J. A. (1953). An unusually large *Fomes*. *Mycologia*, **45,** 622–4.

Heslop-Harrison, J. (1957). The experimental modification of sex expression in flowering plants. *Biol. Rev.* **32,** 38–90.

Heslot, H. (1958). Contribution à l'étude cytogenetique et genetique des Sordariacées. *Rev. Cytol. Biol. vég.* **19,** Suppl. 2, 1–235.

Hesseltine, C. W. (1960). *Gilbertella* gen. nov. (Mucorales). *Bull. Torrey bot. Club*, **87,** 21–30.

Hesseltine, C. W. (1953). A revision of the Choanephoraceae. *Am. Midl. Nat.* **50,** 248–56.

Hesseltine, C. W. (1955). Genera of Mucorales with notes on their synonymy. *Mycologia*, **47,** 344–63.

Hesseltine, C. W. (1957). The genus *Syzygites* (Mucoraceae). *Lloydia*, **20,** 228–37.

Hesseltine, C. W. (1961). Carotenoids in the fungi Mucorales: special reference to Choanephoraceae. *U.S. Dept. Agric. Tech. Bull.* No. 1245, 1–33.

Hesseltine, C. W. & Anderson, P. (1956). The genus *Thamnidium* and a study of the formation of its zygospores. *Am. J. Bot.* **43,** 696–703.

Hesseltine, C. W. & Anderson, P. (1957). Two genera of molds with low temperature requirements. *Bull. Torrey bot. Club*, **84,** 31–45.

Hesseltine, C. W. & Benjamin, C. R. (1957). Notes on the Choanephoraceae. *Mycologia*, **49,** 723–33.

Hesseltine, C. W., Benjamin, C. R. & Mehrotra, B. S. (1959). The genus *Zygorhynchus*. *Mycologia*, **51,** 173–94.

Hesseltine, C. W., Whitehill, A. R., Pidacks, C., Tenhagen, M., Bohonos, M., Hutchings, B. L. & Williams, J. H. (1953). Coprogen, a new growth factor present in dung required by *Pilobolus*. *Mycologia*, **45,** 7–19.

Hewitt, W. B. & Grogan, R. G. (1967). Unusual vectors of plant viruses. *A. Rev. Microbiol.* **21,** 205–24.
Hickman, C. J. (1958). *Phytophthora*—Plant destroyer. *Trans. Br. mycol. Soc.* **41,** 1–13.
Hirsch, H. E. (1950). No brachymeiosis in *Pyronema confluens. Mycologia,* **42,** 301–5.
Hirst, J. M. (1955). The early history of a potato blight epidemic. *Pl. Path.* **2,** 44–50.
Hirst, J. M. & Stedman, O. J. (1960*a*). The epidemiology of *Phytophthora infestans.* I. Climate, ecoclimate and the phenology of disease outbreak. *Ann. appl. Biol.* **48,** 471–88.
Hirst, J. M. & Stedman, O. J. (1960*b*). The epidemiology of *Phytophthora infestans.* II. The source of inoculum. *Ann. appl. Biol.* **48,** 489–517.
Hocking, D. (1967). Zygospore initiation, development and germination in *Phycomyces blakesleeanus. Trans. Br. mycol. Soc.* **50,** 207–20.
Hodgetts, W. J. (1917). On the forcible discharge of spores of *Leptosphaeria acuta. New Phytol.* **15,** 139–46.
Holliday, R. (1961*a*). The genetics of *Ustilago maydis. Genet Res. Camb.* **2,** 204–30.
Holliday, R. (1961*b*). Induced mitotic crossing-over in *Ustilago maydis. Genet. Res. Camb.* **2,** 231–48.
Holm, L. (1959). Some comments on the ascocarps of the Pyrenomycetes. *Mycologia,* **50,** 777–88.
Holm, L. & Müller, E. (1965). Nomina conservanda proposita. II. Proposals in Fungi. *Xylaria* Hill ex Greville. *Regnum veg.* **40,** 13.
Holton, C. S. (1936). Origin and production of morphologic and pathogenic strains of the oat fungi by mutation and hybridisation. *J. agric. Res.* **52,** 311–17.
Holton, C. S. (1959). Genetic controls of host-parasite interactions in smut diseases. In *Plant Pathology: Problems and Progress,* Chapter 15. Editor, Holton, C. S. Madison, U.S.A.: American Phytopathological Society.
Houwink, A. L. & Kreger, D. R. (1954). Observations on the cell wall of yeasts. *Antonie van Leenwenhoek,* **19,** 1–8.
Huber, J. (1958). Untersuchungen zur Physiologie insektentötender Pilze. *Arch. Mikrobiol.* **29,** 257–76.
Hughes, G. C. (1962). Seasonal periodicity of the Saprolegniaceae in the South-Eastern United States. *Trans. Br. mycol. Soc.* **45,** 519–31.
Hughes, S. J. (1953). Conidiophores, conidia, and classification. *Can. J. Bot.* **31,** 577–659.
Hüttig, W. (1931). Über den Einfluss der Temperatur auf die Keimung und Geschlechtsverteilung bei Brandpilzen. *Z. Bot.* **24,** 529–77.
Hunt, J. (1956). Taxonomy of the genus *Ceratocystis. Lloydia,* **19,** 1–58.
Hyde, H. A. & Williams, D. A. (1946). A daily census of *Alternaria* spores caught from the atmosphere at Cardiff in 1942 and 1943. *Trans. Br. mycol. Soc.* **29,** 78–85.
Ingold, C. T. (1939). *Spore Discharge in Land Plants.* 178 pp. Oxford University Press.
Ingold, C. T. (1946*a*). Spore discharge in *Daldinia concentrica. Trans. Br. mycol. Soc.* **29,** 43–51.
Ingold, C. T. (1946*b*). Size and form in agarics. *Trans. Br. mycol. Soc.* **29,** 108–13.
Ingold, C. T. (1948). The water relations of spore discharge in *Epichloe. Trans. Brit. mycol. Soc.* **31,** 277–80.
Ingold, C. T. (1953). *Dispersal in Fungi.* 197 pp. Oxford University Press.
Ingold, C. T. (1954*a*). The ascogenous hyphae in *Daldinia. Trans. Br. mycol. Soc.* **37,** 108–10.
Ingold, C. T. (1954*b*). Fungi and water. *Trans. Br. mycol. Soc.* **37,** 97–107.
Ingold, C. T. (1956). A gas phase in viable fungal spores. *Nature, Lond.* **177,** 1242–3.

Ingold, C. T. (1957). Spore liberation in higher fungi. *Endeavour*, **16,** 78–83.
Ingold, C. T. (1959). Jelly as a water reserve in fungi. *Trans. Br. mycol. Soc.* **42,** 475–8.
Ingold, C. T. (1965). *Spore Liberation.* 210 pp. Oxford University Press.
Ingold, C. T. (1966). Aspects of spore liberation: violent discharge. In *The Fungus Spore*, 113–32. Editor, Madelin, M. F. Colston Papers No. 18. London: Butterworths.
Ingold, C. T. & Cox, V. J. (1955). Periodicity of spore discharge in *Daldinia. Ann. Bot. Lond.* N. S. **19,** 201–9.
Ingold, C. T. & Dann, V. (1968). Spore discharge in fungi under very high surrounding air-pressure, and the bubble-theory of ballistospore release. *Mycologia*, **60,** 285–9.
Ingold, C. T. & Dring, V. J. (1957). An analysis of spore discharge in *Sordaria. Ann. Bot. Lond.* N. S. **21,** 465–7.
Ingold, C. T. & Hadland, S. A. (1959*a*). The ballistics of *Sordaria. New Phytol.* **58,** 46–57.
Ingold, C. T. & Hadland, S. A. (1959*b*). Phototropism and pigment production in *Sordaria* in relation to quality of light. *Ann. Bot. Lond.* N.S. **23,** 425–9.
Ingold, C. T. & Zoberi, M. H. (1963). The asexual apparatus of Mucorales in relation to spore liberation. *Trans. Br. mycol. Soc.* **46,** 115–34.
Ingram, M. (1955). *An Introduction to the Biology of Yeasts.* 273 pp. London: Pitman.
Jackson, H. S. (1931). Present evolutionary tendencies and the origin of life cycles in the Uredinales. *Mem. Torrey bot. Club*, **18,** 1–108.
James, P. W. (1965). A new check-list of British lichens. *Lichenologist*, **3,** 95–153.
Jeffreys, D. B. & Greulach, V. A. (1956). The nature of tropisms of *Coprinus sterquilinus. J. Elisha Mitchell scient. Soc.* **72,** 153–8.
Jenkins, W. A. (1934). The development of *Cordyceps agariciformia. Mycologia*, **26,** 220–43.
Jinks, J. L. (1952). Heterokaryosis: a system of adaptation in wild fungi. *Proc. R. Soc.* B, **140,** 83–99.
Jinks, J. L. & Simchen, G. (1966). A consistent nomenclature for the nuclear status of fungal cells. *Nature, Lond.* **210,** 778–80.
Johns, R. M. & Benjamin, R. K. (1954). Sexual reproduction in *Gonapodya. Mycologia*, **46,** 202–8.
Johnson, T. (1953). Variation in the rusts of cereals. *Biol. Rev.* **28,** 105–57.
Johnson, T. (1960). Genetics of pathogenicity. In *Plant Pathology: An Advanced Treatise*, **2,** 407–59. Editors, Horsfall, J. G. & Dimond, A. E. New York and London: Academic Press.
Johnson, T. (1961). Rust research in Canada and related plant-disease investigations. Publication 1098, Research Branch, Canada Dept. of Agriculture. 1–69.
Johnson, T. W. (1956). *The Genus Achlya: Morphology and Taxonomy.* 180 pp. The University of Michigan Press.
Jones, S. G. (1925). Life-history and cytology of *Rhytisma acerinum* (Pers.) Fries. *Ann. Bot. Lond.* **39,** 41–73.
Kanouse, B. B. (1936). Studies of two species of *Endogone* in culture. *Mycologia*, **28,** 47–62.
Karling, J. S. (1942). *The Plasmodiophorales.* 144 pp. New York: Published by the author.
Karling, J. S. (1944). *Phagomyxa algarum* n. gen., n. sp., an unusual parasite with Plasmodiophoralean and Protomyxean characteristics. *Am. J. Bot.* **31,** 38–52.
Karling, J. S. (1958). *Synchytrium fulgens* Schroeter. *Mycologia*, **50,** 373–5.
Karling, J. S. (1964). *Synchytrium.* 470 pp. New York and London: Academic Press.

Kennedy, L. L. (1958*a*). The genera of the Dacrymycetaceae. *Mycologia*, **50,** 874–95.
Kennedy, L. L. (1958*b*). The genus *Dacrymyces*. *Mycologia*, **50,** 896–915.
Kershaw, K. A. & Alvin, K. L. (1963). *The Observer's Book of Lichens*. 126 pp. London: Warne.
Kevorkian, A. G. (1937). Studies in the Entomophthoraceae. I. Observations on the genus *Conidiobolus*. *J. Agric. Univ. P. Rico*, **21,** 191–200.
Killian, C. (1919). Sur la sexualité de l'ergot de Seigle, le *Claviceps purpurea* (Tulasne). *Bull. Soc. mycol. Fr.* **35,** 182–97.
Kinugawa, K. (1965). On the growth of *Dictyophora indusiata*. II. Relations between the change in osmotic value of expressed sap and the conversion of glycogen to reducing sugar in tissues during receptaculum elongation. *Bot. Mag. Tokyo*, **78,** 171–6.
Kinugawa, K. & Furukawa, H. (1965). The fruit-body formation in *Collybia velutipes* induced by the lower temperature treatment of one short duration. *Bot. Mag. Tokyo*, **78,** 240–4.
Kirby, E. J. M. (1961). Host-parasite relations in the choke disease of grasses. *Trans. Br. mycol. Soc.* **44,** 493–503.
Kitani, Y., Olive, L. S. & El-Ani, A. S. (1962). Genetics of *Sordaria fimicola*. V. Aberrant segregation at the G locus. *Am. J. Bot.* **49,** 697–706.
Klein, D. T. (1960). Interrelations between growth rate and nuclear ratios in heterokaryons of *Neurospora crassa*. *Mycologia*, **52,** 137–47.
Kniep, H. (1919). Untersuchungen über den Antherenbrand (*Ustilago violacea* Pers.). Ein Beitrag zum Sexualitätsproblem. *Z. Bot.* **11,** 275–84.
Koch, W. J. (1951). Studies in the genus *Chytridium*, with observations on a sexually reproducing species. *J. Elisha Mitchell scient. Soc.* **67,** 267–78.
Koch, W. J. (1956). Studies of the motile cells of chytrids. I. Electron microscope observations of the flagellum, blepharoplast and rhizoplast. *Am. J. Bot.* **43,** 811–19.
Koch, W. J. (1958). Studies of the motile cells of chytrids. II. Internal structure of the body observed with light microscopy. *Am. J. Bot.* **45,** 59–72.
Koch, W. J. (1961). Studies of the motile cells of Chytrids. III. Major types. *Am. J. Bot.* **48,** 786–8.
Köhler, E. (1923). Über den derseitigen Stand der Erforschung des Kartoffelkrebses. *Arb. biol. BundAnst. Land -u. Forstw.* **11,** 289–315.
Köhler, E. (1931*a*). Der Kartoffelkrebs und sein Erreger (*Synchytrium endobioticum* (Schilb.) Perc.). *Landw. Jb.* **74,** 729–806.
Köhler, E. (1931*b*). Zur Biologie und Cytologie von *Synchytrium endobioticum* (Schilb.) Perc. *Phytopath. Z.* **4,** 43–55.
Köhler, E. (1956). Zur Kenntniss der Sexualität bei *Synchytrium*. *Ber. dt. bot. Ges.* **69,** 121–7.
Köhler, F. (1935). Genetische Studien an *Mucor mucedo* Brefeld I–III. *Z. indukt Abstamm. -u. Vererblehre*, **70,** 1–54.
Kole, A. P. (1954). A contribution to the knowledge of *Spongospora subterranea* (Wallr.) Lagerh., the cause of powdery scab of potatoes. 65 pp. Thesis. University of Wageningen.
Kole, A. P. (1965*a*). Resting-spore germination in *Synchytrium endobioticum*. *Neth. J. Plant. Path.* **71,** 72–8.
Kole, A. P. (1965*b*). Flagella. In *The Fungi: An Advanced Treatise*, **1,** 77–93. Editors, Ainsworth, G. C. & Sussman, A. S. New York and London: Academic Press.
Kole, A. P. & Gielink, A. J. (1961). Electron microscope observations on the flagella of the zoosporangial zoospores of *Plasmodiophora brassicae* and *Spongospora subterranea*. *Proc. K. ned. Akad. Wet.* C, **64,** 157–61.

Kole, A. P. & Gielink, A. J. (1962). Electron microscope observations on the resting spore germination of *Plasmodiophora brassicae*. *Proc. K. ned. Akad. Wet.* C, **65,** 117–21.

Kole, A. P. & Gielink, A. J. (1963). The significance of the zoosporangial stage in the life cycles of the Plasmodiophorales. *Neth. J. Plant. Path.* **69,** 258–62.

Kramer, C. L. (1961). Morphological development and nuclear behaviour in the genus *Taphrina*. *Mycologia*, **52,** 295–320.

Kreger, D. R. (1954). Observations on cell walls of yeasts and some other fungi by X-ray diffraction and solubility tests. *Biochim. Biophys. Acta*, **13,** 1–9.

Kreger-van Rij, N. J. W. (1964). A taxonomic study of the yeast genera *Endomycopsis*, *Pichia* and *Debaryomyces*. Thesis, Leiden University, V.R.B. Kleine der A 3–4, Gröningen, 1–194.

Kreisel, H. (1961). Die Lycoperdaceae der Deutschen Demokratischen Republik. *Reprium nov. Spec. Regni veg.* **64,** 89–201.

Krenner, J. A. (1961). Studies in the field of microscopic fungi. III. On *Entomophthora aphidis* H. Hoffm. with special regard to the family of the Entomophthoraceae in general. *Acta bot. hung.* **7,** 345–76.

Kuehn, H. H. (1956). Observations on the Gymnoascaceae. III. Developmental morphology of *Gymnoascus reessii*, a new species of *Gymnoascus* and *Eidamella deflexa*. *Mycologia*, **48,** 805–20.

Kuehn, H. H. (1958). A preliminary survey of the Gymnoascaceae. I. *Mycologia*, **50,** 417–39.

Kuehn, H. H. (1959). A preliminary survey of the Gymnoascaceae. II. *Mycologia*, **51,** 665–92.

Kuehn, H. H., Orr, G. F. & Ghosh, G. R. (1964). Pathological implications of Gymnoascaceae. *Mycopath. Mycol. appl.* **24,** 35–46.

Kühner, R. & Romagnesi, H. (1953). *Flore analytique des champignons supérieurs* (*Agarics*, *Bolets*, *Chanterelles*). 556 pp. Paris: Masson.

Kulkarni, U. K. (1963). Initiation of the dikaryon in *Claviceps microcephala* (Wallr.) Tul. *Mycopathologia*, **21,** 19–22.

Kusano, S. (1930*a*). The life-history and physiology of *Synchytrium fulgens* Schroet., with special reference to its sexuality. *Jap. J. Bot.* **5,** 35–132.

Kusano, S. (1930*b*). Cytology of *Synchytrium fulgens* Schroet. *J. Coll. Agric. imp. Univ. Tokyo*, **10,** 347–88.

Kwon, K.-J. & Raper, K. B. (1967). Sexuality and cultural characteristics of *Aspergillus heterothallicus*. *Am. J. Bot.* **54,** 36–48.

Lacoste, L. (1965). Biologie naturelle et culturale du genre *Leptosphaeria* Cesati & de Notaris. Determinisme de la réproduction sexuelle. Thèse de Doctorat és Sciences. Toulouse, 1965. 230 pp.

Laibach, F. (1927). Zytologische Untersuchungen über die Monoblepharideen *Jb. wiss. Bot.* **66,** 596–630.

Laffin, R. J. & Cutter, V. M. (1959), Investigations on the life cycle of *Sporidiobolus johnsonii*. I. Irradiation and cytological studies. *J. Elisha Mitchell scient. Soc.* **75,** 89–96.

Lakon, G. (1963). Entomophthoraceae. *Nova Hedwigia*, **5,** 7–26.

Lange, I. (1966). Das Bewegungsverhalten der Kerne in fusionierten Zellen von *Polystictus versicolor* (L). *Flora*, Abt. A, **156,** 487–97.

Lange, J. E. (1935–40). *Flora Agaricina Danica*. I–V. Copenhagen.

Lange, M. & Hora, F. B. (1963). *Collins Guide to Mushrooms and Toadstools* 257 pp. London: Collins.

Large, E. C. (1952). Surveys for choke (*Epichloe typhina*) in cocksfoot seed crops, 1951. *Pl. Path.* **1,** 23–8.

Large, E. C. (1958). *The Advance of the Fungi*. 488 pp. London: Jonathan Cape.

Last, F. T. (1955). Spore content of air within and above mildew-infected cereal crops. *Trans. Br. mycol. Soc.* **38,** 453–64.
Last, F. T. (1962). Analysis of effects of *Erysiphe graminis* DC on the growth of barley. *Ann. Bot. Lond.* N.S. **26,** 279–89.
Last, F. T. & Deighton, F. C. (1965). The non-parasitic microflora on the surfaces of living leaves. *Trans. Br. mycol. Soc.* **48,** 83–99.
Laundon, G. F. (1967). Terminology in the rust fungi. *Trans. Br. mycol. Soc.* **50,** 189–94.
Leadbeater, G. & Mercer, C. (1956). Zygospores in *Piptocephalis cylindrospora* Bain. *Trans. Br. mycol. Soc.* **39,** 17–20.
Leadbeater, G. & Mercer, C. (1957*a*). Zygospores in *Piptocephalis. Trans. Br. mycol. Soc.* **40,** 109–16.
Leadbeater, G. & Mercer, C. (1957*b*). *Piptocephalis virginiana* sp. nov. *Trans. Br. mycol. Soc.* **40,** 461–71.
Lebeau, J. B. & Hawn, E. J. (1963). Formation of hydrogen cyanide by the mycelial stage of a fairy ring fungus. *Phytopathology,* **53,** 1395–6.
Lentz, P. L. (1954). Modified hyphae of Hymenomycetes. *Bot. Rev.* **20,** 135–99.
Lessie, P. E. & Lovett, J. S. (1968). Ultrastructural changes during sporangium formation and zoospore differentiation in *Blastocladiella. Am. J. Bot.* **55,** 220–36.
Levi, M. P. & Preston, R. D. (1965). A chemical and microscopic examination of the action of the soft-rot fungus *Chaetomium globosum* on Beechwood (*Fagus sylv.*) *Holzforschung,* **19,** 183–90.
Levisohn, I. (1927). Beitrag zur Entwicklungsgesichte und Biologie von *Basidiobolus ranarum* Eidam. *Jb. wiss. Bot.* **66,** 513–55.
Lichtwardt, R. W. (1960). Taxonomic position of the Eccrinales and related fungi. *Mycologia,* **52,** 410–28.
Lilly, V. G. (1966). The effects of sterols and light on the production and germination of *Phytophthora* spores. In *The Fungus Spore,* 259–71. Editor, Madelin, M. F. Colston Papers No. 18. London: Butterworths.
Lilly, V. G. & Barnett, H. L. (1951). *Physiology of the Fungi.* 464 pp. New York: McGraw Hill.
Lindegren, C. C. (1949). *The Yeast Cell, its Genetics and Cytology.* St. Louis, U.S.A.: Educational Publishers.
Lindegren, C. C. & Lindegren, G. (1943). Segregation, mutation, and copulation in *Saccharomyces cerevisiae. Ann. Mo. bot. Gdn.* **30,** 453–68.
Linder, D. H. (1933). The genus *Schizophyllum.* I. Species of the Western Hemisphere. *Am. J. Bot.* **20,** 552–62.
Lingappa, B. T. (1958*a*). Development and cytology of the evanescent prosori of *Synchytrium brownii* Karling. *Am. J. Bot.* **45,** 116–23.
Lingappa, B. T. (1958*b*). The cytology of development and germination of resting spores of *Synchytrium brownii. Am. J. Bot.* **45,** 613–20.
Linnemann, G. (1941). Die Mucorineen Gattung *Mortierella* Coemans. *Pflanzenforschung,* Heft 23, 1–64.
Lissouba, P., Mousseau, J., Rizet, G. & Rossignol, J. L. (1962). Fine structure of genes in the Ascomycete *Ascobolus immersus. Adv. Genet.* **11,** 343–80.
Lodder, J. & Kreger-van Rij. (1952). *The Yeasts: A Taxonomic Study.* 713 pp. Amsterdam: North Holland Publishing Co.
Lodha, B. C. (1964*a*). Studies on coprophilous fungi. I. *Chaetomium. J. Ind. bot. Soc.* **43,** 121–40.
Lodha, B. C. (1964*b*). Studies on coprophilous fungi. II. *Chaetomium. Antonie van Leeuwenhoek.* **20,** 163–7.
Lorenz, F. (1933). Beiträge zur Entwicklungsgeschichte von *Sphaerobolus. Arch. Protistenk.* **81,** 361–98.

Lowry, B. (1951). A morphological basis for classifying the species of *Auricularia*. *Mycologia*, **43**, 351–58.

Lowry, B. (1952). The genus *Auricularia*. *Mycologia*, **44**, 656–92.

Lowry, R. J. & Sussman, A. S. (1958). Wall structure of ascospores of *Neurospora tetrasperma*. *Am. J. Bot.* **45**, 397–403.

Lu, B. C. (1964). Polyploidy in the Basidiomycete *Cyathus stercoreus*. *Am. J. Bot.* **51**, 343–7.

Lu, B. C. & Brodie, H. J. (1964). Preliminary observation of meiosis in the fungus *Cyathus*. *Can. J. Bot.* **42**, 307–10.

Lucas, M. T. & Webster, J. (1967). Conidial states of British species of *Leptosphaeria*. *Trans. Br. mycol. Soc.* **50**, 85–121.

Luttrell, E. S. (1951). Taxonomy of the Pyrenomycetes. *Univ. Mo. Stud.* **24**, (3), 1–120.

Luttrell, E. S. (1955). The ascostromatic ascomycetes. *Mycologia*, **47**, 511–32.

Luttrell, E. S. (1965*a*). Classification of the Loculoascomycetes. *Phytopathology*, **55**, 828–33.

Luttrell, E. S. (1965*b*). Paraphysoids, pseudoparaphyses, and apical paraphyses. *Trans. Br. mycol. Soc.* **48**, 135–44.

Lyr, H. (1953). Zur Kenntniss der Ernahrungsphysiologie der Gattung *Pilobolus*. *Arch. Mikrobiol.* **19**, 402–34.

Lythgoe, J. N. (1958). Taxonomic notes on the genera *Helicostylum* and *Chaetostylum* (Mucoraceae). *Trans. Br. mycol. Soc.* **41**, 135–41.

Lythgoe, J. N. (1961). Effect of light and temperature on growth and development in *Thamnidium elegans* Link. *Trans. Br. mycol. Soc.* **44**, 199–213.

Lythgoe, J. N. (1962). Effect of light and temperature on sporangium development in *Thamnidium elegans* Link. *Trans. Br. mycol. Soc.* **45**, 161–8.

Maas Geesteranus, R. A. (1963*a*). Hyphal structures in Hydnums. II. *Proc. K. ned. Akad. Wet.* (C), **66**, 426–36.

Maas Geesteranus, R. A. (1963*b*). Hyphal structures in Hydnums. IV. *Proc. K. ned. Akad. Wet.* (C), **66**, 447–57.

Macdonald, J. A. (1937). A study of *Polyporus betulinus* (Bull.) Fr. *Ann. appl. Biol.* **24**, 289–310.

Macer, R. C. F. (1959). Pathology. *A. Rept. Plant Breeding Institute*, Cambridge, 1958/9. 60–3.

Macfarlane, I. (1952). Factors affecting the survival of *Plasmodiophora brassicae* Wor. in the soil and its assessment by a host test. *Ann. appl. Biol.* **39**, 239–56.

Macfarlane, I. (1959). A solution-culture technique for obtaining root-hair, or primary, infection by *Plasmodiophora brassicae*. *J. gen. Microbiol.* **18**, 720–32.

Macfarlane, I. & Last, F. T. (1959). Some effects of *Plasmodiophora brassicae* Woron. on the growth of the young cabbage plant. *Ann. Bot. Lond.* N.S. **23**, 547–70.

Machlis, L. (1958*a*). Evidence for a sexual hormone in *Allomyces*. *Physiol. Pl.* **11**, 181–92.

Machlis, L. (1958*b*). A study of sirenin, the chemotactic sexual hormone from the watermold *Allomyces*. *Physiol. Pl.* **11**, 845–54.

Maclean, N. (1964). Electron microscopy of a fission yeast, *Schizosaccharomyces pombe*. *J. Bact.* **88**, 1459–66.

Macleod, D. M. (1963). Entomophthorales infections. In *Insect Pathology: An Advanced Treatise*, **2**, 189–231. Editor, Steinhaus, E. A. New York and London: Academic Press.

Madelin, M. F. (1956). The influence of light and temperature on fruiting of *Coprinus lagopus* Fr. in pure culture. *Ann. Bot. Lond.* N.S. **20**, 467–80.

Malençon, G. (1938). Les truffes européenes: historique, morphogénie, organographie, classification, culture. *Rev. mycol. Paris.* (Mém. hors série), 1–92.

Manier, J. F. (1963). Current status of the knowledge on Trichomycetes. *Archs. Zool. exp. gén.* **102,** 201–10.

Manier, J. F. (1964). Position systematique des Trichomycètes. *Archs. Zool. exp. gén.* **104,** 95–8.

Manocha, M. S. (1965). Fine structure of the *Agaricus* carpophore. *Can. J. Bot.* **43,** 1329–34.

Manocha, M. S. & Shaw, M. (1966). The physiology of host-parasite relations. XVI. Fine structure of the nucleus in rust-infected mesophyll cells of wheat. *Can. J. Bot.* **44,** 669–73.

Manton, I., Clarke, B., & Greenwood, A. D. (1951). Observations with the electron microscope on a species of *Saprolegnia. J. exp. Bot.* **2,** 321–31.

Marchant, R., Peat, A. & Banbury, G. H. (1967). The ultrastructural basis of hyphal growth. *New Phytol.* **66,** 623–9.

Marshall, G. M. (1960*a*). The incidence of certain seed-borne diseases in commercial seed-samples. II. Ergot, *Claviceps purpurea* (Fr.) Tul. in cereals. *Ann. appl. Biol.* **48,** 19–26.

Marshall, G. M. (1960*b*). The incidence of certain seed-borne diseases in commercial seed-samples. IV. Bunt of wheat, *Tilletia caries* (DC) Tul. V. Earcockles of wheat, *Anguina tritici* (Stein.) Filipjer. *Ann. appl. biol.* **48,** 34–8.

Martin, E. (1940). The morphology of *Taphrina deformans. Am. J. Bot.* **27,** 743–51.

Martin, G. W. (1925). Morphology of *Conidiobolus villosus. Bot. Gaz.* **80,** 311–18.

Martin, G. W. (1927). Basidia and spores of the Nidulariaceae. *Mycologia,* **19,** 239–47.

Martin, G. W. (1952). Revision of the North Central Tremellales. *Univ. Iowa. Stud. Nat. Hist.* **19** (3), 1–122.

Martin, G. W. (1957). The Tulasnelloid fungi and their bearing on basidial terminology. *Brittonia,* **9,** 25–30.

Martin, G. W. (1961). Key to the families of fungi. In Ainsworth, G. C. (1961). *Ainsworth & Bisby's Dictionary of the Fungi.* 5th edition, 497–517. Kew, Surrey: Commonwealth Mycological Institute.

Masri, S. S. & Ellingboe, A. H. (1966). Primary infection of wheat and barley by *Erysiphe graminis. Phytopathology,* **56,** 389–95.

Mather, K. & Jinks, J. L. (1958). Cytoplasm in sexual reproduction. *Nature, Lond.* **182,** 1188–90.

McClary, D. O. (1964). The cytology of yeasts. *Bot. Rev.* **30,** 167–225.

McClary, D. O., Williams, M. A., Lindegren, C. C. & Ogur, M. (1957). Chromosome counts in a polyploid series of *Saccharomyces. J. Bact.* **73,** 360–4.

McClintock, B. (1945). *Neurospora.* I. Preliminary observations on the chromosomes of *Neurospora crassa. Am. J. Bot.* **32,** 671–8.

McCranie, J. (1942). Sexuality in *Allomyces cystogenus. Mycologia,* **34,** 209–13.

McEwen, F. L. (1963). *Cordyceps* infections. In *Insect Pathology: An Advanced Treatise,* **2,** 273–90. Editor, Steinhaus, E. A. New York and London: Academic Press.

McIntosh, D. L. (1954). A cytological study of ascus development in *Pyronema confluens* Tul. *Can. J. Bol.* **32,** 440–6.

McKay, R. (1957). The longevity of the oospores of onion downy mildew *Peronospora destructor* (Berk.) Casp. *Scient. Proc. R. Dubl. Soc.* N.S. **27,** 295–307.

McKeen, W. E. (1962). The flagellation, movement and encystment of some Phycomycetous zoospores. *Can. J. Microbiol.* **8,** 897–904.

McKeen, W. E., Mitchell, N. & Smith, R. (1967). The *Erysiphe cichoracearum* conidium. *Can. J. Bot.* **45,** 1489–96.

McKeen, W. E., Smith, R. & Mitchell, N. (1966). The haustorium of *Erysiphe cichoracearum* and the host-parasite interface on *Helianthus annuus*. *Can. J. Bot.* **44,** 1299–306.

McMeekin, D. (1960). The role of the oospores of *Peronospora parasitica* in downy mildew of crucifers. *Phytopathology*, **50,** 93–7.

McMorris, T. C. & Barksdale, A. W. (1967). Isolation of a sex-hormone from the water-mould *Achlya bisexualis*. *Nature, Lond.* **215,** 320–1.

McNabb, R. F. R. (1965). Taxonomic studies in the Dacrymycetaceae. II. *Calocera* (Fries) Fries. *N.Z. J. Bot.* **3,** 31–58.

Meier, H. & Webster, J. (1954). An electron microscope study of zoospore cysts in the Saprolegniaceae. *J. exp. Bot.* **5,** 401–9.

Menzies, B. P. (1953). Studies on the systemic fungus *Puccinia suaveolens*. *Ann. Bot. Lond.* N.S. **17,** 551–68.

Meredith, D. S. (1965). Violent spore release in *Stemphylium botryosum* Wallr. *Pl. Dis. Reptr.* **49,** 1006.

Middleton, J. T. (1943). The taxonomy, host range and geographic distribution of the genus *Pythium*. *Mem. Torrey. bot. Club*, **20,** 1–171.

Middleton, J. T. (1952). Generic concepts in the Pythiaceae. *Tijd. Pl. Ziekt.* **58,** 226–35.

Miller, C. E. (1959). Studies on the life cycle and taxonomy of *Ligniera verrucosa*. *Am. J. Bot.* **46,** 725–9.

Miller, J. H. (1928). Biologic studies in the Sphaeriales. *Mycologia*, **20,** 187–213.

Miller, J. H. (1930). British Xylariaceae. *Trans. Br. mycol. Soc.* **15,** 134–54.

Miller, J. H. (1932*a*). British Xylariaceae. II. *Trans. Br. mycol. Soc.* **17,** 125–35.

Miller, J. H. (1932*b*). British Xylariaceae. III. A revision of specimens in the Herbarium of the Royal Botanic Gardens, Kew. *Trans. Br. mycol. Soc.* **17,** 136–46.

Miller, J. H. (1959). A revision of the classification of the Ascomycetes with special emphasis on the Pyrenomycetes. *Mycologia*, **41,** 99–127.

Miller, J. H. (1961). *A Monograph of the World Species of Hypoxylon*. 158 pp. University of Georgia Press.

Mitchell, M. B. (1965). Characteristics of developing asci of *Neurospora crassa*. *Can. J. Bot.* **43,** 933–9.

Mix, A. J. (1949). A monograph of the genus *Taphrina*. *Kansas Univ. Sci. Bull.* **33,** 3–167.

Moor, H. & Mühlethaler, K. (1963). Fine structure in frozen-etched yeast cells. *J. Cell Biol.* **17,** 609–28.

Moore, E. J. (1963). The ontogeny of the apothecia in *Pyronema domesticum*. *Am. J. Bot.* **50,** 37–44.

Moore, E. J. & Korf, R. P. (1963). The genus *Pyronema*. *Bull. Torrey bot. Club*, **90,** 33–42.

Moore, R. T. (1965). The ultrastructure of fungal cells. In *The Fungi: An Advanced Treatise*, **1,** 95–118. Editors, Ainsworth, G. C. & Sussman, A. S. New York and London: Academic Press.

Moore, R. T. & McAlear, J. H. (1961). Fine structure of mycota. 5. Lomasomes—previously uncharacterised hyphal structures. *Mycologia*, **53,** 194–200.

Moore, R. T. & McAlear, J. H. (1962). Fine structure of mycota. 7. Observations on septa of Ascomycetes and Basidiomycetes. *Am. J. Bot.* **49,** 86–94.

Moore, W. C. (1959). *British Parasitic Fungi*. 430 pp. Cambridge University Press.

Moreau, C. (1953). Les genres *Sordaria* et *Pleurage*, leurs affinités systématiques. *Encyclopédie Mycologique*, **15,** 1–330. Paris: Lechevalier.

Morisset, E. (1963). Recherches sur le Pyrenomycete *Sporormia leporina* Niessl (Pléosporale Sordarioide). *Rev. gén. Bot.* **60,** 69–106.

Morgan, M. T. (1929). Report of alleged ergot poisoning by rye bread in Manchester. *J. Hyg. Camb.* **29,** 51–61.

Mortimer, R. K. & Hawthorne, D. C. (1966). Yeast genetics. *A. Rev. Microbiol.* **20,** 151–68.

Morton, D. J. (1961). Trypan blue and boiling lactophenol for staining and cleaning barley tissues infected with *Ustilago nuda. Phytopathology,* **51,** 27–9.

Moseman, J. G. (1966). Genetics of powdery mildews. *A. Rev. Pl. Pathol.* **4,** 269–90.

Moseman, J. G. & Powers, H. R. (1957). Function and longevity of cleistothecia of *Erysiphe graminis* f. sp. hordei. *Phytopathology,* **47,** 53–6.

Moser, M. (1967). *Die Röhrlinge und Blatterpilze* (*Agaricales*). Kleine Kryptogamenflora. Bd. IIb2. Basidiomyceten II Teil. Editor, Gams, H. Stuttgart: Fischer.

Mosse, B. (1959). The regular germination of resting spores and some observations on the growth requirements of an *Endogone* sp. causing vesicular arbuscular mycorrhiza. *Trans. Br. mycol. Soc.* **42,** 273–86.

Mosse, B. (1963). Vesicular-arbuscular mycorrhiza: an extreme form of fungal adaptation. In *Symbiotic Associations.* 13th *Symp. Soc. gen. Microbiol.* 146–70. Editors, Nutman, P. S. and Mosse, B. Cambridge University Press.

Müller, D. (1954). Die Abschleuderung der Sporen von *Sporobolomyces*—Spiegelhefe—gefilmt. *Friesia,* **6,** 65–74.

Müller, E. & Arx, J. S. von. (1962). *Die Gattungen der didymosporen Pyrenomyceten. Beitr. KryptogFlora Schweiz.* Bd. 11, (2). 1–922.

Müller, E. & Tomasevic, M. (1957). Kulturversuche mit einigen Arten der Gattung *Leptosphaeria* Ces. & de Not. *Phytopath. Z.* **29,** 287–94.

Müller, G. (1964). Zur Kenntniss der Gattung *Endomycopsis* Stelling-Dekker. *Zbl. Bakt.* Abt. 2, **118,** 40–3.

Müller, K. O. (1959). Hypersensitivity. In *Plant Pathology: An Advanced Treatise,* **1,** 469–519. Editors, Horsfall, J. G. & Dimond, A. E. New York and London: Academic Press.

Müller-Kögler, E. (1959). Zur Isolierung und Kultur insektpathogener Entomophthoraceen. *Entomophaga,* **4,** 251–74.

Mullins, J. T. & Raper, J. R. (1965). Heterothallism in biflagellate aquatic fungi: preliminary genetic analysis. *Science, N.Y.* **150,** 1174–5.

Nagai, M. & Takahashi, T. (1962). Electron microscope observations on the zoospores of *Saprolegnia diclina. Trans. mycol. Soc. Japan.* **3,** 19–23.

Nannfeldt, J. A. (1932). Studien über die Morphologie und Systematik der nichtlichenisierten inoperculaten Discomyceten. *Nova Acta R. Soc. Scient. upsal.* IV, **8** (2), 1–368.

Nannfeldt, J. A. (1942). The Geoglossaceae of Sweden. *Ark. Bot.* Bd. 30 A, No. 4, 67 pp.

Nicolson, T. H. (1967). Vesicular-arbuscular mycorrhiza—a universal plant symbiosis. *Sci. Progr. Oxf.* **55,** 561–81.

Nicolson, T. H. & Gerdemann, J. W. (1968). Mycorrhizal *Endogone* species. *Mycologia,* **60,** 313–25.

Nobles, M. K. (1958*a*). A rapid test for extracellular oxidase in cultures of wood-inhabiting Hymenomycetes. *Can. J. Bot.* **36,** 91–9.

Nobles, M. K. (1958*b*). Cultural characters as a guide to the taxonomy and phylogeny of the Polyporaceae. *Can. J. Bot.* **36,** 883–926.

Nobles, M. K. (1965). Identification of cultures of wood-inhabiting Hymenomycetes. *Can. J. Bot.* **43,** 1097–139.

Northcote, D. H. & Horne, R. W. (1952). The chemical composition of the yeast cell wall. *Biochem. J.* **51,** 232–6.

Nyland, G. (1949). Studies on some unusual Heterobasidiomycetes from Washington State. *Mycologia*, **41**, 686–701.

Nyland, G. (1950). The genus *Tilletiopsis*. *Mycologia*, **42**, 487–96.

Ogilvie, L. & Thorpe, I. G. (1966). Black stem rust of wheat in Great Britain. In *Cereal Rust Conferences, Cambridge*, 1964, 172–6. Cambridge: Plant Breeding Institute.

Olive, L. S. (1952). Studies on the morphology and cytology of *Itersonilia perplexans* Derx. *Bull. Torrey bot. Club*, **79**, 126–38.

Olive, L. S. (1953). The structure and behaviour of fungus nuclei. *Bot. Rev.* **19**, 439–586.

Olive, L. S. (1956). Genetics of *Sordaria fimicola*. I. Spore color mutants. *Am. J. Bot.* **43**, 97–107.

Olive, L. S. (1963). Genetics of homothallic fungi. *Mycologia*, **55**, 93–103.

Olive, L. S. (1964). Spore discharge mechanism in Basidiomycetes. *Science, N.Y.* **146**, 542–3.

Olive, L. S. (1965). Nuclear behavior during meiosis. In *The Fungi: An Advanced Treatise*, **1**, 143–61. Editors, Ainsworth, G. C. & Sussman, A. S. New York and London: Academic Press.

Olive, L. S. (1967). The Protostelida—a new order of the Mycetozoa. *Mycologia*, **59**, 1–29.

Olive, L. S. & Stoianovitch, C. (1966). A simple new mycetozoan with ballistospores. *Am. J. Bot.* **53**, 344–9.

Orr, G. F., Kuehn, H. H. & Plunkett, O. A. (1963). The genus *Gymnoascus* Baranetzky. *Mycopath. Mycol. appl.* **21**, 1–18.

Page, R. M. (1962). Studies on the development of asexual reproductive structures in *Pilobolus*. *Mycologia*, **48**, 206–24.

Page, R. M. (1959). Stimulation of asexual reproduction of *Pilobolus* by *Mucor plumbeus*. *Am. J. Bot.* **46**, 579–85.

Page, R. M. (1960). The effect of ammonia on growth and reproduction of *Pilobolus kleinii*. *Mycologia*, **52**, 480–9.

Page, R. M. (1964). Sporangium discharge in *Pilobolus*: a photographic study. *Science, N.Y.* **146**, 925–7.

Page, R. M. & Kennedy, D. (1964). Studies on the velocity of discharged sporangia of *Pilobolus kleinii*. *Mycologia*, **56**, 363–8.

Palmer, J. T. (1955). Observations on Gasteromycetes. 1–3. *Trans. Br. mycol. Soc.* **38**, 317–34.

Papavizas, G. C. & Davey, C. B. (1960). Some factors affecting growth of *Aphanomyces euteiches* in synthetic media. *Am. J. Bot.* **47**, 758–65.

Papazian, H. P. (1950*a*). Physiology of the incompatibility factors in *Schizophyllum commune*. *Bot. Gaz.* **112**, 143–63.

Papazian, H. P. (1950*b*). A convenient method of growing fruiting bodies of a bracket fungus. *Bot. Gaz.* **112**, 138.

Parag, Y. (1965). Genetic investigation into the mode of action of the genes controlling self-incompatibility and heterothallism in Basidiomycetes. In *Incompatibility of Fungi*, 80–98. Editors, Esser, K. & Raper, J. R. Berlin: Springer.

Paravicini, E. (1917). Untersuchungen über das Verhalten der Zellkerne bei der Fortpflanzung der Brandpilze. *Annls. mycol.* **15**, 57–96.

Parker-Rhodes, A. F. (1955). Fairy ring kinetics. *Trans. Br. mycol. Soc.* **38**, 59–72.

Parmeter, J. R., Snyder, W. C. & Reichle, R. E. (1963). Heterokaryosis and variability in plant-pathogenic fungi. *A. Rev. Phytopath.* **1**, 51–76.

Peace, W. R. (1962). *Pathology of Trees and Shrubs with Special Reference to Britain*, 722 pp. Oxford University Press.

Peat, A. & Banbury, G. H. (1967). Ultrastructure, protoplasmic streaming, growth and tropisms of *Phycomyces* sporangiophores. *New Phytol.* **66,** 475–84.

Pegler, D. N. (1966). 'Polyporaceae'—Part I, with a key to British genera. *News Bull. Br. mycol. Soc.* **26,** 14–28.

Pegler, D. N. (1967). 'Polyporaceae'—Part II, with a key to world genera. *Bull. Br. mycol. Soc.* **1,** 17–38.

Pendergrass, W. R. (1950). Studies on the Plasmodiophoraceous parasite, *Octomyxa brevilegniae. Mycologia,* **42,** 279–89.

Perrott, P. E. (1955). The genus *Monoblepharis. Trans. Br. mycol. Soc.* **38,** 247–82.

Perrott, P. E. (1958). Isolation and pure culture of *Monoblepharis. Nature, Lond.* **182,** 1322–4.

Perrott, P. E. (1960). The ecology of some aquatic phycomycetes. *Trans. Br. mycol. Soc.* **43,** 19–30.

Petch, T. (1936). *Cordyceps militaris* and *Isaria farinosa. Trans. Br. mycol. Soc.,* **20,** 216–24.

Petch, T. (1938). British Hypocreales. *Trans. Br. mycol. Soc.* **21,** 243–305.

Petrak, F. (1962). Über die Gattungen *Xylosphaera* Dum. und *Xylosphaeria* Otth. *Sydowia,* **15,** 188–209.

Peyton, G. A. & Bowen, C. C. (1963). The host-parasite interface of *Peronospora manshurica* on *Glycine max. Am. J. Bot.* **50,** 787–97.

Phaff, H. J. (1963). Cell wall of yeasts. *A. Rev. Microbiol.* **17,** 15–30.

Phaff, H. J. & Yoneyama, M. (1961). *Endomycopsis scolyti,* a new heterothallic species of yeast. *Antonie van Leeuwenhoek.* **27,** 196–202.

Piard-Douchez, Y. (1949). Le *Spongospora subterranea* et son action pathogène. *Ann. Sci. nat. Bot.* Sér. **11,** 91–122.

Pidacks, C., Whitehill, A. R., Pruess, L. M. Hesseltine, C. W., Hutchings, B. C., Bohonos, N. & Williams, J. H. (1953). Coprogen, the isolation of a new growth factor required by *Pilobolus* species. *J. Am. chem. Soc.* **75,** 6064–5.

Piehl, A. (1929). The cytology and morphology of *Sordaria fimicola. Trans. Wis. Acad. Sci. Arts. Lett.* **24,** 323–41.

Pittenger, R. H. & Atwood, K. C. (1956). Stability of nuclear proportions during growth of *Neurospora* heterokaryons. *Genetics,* **41,** 227–41.

Pittenger, R. H., Kimball, A. W. & Atwood, K. C. (1955). Control of nuclear ratios in *Neurospora* heterokaryons. *Am. J. Bot.* **42,** 954–8.

Plempel, M. (1957). Die Sexualstoffe der Mucoraceae. *Arch. Mikrobiol.* **26,** 151–74.

Plempel, M. (1963). Die chemischen Grundlagen der Sexualreaktion bei Zygomyceten. *Planta,* **59,** 492–508.

Plunkett, B. E. (1953). Nutritional and other aspects of fruit body formation in pure cultures of *Collybia velutipes* (Curt.) Fr. *Ann. Bot. Lond.* N.S. **17,** 193–217.

Plunkett, B. E. (1956). The influence of factors of the aeration complex and light upon fruit-body form in pure culture of an Agaric and a Polypore. *Ann. Bot. Lond.* N.S. **20,** 563–86.

Plunkett, B. E. (1958). Translocation and pileus formation in *Polyporus brumalis. Ann. Bot. Lond.* N.S. **22,** 237–49.

Plunkett, B. E. (1961). The change of tropism of *Polyporus brumalis* stipes and the effect of directional stimuli on pileus differentiation. *Ann. Bot. Lond.* N.S. **25,** 206–23.

Poitras, A. W. (1955). Observations on asexual and sexual reproductive structures of the Choanephoraceae. *Mycologia,* **47,** 702–13.

Pontecorvo, G. (1956). The parasexual cycle in fungi. *A. Rev. Microbiol.* **10,** 393–400.

Popp, W. (1958). An improved method of detecting loose-smut mycelium in whole embryos of wheat and barley. *Phytopathology,* **48,** 641–3.

Prince, A. E. (1943). Basidium formation and spore discharge in *Gymnosporangium nidus-avis*. *Farlowia* **1,** 79–93.

Prusso, D. C. & Wells, K. (1967). *Sporobolomyces roseus*. I. Ultrastructure. *Mycologia*, **59,** 337–48.

Quispel, A. (1959). Lichens. In *Encyclopedia of Plant Physiology*, **11,** 577–604. Editor, Ruhland, W. Berlin: Springer.

Quantz, L. (1943). Untersuchungen über die Ernährungsphysiologie einiger niederer Phycomyceten (*Allomyces kniepii*, *Blastocladiella variabilis* und *Rhizophlyctis rosea*). *Jb. wiss. Bot.* **91,** 120–68.

Ragab, M. A. (1953). Hyphal systems of *Auriscalpium vulgare*. *Bull. Torrey bot. Club*, **80,** 21–5.

Raggi, V. (1967). Changes in peach trees (cv. Red Haven) attacked by *Taphrina deformans*, with particular reference to nitrogen metabolism in infected and non-infected leaves. *Can. J. Bot.* **45,** 459–77.

Ramsbottom, J. (1953). *Mushrooms and Toadstools. A Study of the Activities of Fungi*. 306 pp. London: Collins.

Ramsbottom, J. & Stephens, F. L. (1935). *Neurospora* in Britain. *Trans. Br. mycol. Soc.* **19,** 215–20.

Raper, J. R. (1939). Sexual hormones in *Achlya*. I. Indicative evidence of a hormonal coordinating mechanism. *Am. J. Bot.* **26,** 639–50.

Raper, J. R. (1950). Sexual hormones in *Achlya*. VII. The hormonal mechanism in homothallic species. *Bot. Gaz.* **112,** 1–24.

Raper, J. R. (1952). Chemical regulation of the sexual processes in the Thallophytes. *Bot. Rev.* **18,** 447–545.

Raper, J. R. (1954). Life cycles, sexuality, and sexual mechanisms in the fungi. In *Sex in Microorganisms*, 42–81. Editor, Wenrich, D. H. Washington: American Association for the Advancement of Science.

Raper, J. R. (1957). Hormones and sexuality in lower plants. *Symp. Soc. exp. Biol.* **11,** 143–65.

Raper, J. R. (1959). Sexual versatility and evolutionary processes in fungi. *Mycologia*, **51,** 107–24.

Raper, J. R. (1966). *Genetics of Sexuality in Higher Fungi*. 283 pp. New York and London: Ronald Press Company.

Raper, J. R., Krongelb, G. S. & Baxter, M. G. (1958). The number and distribution of incompatibility factors in *Schizophyllum commune*. *Amer. Nat.* **92,** 221–32.

Raper, J. R. & Miles, P. G. (1958). The genetics of *Schizophyllum commune*. *Genetics, Princeton*, **43,** 530–46.

Raper, K. B. (1952). A decade of antibiotics in America. *Mycologia*, **44,** 1–59.

Raper, K. B. (1957). Nomenclature in *Aspergillus* and *Penicillium*. *Mycologia*, **49,** 644–62.

Raper, K. B. & Fennell, D. I. (1952). Homothallism vs heterothallism in the *Penicillium luteum* series. *Mycologia*, **44,** 101–11.

Raper, K. B. & Fennell, D. I. (1965). *The genus Aspergillus*. 686 pp. Baltimore: Williams and Wilkins.

Raper, K. B. & Thom, C. (1949). *A Manual of the Penicillia*. 875 pp. London: Baillière, Tindall and Cox.

Reichle, R. E. & Fuller, M. S. (1967). The fine structure of *Blastocladiella emersonii* zoospores. *Am. J. Bot.* **54,** 81–92.

Reiff, F., Kautzmann, R., Luers, H. & Lindemann, M. (1960). Editors. *Die Hefen*. Bd. 1. *Die Hefen in der Wissenschaft*. 1024 pp. Nürnberg: Hans Carl.

Reijnders, A. F. M. (1963). *Les problèmes du développment des carpophores des Agaricales et de quelques groupes voisins*. 412 pp. The Hague: Junk.

Reischer, H. S. (1951*a*). Growth of Saprolegniaceae in synthetic media. I. Inorganic nutrition. *Mycologia*, **43**, 142–55.

Renaud, F. L. & Swift, H. (1964). The development of basal bodies and flagella in *Allomyces arbusculus*. *J. Cell. Biol.* **23**, 339–54.

Remacle, J. & Moreau, C. (1962). Le *Pleurage tetraspora* (Wint.) Griffiths est-il une forme à asques tetrasporés du *Pleurage curvula* (de Bary) Kuntze. *Rev. mycol. Paris*, **27**, 213–17.

Reeves, F. (1967). The fine structure of ascospore formation in *Pyronema domesticum*. *Mycologia*, **59**, 1018–33.

Riddle, L. W. (1906). On the cytology of the Entomophthoraceae. *Proc. Am. Acad. Arts. Sci.* **42,** 177–200.

Rifai, M. & Webster, J. (1966). Culture studies on *Hypocrea* and *Trichoderma*. III. *H. lactea* (= *H. citrina*) and *H. pulvinata*. *Trans. Br. mycol. Soc.* **49,** 297–310.

Rishbeth, J. (1952). Control of *Fomes annosus* Fr. *Forestry*, **25,** 41–50.

Rishbeth, J. (1961). Inoculation of pine stumps against infection by *Fomes annosus*. *Nature, Lond.* **191,** 826–7.

Ritchie, D. (1937). The morphology of the perithecium of *Sordaria fimicola*. *J. Elisha Mitchell scient. Soc.* **53,** 334–42.

Ritchie, D. (1948). The development of *Lycoperdon oblongisporum*. *Am. J. Bot.* **35,** 215–19.

Robertson, J. & Ashby, H. T. (1928). Ergot poisoning among rye bread consumers. *Brit. med. J.* **3503,** 302–3.

Robinow, C. F. (1957*a*). The structure and behavior of the nuclei in spores and growing hyphae of Mucorales. I. *Mucor hiemalis* and *Mucor fragilis*. *Can. J. Microbiol.* **3,** 771–89.

Robinow, C. F. (1957*b*). The structure and behavior of the nuclei in spores and growing hyphae of Mucorales. II. *Phycomyces blakesleeanus*. *Can. J. Microbiol.* **3,** 791–8.

Robinow, C. F. (1962). Some observations on the mode of division of somatic nuclei of *Mucor* and *Allomyces*. *Arch. Mikrobiol.* **42,** 369–77.

Robinow, C. F. & Bakerspigel, A. (1965). Somatic nuclei and forms of mitosis in fungi. In *The Fungi: An Advanced Treatise*, **1,** 119–42. Editors, Ainsworth, G. C. & Sussman, A. S. New York and London: Academic Press.

Robinow, C. F. & Marak, J. (1966). A fiber apparatus in the nucleus of the yeast cell. *J. Cell Biol.* **29,** 129–51.

Roelofsen, P. A. (1950). The origin of spiral growth in *Phycomyces* sporangiophores. *Recl. Trav. bot. néerl.* **42,** 73–110.

Roelofsen, P. A. (1959). *The Plant Cell Wall*. Handbuch der Pflanzenanatomie, III (4). Berlin: Gebrüder Borntraeger.

Roelofsen, P. A. & Hoette, I. (1951). Chitin in the cell wall of yeasts. *Antonie van Leeuwenhoek* **17,** 27–43.

Romagnesi, H. (1944). La cystide chez les Agaricacées. *Rev. mycol. Paris*, **9,** (Suppl.) 4–21.

Roper, J. R. (1966). Mechanisms of inheritance. 3. The parasexual cycle. In *The Fungi: An Advanced Treatise*, **2,** 589–617. Editors, Ainsworth, G. C. & Sussman, A. New York and London: Academic Press.

Rosinski, M. A. (1961). Development of the ascocarp of *Ceratocystis ulmi*. *Am. J. Bot.* **48,** 285–93.

Sahtiyanci, S. (1962). Studien über einige wurzelparasitäre Olpidiaceen. *Arch. Mikrobiol.* **41,** 187–228.

Salaman, R. N. (1949). *The History and Social Influence of the Potato*. 685 pp. Cambridge University Press.

Salvin, S. B. (1941). Comparative studies on the primary and secondary zoospores of the Saprolegniaceae. I. Influence of temperature. *Mycologia*, **33**, 592–600.

Salvin, S. B. (1942*a*). Factors controlling sporangial type in *Thraustotheca primoachlya* and *Dictyuchus achlyoides*. I. *Am. J. Bot.* **29**, 97–104.

Salvin, S. B. (1942*b*). Preliminary report on the intergeneric mating of *Thraustotheca clavata* and *Achlya flagellata*. *Am. J. Bot.* **29**, 674–6.

Sampson, K. (1933). The systemic infection of grasses by *Epichloe typhina* (Pers.) Tul. *Trans. Br. mycol. Soc.* **18**, 30–47.

Sampson, K. (1939). *Olpidium brassicae* (Wor.) Dang. and its connection with *Asterocystis radicis* de Wildeman. *Trans. Br. mycol. Soc.* **32**, 199–205.

Sansome, E. R. (1946). Heterokaryosis, mating-type factors, and sexual reproduction in *Neurospora*. *Bull. Torrey bot. Club*, **73**, 397–409.

Sansome, E. (1961). Meiosis in the oogonium and antheridium of *Pythium debaryanum* Hesse. *Nature, Lond.* **191**, 827–8.

Sansome, E. (1963). Meiosis in *Pythium debaryanum* Hesse and its significance in the life history of the Biflagellatae. *Trans. Br. mycol. Soc.* **46**, 63–72.

Sansome, E. & Harris, B. J. (1962). Use of camphor-induced polyploidy to determine the place of meiosis in fungi. *Nature, Lond.* **196**, 291–2.

Santesson, R. (1953). The new systematics of lichenized fungi. *Proc. 7th Internat. Bot. Congr.* 809–10.

Savile, D. B. O. (1965). Spore discharge in Basidiomycetes: a unified theory. *Science, N.Y.* **147**, 165–6.

Savory, J. G. (1954). Breakdown of timber by Ascomycetes and Fungi Imperfecti. *Ann. appl. Biol.* **41**, 336–47.

Savory, J. G. & Pinion, L. C. (1958). Chemical aspects of decay of beech wood by *Chaetomium globosum*. *Holzforschung*, **12**, 99–103.

Schneider, A. (1956). Sur les asques de *Taphrina* (Hemiascomycetes). *C. r. Acad. Sci. Paris*, **243**, 2139–42.

Schnathorst, W. C. (1965). Environmental relationships in the powdery mildews. *A. Rev. Phytopath.* **3**, 343–66.

Schremmer, F. (1963). Wechselbeziehungen zwischen Pilzen und Insecten. Beobachtungen an der Stinkmorchel, *Phallus impudicus* L. ex Pers. *Öst. bot. Z.* **110**, 380–400.

Schweizer, G. (1947). Über die Kultur von *Empusa muscae* Cohn und anderen Entomophthoraceen auf kalt sterilisierten nährboden. *Planta*, **35**, 132–76.

Scott, G. D. (1956). Further investigation of some lichens for fixation of nitrogen. *New Phytol.* **55**, 111–19.

Scott, W. W. (1956). A new species of *Aphanomyces*, and its significance in the taxonomy of the watermolds. *Va. J. Sci.* **7**, N.S. 170–5.

Scott, W. W. (1961). A monograph of the genus *Aphanomyces*. *Va. Agric. Exp. Sta. Tech. Bull.* **151**, 95 pp.

Scott, W. W. & O'Bier, A. H. (1962). Aquatic fungi associated with diseased fish and their eggs. *Progve Fish Cult.* **24**, 3–15.

Scurfield, G. (1967). Fine structure of the cell walls of *Polyporus myllitae* Cke. et Mass. *J. Linn. Soc.* (*Bot.*) **60**, 159–66.

Seth, H. K. (1967). Studies on the genus *Chaetomium*. I. Heterothallism. *Mycologia*, **59**, 580–4.

Shanor, L. (1936). The production of mature perithecia of *Cordyceps militaris* (Linn.) Link in laboratory culture. *J. Elisha Mitchell scient. Soc.* **52**, 99–103.

Shantz, H. L. & Piemeisel, R. L. (1917). Fungus fairy rings in Eastern Colorado and their effects on vegetation. *J. agric. Res.* **11**, 191–245.

Shatkin, A. J. & Tatum, E. L. (1959). Electron microscopy of *Neurospora crassa* mycelia. *J. biophys. biochem. Cytol.* **6**, 423–6.

Shatla, M. N., Yang, C. Y. & Mitchell, J. E. (1966). Cytological and fine structure studies of *Aphanomyces euteiches*. *Phytopathology*, **56**, 923–8.

Shear, C. L. & Dodge, B. O. (1927). Life histories and heterothallism of the red bread-mold fungi of the *Monilia sitophila* group. *J. agric. Res.* **34**, 1019–42.

Shoemaker, R. A. (1955). Biology cytology and taxonomy of *Cochliobolus sativus*. *Can. J. Bot.* **33**, 562–76.

Shropshire, W. (1963). Photoresponses of the fungus *Phycomyces*. *Physiol. Rev.* **43**, 38–67.

Siegel, M. (1958). Zur Ökologie des Anemonbecherlings. *Z. Pilzk.* **24**, 18–19.

Silver-Dowding, E. (1955). *Endogone* in Canadian rodents. *Mycologia*, **47**, 51–7.

Singer, R. (1961). *Mushrooms and Truffles. Botany, Cultivation and Utilization.* 272 pp. London: Leonard Hill.

Singer, R. (1962). *The Agaricales in Modern Taxonomy*. 915 pp. Weinheim: Cramer.

Singer, R. & Smith, A. H. (1960). Studies on Secotiaceous fungi. XI. The Astrogastraceous series. *Mem. Torrey bot. Club*, **21**, 1–112.

Singleton, J. R. (1953). Chromosome morphology and the chromosome cycle in the ascus of *Neurospora crassa*. *Am. J. Bot.* **40**, 125–44.

Sjöwall, von M. (1945). Studien über Sexualität, Vererbung und Zytologie bei einiger diozischen Mucoraceen. *Akad. Abhandl. Lund. Gleerupska Univ. Bokhandeln.* 1–97.

Sjöwall, von M. (1946). Über die zytologischen Verhaltnisse in den Keimschlauchen von *Phycomyces blakesleeanus* und *Rhizopus nigricans*. *Bot. Not.* 1946, 331–4.

Skucas, G. P. (1966). Structure and composition of zoosporangial discharge papillae in the fungus *Allomyces*. *Am. J. Bot.* **53**, 1006–11.

Smith, A. L. (1918). *A Monograph of British Lichens*. Part I, 519 pp. British Museum.

Smith, A. L. (1921). *A Handbook of the British Lichens*. 158 pp. British Museum.

Smith, A. L. (1926). *A Monograph of British Lichens*. Part II, 447 pp. British Museum.

Smith, D. C. (1961). The physiology of *Peltigera polydactyla* (Neck.) Hoffm. *Lichenologist*, **1**, 209–26.

Smith, D. C. (1962). The biology of lichen thalli. *Biol. Rev.* **37**, 537–70.

Smith, D. C. (1963). Experimental studies of lichen physiology. In *Symbiotic Associations*, 13th *Symp. Soc. gen. Microbiol.*, 31–50, Editors, Nutman, P. S. and Mosse, B. Cambridge University Press.

Smith, D. C., Muscatine, L. & Lewis, D. (1969). Carbohydrate movement from autrophs to heterotrophs in parasitic and mutalistic symbiosis. *Biol. Rev.* **44**, 17–90.

Snider, P. J. (1959). Stages of development in rhizomorphic thalli of *Armillaria mellea*. *Mycologia*, **51**, 693–707.

Snider, P. J. (1965). Incompatibility and nuclear migration. In *Incompatibility in Fungi*, 52–70, Editors, Esser, K., & Raper, J. R. Berlin: Springer.

Somers, C. E., Wagner, R. P. & Hsu, T. G. (1960). Mitosis in nuclei of *Neurospora crassa*. *Genetics*, **45**, 801–10.

Somers, E. & Horsfall, J. G. (1966). The water content of powdery mildew conidia. *Phytopathology*, **56**, 1031–5.

Sommer, N. F. (1961). Production by *Taphrina deformans* of substances stimulating cell elongation and division. *Physiol. Pl.* **14**, 460–9.

Spaeth, H. (1957). Über *Sclerotinia tuberosa*. *Schweiz. Z. Pilzk.* **23**, 20–1.

Sparrow, F. K. (1939). *Monoblepharis taylori*, a remarkable soil fungus from Trinidad. *Mycologia*, **31**, 737–8.

Sparrow, F. K. (1957). A further contribution to the Phycomycete flora of Great Britain. *Trans. Br. mycol. Soc.* **40**, 523–35.

Sparrow, F. K. (1958). Interrelationships and phylogeny of the aquatic Phycomycetes. *Mycologia*, **50**, 797–813.

Sparrow, F. K. (1960). *Aquatic Phycomycetes.* 2nd edition. 1187 pp. The University of Michigan Press.

Springer, M. E. (1945). A morphologic study of the genus *Monoblepharella. Am. J. Bot.* **32,** 259–69.

Srinivasan, M. C. Narasimhan, M. J. & Thirumalachar, M. J. (1964). Artificial culture of *Entomophthora muscae* and morphological aspects for differentiation of the genera *Entomophthora* and *Conidiobolus. Mycologia,* **56,** 683–91.

Srinivasan, M. C. & Thirumalachar, M. J. (1964). On the identity of *Entomophthora coronata. Mycopath. Mycol. appl.* **24,** 294–6.

Stakman, E. C. & Christensen, C. M. (1946). Aerobiology in relation to plant disease. *Bot. Rev.* **12,** 205–53.

Stakman, E. C. & Harrar, J. G. (1957). *Principles of Plant Pathology.* 581 pp. New York: Ronald Press Company.

Stakman, E. C., Levine, M. N. & Cotter, R. U. (1930). Origin of physiologic forms of *Puccinia graminis* through hybridization and mutation. *Scient. Agric.* **10,** 707–20.

Stamps, D. J. (1953). Oospore production in paired cultures of *Phytophthora* species. *Trans. Br. mycol. Soc.* **36,** 255–9.

Stanier, R. Y. (1942). The culture and nutrient requirements of a chytridiaceous fungus. *J. Bact.* **43,** 499–520.

Staples, R. C. & Wynn, W. K. (1965). The physiology of uredospores of the rust fungi. *Bot. Rev.* **31,** 537–64.

Stiegel, S. ((1952). Über Erregungsvorgänge bei der Einwirkung von photischen und mechanischen Reizen auf *Coprinus* Fruchtkörper. *Planta,* **40,** 301–12.

Strickmann, E. & Chadefaud, M. (1961). Recherches sur les asques et les périthèces des *Nectria* et reflexions sur l'évolution des Ascomycètes. *Rev. gén. Bot.* **68,** 725–70.

Stuart, M. R. & Fuller, H. T. (1968). Mycological aspects of diseased Atlantic salmon. *Nature, Lond.* **217,** 90–2.

Suminoe, K. & Dukmo, H. (1963). The life cycles of *Schizosaccharomyces* with reference to the mode of conjugation and ascospore formation. *J. gen. appl. Microbiol. Tokyo,* **9,** 243–7.

Sussman, A. S. (1957). Physiological and genetic adaptability in the fungi. *Mycologia,* **49,** 29–43.

Sussman, A. S. (1966). Types of dormancy as represented by conidia and ascospores of *Neurospora.* In *The Fungus Spore,* 235–56. Editor, Madelin, M. F. Colston Papers No. 18. London: Butterworths.

Sussman, A. S. & Halvorson, H. O. (1966). *Spores: their Dormancy and Germination.* 354 pp. New York and London: Harper and Row.

Swinbank, P., Taggart, J. & Hutchinson, S. A. (1964). The measurement of electrostatic charges on spores of *Merulius lacrymans* (Wulf.) Fr. *Ann. Bot. Lond.* N.S. **28,** 239–49.

Szaniszlo, P. J. (1965). A study of the effect of light and temperature on the formation of oogonia and oospheres in *Saprolegnia diclina. J. Elisha Mitchell scient. Soc.* **81,** 10–15.

Talbot, P. H. B. (1954). Micromorphology of the lower Hymenomycetes. *Bothalia,* **6,** 249–99.

Tatum, E. L. (1946). *Neurospora* as a biochemical tool. *Fedn. Proc. Fedn. Am. Socs. exp. Biol.* **5,** 362–5.

Teixeira, A. R. (1962). The taxonomy of the Polyporaceae. *Biol. Rev.* **37,** 51–81.

Teixeira, A. R. & Furtado, J. S. (1963). Anatomical studies on *Amauroderma regulicolor* (Berk. ex Cke.) Murrill. *Rickia, Archos. Bot. Est. S. Paulo,* **2,** 17–23.

Teter, H. E. (1944). Isogamous sexuality in a new strain of *Allomyces. Mycologia*, **36,** 194–210.

Thaxter, R. (1922). A revision of the Endogonaceae. *Proc. Am. Acad. Arts Sci.* **57,** 291–350.

Thirumalachar, M. J. & Dickson, J. G. (1949). Chlamydospore germination, nuclear cycle and artificial culture of *Urocystis agropyri* on red top. *Phytopathology*, **39,** 333–9.

Tiffney, W. N. (1939*a*). The host range of *Saprolegnia parasitica. Mycologia*, **31,** 310–21.

Tiffney, W. N. (1939*b*). The identity of certain species of the Saprolegniaceae parasitic to fish. *J. Elisha Mitchell. scient. Soc.* **55,** 134–51.

Tomlinson, J. A. (1958*a*). Crook root of watercress. II. The control of the disease with zinc-fritted glass and the mechanism of its action. *Ann. appl. Biol.* **46,** 608–21.

Tomlinson, J. A. (1958*b*). Crook root of watercress. III. The causal organism *Spongospora subterranea* (Wallr.) Lagerh. f. sp. *nasturtii* f. sp. nov. *Trans. Br. mycol. Soc.* **41,** 491–8.

Trotter, M. J. & Whisler, H. C. (1965). Chemical composition of the cell wall of *Amoebidium parasiticum. Can. J. Bot.* **43,** 869–976.

Tschierpe, H. J. (1959). Die Bedeutung des Kohlendioxyds für den Kulturchampignon. *Gartenbauwiss.* **24** (1), 18–75.

Tschierpe, H. J. & Sinden, J. W. (1964). Weitere Untersuchungen über die Bedeutung von Kohlendioxyd für die Fructifikation des Kulturchampignons, *Agaricus campestris* var. *bisporus* (L) Lge. *Arch. Mikrobiol.* **49,** 405–25.

Tubaki, K. (1958). Studies on the Japanese Hyphomycetes. V. Leaf and stem group with a discussion of the classification of Hyphomycetes and their perfect stages. *J. Hattori Bot. Lab.* **20,** 142–244.

Tucker, C. M. (1931). Taxonomy of the genus *Phytophthora* de Bary. *Res. Bull. Mo. agric. exper. sta.* **153,** 1–208.

Tveit, M. (1953). Control of a seed-borne disease by *Chaetomium cochlioides* Pall., under natural conditions. *Nature, Lond.* **172,** 39.

Tveit, M. (1956). Isolation of a chetomin-like substance from oat seedlings infested with *Chaetomium cochlioides. Acta Agric. scand.* **6,** 13–16.

Tveit, M. & Moore, M. B. (1964). Isolates of *Chaetomium* that protect oats from *Helminthosporium victoriae. Phytopathology*, **44,** 686–9.

Tveit, M. & Wood, R. K. S. (1955). The control of *Fusarium* blight in oat seedlings with antagonistic species of *Chaetomium. Ann. appl. Biol.* **43,** 538–52.

Unestam, T. (1965). Studies on the crayfish plague fungus *Aphanomyces astaci*. I. Some factors affecting growth *in vitro. Physiol. Pl.* **18,** 483–505.

Vanterpool, T. C. (1959). Oospore germination in *Albugo candida. Can. J. Bot.* **37,** 169–71.

Varitchak, B. (1931). Contribution à l'étude du développement des Ascomycetes. *Botaniste*, Sér. **23,** 1–182.

Vishniac, H. S. & Nigrelli, R. F. (1957). The ability of the Saprolegniaceae to parasitise platyfish. *Zoologica*, **42,** 131–4.

Vitols, E., North, R. J. & Linnane, A. W. (1961). Studies on the oxidative metabolism of *Saccharomyces cerevisiae*. I. Observations on the fine structure of the yeast cell. *J. Biochem. biophys. Cytol.* **9,** 689–99.

Vujičić, R. & Colhoun, J. (1966). Asexual reproduction in *Phytophthora erythroseptica. Trans. Br. mycol. Soc.* **49,** 245–54.

Wagner, F. (1963). Der Gerstenflugbrand (*Ustilago nuda*) ist chemisch bekämpfbar. *Pflanzenarzt*, **16,** 82–3.

Waksman, S. A., & Bugie, E. (1944). Chaetomin, a new antibiotic substance produced by *Chaetomium cochliodes*. I. Formation and properties. *J. Bact.* **48,** 527–30.

Walker, L. B. (1920). Development of *Cyathus fascicularis*, *C. striatus*, and *Crucibulum vulgare*. *Bot. Gaz.* **70,** 1–24.

Walker, L. B. (1923). Some observations on the development of *Endogone malleola* Hark. *Mycologia*, **15,** 245–57.

Walker, L. B. (1927). Development and mechanism of discharge in *Sphaerobolus iowensis* n. sp. and *S. stellatus* Tode. *J. Elisha Mitchell scient. Soc.* **42,** 151–78.

Walkey, D. G. A. & Harvey, R. (1966*a*). Studies of the ballistics of ascospores. *New Phytol.* **65,** 59–74.

Walkey, D. G. A. & Harvey, R. (1966*b*). Spore discharge rhythms in pyrenomycetes. *Trans. Br. mycol. Soc.* **49,** 583–92.

Walsh, J. H. & Harley, J. L. (1962). Sugar absorption by *Chaetomium globosum*. *New Phytol.* **61,** 299–313.

Wang, D. T. (1932). Observations cytologiques sur l'*Ustilago violacea* (Pers.) Fuckel. *C. r. hebd. Séanc. Acad. Sci., Paris*, **195,** 1417–18.

Wang, D. T. (1934). Contribution à l'étude des Ustilaginées. (Cytologie du parasite et pathologie de la cellule hôte.) *Botaniste*, **26,** 540–672.

Warcup, J. H. (1951). Studies on the growth of Basidiomycetes in soil. *Ann. Bot. Lond.* N.S. **15,** 305–17.

Warcup, J. H. & Talbot, P. H. B. (1962). Ecology and identity of mycelia isolated from soil. *Trans. Br. mycol. Soc.* **45,** 495–518.

Ward, E. W. B. & Ciurysek, K. W. (1962). Somatic mitosis in *Neurospora crassa*. *Am. J. Bot.* **49,** 393–9.

Ward, E. W. B. & Thorn, G. D. (1965). The isolation of a cyanogenic fraction from the fairy ring fungus *Marasmius oreades* Fr. *Can. J. Bot.* **43,** 997–8.

Waterhouse, G. M. (1956). The genus *Phytophthora*. Diagnoses (or descriptions) and figures from original papers. *Misc. Publ. C.M.I.* **12,** 120 pp.

Waterhouse, G. M. (1963). Key to the species of *Phytophthora* de Bary. *Mycol. Pap. C.M.I.* **92,** 22 pp.

Waterhouse, G. M. (1967). Key to *Pythium* Pringsheim. *Mycol. Pap. C.M.I.* **109,** 15 pp.

Waterhouse, G. M. (1968). The genus *Pythium* Pringsheim. Diagnoses (or descriptions) and figures from the original papers. *Mycol. Pap. C.M.I.* **110,** 71 pp.

Watson, I. A. (1957). Further studies on the production of new races from mixtures of races of *Puccinia graminis* var *tritici* on wheat seedlings. *Phytopathology*, **47,** 510–12.

Webster, J. (1964). Culture studies on *Hypocrea* and *Trichoderma*. I. Comparison of perfect and imperfect states of *Hypocrea gelatinosa*, *H. rufa* and *Hypocrea* sp. 1. *Trans. Br. mycol. Soc.* **47,** 75–96.

Webster, J. & Dennis, C. (1967). The mechanism of sporangial discharge in *Pythium middletonii*. *New Phytol.* **66,** 307–13.

Wehmeyer, L. E. (1926). A biologic and phylogenetic study of stromatic Sphaeriales. *Am. J. Bot.* **13,** 575–645.

Wehmeyer, L. E. (1955). Development of the ascostroma in *Pleospora armeriae* of the *Pleospora herbarum* complex. *Mycologia*, **47,** 821–34.

Wehmeyer, L. E. (1961). *A World Monograph of the Genus* Pleospora *and its Segregates*. 451 pp. University of Michigan Press.

Weijer, J., Koopmans, A. & Weijer, D. L. (1965). Karyokinesis of somatic nuclei of *Neurospora crassa*. III. The juvenile and maturation cycles (Feulgen and crystal violet staining). *Can. J. Genet. Cytol.* **7,** 140–63.

Weijer, J. & Weisberg, S. H. (1966). Karyokinesis in the somatic nuclei of *Asper-*

gillus. I. The juvenile chromosome cycle (Feulgen staining). *Canad. J. Genet. Cytol.* **8,** 361–74.
Wells, K. (1964*a*). The basidia of *Exidia nucleata*. I. Ultrastructure. *Mycologia*, **56,** 327–41.
Wells, K. (1964*b*). The basidia of *Exidia nucleata*. II. Development. *Am. J. Bot.* **51,** 360–70.
Wells, K. (1965). Ultrastructural features of developing and mature basidia and basidiospores of *Schizophyllum commune*. *Mycologia*, **57,** 236–61.
Wessels, J. G. H. (1965). Morphogenesis and biochemical processes in *Schizophyllum commune* Fr. *Wentia*, **13,** 1–113.
Western, J. H. & Cavett, J. J. (1959). The choke disease of cocksfoot (*Dactylis glomerata*) caused by *Epichloe typhina* (Fr.) Tul. *Trans. Br. mycol. Soc.* **42,** 298–307.
Weyen, A van der (1954). L'évolution nucléaire et les hyphes ascogènes chez *Chaetomium globosum* Kunze. Cellule, **56,** 211–26.
Whetzel, H. H. (1945). A synopsis of the genera and species of the Sclerotiniaceae, a family of stromatic inoperculate Discomycetes. *Mycologia*, **37,** 648–714.
Whetzel, H. H. (1946). The cypericolous and juncicolous species of *Sclerotinia. Farlowia*, **2,** 385–437.
Whiffen, A. J. (1944). A discussion of taxonomic criteria in the Chytridiales. *Farlowia*, **1,** 583–97.
Whisler, H. C. (1966). Host-integrated development in the Amoebidiales. *J. Protozool.* **13,** 183–8.
Whitehouse, H. L. K. (1949*a*). Heterothallism and sex in the fungi. *Biol. Rev.* **24,** 411–47.
Whitehouse, H. L. K. (1949*b*). Multiple-allelomorph heterothallism in the fungi. *New Phytol.* **48,** 212–44.
Whitehouse, H. L. K. (1951). A survey of heterothallism in the Ustilaginales. *Trans. Br. mycol. Soc.* **34,** 340–55.
Whiteside, W. C. (1961). Morphological studies in the Chaetomiaceae. I. *Mycologia*, **53,** 512–23.
Whitten, R. R. & Swingle, R. U. (1964). The Dutch elm disease and its control. *U.S. Dept. Agric. Information Bull.* No. 193, 1–12.
Wicker, M. (1962). Le mycélium et les périthèces dans une souche de la Sordariale *Pleurage minuta* (Fuck.) Ktze. *Bull. Soc. mycol. Fr.* **78,** 291–326.
Wickerham, L. J. & Burton, K. A. (1954). A clarification of the relationships of *Candida guilliermondii* to other yeasts by a study of their mating types. *J. Bact.* **68,** 594–7.
Wickerham, L. J. & Duprat, J. (1945). A remarkable fission yeast, *Schizosaccharomyces versatilis* nov. sp. *J. Bact.*, **50,** 597–607.
Wickerham, L. J. Lockwood, L. B., Pettijohn, O. G. & Ward, G. E. (1944). Starch hydrolysis and fermentation by the yeast *Endomycopsis fibuliger*. *J. Bact.* **48,** 413–27.
Widra, A. & Delamater, E. D. (1955). The cytology of meiosis in *Schizosaccharomyces octosporus*. *Am. J. Bot.* **42,** 423–35.
Wieben, M. (1927). Die Infektion, die Myzelüberwinterung und die Kopulation bei Exoasceen. *Forschn Geb. PflKrankh. Berl.* **3,** 139–76.
Williams, P. G., Scott, K. J., & Kuhl, J. L. (1966). Vegetative growth of *Puccinia graminis* f. sp. *tritici in vitro*. *Phytopathology*, **56,** 1418–19.
Williams, P. H. (1966). A cytochemical study of hypertrophy in clubroot of cabbage. *Phytopathology*, **56,** 521–4.
Williams, P. H. & McNabola, S. S. (1967). Fine structure of *Plasmodiophora brassicae* in sporogenesis. *Can. J. Bot.* **45,** 1665–9.

Williams, P. H. & Yukawa, Y. B. (1967). Ultrastructural studies on the host-parasite relations of *Plasmodiophora brassicae*. *Phytopathology*, **57,** 682–7.

Williams, S. T., Gray, T. R. G. & Hitchen, P. (1965). Heterothallic formation of zygospores in *Mortierella marburgensis*. *Trans. Br. mycol. Soc.* **48,** 129–33.

Willoughby, L. G. (1956). Studies on soil chytrids. I. *Rhizidium richmondense* sp. nov. and its parasites. *Trans. Br. mycol. Soc.* **39,** 125–41.

Willoughby, L. G. (1958). Studies on soil chytrids. III. On *Karlingia rosea* Johanson and a multi-operculate chytrid parasitic on *Mucor*. *Trans. Br. mycol. Soc.* **41,** 309–19.

Willoughby, L. G. (1962*a*). The fruiting behaviour and nutrition of *Cladochytrium replicatum* Karling. *Ann. Bot. Lond.* N.S. **26,** 13–36.

Willoughby, L. G. (1962*b*). The occurrence and distribution of reproductive spores of Saprolegniales in fresh water. *J. Ecol.* **50,** 733–59.

Wilsenach, R. & Kessel, M. (1965*a*). The role of lomasomes in wall formation in *Penicillium vermiculatum*. *J. gen. Microbiol.* **40,** 401–4.

Wilsenach, R. & Kessel, M. (1965*b*). On the fuction and structure of the septal pore of *Polyporus rugulosus*. *J. gen. Microbiol.* **40,** 397–400.

Wilson, C. M. (1952). Meiosis in *Allomyces*. *Bull. Torrey bot. Club*, **79,** 139–60.

Wilson, I. M. (1952). The ascogenous hyphae of *Pyronema confluens*. *Ann. Bot. Lond.* N.S. **16,** 321–39.

Wilson, J. F., Garnjobst, L. & Tatum, E. L. (1961). Heterokaryon incompatibility in *Neurospora crassa*—micro-injection studies. *Am. J. Bot.* **48,** 299–305.

Wilson, M. & Henderson, D. M. (1966). *British Rust Fungi*. 384 pp. Cambridge University Press.

Windisch, S. & Bautz, E. (1960). Die innere Bau der Hefezelle. In *Die Hefen*. I. *Die Hefen in der Wissenschaft*. Editors, Reiff, F., Kautzmann, R., Luers, H. & Lindemann, M. Nürnberg: H. Carl.

Winge, Ö. & Lausten, O. (1937). On two types of spore germination, and on genetic segregations in *Saccharomyces*, demonstrated through single spore cultures *C. r. Trav. Lab. Carlsberg phys.* **22,** 99–116.

Winge, Ö. & Roberts, C. (1949). A gene for diploidization in yeasts. *C. r. Trav. Lab. Carlsberg. phys.* **24,** 341–6.

Winge, Ö. & Roberts, C. (1958). Yeast genetics. In *The Chemistry and Biology of Yeasts*, 123–56. Editor, Cook, A. H. New York: Academic Press.

Wolf, E. (1954). Beitrag zur Systematik der Gattung *Mortierella* und *Mortierella*-Arten als Mykorrhizapilze bei Ericaceen. *Zbl. Bakt.* Abt. 2, **107,** 21–4; 523–48.

Wood, J. L. (1953). A cytological study of ascus development in *Ascobolus magnificus* Dodge. *Bull. Torrey bot. Club*, **80,** 1–15.

Woodham-Smith, C. (1962). *The Great Hunger. Ireland 1845–9*. London: Hamish Hamilton.

Woodward, R. C. (1927). Studies on *Podosphaera leucotricha* (Ell. & Ev.). Salm. I. The mode of perennation. *Trans. Br. mycol. Soc.* **12,** 173–204.

Wormald, H. (1921). On the occurrence in Britain of the ascigerous stage of a brown rot fungus. *Ann. Bot. Lond.* **35,** 125–35.

Wormald, H. (1954). The brown rot diseases of fruit trees. *Min. Agric. Tech. Bull.* No. 3, 113 pp.

Woycicki, Z. (1927). Über die Zygotenbildung bei *Basidiobolus* ranarum Eidam. II. *Flora*, **122,** 159–66.

Yarwood, C. E. (1941). Diurnal cycle of ascus maturation of *Taphrina deformans*. *Am. J. Bot.* **28,** 355–7.

Yendol, W. G. & Paschke, J. D. (1965). Pathology of an *Entomophthora* infection in the Eastern subterranean termite, *Reticulitermes flavipes* (Kollar). *J. invert. Pathol.* **7,** 414–22.

Yerkes, W. D. & Shaw, C. G. (1959). Taxonomy of the *Peronospora* species on Cruciferae and Chenopodiaceae. *Phytopathology*, **49,** 499–507.
Yu, C. C. (1954). The culture and spore germination of *Ascobolus* with emphasis on *A. magnificus*. *Am. J. Bot.* **41,** 21–30.
Yuill, E. (1950). The numbers of nuclei in conidia of *Aspergilli*. *Trans. Br. mycol. Soc.* **33,** 324–31.
Zeller, S. M. (1949). Keys to the orders, families, and genera of the Gasteromycetes. *Mycologia*, **41,** 36–58.
Ziegler, A. W. (1953). Meiosis in the Saprolegniaceae. *Am. J. Bot.* **40,** 60–6.
Zoberi, M. H. (1961). Take-off of mould spores in relation to wind speed and humidity. *Ann. Bot. Lond.* N.S. **25,** 53–64.
Zycha, H. (1935). Mucorineae. Kryptogamenflora der Mark Brandenburg, 6a, 1–264.

Index